Trends in Mathematics is a series devoted to the publication of volumes arising from conferences and lecture series focusing on a particular topic from any area of mathematics. Its aim is to make current developments available to the community as rapidly as possible without compromise to quality and to archive these for reference.

Proposals for volumes can be sent to the Mathematics Editor at either

Birkhäuser Verlag
P.O. Box 133
CH-4010 Basel
Switzerland

or

Birkhäuser Boston Inc.
675 Massachusetts Avenue
Cambridge, MA 02139
USA

Material submitted for publication must be screened and prepared as follows:

All contributions should undergo a reviewing process similar to that carried out by journals and be checked for correct use of language which, as a rule, is English. Articles without proofs, or which do not contain any significantly new results, should be rejected. High quality survey papers, however, are welcome.

We expect the organizers to deliver manuscripts in a form that is essentially ready for direct reproduction. Any version of $\TeX$ is acceptable, but the entire collection of files must be in one particular dialect of $\TeX$ and unified according to simple instructions available from Birkhäuser.

Furthermore, in order to guarantee the timely appearance of the proceedings it is essential that the final version of the entire material be submitted no later than one year after the conference. The total number of pages should not exceed 350. The first-mentioned author of each article will receive 25 free offprints. To the participants of the congress the book will be offered at a special rate.

Infinite Length Modules

Henning Krause
Claus Michael Ringel
Editors

Springer Basel AG

Editors' address:

Fakultät für Mathematik
Universität Bielefeld
D-33501 Bielefeld
Germany

2000 Mathematical Subject Classification 16-06; 16GXX, 16D90, 16D80

A CIP catalogue record for this book is available from
the Library of Congress, Washington D.C., USA

Deutsche Bibliothek Cataloging-in-Publication Data

Infinite length modules / Henning Krause ; Claus Michael Ringel ed.. - Basel ; Boston ; Berlin :
Birkhäuser, 2000
(Trends in mathematics)

ISBN 978-3-0348-9562-0 ISBN 978-3-0348-8426-6 (eBook)
DOI 10.1007/978-3-0348-8426-6

Printed on acid-free paper produced from chlorine-free pulp. TCF ∞

9 8 7 6 5 4 3 2 1

CONTENTS

Preface vii

* * *

C.M. Ringel: Infinite length modules. Some Examples as Introduction 1

* * *

P.C. Eklof: Modules with strange decomposition properties 75

A. Facchini: Failure of the Krull-Schmidt theorem
for artinian modules and serial modules 89

K.I. Pimenov, A.V. Yakovlev: Artinian modules over a matrix ring 101

R. Göbel: Some combinatorial principles for solving algebraic problems 107

* * *

T.H. Lenagan: Dimension theory of noetherian rings 129

V. Bavula: Krull, Gelfand-Kirillov, Filter, Faithful
and Schur dimensions 149

A. Martsinkovsky: Cohen-Macaulay modules and approximations 167

* * *

N.J. Kuhn: The generic representation theory of finite fields.
A survey of basic structures 193

G.M.L. Powell: On artinian objects in the category of functors
between $\mathbb{F}_2$-vector spaces 213

L. Schwartz: Unstable modules over the Steenrod algebra, functors,
and the cohomology of spaces 229

* * *

D.J. Benson: Infinite dimensional modules for finite groups 251

J. Rickard: Bousfield localization for representation theoretists 273

J.F. Carlson: The thick subcategory generated by the trivial module 285

* * *

A. Schofield: Birational classification of moduli spaces 297

G. Zwara: Tame algebras and degenerations of modules 311

R. Bautista: On some tame and discrete families of modules 321

* * *

B. Huisgen-Zimmermann: Purity, algebraic compactness,
direct sum decompositions, and representation type 331

M. Prest: Topological and geometrical aspects of the Ziegler spectrum 369

H. Krause: Finite versus infinite dimensional representations.
A new definition of tameness 393

H. Lenzing: Invariance of tameness under stable equivalence:
Krause's theorem 405

J. Schröer: The Krull-Gabriel dimension of an algebra.
Open problems and conjectures 419

S.O. Smalø: Homological differences between
finite and infinite dimensional representations of algebras 425

PREFACE

This volume presents the invited lectures of a conference devoted to *Infinite Length Modules*, held at Bielefeld, September 7–11, 1998. Some additional surveys have been included in order to establish a unified picture. The scientific organization of the conference was in the hands of K. Brown (Glasgow), P.M. Cohn (London), I. Reiten (Trondheim) and C.M. Ringel (Bielefeld).

The conference was concerned with the role played by modules of infinite length when dealing with problems in the representation theory of algebras. The investigation of such modules always relies on information concerning modules of finite length, for example simple modules and their possible extensions. But the converse is also true: recent developments in representation theory indicate that a full understanding of the category of finite dimensional modules, even over a finite dimensional algebra, requires consideration of infinite dimensional, thus infinite length, modules. For instance, the important notion of tameness uses one-parameter families of modules, or, alternatively, generic modules and they are of infinite length. If one tries to exhibit a presentation of a module category, it turns out to be essential to take into account the indecomposable modules which are algebraically compact, or, equivalently, pure injective. Specific methods have been developed over the last few years dealing with such special situations as group algebras of finite groups or noetherian rings, and there are surprising relations to topology and geometry. The conference outlined the present state of the art. Its aim was to stimulate further development by bringing together experts from quite different schools.

The contributions in this volume reflect the directions of the different lecture series. In dealing with general modules, one has to restrict to classes which are still well-behaved and which include the typical modules to be considered. The most important class in this respect is that of the algebraically compact modules, a fact which is supported by investigations from other part of mathematics: mathematical logic, algebraic topology and geometry. Of particular interest is the set of isomorphism classes of indecomposable algebraically compact modules. This set carries a topology and one deals in this way with what is called the Ziegler spectrum (Huisgen-Zimmermann, Prest, Krause, Lenzing, Schröer, Smalø). The use of infinite dimensional representations in the representation theory of finite dimensional algebras is often motivated by geometric concepts which arise if one studies the variety of modules of some fixed dimension. For example, the definitions of tameness and wildness are based on the possibility to parametrize such representations, using generic modules or lattices over orders. Further topics in this context include degenerations and moduli spaces (Schofield, Zwara, Bautista).

In the representation theory of noetherian rings there has been a rather long tradition dealing with arbitrary modules. For instance, many different concepts of "dimension" have been introduced, the most prominent being Krull-Gabriel dimension and Gelfand-Kirillov dimension (Lenagan, Bavula). Of similar importance are homological methods which have their origin in commutative algebra (Martsinkovsky). These are concepts which still have to be pursued in more general settings. For example, passing from the representations of a finite dimensional algebra to the category of functors from finite dimensional representations to abelian groups involves the study of a ring with several objects which is usually not noetherian. In fact, it is noetherian if and only if the finite dimensional algebra in question is of finite representation type.

The general module theory (for example as developed for abelian groups) provides many examples of pathological behaviour which seem to be important. Of particular interest are methods for constructing modules with (nearly) prescribed endomorphism rings and thus modules with strange decomposition properties. This leads to examples for the failure of the Krull-Remak-Schmidt property for infinite length modules (Eklof, Göbel, Facchini, Pimenov and Yakovlev).

A special setting where the study of infinite length modules has turned out to be very fruitful is the representation theory of finite groups. The reason for this is the use of methods from stable homotopy theory. In fact, there is a strong analogy between the stable module category of a finite group and the stable homotopy category of CW-spectra. Usually, the study of finite spectra involves infinite spectra as well, for example via Bousfield localization. The corresponding objects in representation theory are the so-called idempotent modules which are of finite length only in trivial cases. It seems that the passage between representations and cohomology of a finite group, via methods from stable homotopy theory, depends in most cases on infinite length modules (Benson, Carlson, Rickard).

In generic representation theory one studies the representations of the finite general linear groups. In this setting representations of infinite length arise quite naturally when one collects the representations of all finite linear groups over a fixed (finite) field in a single category. There are several reasons for this approach. For example, generic representation theory helps to understand the unstable modules over the Steenrod algebra which are relevant in algebraic topology. Further fields of application include the representation theory of the symmetric groups, Schur algebras, and Topological Hochschild Homology (Kuhn, Schwartz, Powell).

The conference was a "Euroconference", with special support by the European Commission. A joint research project of algebraists from the universities of Antwerp, Bielefeld, Essen, Leeds, Paris VI and Trondheim on "Invariants and Representations of Algebras" has been supported from 1991 to 1997 by the European Union programmes "Science" and "Human Capital and Mobility". This project was coordinated by Mme M.-P. Malliavin (Paris VI). Later, algebraists from the universities of Edinburgh, Ioannina, Murcia and Toruń joined the collaboration. This network has received funds from the European Commission in order to organize four conferences as part of the programme "Training and Mobility of Researchers". The first one, held at Essen in 1997, dealt with *Computational Methods for Representations of Groups and Algebras.* The second one had the title *Interactions of Ring and Module Theory* and was organized in 1998 at the University of Murcia. The Bielefeld conference was the third in this series, and a fourth one has been held in 1999 at the University of Ioannina, devoted to *Homological Methods in Representation Theory.*

The editors, in the name of all participants of the meeting, thank the European Community for its financial support. Nevertheless, it has to be mentioned that the financial and bureaucratic regulations of the EU cause a lot of difficulties. These constraints are in conflict with good scientific traditions, in particular the proclivity to focus all the attention on the nationality of the speakers and the participants. Without the additional support of the Sonderforschungsbereich 343, the Ministerium für Wissenschaft und Forschung des Landes Nordrhein-Westfalen and the Westfälisch-Lippische Universitäts-Gesellschaft it would not have been possible to present all the topics in a satisfactory way. We hope that this volume reflects, as did the meeting, the truly international character of present research, not restricted to narrow boundaries of countries or communities.

Finally, it is a pleasure to thank the authors and referees of the papers in this volume for their efforts and enthusiasm in helping to make this a broad collection reflecting the interaction between various topics and disciplines.

Bielefeld, October 30, 1999 H. Krause, C.M. Ringel

Trends in Mathematics, © 2000 Birkhäuser Verlag Basel/Switzerland

INFINITE LENGTH MODULES. SOME EXAMPLES AS INTRODUCTION.

CLAUS MICHAEL RINGEL

The aim of this introduction is to outline the general setting and to exhibit some examples, in order to show interesting features of infinite length modules, but also to point out the relevance of these features in representation theory. We try to present examples as explicit as possible, in contrast to the quite common attitude of being satisfied with the mere existence, an attitude which indicates the desire to rate such features as unpleasant and to avoid them. In contrast, the phenomena we deal with should be considered as typical and as exciting.

The main references to be quoted are the books by Jensen and Lenzing [JL] and Prest [P1], but also volume 2 of Faith [Fa2]. The reader will realize that we follow closely the path of the Trondheim lectures of Crawley-Boevey [CB4] and we have to admit that we are strongly indebted to his mathematical insight. The text is written to be accessible even for a neophyte. We hope that some of the considerations, in particular several examples, are of interest for a wider audience, as we try to present as easy as possible some gems which seem to be hidden in the literature. Part of the text may be rated as "descriptive mathematics", definitely not fitting into the usual pattern of "definition – theorem – proof", but we hope that the topics presented in this way will be illuminating and will provide a better understanding of some problems in representation theory.

General conventions: Always, k will be a field. We denote by $\subseteq$ an inclusion of sets, and we write $\subset$ in case the inclusion is proper. Recall that a quiver $Q = (Q_0, Q_1, s, t)$ consists of a set Q_0 (its elements are called vertices) and a set Q_1 (of arrows); for every arrow $\alpha \in Q_1$, $s(\alpha)$ is its starting point, $t(\alpha)$ its end point.

If R is a ring, we consider mainly left modules and we denote by $\operatorname{Mod} R$ the category of all (left) R-modules, by $\operatorname{mod} R$ the full subcategory of all finitely presented ones. The module ${}_RR$ will be called the regular representation of R.

We denote by $\mathbb{N}$ the set of natural numbers $\{0, 1, 2, \dots\}$, by $\mathbb{Z}$ the integers, and by $\mathbb{Q}$ the rational numbers. For any natural number $n \geq 2$, let $\mathbb{Z}_{(n)}$ be the set of rational numbers $\frac{a}{b}$ such that 1 is the only positive integer which divides both b and n.

The author is indebted to P.N. Ánh, P. Dräxler, L. Hille, H. Krause and J. Schröer for helpful comments concerning the presentation.

Contents.

Finite length modules. 2
1. Modules in general. 6
2. The categorical setting. 16
3. Well-behaved modules. 30
4. Finite-dimensional algebras. 36
5. Hammocks and quilts. 51
Epilogue. 68
References. 70

Finite length modules

Composition series. Let R be a ring and M an R-module. A *composition series* of M is a finite sequence of submodules

$$(*) \qquad 0 = M_0 \subset M_1 \subset \cdots \subset M_{n-1} \subset M_n = M$$

which cannot be refined (this means that if N is a submodule of M with $M_{i-1} \subseteq N \subseteq M_i$, then $M_{i-1} = N$ or $N = M_i$). Of course, a sequence $(*)$ of submodules is a composition series if and only of all the factors M_i/M_{i-1}, for $1 \le i \le n$ are simple R-modules. Given a composition series $(*)$ of M, then the modules M_i/M_{i-1} are called the *composition factors* of M and $n = |M|$ is said to be its *length.* If M has a composition series, then M is said to be *of finite length.* The **Theorem of Jordan-Hölder** asserts that *the length of a finite length module is uniquely determined,* and that also the composition factors are unique in the following sense: *Given a second composition series*

$$0 = M'_0 \subset M'_1 \subset \cdots \subset M'_{n-1} \subset M'_n = M,$$

there is a permutation σ of $\{1, 2, \dots, n\}$ such that $M'_i/M'_{i-1} \simeq M_{\sigma(i)}/M_{\sigma(i)-1}$. We can phrase this uniqueness assertion also as follows: *Given a simple R-module S and a composition series $(*)$, then the number of indices i with $M_i/M_{i-1} \simeq S$ only depends on M and S and not on the chosen composition series $(*)$;* this number is called the *Jordan-Hölder multiplicity of S in M*. For further reference, we have given in detail the definition of a finite length module, even though the book exhibited here — as the title makes clear — is devoted to modules which are just **not** of finite length. But the aim of the book is, on the one hand, the comparison of properties of finite length modules with those of general ones, and, on the other hand, to stress the need for using infinite length modules in situations which seem to be confined to finite length modules.

The term *finite length module* obviously refers to the existence of a composition series and to its length. But it is worthwhile to note that there exists a characterization of these modules which does not refer to the actual length and which does not give any hint at all what the length of such a module M could be: *The module M has finite length if and only if M is both artinian and noetherian.* (Recall that M is said to be *noetherian* - or to satisfy the ascending chain condition for submodules - provided for every chain of submodules $M_1 \subseteq M_2 \subseteq \cdots \subseteq M_i \subseteq M_{i+1} \subseteq \cdots$ there exists some number n with $M_n = M_{n+1}$; similarly, M is said to be *artinian* - or to satisfy the descending chain condition for submodules - provided for any chain of submodules $M_1 \supseteq M_2 \supseteq \cdots \supseteq M_i \supseteq M_{i+1} \supseteq \cdots$ there exists some number n with $M_n = M_{n+1}$. The properties of being artinian and noetherian are dual to each other, but there is a characterization of the noetherian modules for which no neat dual property can be named: a module M is noetherian if and only if all submodules are finitely generated.)

Let us consider the interrelation between the various composition series of a finite length module M. Starting with a composition series $(*)$, one looks at two consecutive composition factors M_i/M_{i-1}, M_{i+1}/M_i and the corresponding length 2 module M_{i+1}/M_{i-1}. In case the exact sequence

$$0 \to M_i/M_{i-1} \to M_{i+1}/M_{i-1} \to M_{i+1}/M_i \to 0$$

splits, we find some submodule $M_i' \neq M_i$ with $M_{i-1} \subset M_i' \subset M_{i+1}$, and we obtain a different composition series

$$0 = M_0 \subset \cdots \subset M_{i-1} \subset M_i' \subset M_{i+1} \subset \cdots \subset M_n = M.$$

By a sequence of such changes, we obtain from the given composition series any other composition series (Proof: Given two composition sequences $(M_i)_{0 \le i \le n}$ and $(M_i')_{0 \le i \le n}$ of a module M, we may assume that $M_{n-1} \neq M_{n-1}'$, otherwise use induction. Take any composition series $(M_i'')_{0 \le i \le n-2}$ of $M_{n-1} \cap M_{n-1}'$. The submodules M_i and M_i'' provide two composition series for M_{n-1}, and similarly, the submodules M_i'' and M_i' provide two composition series of M_{n-1}'. It remains to use induction.)

Direct decompositions. Any finite length module can be written as the direct sum of a finite number of indecomposable[1] modules. Now, it is of great importance that such a decomposition is essentially unique, this follows from the following fact: *The endomorphism ring of an indecomposable module of finite length is a local ring* (recall that a ring R is local provided (it is non-zero and) the set of non-invertible elements of R is an ideal). The only idempotents of a local ring are 0 and 1; thus a module with a local endomorphism ring necessarily is indecomposable. The uniqueness statement is known as the **Theorem of Krull-Remak-Schmidt**[2]: *Let $M = \bigoplus_{i=1}^m M_i = \bigoplus_{j=1}^n N_j$, such that all the modules M_i have local endomorphism rings, $1 \le i \le m$, and the modules N_j are indecomposable, $1 \le j \le n$.*

[1] A module M is said to be *indecomposable* provided M is non-zero and there is no direct sum decomposition $M = M_1 \oplus M_2$ with both M_1, M_2 non-zero.

[2] This theorem sometimes is referred just as the Krull-Schmidt Theorem, and

Then $n = m$ and there is a permutation σ with $N_j \simeq M_{\sigma(j)}$ for all $1 \le j \le m$. Actually, the corresponding result for arbitrary (not necessarily finite) direct sums is also true, this is the **Theorem of Azumaya:** *Let $M = \bigoplus_{i\in I} M_i = \bigoplus_{j\in J} N_j$, such that all the module M_i have local endomorphism rings, $i \in I$, and the modules N_j are indecomposable, $j \in J$. Then there is a bijection $\sigma\colon J \to I$ such that $N_j \simeq M_{\sigma(j)}$ for all $j \in J$.*

The Krull-Remak-Schmidt theorem yields a reduction of the study of finite length modules to the indecomposable ones, and one of the main tasks of present representation theory is to describe the shape of indecomposable modules of finite length, to look for invariants which allow to read off their properties. In addition to the knowledge of the simple R-modules S_i, one needs to know the possible extensions of these modules, thus the extension groups $\mathrm{Ext}^1_R(S_i, S_j)$, for all simple R-modules S_i, S_j. This knowledge is collected in the *quiver* of R, its vertices are the isomorphism classes $[S_i]$ of the simple R-modules, and one draws an arrow $[S_i] \to [S_j]$ provided $\mathrm{Ext}^1_R(S_i, S_j) \neq 0$. Actually, in case we deal with a k-algebra and $\mathrm{End}(S_i) = k = \mathrm{End}(S_j)$, one will draw not just one arrow, but d arrows $[S_i] \to [S_j]$, where d is the k-dimension of $\mathrm{Ext}^1_R(S_i, S_j)$; in general one may endow the arrow $[S_i] \to [S_j]$ with the $\mathrm{End}(S_i)$-$\mathrm{End}(S_j)$-bimodule $\mathrm{Ext}^1_R(S_i, S_j)$.

Serial modules. Since ancient times, the modules whose submodule lattice is a chain, the so called *serial* (or uniserial) modules, have attracted a lot of interest, these are those modules M such that any pair N_1, N_2 of submodules is comparable: we have $N_1 \subseteq N_2$ or $N_2 \subset N_1$.

First of all, for some quite important rings, all the indecomposable finite length modules are serial: this includes the two most prominent commutative rings, the integers and the polynomial ring in one variable (and the necessity to use their finite length modules should be out of doubt), more generally, all Dedekind rings, but also, in the non-commutative realm, the full rings of upper triangular matrices over some field. A second reason is a very trivial one: any indecomposable module of length 2 is serial, and the R-modules of length 2 provide the information exhibited in the quiver of R. However, a general ring, even a general finite-dimensional k-algebra may have few serial modules: for example, consider the exterior algebra $\Lambda_n(k)$ (with generators $X_1, \dots, X_n$ and relations $X_iX_j + X_jX_i = 0$ and $X_i^2 = 0$ for all i, j). Then the only serial modules are the indecomposable modules of length 1 and 2. In particular, if J is the radical of $\Lambda_n(k)$, then J^2 annihilates all the serial modules. Thus, the structure of the algebra is not determined by the serial modules.

Uniform and couniform modules. Recall that a module M is said to be *uniform* provided M is non-zero, and the intersection of any two non-zero sub-

---

we also did so in earlier publications. But one should note that already in 1911 Remak was aware of this kind of results, whereas the relevant papers by Krull and Schmidt are from 1925 and 1928, respectively. The Germans took his life (he died in Auschwitz in 1942), we should not take also his mathematical insight.

modules of M is non-zero again. Thus, M is uniform if and only if M is non-zero and any non-zero submodule of M is indecomposable, again. An injective indecomposable module always has to be uniform, but usually there will exist many indecomposable modules which are not uniform. A module is uniform if and only if its injective envelope is indecomposable. We obtain all uniform R-modules for a ring R as follows: start with all cyclic uniform R-modules U_i, and take all non-zero submodules of their injective envelopes; in this way, we see that the isomorphism classes of uniform modules form a set. There is the dual notion of a couniform module: M is said to be *couniform* provided the following condition is satisfied: if M_1, M_2 are submodules with $M = M_1 + M_2$, then $M = M_1$ or $M = M_2$. Serial modules are both uniform and couniform. Recall that a submodule N of the module M is said to be *essential* provided the only submodule $U \subseteq M$ with $N \cap U = 0$ is the zero module $U = 0$. Dually, N is said to be *small* in M provided the only submodule $V \subseteq M$ with $N + V = M$ is $V = M$ itself. A non-zero module is uniform if and only if every non-zero submodule is essential; it is couniform if any only if every proper submodule is small.

A module M with an essential simple submodule is uniform. In this case, there is a non-zero submodule M_1 which is contained in any non-zero submodule. More generally, one may be interested in modules which have a chain of non-zero submodules $M = M_0 \supseteq M_1 \supseteq M_2 \supseteq \cdots$ indexed by the natural numbers such that any non-zero submodule U of M contains at least one of the modules M_i (and then almost all). We say that such a module is $\mathbb{N}$-*uniform*. If R is a Dedekind ring with only countably many prime ideals (for example $R = \mathbb{Z}$, or $R = k[T]$ where k is a countable field), then the module ${}_RR$ is $\mathbb{N}$-uniform. (Proof: Let $P_1, P_2, \dots$ be all the prime ideals and form $M_i = P_1^{i-1}P_2^{i-2}\cdots P_{i-1}^{1} = \prod_{1\leq j<i} P_j^{i-j}$. Then these ideals form a chain, and any non-zero ideal contains one of them.)

Diamonds. An important class of modules which are both uniform and couniform should be mentioned: We call a module D a *diamond* provided it has a simple essential submodule and also a small maximal submodule (such modules look really quite like diamonds, and their value in ring and module theory is that of diamonds). Let us show that for a semiperfect ring, the direct sum of all diamonds, one from each isomorphism class, is a faithful module: *For any ideal I of a semiperfect ring R, let $\mathcal{D}(I)$ be the set of diamonds annihilated by I. Then the intersection of the annihilators of the diamonds in $\mathcal{D}(I)$ is I.* (Proof: Let I' be the intersection of the annihilators of all the modules in $\mathcal{D}(I)$. Of course, $I \subseteq I'$. Let a be an element of $R \setminus I$, we show that it does not belong to I'. Choose a left ideal L with $I \subseteq L \subset R$ such that $a \notin L$, and maximal with this property, here we use the axiom of choice. It is well-known and easy to see that R/L has a unique minimal submodule, namely the left module generated by the residue class of a, and this submodule is contained in any non-zero submodule. Since R is semiperfect, any module is a sum of local modules, in particular this is true for R/L. Let $R/L = \sum M_i$, with local modules M_i. All these modules M_i are diamonds. Since $a \cdot (1 + L) = a + L \neq L$, we see that R/L is not annihilated by a. It follows that there exists some i such that M_i is not annihilated by a. On the other hand, R/L and therefore all modules M_i are annihilated by I. This shows that a does not belong to I'.)

Let us assume for a moment that R is a finite-dimensional algebra (or, more generally, an artin algebra). Note that all indecomposable modules of length 2 are diamonds, thus it is quite usual to have infinitely many isomorphism classes of diamonds. In case there are only finitely many isomorphism classes of indecomposable R-modules of finite length, the algebra R is said to be *representation-finite.* For a representation-finite algebra the shape of the possible diamonds is very restricted[2]: For example, any semisimple subfactor of a diamond is of length at most 3. Also, for a representation-finite algebra R, diamonds D, D' with the same Loewy length, and with isomorphic top and isomorphic socle are isomorphic[3].

1. Modules in general.

Let us consider now an arbitrary R-module M, also without any restriction on the ring R.

Composition factors. Note that any module M has sufficiently many simple subfactors, in the following sense: given a non-zero element $m \in M$, there are submodules $M'' \subseteq M' \subseteq M$ with M'/M'' simple, such that $m \in M' \setminus M''$.

Generalizing the concept of a composition series, one may look for chains of submodules M_i of M indexed by $i \in I$, where I is one of the sets $\mathbb{N}, -\mathbb{N}, \mathbb{Z}$, with $M_{i-1} \subset M_i$, such that the factors M_i/M_{i-1} are simple, for all i, with $\bigcap_{i\in I} M_i = 0$ and $\bigcup_{i\in I} M_i = M$.

First, consider the case of an ascending chain, thus $I = \mathbb{N}$: in this case, the corresponding factors are again uniquely determined, as in the Jordan-Hölder-Theorem: Assume that there is given a second chain M'_i, $i \in \mathbb{N}$, with the corresponding properties. Consider the index $j \in \mathbb{N}$. The submodule M_j is contained

[2] Let us assume that R is a k-algebra, where k is an algebraically closed field. Using covering theory [GR], we can assume that R is representation-directed. In addition, we may assume that we deal with a faithful diamond D. But then D is both projective and injective and R is the incidence algebra of a finite poset with one minimal and one maximal element. The possible cases are well-known: For example, if there are more than 13 simple R-modules, then there are just seven possibilities labelled (Bo1), (Bo15), (Bo16), (Bo17), (Bo19), (Bo20), (Bo21) in [R4]. Note that all the Jordan-Hölder multiplicities of such a module D are equal to 1 and one obtains in this way a maximal positive root of the corresponding quadratic form. Conversely, Dräxler [Dl] has observed that any representation-directed k-algebra which has a maximal positive root with all coefficients equal to 1 is obtained from an algebra with a faithful indecomposable projective-injective module by a change of orientation.

[3] In case there do exist non-isomorphic diamonds D, D' with the same Loewy length, isomorphic top and isomorphic socle, then it is easy to show that R is even of "strongly unbounded representation type".

in $M = \bigcup M'_j$, and since M_j is finitely generated, it is contained already in some $M'_{j'}$; but $M'_{j'}$ is a finite length module, thus the composition factors of M_j occur as composition factors of $M'_{j'}$. It follows that for any simple module S, the number of indices i with $M_i/M_{i-1} \simeq S$ is smaller or equal to the number of factors $M'_i/M'_{i-1} \simeq S$. By symmetry, these numbers actually are equal, and this is what we wanted to show.

In contrast, for descending chains ($I = -\mathbb{N}$) we cannot expect a corresponding assertion, as already the case $R = \mathbb{Z}$ shows: It is easy to write down all possible descending chains with simple factors: just pick any sequence $p_1, p_2, \ldots$ of prime numbers and take $M_{-i} = p_1 \cdots p_i\mathbb{Z}$. Then $M_{-i}/M_{-i-1} \simeq \mathbb{Z}/\mathbb{Z}p_{i+1}$. For example, we can take the constant sequences $2, 2, \ldots$ or $3, 3, \ldots$; they will not have any common factor. We also should note another fact: both $\mathbb{Z}$ and its p-adic completion $\hat{\mathbb{Z}}_p$ have descending chains M_i, $i \in -\mathbb{N}$ with all the factors $M_i/M_{i-1} \simeq \mathbb{Z}/\mathbb{Z}p$; but the sets $\mathbb{Z}$ and its p-adic completion $\hat{\mathbb{Z}}_p$ have different cardinalities. Of course, similar features occur for the case $I = \mathbb{Z}$. Here another example: again, take $R = \mathbb{Z}$ and consider the $\mathbb{Z}$-module $\mathbb{Q}$. Given a chain M_i $i \in \mathbb{Z}$ of submodules with simple factors such that $\bigcap_i M_i = 0$ and $\bigcup_i M_i = \mathbb{Q}$, we obtain via $M_i/M_{i-1} \simeq \mathbb{Z}/\mathbb{Z}p_i$ a family p_i of prime numbers indexed by $i \in \mathbb{Z}$ such that for any prime number p and any natural number n, there is an index $i \geq n$ with $p_i = p$. And conversely, any such indexed family of prime numbers occurs in this way.

Direct decompositions. Concerning direct decompositions of modules, quite surprising phenomena are possible:

- **Failure of cancelation.** The existence of indecomposable modules M_0, M_1, M_2 such that $M_0 \oplus M_1 \simeq M_0 \oplus M_2$, but $M_1 \not\simeq M_2$.
- **Failure of the Schröder-Bernstein property.** The existence of non-isomorphic modules M, N with monomorphisms $M \to N$ and $N \to M$.
- **Non-uniqueness of roots.** The existence of non-isomorphic modules M, N such that with $M^s \simeq N^s$ for some $s \geq 2$.
- **Isomorphism of specified powers.** The existence of a module M such that $M^s \simeq M^t$ for natural numbers s, t if and only if $s \cong t$ modulo some fixed number n.
- **Decompositions into an arbitrary finite number of indecomposables.** The existence of a module M which can be written as the direct sum of t indecomposable modules for any $t \geq 2$, but not as the direct sum of infinitely many non-zero modules.
- **Superdecomposability.** The existence of a non-zero module M such that no non-zero direct summand of M is indecomposable (thus M itself is not indecomposable, and if we take any non-trivial direct decomposition $M = M_1 \oplus M_2$, then both direct summands M_1, M_2 can be further decomposed).
- **Finiteness of direct decompositions, but any direct decomposition involves decomposable modules.** The existence of a module M with the following property: If $M = \bigoplus_{i \in I} M_i$ is a direct decomposition with non-zero

modules M_i, then the index set I is finite, but at least one of the modules M_i is decomposable.

- **Existence of large indecomposables.** The existence of indecomposable modules which have arbitrarily large cardinality.

The free algebra $k\langle X, Y\rangle$ in two variables has such modules, but many other algebras also. Sometimes, such modules are called "pathological", but this term is misleading: what is called "pathological" seems to be the general behaviour of modules, and why should we think of a general module to be pathological? Let us exhibit some examples which are easy to comprehend and to remember. In addition, we want to formulate some consequences.

Modules and their endomorphism rings. Properties concerning the possible direct decompositions of a module M are reflected in its endomorphism ring $\mathrm{End}(M)$, more precisely in the set of idempotents in $\mathrm{End}(M)$. Namely, the direct sum decompositions $M = M_1 \oplus M_2$ correspond bijectively to the idempotents $e \in \mathrm{End}(M)$ (given such an idempotent e, let $M_1 = eM$ and $M_2 = (1-e)M$). Thus, the study of modules with prescribed behaviour with respect to direct decompositions is part of the question what kind of rings can be realized as endomorphism rings of modules. This topic will be discussed in detail by Eklof in his contribution [E].

Indecomposable modules. Let us start with the indecomposable modules themselves. Of course, a non-zero module M is indecomposable if and only if the only idempotents in $\mathrm{End}(M)$ are the elements 0 and 1.

One of the assumptions in the theorem of Krull-Remak-Schmidt is that one deals with modules M_i with local endomorphism rings. It should be kept in mind that the conclusion of the theorem may hold also in case some of the endomorphism rings are not local. For example, the finitely generated abelian groups are direct sums of indecomposable torsion groups (they are of finite length, thus have local endomorphism rings) and copies of $\mathbb{Z}$ (here the endomorphism ring is $\mathbb{Z}$ again, thus a ring which is **not** local). Given a module M, the full force of $\mathrm{End}(M)$ being local is expressed in the exchange property [CJ,Wa]: If there is a split embedding $M \subset N$ and $N = \bigoplus_{i\in I} N_i$ with arbitrary modules N_i, then there exist submodules $N_i' \subseteq N_i$ such that $N = M \oplus \left(\bigoplus_{i\in I} N_i'\right)$ (and these submodules N_i' are necessarily direct summands of N_i).

There do exist indecomposable modules whose endomorphism rings are not local, but still semilocal; such modules are discussed by Facchini [Fc3] and Pimenov and Yakovlev [PY]; they share some of the properties of modules with local endomorphism rings. An artinian indecomposable module always has a semilocal endomorphism ring [CD]. Let us quote from [PY] an example of an artinian modules which is cyclic, but not of finite length[4]: Take the ring $R = \begin{bmatrix} \mathbb{Q} & 0 \\ \mathbb{Q} & \mathbb{Z} \end{bmatrix}$ and

[4] Observe that for R being commutative, cyclic artinian modules always are

consider the indecomposable projective module P and its submodules U and V, where

$$P = \begin{bmatrix} \mathbb{Q} \\ \mathbb{Q} \end{bmatrix} \supset V = \begin{bmatrix} 0 \\ \mathbb{Q} \end{bmatrix} \supset U = \begin{bmatrix} 0 \\ \mathbb{Z}_{(p)} \end{bmatrix},$$

where p is a prime number. We are interested in the factor module P/U. First of all, as a factor module of P, this is a cyclic module. Second, it is artinian and of infinite length: consider the filtration $0 \subset V/U \subset P/U$. Here, $(P/U)/(V/U) = P/V$ is a simple R-module, whereas V/U is annihilated by P (this is a twosided ideal), thus an R/P-module and $R/P = \mathbb{Z}$, and actually, V/U is just a Prüfer group. Also, it is easy to check that $\mathrm{End}(P/U)$ is equal to $\mathbb{Z}_{(p)}$. — We may do the same construction, replacing p by a product pq of two different prime numbers, thus we consider

$$U' = \begin{bmatrix} 0 \\ \mathbb{Z}_{(pq)} \end{bmatrix}.$$

Again, P/U' is cyclic and artinian (V/U' is the direct sum of two Prüfer groups), and this time $\mathrm{End}(P/U') = \mathbb{Z}_{(pq)}$, thus we obtain a semilocal ring which is not local[5]. Note that the semilocal ring $\mathrm{End}(P/U')$ has precisely two maximal ideals and is a domain; in particular, it is not semiperfect. In the same way, replacing p by a product of n pairwise different prime numbers, we obtain a cyclic artinian module whose endomorphism ring is a semilocal domain with precisely n maximal ideals.

Large indecomposables. Using transfinite induction, Fuchs (1959) has constructed large indecomposable abelian groups, namely groups whose cardinality is any cardinal number less than the first strongly inaccessible cardinal number (a cardinal number λ is said to be strongly inaccessible, provided first: $\lambda > \aleph_0$, second: $2^\mu < \lambda$ for every cardinal number $\mu < \lambda$, and third: $\sum_{i \in I} \mu_i < \lambda$, whenever I is an index set of cardinality smaller than λ and also all $\mu_i < \lambda$). These strongly inaccessible cardinal numbers are huge, and their existence is independent of the usual axioms of set theory. On the other hand, in 1973, Shelah was able to remove this cardinality restriction. Shelah's methods will be presented in this volume by Göbel [Gö]. To quote from his introduction: these are "simple, but

---

of finite length. The reason is that in the commutative case, any cyclic module is really a factor ring, and artinian rings always are noetherian. Let us stress the importance of understanding cyclic artinian modules in case one is interested in artinian modules in general: after all, every module is the sum of its cyclic submodules, thus an artinian one is the sum of artinian cyclic modules. The usual predominance of commutative ring theory is here a clear source for misdirection: the possible existence of non-trivial artinian modules is one of the intrinsic features of non-commutative algebra.

[5] In [Wi], 31.14, Wisbauer claims that the endomorphism ring of an indecomposable artinian module is local.

clever counting arguments"; they are put together in "Shelah's Black Box" and designed for applications in different areas of mathematics.

Formatted modules. We suggest to call a module M *formatted* provided it contains an essential submodule which is a direct sum of uniform modules. Note that a module M is formatted if and only if every non-zero submodule of M contains a uniform submodule (for the proof, use the lemma of Zorn). As a consequence, the class of formatted modules is closed under submodules.

Let M be a formatted module, thus there is an essential submodule $\bigoplus_{i\in I} U_i$ with uniform modules U_i. Note that the cardinality of I is an invariant of M, which may be called its *uniform dimension* (or *Goldie dimension*). (Proof: Consider the set $\mathcal{U}$ of uniform submodules of an module. We call $Q \in \mathcal{U}$ dependent on $P_i (i \in I)$ provided Q intersects $\sum P_i$ non trivially. A family in $\mathcal{U}$ is said to be independent, if no element is dependent on the rest. With these definitions, the set $\mathcal{U}$ satisfies the axioms of an abstract dependence relation [Cn]: First, any element of a set is dependent on the set. Second: If Z is dependent on the independent set Y and each element of Y is dependent on X, then Z is dependent on X. And finally, the exchange property: If Y is dependent on the set $\{X_1, \ldots, X_n\}$, but not on $\{X_2, \ldots, X_n\}$, then X_1 is dependent on $\{Y, X_2, \ldots, X_n\}$. For such a dependence relation $\mathcal{U}$, the maximal independent subsets of $\mathcal{U}$ are precisely the minimal spanning sets [Cn, 1.4.1] (a spanning set is a subset of $\mathcal{U}$ such that every element of $\mathcal{U}$ is dependent on it) — such a set is called a basis, and it is a general assertion that the cardinality of a basis is an invariant.)

It is easy to construct indecomposable modules of infinite uniform dimension. Consider the subring $R = \begin{bmatrix} k & 0 \\ k[T] & k \end{bmatrix}$ of the ring of 2×2 matrices over $k[T]$. There are up to isomorphism two indecomposable projective R-modules: one is simple, the other one is a local module of infinite uniform dimension.

In noetherian ring theory, the modules of finite uniform dimension play a decisive role, these are just those modules M which do not contain an infinite direct sum of non-zero submodules, as one easily verifies. As a trivial consequence, one sees that any noetherian module has finite uniform dimension, thus is an essential extension of a finite direct sum of uniform modules. More generally, any module M which is generated by noetherian modules is formatted[6]. (Proof: Assume that M is generated by noetherian submodules M_i with $i \in I$. Note that the sum of two noetherian submodules is again noetherian, thus we may assume that the submodules M_i form a directed family. If U is any non-zero submodule of M, then $U = U \cap M = U \cap \sum M_i = \sum (U \cap M_i)$ yields some $i \in I$ with $U \cap M_i$ non-zero. Now, $U \cap M_i$ is noetherian, since it is a submodule of M_i, thus $U \cap M_i$ contains some uniform submodule. This shows that U has a uniform submodule.)

As a consequence, given a left noetherian ring R, then any R-module is formatted. The same is true in case R is semi-artinian, since the semi-artinian

[6] Warning: In general, the sum of two formatted submodules does not have to be formatted, an example will be given below.

rings are characterized by the property that every non-zero module has a simple submodule.

Discrete modules. Next, we want to discuss the question of the existence of indecomposable direct summands. A module M is said to be *discrete* provided every non-zero direct summand of M has an indecomposable direct summand. As we will see below, for injective modules the formatted ones are just the discrete ones. For non-injective modules, being formatted and being discrete are completely different properties: For an artinian ring T, all the R-modules are formatted, but any strictly wild finite-dimensional algebra has many superdecomposable modules. This shows that a formatted module does not have to be discrete. Conversely, it is easy to construct local modules which do not contain uniform submodules (this provides an indecomposable, thus discrete module which is not formatted): consider the matrix algebra $R = \begin{bmatrix} k & 0 \\ k\langle X,Y\rangle & k\langle X,Y\rangle \end{bmatrix}$, where $k\langle X,Y\rangle$ is the free algebra in two generators. The indecomposable projective R-module given by the first column is local and has no uniform submodule. Namely, observe that for any element $a \in k\langle X,Y\rangle$, the two elements Xa and Ya of $k\langle X,Y\rangle$ generate submodules with zero intersection:

$$k\langle X,Y\rangle Xa \cap k\langle X,Y\rangle Ya = 0.$$

Superdecomposable modules. A module M is said to be *superdecomposable* provided no direct summand of M is indecomposable. Note that the class of superdecomposable modules is closed under direct summands.

Superdecomposable modules may look surprising at the first sight, but there are at least two natural examples to be mentioned. The first is the injective envelope I of the regular representation ${}_{k\langle X,Y\rangle}k\langle X,Y\rangle$ of the free algebra $k\langle X,Y\rangle$ in two variables. Given a non-zero direct summand N of I, the intersection $N \cap k\langle X,Y\rangle$ is non-zero. Take an element a in this intersection, then the injective hull of $k\langle X,Y\rangle Xa$ is a proper non-zero direct summand of N. This shows that I is superdecomposable.

Another example of a superdecomposable module is the regular representation of the ring $\mathcal{C}(X)$ of all continuous functions $X \to \mathbb{R}$, where X is the Cantor discontinuum (or any totally disconnected topological space X without isolated points). As for any ring R, the direct summands of the regular representation ${}_RR$ are of the form Re, where e is an idempotent in R. The idempotents e of R correspond bijectively to the subsets of X which are both open and closed. Note that we assume that X is totally disconnected and that there are no isolated points. Thus, if X' is any non-empty subset of X which is open and closed, then X' can be written (in many ways) as the disjoint union of subsets which again are open and closed.

Finiteness of direct decompositions, but any direct decomposition involves decomposable modules. Let us stress that there do exist superdecomposable modules which cannot be written as infinite direct sums of non-zero modules. Actually, both examples presented above can be quoted. First, consider the injective envelope I of the regular representation ${}_RR$ of the free algebra $R = k\langle X,Y\rangle$ in two variables, and assume $I = \bigoplus_{j\in J} I_j$. The unit element 1_R belongs to a finite direct sum $\bigoplus_{j\in J'} I_j$, where J' is a finite subset of J, but then also $R = R\cdot 1_R$ is contained in $\bigoplus_{j\in J'} I_j$, thus also I. Similarly, consider $R = \mathcal{C}(X)$, where X is totally disconnected without isolated points. If we assume in addition that X is compact, then again we see that ${}_RR$ cannot be written as the direct sum of infinitely many non-zero submodules.

But there are also discrete modules with only finite direct decompositions, but such that any direct decomposition involves decomposable modules. Consider again the subring $R = \begin{bmatrix} k & 0 \\ k[T] & k \end{bmatrix}$ of the ring of 2×2 matrices over $k[T]$ and let $P(1)$ be the indecomposable projective R-module which is not simple. We denote by I the injective envelope of $P(1)$. Since the socle U of $P(1)$ is an essential submodule, it follows that U is essential in I. Also, $\operatorname{Hom}(I/U, I) = 0$, thus the intersection with U yields a bijection between the direct decompositions $I = I' \oplus I''$ and the direct decompositions of U. In particular, we see that I is a discrete module. On the other hand, assume there is given a countable direct decomposition $I = \bigoplus_i I_i$ with non-zero submodules I_i. Note that $I_i \cap U \neq 0$, thus $I_i \cap U$ contains a simple submodule S_i. Of course, all these submodules S_i are isomorphic to the simple projective module $P(2)$, thus $\bigoplus_i S_i$ is isomorphic to U. In this way, we obtain an injective map $U \to \bigoplus_i S_i \subset \bigoplus_i I_i = I$. Using the injectivity of I, we obtain an extension of f to a homomorphism $f'\colon P(1) \to \bigoplus_i I_i$. But $P(1)$ is cyclic, thus the image of f' is contained in a finite direct sum, a contradiction. This shows that any direct decomposition of I has to be finite.

Comparison of different direct decompositions. Let us return to the theorem of Krull-Remak-Schmidt: If M is a module whose endomorphism ring is local, one has the following nice situation: if M occurs as a direct summand of a module M' and M' can be decomposed in the form $M' = M_1' \oplus M_2'$, then M is isomorphic to a direct summand of M_1 or M_2. This means that such a module M behaves like a prime element when we consider the relation of being a direct summand in analogy to the relation that one element divides another element in a commutative ring.

We are going to exhibit some typical ways for obtaining modules M_1, M_2, N_1, N_2 with an isomorphism $M_1 \oplus M_2 \simeq N_1 \oplus N_2$. The aim is to provide examples of indecomposable modules M_1, N_1, N_2 such that M_1 is a direct summand of $N_1 \oplus N_2$, but not isomorphic to any one of the modules N_1, N_2.

(1) Let M be a projective module with submodules N_1, N_2 such that $N_1 + N_2 = M$. Let $M' = N_1 \cap N_2$. Then the inclusion maps ι give rise to an exact sequence

$$0 \longrightarrow M' \xrightarrow{\begin{bmatrix} \iota \\ -\iota \end{bmatrix}} N_1 \oplus N_2 \xrightarrow{[\iota \ \ \iota]} M \longrightarrow 0,$$

and, since M is projective, the sequence splits. Thus

$$M \oplus M' \simeq N_1 \oplus N_2.$$

We consider the special case where R is an integral domain and consider the regular representation $M = {}_RR$. Given non-zero ideals N_1, N_2, then also $M' = N_1 \cap N_2$ is non-zero (an integral domain is a uniform module), thus all the modules M, M', N_1, N_2 are indecomposable.

Here are some typical cases. We may take as N_1, N_2 any pair of two different maximal ideals of an integral domain R, then clearly $N_1 + N_2 = R$. For example, let $R = k[X,Y]$ be the polynomial ring in two variables, $N_1 = RX + RY$, and $N_2 = RX + R(Y-1)$. The submodules N_1, N_2 are not cyclic; in particular, they are not isomorphic to ${}_RR$. Thus we see that ${}_RR$ is an indecomposable direct summand of $N_1 \oplus N_2$, and not isomorphic to N_1 or N_2.

Second, an example suggested by Mazorchuk: Let R be the subring of the polynomial ring $k[X]$ generated by X^2 and X^3. Let $N_1 = R(X^2+1)$ and $N_2 = RX^2 + RX^3$. Again we see that $N_1 + N_2 = R$. Note that $N_1 \cap N_2 = R(X^4 + X^2)$, in particular, both N_1 and $N_1 \cap N_2$ are principal ideals, thus isomorphic, as R-modules, to ${}_RR$. The isomorphism $M \oplus M' \simeq N_1 \oplus N_2$ can be rewritten in the form

$${}_RR \oplus {}_RR \simeq {}_RR \oplus N_2,$$

and N_2 is not cyclic. A typical instance where cancelation fails.

(2) Consider a module with submodules $0 \subset U \subset M' \subset M$. There is the following exact sequence:

$$0 \longrightarrow M' \xrightarrow{\begin{bmatrix} \iota \\ -\pi' \end{bmatrix}} M \oplus M'/U \xrightarrow{[\pi \ \ \iota']} M/U \longrightarrow 0,$$

again inclusion maps are denoted by ι, projection maps by π. Let us assume that the module M' is both uniform and couniform, and, in addition[7], that there is a monomorphism $f\colon M \to M'$ and an epimorphism $g\colon M'/U \to M'$. Under these assumptions, the sequence splits, thus

$$M \oplus M'/U \simeq M' \oplus M/U.$$

(Proof: Consider the endomorphism $h = f \circ \iota - g \circ \pi'$ of M'. First, let us show that h is a monomorphism. Note that $\mathrm{Ker}(h) \cap \mathrm{Ker}(g \circ \pi')$ is contained in the

[7] These additional assumptions imply that the given module M is isomorphic to a proper submodule as well as a proper factor module of itself (note that an artinian module is never isomorphic to any of its proper submodules, a noetherian module is never isomorphic to any of its proper factor modules; thus, the modules we are dealing with are neither artinian nor noetherian).

kernel $\operatorname{Ker}(f\circ\iota)$ and $\operatorname{Ker}(f\circ\iota)=0$. Since M' is uniform, and $\operatorname{Ker}(g\circ\pi')$ is non-zero, it follows that $\operatorname{Ker}(h)=0$. Second, $\operatorname{Im}(f\circ\iota)$ is a proper submodule of M', since both f and ι are monomorphism and at least ι is a proper inclusion. Now, $\operatorname{Im}(f\circ\iota)+\operatorname{Im}(h)\supseteq\operatorname{Im}(g\circ\pi')=M'$. Since M' is couniform, it follows that $\operatorname{Im}(h)=M'$. This shows that $h=[f\ \ g]\circ\begin{bmatrix}\iota\\-\pi'\end{bmatrix}$ is an isomorphism, thus the sequence splits.)

The latter construction is the one used by Facchini in order to show that the Krull-Remak-Schmidt property does not hold for finitely presented modules over serial rings [Fc1,Fc2,Fc3], in this way solving a problem raised by Warfield. Consider the ring

$$R=\begin{bmatrix}\mathbb{Z}_{(p)} & p\mathbb{Z}_{(p)} & 0 & 0\\ \mathbb{Z}_{(p)} & \mathbb{Z}_{(p)} & 0 & 0\\ \mathbb{Q} & \mathbb{Q} & \mathbb{Z}_{(q)} & q\mathbb{Z}_{(q)}\\ \mathbb{Q} & \mathbb{Q} & \mathbb{Z}_{(q)} & \mathbb{Z}_{(q)}\end{bmatrix}$$

where p,q are different primes. Denote by $P(i)$, for $1\le i\le 4$, the indecomposable projective R-module given by the ith column, let $V=\begin{bmatrix}0\\0\\\mathbb{Q}\\\mathbb{Q}\end{bmatrix}$ and note the various inclusions of these modules: all are submodules of $P(1)$. Here, on the left, is the complete submodule lattice of $P(1)$, this module $P(1)$ is a serial module. On the right, the module $M=P(1)/P(4)$ is depicted and we have marked its submodules $U=P(3)/P(4)$ and $M'=P(2)/P(4)$:

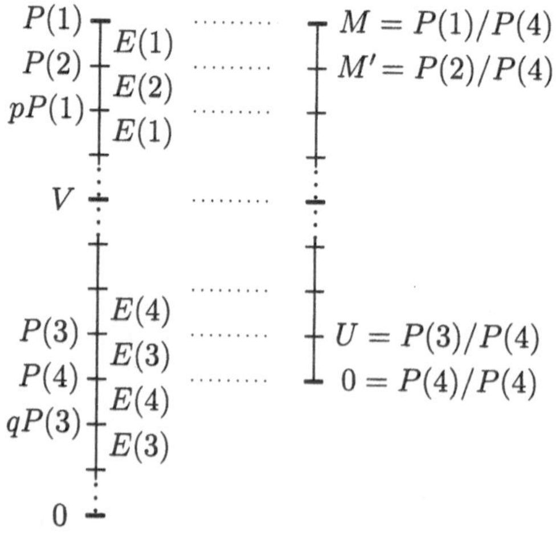

The following observation is of importance: The scalar multiplication by p maps $P(1)$ onto the submodule $pP(1)$, with $pP(1)\subset P(2)\subset P(1)$, and the factors $E(1)=P(1)/P(2)$ and $E(2)=P(2)/pP(1)$ are simple. However $pP(3)=P(3)$

and $pP(4) = P(4)$, since p is invertible in $\mathbb{Z}_{(q)}$. Similarly, $qP(1) = P(1)$, $qP(2) = P(2)$ and $qP(3) \subset P(4) \subset P(3)$ with simple factors $E(3) = P(3)/P(4)$ and $E(4) = P(4)/qP(3)$. As monomorphism $f\colon M \to M'$ we take the multiplication by p. As epimorphism $g\colon M'/U \to M'$ we compose the isomorphism $M'/U = P(2)/P(3) \to P(2)/qP(3)$ given by the multiplication with q and the canonical projection $P(2)/qP(3) \to P(2)/P(4)$. In this way, we see that all the requirements needed are fulfilled and $M \oplus M'/U \simeq M' \oplus M/U$. The modules occurring in these two direct decompositions have the following form:

$$
\begin{array}{cccc}
M & M'/U & M' & M/U \\[2pt]
\begin{array}{c} E(1) \\ E(2) \\ E(1) \\ \vdots \\ E(3) \\ E(4) \\ E(3) \end{array} &
\begin{array}{c} E(2) \\ E(1) \\ E(2) \\ \vdots \\ E(4) \\ E(3) \\ E(4) \end{array} &
\begin{array}{c} E(2) \\ E(1) \\ E(2) \\ \vdots \\ E(3) \\ E(4) \\ E(3) \end{array} &
\begin{array}{c} E(1) \\ E(2) \\ E(1) \\ \vdots \\ E(4) \\ E(3) \\ E(4) \end{array}
\end{array}
$$

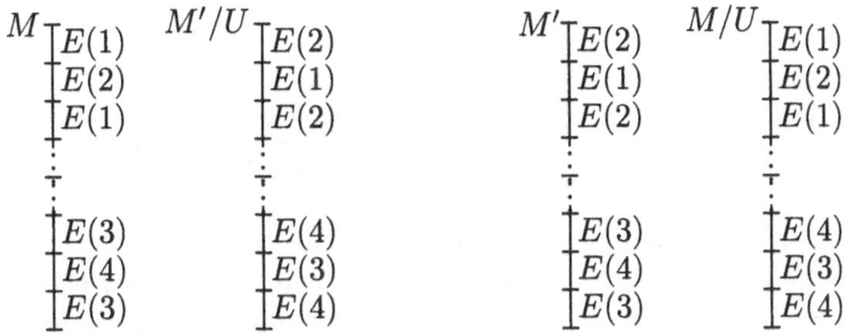

What we see is a sort of partial exchange, we may describe it as follows: if we keep the upper (noetherian) parts of the modules, then the lower (artinian) parts are exchanged. Remember that for all these considerations, the essential ingredient is the existence of a proper monomorphism $M \to M$ and a proper epimorphism $M \to M$; here the scalar multiplications by p and by q serve this purpose[8]. It should also be remarked that the endomorphism ring of our module M is just $\mathbb{Z}_{(pq)}$. In general [Fc3], the endomorphism ring of a serial module is semilocal with at most two maximal ideals, one consists of all endomorphisms which are not injective, the other of all endomorphisms which are not surjective, and any proper one-sided ideal is contained in at least one of these ideals (these ideals may coincide, as one knows from the indecomposable modules of finite length, but in general they will not, a typical example is the module M considered above).

The isomorphism property for specified powers. Consider a module with the isomorphism property for specified powers, say for $n = 2$. In particular, this means that $M \not\simeq M \oplus M$, but $M \simeq M \oplus M \oplus M$. Two consequences should be mentioned: Consider the modules M and $N = M \oplus M$. *The modules M, N are non-isomorphic and have the following properties:* First, *M is a direct summand of N and N is a direct summand of M* (failure of the Schröder-Bernstein property). Second, *the modules $M \oplus M$ and $N \oplus N$ are isomorphic* (non-uniqueness

---

[8] Note that the existence of the scalar multiplications by p and by q has the consequence that all serial modules with top $E(1)$ and socle $E(3)$ are isomorphic. It follows that the ring R has only finitely many isomorphism classes of serial modules of infinite length. In contrast, if we deal with the similarly defined 4×4 matrix ring using only one prime number $p = q$, then there are infinitely many isomorphism classes of serial modules of infinite length.

of roots). In case we deal with abelian groups, thus with $\mathbb{Z}$-modules, the question whether non-isomorphic modules with one of these properties do exist, was raised by Kaplansky [Ka] in 1954 as his two famous *Test Problems;* the existence of non-isomorphic abelian groups M and N such that $M \oplus M$ and $N \oplus N$ are isomorphic was shown by Jónsson in 1957, that of non-isomorphic abelian groups M and N such that M is a direct summand of N and N is a direct summand of M was shown by Sąsiada in 1961. In 1964, Corner exhibited an abelian group M which is not isomorphic to $M \oplus M$, but isomorphic to $M \oplus M \oplus M$. We refer to Eklof's contribution [E] in this volume which focuses the attention precisely to this problem.

The non-uniqueness of roots is discussed in detail in the very nice booklet by Lam [La] which just has appeared. He starts with a detailed discussion of the isomorphism $P \oplus P \simeq R \oplus R$, where $R = \mathbb{Z}[\sqrt{-5}]$ and P is the ideal of R generated by 2 and $1 + \sqrt{-5}$. Lam presents several additional examples, but also many affirmative results valid under suitable conditions on the ring or the modules and he outlines the relationship to problems in number theory, K-theory and the theory of operator algebras. He also includes related results concerning the cancelation problem.

2. The categorical setting

For the problems to be discussed here, it seems to be appropriate to consider not only modules over a ring with 1, but more generally over a ring R which has sufficiently many idempotents: this means that $R = \bigoplus_{i,j \in I} e_i R e_j$ for some set of idempotents e_i, $i \in I$ (and here we should not be too fuzzy about the word "set"). Of course, this means that R is really just an additive category (preferably with a set of objects), and the category of R-modules is the category of all contravariant functors to abelian groups (this point of view is often expressed by Mitchell's formulation that we deal with a "ring with several objects"). The advantage of this slight generalization is the following: starting with such a ring with several objects, thus with an additive category $\mathcal{A}$, the functor category $\operatorname{Mod}\mathcal{A}$ of all controvariant functors from $\mathcal{A}$ to $\operatorname{Mod}\mathbb{Z}$ again is an additive category, and we may repeat this process. For set theoretical reasons, it will be necessary to restrict the size of the modules which we consider, or to work with increasing universes, but this should not lead to confusion.

Additive categories arise very frequently in mathematics and all the questions which usually are asked when dealing with a ring (a commutative one as it is considered in number theory or in algebraic geometry, or the non-commutative versions which are now very popular under the name of quantizations) can and have to be asked in this more general setting. One of the main questions concerns the description of such a ring or such a category, by generators and relations. If we consider a representation-finite k-algebra R, where k is an algebraically closed field of characteristic different from 2, it is the Auslander-Reiten quiver Γ_R which produces such a presentation: the generators are given by the vertices and the arrows of Γ_R, thus by the indecomposable R-modules and the irreducible maps between them, the relations are just the mesh relations. Also for a general, not necessarily representation-finite k-algebra R, the Auslander-Reiten quiver Γ_R describes at least part of the category of R-modules, the next section will be devoted to discuss some examples.

When dealing in this way with a module category as a sort of ring, the individual modules are just mere elements of this ring (note that a category is just a set or class of "maps", these are the elements of the category; in this interpretation, the objects of the category are the identity maps, thus special idempotents.) On the other hand, many of the properties of a module which are of interest are really categorical properties, they can be recovered from the module category. In particular, this holds true for the property of having finite or infinite length, at least as long as we work with the full module category.

Ideals. When dealing with a ring, one of the basic concepts is that of an ideal and the corresponding factor ring. Clearly, dealing with an additive category $\mathcal{A}$, there also is the notion of an ideal $\mathcal{I}$ in $\mathcal{A}$ and of the corresponding factor category[9] $\mathcal{A}/\mathcal{I}$. In particular, given a class $\mathcal{U}$ of objects in $\mathcal{A}$ (or, equivalently, a full subcategory), we may consider the ideal $\langle\mathcal{U}\rangle$ given by all maps which factor through a finite direct sum of objects in $\mathcal{U}$. Such an ideal is idempotent since it is generated by idempotent elements (the identity maps 1_U, where U is a finite direct sum of objects in $\mathcal{U}$). But note that not all idempotent ideals have to be of this form, as we will see below.

As for a ring, we may speak of the Jacobson radical $\mathrm{rad}(\mathcal{A})$ of an additive category $\mathcal{A}$: it consists of all homomorphisms $f\colon A \to A'$ such that for any homomorphism $g\colon A' \to A$, the endomorphism $1 + g \circ f$ is invertible (this just means that $\mathrm{Hom}(A', A) \circ f$ is contained in the Jacobson radical of $\mathrm{End}(A)$).

If we consider idempotent ideals in an additive category $\mathcal{A}$, two different kinds can be distinguished. First of all, the ideal $\mathcal{I}$ may be generated by idempotent elements, thus "by objects", as mentioned above. But second, it is also easy to construct idempotent ideals which contain only nilpotent elements. We stress that the only idempotents contained in the Jacobson radical of $\mathcal{A}$ are the zero endomorphisms. In case every object of $\mathcal{A}$ is a finite direct sum of objects with local endomorphism rings, then any ideal of $\mathcal{A}$ which properly contains $\mathrm{rad}(\mathcal{A})$ will also contain non-trivial idempotents; thus, in his case $\mathrm{rad}(\mathcal{A})$ is the largest

---

[9] When starting with an abelian category $\mathcal{A}$, the symbol / sometimes also is used in order to denote quotient categories with respect to a Serre subcategory; this is a completely different construction and should not be confused. Well-known situations of forming the factor categories $\mathcal{A}/\mathcal{J}$ are the categories $\underline{\mathrm{mod}}R = \mathrm{mod}\,R/\langle\mathcal{P}\rangle$ and $\overline{\mathrm{mod}}R = \mathrm{mod}\,R/\langle\mathcal{I}\rangle$ where $\langle\mathcal{P}\rangle$ is the ideal of all homomorphisms which factor through a projective module, and $\langle\mathcal{I}\rangle$ is that of all homomorphisms which factor through an injective module. For R selfinjective, these categories coincide and are of special importance.

ideal without non-trivial idempotents. The Jacobson radical $\mathrm{rad}(\mathcal{A})$ never contains non-trivial idempotents, but it may contain non-trivial idempotent ideals. Actually, we may define by transfinite induction the power $\mathrm{rad}^{\alpha}(\mathcal{A})$ for any ordinal number α, and then the transfinite radical[10] $\mathrm{rad}^{\infty}(\mathcal{A}) = \bigcap_{\alpha} \mathrm{rad}^{\alpha}(\mathcal{A})$, where α runs through all ordinal numbers, see [P1,Sc1,Sc3]. This transfinite radical $\mathrm{rad}^{\infty}(\mathcal{A})$ is an idempotent ideal and it contains all the idempotent ideals of $\mathcal{A}$ which are contained in $\mathrm{rad}(\mathcal{A})$. An easy way for obtaining an idempotent ideal is the following: select a class $\mathcal{H}$ of homomorphisms such that any $h \in \mathcal{H}$ can be written in the form $h = g \circ h' \circ g' \circ h'' \circ g''$ with $h', h'' \in \mathcal{H}$ and arbitrary maps g, g', g'', then $\mathcal{H}$ generates an idempotent ideal.

Let us mention two typical situations where it is easy to specify such classes $\mathcal{H}$ inside $\mathrm{rad}(\mathcal{A})$. First of all, consider a tubular algebra [R6]. Recall that there are a preprojective component and a preinjective component, and that the remaining components form tubular families $\mathcal{T}_{\gamma}$ with γ a non-negative rational number or the symbol ∞. Let $\mathcal{H}$ be the set of all homomorphisms $h\colon A \to A'$, where A belongs to a family $\mathcal{T}_{\gamma}$, A' to $\mathcal{T}_{\gamma'}$ and $\gamma < \gamma'$. Choose γ'' with $\gamma < \gamma'' < \gamma'$ and note that $\mathcal{T}_{\gamma''}$ is a separating tubular family: this shows that h can be factored through a module in $\mathcal{T}_{\gamma''}$. Note that the idempotent ideal generated by $\mathcal{H}$ has the following property: any element of this ideal is the sum of elements f with square zero.

Second, consider a non-domestic string algebra, see [Sc3]. There are arrows $\alpha, \beta, \gamma, \delta$ and words of the form $u\gamma^{-1}v, u\delta w$ which both start with the letter α and end in the letter β^{-1}. We fix $\alpha, \beta, \gamma, \delta$ and v, w, and consider the set $\mathcal{U}$ of all words which start with α and such that both words $u' = u\gamma^{-1}v$ and $u'' = u\delta w$ exist. Note that with u also $u'u$ and $u'u''u$ belong to $\mathcal{U}$. Let $\mathcal{H}$ be the set of canonical maps $h_u\colon M(u') \to M(u'')$ with image $M(u)$, where $u \in \mathcal{U}$. Given $u \in \mathcal{U}$, we can factorize h_u as follows:

$$M(u') \xrightarrow{g''} M(u'u') \xrightarrow{h_{u'u}} M(u'u'') \xrightarrow{g'} M(u'u''u') \xrightarrow{h_{u'u''u}} M(u'u''u'') \xrightarrow{g} M(u''),$$

here g'', g' are the canonical inclusions, and g is the canonical map with image $M(u)$.

On the other hand, it has been conjectured (by Prest and others) that for a domestic algebra R, the transfinite radical $\mathrm{rad}(\mathrm{mod}\, R)$ is zero, thus that the only idempotent ideal inside $\mathrm{rad}(\mathrm{mod}\, R)$ is the zero ideal. This conjecture has been verified by Skowroński for strongly simply connected algebras [Sk] and by Schröer for string algebras.

Ideals and "ideal objects". Recall that the concept of an ideal in a ring R goes back to Kummer. Consider the ring of algebraic integers in a number field, it is what now is called a Dedekind ring, but in contrast to say the

[10] One should be careful to distinguish the transfinite radical rad^{∞} and the "infinite" radical $\mathrm{rad}^{\omega} = \bigcap_{n \in \mathbb{N}} \mathrm{rad}^{n}$; often the latter also is denoted by rad^{∞}.

integers $\mathbb{Z}$, it need not be a principal ideal domain. In a principal ideal domain, every non-zero element can be written as a product of prime elements and such a factorization is essentially unique. Such a result is not valid for arbitrary Dedekind rings, already in $\mathbb{Z}[\sqrt{-5}]$ we have two completely different factorizations $6 = 2 \cdot 3 = (1+\sqrt{-5})(1-\sqrt{-5})$. To overcome this difficulty, Kummer introduced his "ideal numbers" and he obtained an essentially unique factorization of non-zero elements (even of arbitrary "ideal numbers") as products of "ideal numbers". In the terminology of Dedekind which is used today, instead of considering an element, we look at the corresponding principal ideal generated by the element, but we also take into account ideals which are not principal ideals. As it turns out, in a Dedekind ring, the set of non-zero ideals is a free commutative semigroup, the free generators are the non-zero prime ideals. As we have mentioned, ideals serve as a generalization of the principal ideals. One may try to associate to an ideal I of a ring R an "ideal element" s which generates I; of course, in case I is not a principal ideal of R, then s cannot be an element of R, but it may be an element of some overring R' of R. We may reverse these considerations: let $R \subset R'$ be commutative rings, and let $s \in R'$ be a non-zero element. The intersection of the principal ideal of R' generated by s with R yields an ideal of R which usually will not be principal.

In the context of categories, Krause has followed these ideas: starting with a category $\mathcal{C}'$ and a subcategory $\mathcal{C}$, we may intersect any ideal of $\mathcal{C}'$ with $\mathcal{C}$ and we will obtain an ideal of $\mathcal{C}$. In particular, we may start with an object S in $\mathcal{C}'$ and take the ideal of $\mathcal{C}'$ generated by 1_S, this is just the class of all homomorphisms which factor through a direct sum of copies of S; this is an idempotent ideal of $\mathcal{C}'$, however the intersection of this ideal with $\mathcal{C}$ does not have to be idempotent again: Actually, it may be nilpotent: take $\mathcal{C} = \operatorname{mod} R$, $\mathcal{C}' = \operatorname{Mod} R$, where R is a tame hereditary finite-dimensional algebra and let S be the generic module of infinite length. Of course, the ideal of $\operatorname{Mod} R$ generated by 1_S is idempotent, but its intersection $\mathcal{I}$ with $\operatorname{mod} R$ satisfies $(\mathcal{I})^2 = 0$. Krause [K6,K7] has introduced the concept of fp-idempotence for ideals, and it turns out that such ideals as $\mathcal{I}$ are fp-idempotent. We may say that $\mathcal{I}$ is a sort of shadow of the generic module S inside the category $\operatorname{mod} R$; the object S itself is not visible in $\operatorname{mod} R$, but its shadow is. Conversely, starting from the category $\mathcal{C}$, there may exist ideals $\mathcal{I}$ in $\mathcal{C}$ which can be considered as shadows of objects outside of the category, of "ideal objects": ideals which can be obtained as the intersection $\langle S \rangle_{\mathcal{C}'} \cap \mathcal{C}$ where S is an object of some category $\mathcal{C}' \supset \mathcal{C}$.

Categorical equivalences. Concerning the categorical approach, a further view point should be stressed: categorical equivalences show that a given category may be realized in different ways. The paper [PY] by Pimenov and Yakovlev provides a particularly nice example of two different realizations of a category. They consider abelian groups and maps between them. Let $\mathcal{T}$ be the category of all torsionfree groups and as second category take the category[11] $\mathcal{R}$ of surjective group homomorphisms $f\colon M \to N$, where M is torsionfree divisible and N is torsion (and also divisible, since N is a factor group of the divisible group M). It is easy to see that the categories $\mathcal{T}$ and $\mathcal{R}$ equivalent: given a torsionfree group F, let $I(F)$ be its injective hull, then the canonical projection $I(F) \to I(F)/F$ belongs to $\mathcal{R}$; conversely, given an object (f) in $\mathcal{R}$, then $\mathrm{Ker}(f)$ is a torsionfree group. In this way we obtain mutually inverse equivalences. Note that the functor $\mathcal{R} \to \mathcal{T}$ which attaches to (f) its kernel $\mathrm{Ker}(f)$ is faithful, since any homomorphism between torsionfree abelian groups has a unique extension to their injective envelopes.

The category $\mathcal{R}$ can be identified with the the category of those representations[12] $(f) = (M, N, f)$ of the bimodule ${}_{\mathbb{Z}}\mathbb{Q}_{\mathbb{Q}}$, for which f is surjective. Note that a general representation (M, N, f) of ${}_{\mathbb{Z}}\mathbb{Q}_{\mathbb{Q}}$ is the direct sum of $(M, f(M), f)$ (which belongs to $\mathcal{R}$) and of $(0, N/f(M), 0)$; here we use that with M also $f(M)$ is divisible, thus $f(M)$ is a direct summand of N. As we know, representations of the bimodule ${}_{\mathbb{Z}}\mathbb{Q}_{\mathbb{Q}}$ are just R-modules, where $R = \begin{bmatrix} \mathbb{Q} & 0 \\ \mathbb{Q} & \mathbb{Z} \end{bmatrix}$. In $\mathrm{Mod}\, R$, the full subcategory $\mathcal{R}$ is the class of all torsion modules for a split torsion pair, the torsionfree modules being those of the form $(0, G, 0)$, where G is an arbitrary abelian group.

Let $\mathcal{T}'$ be the full subcategory of all torsionfree groups F of finite rank such that $pF = F$ for almost all primes p. Let $\mathcal{R}'$ be the full subcategory of $\mathcal{R}$ consisting of all artinian R-modules. Observe that an R-module (M, N, f) is artinian if and only if both M and N are artinian abelian groups, thus if we assume that f is surjective, (M, N, f) is artinian in $\mathrm{Mod}\, R$ if and only if M is of finite rank and N is the direct sum of finitely many Prüfer groups. Under the equivalences described above, these subcategories $\mathcal{T}'$ and $\mathcal{R}'$ correspond to each other. There is nothing strange about the equivalence of the subcategory $\mathcal{T}$ of $\mathrm{Mod}\, \mathbb{Z}$ and the subcategory $\mathcal{R}$ of $\mathrm{Mod}\, R$, or also about the equivalence of the categories $\mathcal{T}'$ and $\mathcal{R}'$, but nevertheless it is worthwhile to contemplate. By definition, all the R-modules in $\mathcal{T}'$ are artinian, whereas the only artinian $\mathbb{Z}$-module in $\mathcal{R}'$ is the zero module. Also, the artinian property of the modules in $\mathcal{T}'$ does not seem to be related to a dual property for the modules in $\mathcal{R}'$ such as being noetherian: some of the $\mathbb{Z}$-modules in $\mathcal{R}'$ are noetherian, most of them not.

Recall that the category $\mathcal{T}'$ of all torsionfree abelian groups of finite rank was

---

[11] We write the objects in $\mathcal{R}$ in the form (f), where f is a homomorphism as indicated; given two homomorphisms $f\colon M \to N$ and $f'\colon M' \to N'$, the maps $(\alpha, \beta)\colon (f) \longrightarrow (f')$ in $\mathcal{R}$ are as usual pairs, with $\alpha : M \to M'$ and $\beta\colon N \to N'$ such that $\beta \circ f = f' \circ \alpha$.

[12] If R, S are rings and ${}_S X_R$ is a bimodule, a *representation* of ${}_S X_R$ is by definition of the form $(f) = ({}_R M, {}_S N, f\colon {}_S X \otimes_R M \to {}_S N)$, with f being S-linear, but this is nothing else than a C-module, where $C = \begin{bmatrix} R & 0 \\ X & S \end{bmatrix}$.

the main playing ground for considering the failure of the Krull-Remak-Schmidt property, since papers by Jónsson (1945, 1957) and Corner (1961). Many different constructions are known, and via the stated equivalence they carry over to the category $\mathcal{R}'$, thus one obtains in this way many examples of artinian modules which do not satisfy the Krull-Remak-Schmidt property. Note that Krull raised the question whether this property holds for artinian modules in 1932, and the first answer to this question is only from 1995, see the paper [FHLV] by Facchini, Herbera, Levy and Vámos. Let us repeat: If we assume the knowledge on $\mathcal{T}'$ established a long time ago, we obtain the corresponding assertions for $\mathcal{R}'$ using the equivalence of the categories. However, it may be reasonable to revert these considerations. If one analyzes the usual constructions made in $\mathcal{T}'$ for obtaining a torsionfree group F of finite rank, one observes that these are really constructions starting with a finite direct sum of copies of $\mathbb{Q}$, thus with $I(F)$, and prescribing a direct sum of Prüfer modules as factor module: the group F is given by a minimal injective resolution, thus by an object in $\mathcal{R}'$.

It may be sufficient to discuss just one example, the construction of a torsionfree group $F_1 \oplus F_2 \simeq F_3 \oplus F_4 \oplus F_5$, with indecomposable direct summands F_i such that F_1, F_2, F_3 have rank 2 (and F_4, F_5 rank 1), see Fuchs [Fu2] Theorem 90.1. We use the following notation: The factor group $\mathbb{Q}/\mathbb{Z}$ is the direct sum of all Prüfer groups P_p, p a prime number, each occurring with multiplicity one. Denote by π_p the composition of the canonical projections $\mathbb{Q} \to \mathbb{Q}/\mathbb{Z} \to P_p$. We also will need the map $p\pi_p\colon \mathbb{Q} \to P_p$; note that its kernel $\operatorname{Ker}(p\pi_p)$ contains $\operatorname{Ker}(\pi_p)$ with index p. For $1 \le i \le 3$, let F_i be the kernel of a map

$$f_i\colon \mathbb{Q}^2 \to P_2 \oplus P_3 \oplus P_5 \oplus P_5 \oplus P_7 \oplus P_7,$$

namely of

$$f_1 = \begin{bmatrix} \pi_2 & 0 \\ 0 & \pi_3 \\ \pi_5 & \pi_5 \\ 5\pi_5 & 0 \\ \pi_7 & 0 \\ 0 & \pi_7 \end{bmatrix}, \qquad f_2 = \begin{bmatrix} \pi_2 & 0 \\ 0 & \pi_3 \\ \pi_5 & 0 \\ 0 & \pi_5 \\ \pi_7 & \pi_7 \\ 7\pi_7 & 0 \end{bmatrix}, \qquad f_3 = \begin{bmatrix} \pi_2 & 0 \\ 0 & \pi_3 \\ \pi_5 & \pi_5 \\ 5\pi_5 & 0 \\ \pi_7 & \pi_7 \\ 7\pi_7 & 0 \end{bmatrix}.$$

The additional groups F_4 and F_5 are the kernels of the maps

$$f_4 = \begin{bmatrix} \pi_2 \\ \pi_5 \\ \pi_7 \end{bmatrix} : \mathbb{Q} \to P_2 \oplus P_5 \oplus P_7, \quad \text{and} \quad f_5 = \begin{bmatrix} \pi_3 \\ \pi_5 \\ \pi_7 \end{bmatrix} : \mathbb{Q} \to P_3 \oplus P_5 \oplus P_7,$$

respectively. Of course, the map f_i is the minimal injective resolution of F_i and belongs to $\mathcal{R}'$. Thus, we really deal with indecomposable artinian R-modules $(f_1), \dots, (f_5)$ and with an isomorphism[13] $(f_1) \oplus (f_2) \simeq (f_3) \oplus (f_4) \oplus (f_5)$.

[13] Matrices which yield an isomorphism can be calculated easily; or see [R13].

The example presented here uses four prime numbers, but actually it is sufficient to work with a single prime. Namely, as Butler has pointed out, for any prime $p \geq 5$, there do exist torsionfree abelian groups F of finite rank which do not satisfy the Krull-Remak-Schmidt property, such that $qF = F$ for all primes $q \neq p$, see [Ar] 2.15. Of course, such examples give rise to corresponding artinian modules over the ring $\begin{bmatrix} \mathbb{Q} & 0 \\ \mathbb{Q} & \mathbb{Z}_{(p)} \end{bmatrix}$. Note that this ring is a very well-behaved ring: all its indecomposable projective (left or right) modules are serial.

A slight modification of the Pimenov-Yakovlev construction allows to replace the rings considered by local rings [R13], and we obtain in this way examples of artinian modules even over a local ring which do not satisfy the Krull-Remak-Schmidt property.

Categorical equivalences, again. We have seen above an interesting example of realizing a category in two different ways, once as a full subcategory of the category of abelian groups, once as modules over some non-commutative ring. But the most important setting for using a categorical equivalence concerns a process which may be called projectification. It is the following quite trivial, but very effective procedure: Given any module M, its endomorphism ring $\operatorname{End}(M)$ is also the endomorphism ring of a projective module, just take the regular representation of E^{op}, where $E = \operatorname{End}(M)$; the categorical version of this statement is: Given an R-module M with endomorphism ring $E = \operatorname{End}(M)$, let $\operatorname{add} M$ be the additive closure, this is the full subcategory of $\operatorname{Mod} R$ consisting of all direct summands of finite direct sums of copies of M. Then this category is equivalent to the category $\operatorname{pro} E^{\mathrm{op}}$ of all finitely generated projective E^{op}-modules, an equivalence is given by the functor $\operatorname{Hom}_R(M, -)$.

A related result has to be mentioned here, a theorem due to Swan [Sw,Ba]: *Let X be a compact Hausdorff space and denote by $C_0(X)$ the ring of continuous functions $X \to \mathbb{R}$. Then the category of $\mathbb{R}$-vector bundles on X is equivalent to the category of finitely generated projective $C_0(X)$-modules,* an equivalence is given by sending the $\mathbb{R}$-vector bundle E to the $C_0(X)$-module $\Gamma(E)$ of all continuous sections of E. Note that this result provides a bridge to topology and differential geometry. Note that the most prominent vector bundles are the tangent bundles of differential manifolds, thus questions concerning vector fields (these are the sections of the tangent bundle) can be reformulated in terms of projective modules. The analogy between finitely generated projective modules and vector bundles can be an important source for inspiration, it provides a nice geometrical model for questions concerning projective modules.

There is a similar bridge to number theory: many problems about rings of integers in number fields concern the structure of their finitely generated projective modules. In particular, the ideal class group of a Dedekind ring can be interpreted as the set of isomorphism classes of projective modules of rank 1 with respect to the tensor product (for a general commutative ring, the latter group is called its Picard group).

We end this section with some general considerations concerning the role of simple objects and of finite length objects in abelian categories.

Abelian categories with no finite length modules. First of all, we have to stress that *an abelian category may not have simple objects at all,* a typical example $\mathcal{A}$ can be constructed as follows: As set of indecomposable objects take the set of open (non-empty) intervals $(a,b) = \{q \in \mathbb{Q} \mid a < q < b\}$ in $\mathbb{Q}$, let $\operatorname{Hom}_{\mathcal{A}}((a,b),(c,d)) = k$ if $a \leq c < b \leq d$ and 0 otherwise; as composition take the multiplication in k. By adding formal (finite) direct sums (see for example [GR], p.18), we obtain an additive category $\bigoplus \mathcal{A}$ which easily can be shown to be abelian: just observe that it is sufficient to consider a finite sequence $c_0 < c_1 < \cdots < c_m$ of rational numbers and the full subcategory of direct sums of objects of the form (c_i, c_j) with $i < j$. This subcategory is equivalent to the module category of the ring of upper triangular matrices over k.

Of course, a Grothendieck category[14], in particular the module category $\operatorname{Mod} R$ over a ring R, always has sufficiently many simple objects. There do exist non-trivial examples of rings R where all simple modules are injective. A ring R with all simple modules injective is called a *V-ring*. Cozzens [Cz,Fa2] has constructed rings of differential polynomials and also twisted polynomial rings which are V-rings, but have zero socle.

Objects of finite Loewy length. *Let $\mathcal{A}$ be an abelian category, let $S_1, \ldots, S_n$ be pairwise non-isomorphic simple objects in $\mathcal{A}$. Let $\mathcal{S}(S_1, \ldots, S_n)$ be the class of semisimple objects in $\mathcal{A}$ which are (finite or infinite) direct sums of copies of $S_1, \ldots, S_n$. If t is a natural number, let $\mathcal{A}(S_1, \ldots, S_n; t)$ be the set of objects in $\mathcal{A}$ which have a filtration of length t with factors in $\mathcal{S}(S_1, \ldots, S_n)$. Then $\mathcal{A}(S_1, \ldots, S_n; t)$ is equivalent to the module category of a semiprimary ring R. If J is the radical of R, then $J^t = 0$ and R/J is isomorphic to the endomorphism ring of $\bigoplus_{i=1}^n S_i$.* Proof: For every object S_i we can construct its relative projective cover P_i in $\mathcal{A}' = \mathcal{A}(S_1, \ldots, S_n; t)$. In this way, we clearly obtain a progenerator $P = \bigoplus_{i=1}^n P_i$ for $\mathcal{A}'$. Let R be the endomorphism ring of P.

Note that all the finite length objects are recovered in this way: If M has length at most t and if all the composition factors of M belong to the set $\{S_1, \ldots, S_n\}$, then M belongs to $\mathcal{A}(S_1, \ldots, S_n; t)$.

Decomposing projective objects in an abelian category. Let us stress that for general abelian categories we cannot expect any structure theory. In case we consider a Grothendieck category, the assertions concerning projective objects and those concerning injective objects are very different. Of course, dealing with the dual of a Grothendieck category, we obtain examples with the opposite features.

Let $\mathcal{C}$ be a Grothendieck category. Note that $\mathcal{C}$ may not have enough projective modules (example: the category of all abelian p-groups); there may be

[14] For the definition, see for example [Fa1].

enough projectives, but not enough projective covers (example: $\operatorname{Mod}\mathbb{Z}$); an indecomposable projective object P does not have to be couniform (example again: ${}_RR$ for $R=\mathbb{Z}$), and this may happen even if the radical of P is superfluous in P (example: take ${}_RR$, where $R=\mathbb{Z}_{(pq)}$ is obtained from $\mathbb{Z}$ by localizing at the product of two different primes p,q). If $\mathcal{C}=\operatorname{Mod}R$ for some ring R, then there are enough projectives, but all the other anomalies mentioned occur already in module categories. On the other hand there is **Kaplansky's Theorem:** *Every projective module is a direct sum of countably generated projective modules.* Thus, we obtain a strong bound on the size of indecomposable projective modules.

For an integral domain R, the projective modules of rank 1 play an important role; as we have mentioned, the set of isomorphism classes with respect to the tensor product is called the Picard group $\operatorname{Pic}R$ of R. Of course, all these projective rank 1 modules are indecomposable. There do exist also indecomposable projective modules of rank greater than 1. The typical example of an integral domain R which has such a module is the coordinate algebra $\mathbb{R}[x_1,x_2,x_3]/\langle x_1^2+x_2^2+x_3^2-1\rangle$ of the sphere S^2, see [Sw]: Let P be the kernel of the homomorphism $\phi\colon R^3\to R$ defined by $\phi(r_1,r_2,r_3)=\sum r_ix_i$. Note that ϕ is surjective, thus it splits and therefore $R^3\simeq P\oplus R$. This shows that P is projective and has rank 2. In case P would be decomposable, one would have $P\simeq R^2$, since the ring R is known to be a principal ideal domain. But an isomorphism $P\simeq R^2$ would provide a continuous vector field on S^2 which nowhere vanishes, impossible.

Decomposing injective objects in an abelian category. Consider now the injectives in a Grothendieck category $\mathcal{C}$. There are always sufficiently many injective objects, even sufficiently many injective envelopes. Indecomposable injective objects are always uniform. In case $\mathcal{C}=\operatorname{Mod}R$ for some ring R, we obtain all indecomposable injective modules as injective envelopes of uniform cyclic modules. In contrast to the case of projective modules, we cannot expect to be able to write all the injective modules as direct sums of countably generated modules (example: the injective envelope of ${}_RR$, where R is the polynomial ring $k[T]$ in one variable is given by the field $k(T)$ of rational functions, and if k is an uncountable field, then ${}_{k[T]}k(T)$ is not countably generated), and not even of modules which are generated by λ elements, where λ is a fixed cardinal number (this is only possible for left noetherian rings, by the Faith-Walker theorem mentioned already). Any non-zero ring has indecomposable injective modules (since it has uniform modules, namely at least the simple modules), but, as we have noted already, there are rings R which also have non-zero superdecomposable injective modules, for example the free algebra $k\langle X,Y\rangle$ in two variables.

Let I be an injective module. Then
(i) *I is indecomposable if and only if I is uniform.*
(ii) *I is discrete if and only if I is formatted.*
(iii) *I is superdecomposable if and only if I has no uniform submodule.*

Here, the conditions mentioned left deal with the behaviour with respect to direct decompositions, the right ones with the submodule structure, namely the uniform submodules of the module. The direct sum conditions are properties which concern the endomorphism rings, thus they remain valid when we apply a functor which preserves endomorphism rings. On the other hand, the submodule conditions are preserved under submodules as follows: a non-zero submodule of a uniform module is uniform; any submodule of a formatted module is formatted; and finally, if a module has no uniform submodule, then the same is true for any of its submodules. (Let us sketch the proof of (ii). First, assume that the injective module I is formatted and consider a direct decomposition $I = I' \oplus I''$ with $I' \neq 0$. Since I is formatted, I' contains a uniform submodule, say U. But the injective envelope of U is a direct summand of I. This shows that I is discrete. Assume conversely that I is discrete and let N be a non-zero submodule. Then the injective envelope I' of N is a direct summand of I and by assumption, I' has an indecomposable direct summand, say J. Now $J \cap N \neq 0$, since J is a non-zero submodule of I' and N is essential in I'. Since J is indecomposable injective, its submodule $J \cap N$ is uniform. This shows that I is formatted.)

Theorem of Gabriel and Oberst. *Any injective module I is the direct sum of a discrete module I_1 and a superdecomposable module I_2. If $I = I_1' \oplus I_2'$ is a second decomposition with I_1' discrete and I_2' superdecomposable, then $I = I_1 \oplus I_2'$.* (In particular, the modules I_1 and I_1' are isomorphic, and similarly, the modules I_2 and I_2' are isomorphic.) The usual discussions of the Gabriel-Oberst-Theorem invoke so called spectral categories [GO] (they are obtained by factoring out from the category of all injective R-modules the ideal of all maps $f\colon I \to I'$ which vanish on an essential submodule). The use of spectral categories is illuminating, but it seems also misleading[15]. Let us sketch a direct and elementary proof. Let I be an injective module. Using the lemma of Zorn, choose a submodule I' of I which is a direct sum of uniform modules and such that there does not exist a uniform submodule U of I with $I' \cap U = 0$. Let I_1 be an injective envelope of I', thus I_1 is discrete. Since I is injective, we may assume that I_1 is a submodule of I, thus there is a direct decomposition $I = I_1 \oplus I_2$. Since I_2 has no uniform submodule, it is superdecomposable. Now assume that there is given a second decomposition $I = I_1' \oplus I_2'$ with I_1' discrete and I_2' superdecomposable. The intersection $I_1 \cap I_2'$ is a submodule of I_1, thus formatted, and a submodule of I_2, thus also superdecomposable, and therefore zero. It follows that $I_1 + I_2'$ is a direct sum; it is an injective module, thus there is a submodule C with $I = I_1 \oplus I_2' \oplus C$. Both I_2 and $I_2' \oplus C$ are direct complements for I_1, thus they are isomorphic and C is isomorphic to a submodule of I_2. This implies that C is superdecomposable. On the other hand, both I_1' and $I_1 \oplus C$ are direct complements for I_2', thus isomorphic and therefore C is isomorphic to a submodule of I_1', thus formatted.

[15] For example, based on the use of spectral categories, the book by Jensen-Lenzing [JL,8.24] asserts that the maximal discrete direct summand I_1 of I is uniquely determined, Prest [P1, Corollary 4.A14] even claims that both summands I_1, I_2 are unique, in contrast to examples which we are going to present.

Since C is both formatted and superdecomposable, we see that $C = 0$. This completes the proof.

In the decomposition $I = I_1 \oplus I_2$ of an injective module I, with I_1 discrete and I_2 superdecomposable, none of the direct summands I_1 and I_2 has to be unique, as the following examples show. It is sufficient to find a ring R, a discrete injective R-module I_1 and a superdecomposable injective R-module I_2 such that $\mathrm{Hom}(I_1, I_2) \neq 0$ or $\mathrm{Hom}(I_2, I_1) \neq 0$ (since the graph $G(f)$ of an R-linear map $f\colon M \to M'$ is an R-submodule of $M \oplus M'$ which satisfies $M \oplus M' = G(f) \oplus M'$). For an example with $\mathrm{Hom}(I_1, I_2) \neq 0$, consider the quiver with two vertices, say labeled 1 and 2, with two loops x, y at the vertex 1 and one arrow $1 \to 2$.

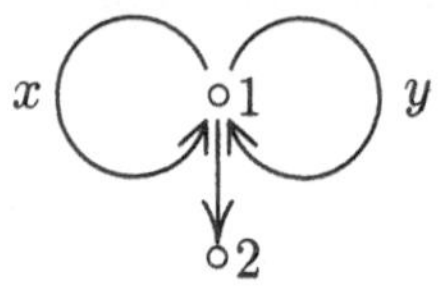

Consider the indecomposable projective representation $P(1)$ corresponding to the vertex 1, it has a k-basis given by the set of paths starting in 1. The socle N of $P(1)$ is an infinite direct sum of copies of the simple module $S(2)$ corresponding to the vertex 2. Since N is essential in $P(1)$, we see that $P(1)$ is formatted, whereas $P(1)/N$ is superdecomposable. Take for I_1 the injective envelope of $P(1)$ and for I_2 the injective envelope of $P(1)/N$. The canonical map $P(1) \to P(1)/N$ induces a non-zero map $I_1 \to I_2$. Thus we see[16] that in $I_1 \oplus I_2$, there are other maximal discrete submodules than I_1.

As an example with $\mathrm{Hom}(I_2, I_1) \neq 0$, we take the opposite quiver:

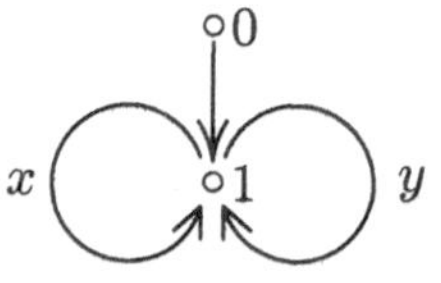

Also here, the injective envelope I_2 of $P(1)$ is superdecomposable, and now it maps onto the injective module $I_1 = S(0)$. This shows that in $I_1 \oplus I_2$, there are other maximal superdecomposable submodules than I_2. Actually, given an arbitrary ring R and I_2 a non-zero superdecomposable R-module, we obtain a corresponding example: consider a simple subfactor S of I_2 and its injective envelope I_1. Clearly, I_1 is discrete and $\mathrm{Hom}(I_2, I_1) \neq 0$. This shows that in $I_1 \oplus I_2$, there are several different maximal superdecomposable submodules.

[16] This ring R provides also an example of a sum of two formatted modules which is not formatted: Consider $M = P(1) \oplus P(1)/N$. Note that both modules $P(1)$ and $P(1)/N$ are cyclic, say generated by elements a and $b = a + N$, respectively. Then M is generated by the elements $(a, 0)$ and (a, b). The submodules generated by $(a, 0)$ and by (a, b) both are isomorphic to $P(1)$, thus formatted, but M is not formatted, since its submodule $P(1)/N$ has no uniform submodules.

One may ask under what conditions on a ring R every discrete injective R-module actually is a direct sum of uniform modules. In case all the injective modules are discrete, this has been answered by the **Matlis–Papp theorem**: *A ring R is left noetherian if and only if every left module is a direct sum of uniform modules.* We should mention also the following characterization of left noetherian rings. **Faith-Walker theorem.** *A ring R is left noetherian if and only if each injective module is a direct sum of modules, each being generated by λ elements, where λ is some fixed cardinal number.* For both results we may refer to [Fa2].

Diamond categories. Let us assume that $\mathcal{C}$ is an abelian category. We say that $\mathcal{C}$ is a *diamond category* provided first, every simple object has a projective cover and an injective envelope, and second every object C has an essential subobject C' which is semisimple and of finite length, and dually, a superfluous subobject C'' such that C/C'' is semisimple and of finite length. Of course, the subobject C' will be called the socle of C and denoted by $\operatorname{soc} C$, the subobject C'' is called the radical of C and denoted by $\operatorname{rad} C$. Let us state some properties of a diamond category $\mathcal{C}$. First of all, it follows easily that every object has a projective cover and an injective envelope. Next, *any object in $\mathcal{C}$ is a finite direct sum of indecomposable objects, and indecomposable objects have local endomorphism rings.* In particular, the Krull-Remak-Schmidt Theorem holds in $\mathcal{C}$. Also, a projective object is indecomposable if and only if its radical is a maximal subobject; an injective object is indecomposable if and only if its socle is simple.

A set of subfactors $C_i' \subset C_i$ of an object C is said to *cover* C provided for any subfactor $U' \subset U$, there exists an index i such that $U \cap C_i \not\subseteq U' + C_i'$. *Any object C in a diamond category has finitely many subfactors $C_i' \subset C_i \subseteq C$ which cover C, such that any object C_i/C_i' is a diamond* (this explains the name). (For the proof, take a projective cover $(p_a)_a\colon \bigoplus_{a=1}^n P_a \to C$ with indecomposable projective objects P_a, and an injective envelope $(u_b)_b\colon C \to \bigoplus_{b=1}^m I_b$ with indecomposable injective objects I_b. Consider a pair $i = (a, b)$ such that the composition $u_b \circ p_a$ is non-zero. Let C_i be the image of p_a in C, let C_i' be the intersection of C_i with the kernel of u_b. Then $C_i' \subset C_i$ and C_i/C_i' is a diamond, since it is isomorphic to a factor object of P_a and to a subobject of I_b. In order to see that these subfactors cover C, take a subfactor $U' \subset U$ in C.)

Given an object C in a diamond category $\mathcal{C}$, we may define its socle sequence $\operatorname{soc}_i C$ by $\operatorname{soc}_0 C = 0$ and $\operatorname{soc}_{i+1} C/\operatorname{soc}_i C = \operatorname{soc}(C/\operatorname{soc}_i C)$, for all natural numbers i. Similarly, let $\operatorname{rad}^0 C = C$ and $\operatorname{rad}^{i+1} C = \operatorname{rad}(\operatorname{rad}^i C)$, for all i. Note that all the objects $\operatorname{soc}_i C$ and $C/\operatorname{rad}^i C$ are of finite length. As a consequence, all objects in a diamond category which are of finite Loewy length are of finite length.

Starting with a diamond category $\mathcal{C}$, we may add formal direct limits in order to obtain a Grothendieck category $\varinjlim \mathcal{C}$. Note that an object of $\mathcal{C}$ may have in $\varinjlim \mathcal{C}$ subfactors $C'' \subset C' \subset C$ such that C'/C'' is an infinite direct sum of simple objects.

If C is an object of $\mathcal{C}$ which is not of finite length, then all the inclusions occurring in the socle sequence and in the radical sequence of C

$$0 \subset \operatorname{soc} C \subset \operatorname{soc}_2 C \subset \cdots \qquad \text{and} \qquad \cdots \subset \operatorname{rad}^2 C \subset \operatorname{rad} C \subset C$$

are proper. If we take the union $\bigcup_i \operatorname{soc}_i C$ in $\varinjlim \mathcal{C}$, we obtain an artinian object. Similarly, we may consider the intersection $C' = \bigcap_i \operatorname{rad}^i C$ in $\varinjlim \mathcal{C}$, the corresponding factor object C/C' will be a noetherian object. Note that for C not of finite length, C/C' as well as $\bigcup_i \operatorname{soc}_i C$ never will belong to $\mathcal{C}$ itself.

Functor categories. If $\mathcal{C}$ is an additive category, recall that $\operatorname{Mod}\mathcal{C}$ denotes the category of all additive contravariant functors from the category $\mathcal{C}$ into the category of all abelian groups, and $\operatorname{mod}\mathcal{C}$ the full subcategory of all finitely presented functors (a functor F is said to be finitely presented provided there is an exact sequence of the form $\operatorname{Hom}_{\mathcal{C}}(-,C) \to \operatorname{Hom}_{\mathcal{C}}(-,C') \to F \to 0$, with objects C, C' in $\mathcal{C}$). Of course, in case R is a ring and $\operatorname{pro} R$ is the category of all finitely generated projective R-modules, then $\operatorname{Mod} R = \operatorname{Mod}\operatorname{pro} R$ and $\operatorname{mod} R = \operatorname{mod}\operatorname{pro} R$.

Consider now the case of $\mathcal{C}$ being a *dualizing k-variety*, where k is a commutative artinian ring (for example a field), as studied by Auslander and Reiten [AR1]. This means that first of all $\mathcal{C}$ is a k-category with all Hom-sets being finitely generated k-modules, and second, that for every contravariant functor F in $\operatorname{mod}\mathcal{C}$, also the covariant functor $D(F)$ is finitely presented, and dually, that for every finitely presented covariant functor G on $\mathcal{C}$, also the contravariant functor $D(G)$ is finitely presented; here D is the duality with respect to the minimal injective cogenerator I for k, thus $D(F)(C) = \operatorname{Hom}_k(F(C), I)$ and $D(G)(C) = \operatorname{Hom}_k(G(C), I)$. For example, for any artin algebra R, the category $\operatorname{pro} R$ is such a dualizing k-variety. The first result to remember is: **Theorem (Auslander-Reiten).** *If $\mathcal{C}$ is a dualizing k-variety, then so is* $\operatorname{mod}\mathcal{C}$. ([AR1], 2.6). As a consequence, we may iterate the construction, and consider the sequence $\mathcal{C}, \operatorname{mod}\mathcal{C}, \operatorname{mod}\operatorname{mod}\mathcal{C}, \ldots$ and so on. We obtain a sequence of categories which are getting larger and larger, but in some sense more and more well behaved. Note the following: if we start with a representation-finite algebra R, then $\operatorname{mod} R = \operatorname{pro} A(R)$, where $A(R)$ is the Auslander algebra of R, and, $A(R)$ is well-behaved both with respect to global dimension and dominant dimension (it has global dimension at most 2 and dominant dimension at least 2). Also, the category $\mathcal{C}$ does not have to be abelian, however $\operatorname{mod}\mathcal{C}$ always will be. The second result to remember: **Theorem (Auslander-Reiten).** *If $\mathcal{C}$ is a dualizing k-variety, then* $\operatorname{mod}\mathcal{C}$ *is a diamond category.* ([AR1], 3.7).

An additive category which is equivalent to a category of the form $\operatorname{mod}\mathcal{C}'$ will be said to have *Auslander dimension* at least 1. Inductively, we may say that $\mathcal{C}$ has Auslander dimension at least $n+1$ provided $\mathcal{C}$ is equivalent to a category of the form $\operatorname{mod}\mathcal{C}'$ where $\mathcal{C}'$ has Auslander dimension at least n. Note that for any additive category $\mathcal{C}'$, the category $\operatorname{pro}\operatorname{mod}\mathcal{C}'$ is just the closure of $\mathcal{C}'$ under finite direct sums and direct summands. Thus, if $\mathcal{C}$ is equivalent to $\operatorname{mod}\mathcal{C}'$, and if

$\mathcal{C}'$ has finite direct sums and split idempotents, then we can recover $\mathcal{C}'$ as the full subcategory $\operatorname{pro}\mathcal{C}'$ of all projective objects in $\mathcal{C}$. If the Auslander dimension of $\mathcal{C}$ is at least n, we can apply n times pro to $\mathcal{C}$. In particular, if $\mathcal{C} = \operatorname{mod}\operatorname{mod}\mathcal{B}$, and $\mathcal{B}$ has finite sums and split idempotents, then we get $\mathcal{B} = \operatorname{pro}\operatorname{pro}\mathcal{C}$.

Let us assume that $\mathcal{B}$ is a dualizing k-variety with finite direct sums and split idempotents, and let $\mathcal{C} = \operatorname{mod}\operatorname{mod}\mathcal{B}$. Then we can recover $\mathcal{B}$ from $\mathcal{C}$ also in a different way, namely as the full subcategory of $\mathcal{C}$ of all objects which are both projective and injective. The indecomposable objects in $\operatorname{mod}\operatorname{mod}\mathcal{C}$ which are both projective and injective may be compared with the "hammocks" as considered by S. Brenner: The name "hammock" was introduced by her when she considered $\mathcal{C} = \operatorname{mod}\operatorname{mod} R$ for a representation-finite algebra R aiming at a combinatorial characterization of these hammock functors $\mathcal{H}(S) = \operatorname{Hom}(-, I(S)) = D\operatorname{Hom}(P(S), -)$, here $I(S)$ is the injective envelope of a simple module S and $P(S)$ its projective cover. It seems to be of great importance to study these functors $\operatorname{Hom}(-, I(S)) = D\operatorname{Hom}(P(S), -)$ not only in the case of a representation-finite algebra, but in general. We also have to refer to Tachikawa [T] who emphasized the importance of the objects in $\operatorname{mod}\operatorname{mod} R$ which are projective as well as injective.

The dualizing k-varieties are the proper setting for the Auslander-Reiten theory. *If $\mathcal{C}$ is a dualizing k-variety, then the category* $\operatorname{mod}\mathcal{C}$ *has Auslander-Reiten sequences*[17], and if $0 \to X \to Y \to Z \to 0$ is such a sequence, then $X = \tau Z$ is calculated as usual as the "dual of the transpose".

The categories to be considered later are of the form $\operatorname{mod} R$ or $\operatorname{mod}\operatorname{mod} R$, where R is a finite-dimensional k-algebra with k a field, thus they are dualizing k-varieties and therefore diamond categories. The combinatorial flavour of the representation theory of such algebras R is due to this fact. In case R is not representation-finite, the assertion that $\operatorname{mod}\operatorname{mod} R$ is a diamond category has not yet found the appreciation which it deserves, even though many classical facts from Auslander-Reiten theory find a natural interpretation in this setting. Only in very special cases, the structure of finitely presented functors on $\operatorname{mod} R$, thus of objects in the category $\operatorname{mod}\operatorname{mod} R$, has been analyzed in detail. The first question to be raised concerns the possible serial objects. Finitely presented functors which are serial have been studied by Auslander and Reiten in [AR2]. Some typical examples of such functors will be seen below.

---

[17] A general diamond category $\mathcal{C}$ may not have sufficiently many Auslander-Reiten sequences, even if every object in $\mathcal{C}$ has finite length. A typical example has been exhibited in [R2]: Let F be a field with a derivation δ and consider the F-F-bimodule ${}_FM_F$ where ${}_FM = {}_FF \oplus {}_FF$ and such that the action by F on the right is given by $(a,b)c = (ac + b\delta(c), bc)$ for all $a, b, c \in F$. The category of representations of ${}_FM_F$ of finite length is a diamond category. The finite direct sums of indecomposable representations of ${}_FM_F$ of even length form an abelian subcategory $\mathcal{R}_0$. If (F, δ) is differentially closed, there are just two indecomposable representations of length two. One of them is projective and injective in $\mathcal{R}_0$ and does not occur in any Auslander-Reiten sequence.

3. Well-behaved modules

We have seen that for most of the rings there is an abundance of different types of modules so that it will be a waste of time to try to deal with all of them. Even if we restrict to finite length modules, Maurice Auslander strongly argued against the usual classification procedures many mathematicians are fond of: he stressed that it should be of greater interest to deal with some classes of modules with specific and important properties, than to publish large lists of normal forms no-one is going to use. Of course, up to now such lists have been put forward only in tame cases (and some people believe that the notion of wildness just excludes the possibility of producing complete lists) and there it seems that really all the indecomposable modules have specific and presumably also important properties. Thus Auslander's argument could be rephrased in these cases better as follows: it may be of little interest to exhibit lists of modules unless one cannot figure out their properties. But the by now usual procedure of determining complete Auslander-Reiten components or even families of such components as well as their global behaviour aims at a reasonable description of such module categories and this seems to be an endeavor which has to be appreciated. Too little is known at the moment about any wild module category for arguing in real favour for a classification program, but there do exist dreams about a "very well-behaved wild world": the Kac conjecture [Kc] that the module varieties of wild hereditary algebras have a cellular decomposition with affine strata is part of it. What one may hope for are, on the one hand, discrete invariants which fit into some concise combinatorial picture, and, on the other hand, for any given set of such invariants, a nice, hopefully even rational variety which describes the modules with these data. In this way, it may turn out that all the finite length modules are considered to be of interest and of importance, at least when considered in the natural setting of their relatives, as part of such a family with a fixed set of combinatorial data. It is one of the main reason for the present book to stress that a description of the category of all finite length modules quite naturally has to rely on incorporating infinite length modules, but clearly only some of them. It should be out of question that we have to be very restrictive about the infinite length modules which we are going to involve.

The algebraically compact modules. There is a natural choice of the class of modules to be considered, a choice which can be justified both by usual algebraic arguments, but also by mathematical logic. Recall that there is an overlap between the interests of algebraist and logicians. There are many important questions handled in a different, but quite parallel way by algebraists and logicians. Many notions and constructions both in algebra and in logic stem from difficulties which have been encountered in the 19th century before the set theoretical foundation of modern mathematics was laid. To overcome such difficulties, different, and sometimes incompatible remedies have been found. For example, in first order logic, the notion of a logic "with equality" just tries to formalize a specific way of dealing with factor objects. The whole model theory should be considered as part of algebra, but the conflicting terminologies make it difficult for algebraists and logicians to communicate. Some attempts have been made by Ziegler, and especially by Prest and Herzog in order to overcome these difficulties and to provide some reasonable dictionary in order to translate questions and results from one language to the other. Unfortunately, as it is usual with dictionaries, the scope of notions in the languages to be compared often are not really compatible, so that convenient groupings in one language may look slightly artificial in the other, at least on the first sight.

One of the basic notions presented in this book is a concept which can easily be described in both languages, on the first sight in completely different ways (but a deeper understanding relates the two definitions quite clearly), namely that of a module which is *algebraically compact* or, what is the same, *pure injective.* The two denominations already point out the two different roles these modules play.

Algebraic compactness refers to solving systems of linear equations. Such a system consists of equations indexed by the elements i of a set I, the equations are of the form

$$(*) \qquad \sum\nolimits_{j\in J} r_{ij}x_j = a_i,$$

where a_i are elements of a given R-module M, the elements r_{ij} belong to the ring R, for $i \in I, j \in I$, and one assumes that for given $i \in I$, almost all r_{ij} are supposed to be zero (so that forming the sum $\sum_{j\in J} r_{ij}x_j$ makes sense in this algebraic context). In principle, the x_j, $j \in J$ are variables; a solution of this system of equations consists of elements $x_j \in M$ such that all the equations $(*)$ are satisfied. Such a system of equations is said to be finitely solvable provided for any finite subset $I' \subseteq I$, there exists a solution for the equations $(*)$ with $i \in I'$. The R-module M is said to be *algebraically compact* provided any system of linear equations which is finitely solvable has a solution.

On the other hand, pure injectivity means "relative injectivity with respect to pure embeddings". Recall that forming tensor products does not respect monomorphisms: a typical example is the map $f\colon \mathbb{Z} \to \mathbb{Z}$ given by the multiplication with 2, this is a monomorphism, but if we tensor f with $\mathbb{Z}/2$, we obtain the zero map (of course, the tensor product of $\mathbb{Z}$ with $\mathbb{Z}/2$ is just $\mathbb{Z}/2$ and $1_{\mathbb{Z}/2}\otimes f\colon \mathbb{Z}/2 \to \mathbb{Z}/2$ is still the multiplication by 2, but this is now the zero map). For any ring R, a homomorphism $f\colon M \to N$ of R-modules (as usual, this means left R-modules) is said to be a *pure monomorphism* provided $1_X \otimes_R f\colon X \otimes_R M \to X \otimes_R N$ is a monomorphism, for any right R-module X_R. This concept of purity will be discussed in detail by Huisgen-Zimmermann in [H]. The R-module M is said to be *pure injective* provided the only pure monomorphisms $M \to N$ (with arbitrary N) are the split monomorphisms.

The two names of "algebraic compactness" and "pure injectivity" show the main directions of interest, but actually, these modules can be characterized in many additional ways (and some of these could equally well serve as the source of naming these modules). For a very elegant and concise treatment of these modules including seven characterizations we refer to the book of Jensen and Lenzing [JL], Chapter 7. Chapter 8 of the same book provides a corresponding account dealing with the Σ-algebraically compact modules: these are those modules M, for which any direct sum of copies of M is again algebraically compact.

It seems to be of interest to quote at least the following characterization: *A module M is algebraically compact if and only if for any index set I the summation map $\bigoplus_I M \to M$ can be extended to a map $\prod_I M \to M$.* The importance of this characterization is due to the fact that it has the following consequence: **Corollary.** *Let F be an additive functor which commutes with direct sums and direct products. If M is algebraically compact, then also $F(M)$ is algebraically compact.* This is a convenient tool for showing that modules are algebraically compact. For example, given a ring homomorphism $\varphi\colon R \to S$, any S-module M may be considered as an R-module via $r \cdot m = \varphi(r)m$ for $m \in M$, $r \in R$, this yields a functor $F\colon \operatorname{Mod} S \to \operatorname{Mod} R$ which does not change the underlying sets, thus it commutes with direct sums and direct products. As a consequence, we see that any algebraically compact S-module is also algebraically compact when considered as R-module. In particular, if I is an ideal of R, and M is an R-module which is annihilated by I, then M is algebraically compact as an R-module if and only if it is algebraically compact when considered as an R/I-module. This shows that all injective R/I-modules are algebraically compact R-modules.

Algebraically compact modules as injective objects in a Grothendieck category. According to Gruson and Jensen [GJ], the category of algebraically compact modules is equivalent to the category of injective objects in some Grothendieck category – a Grothendieck category which is usually far away from module categories of rings with finiteness conditions. The equivalence is given as follows: Let R be a ring and $\operatorname{mod}(R^{\mathrm{op}})$ the category of finitely presented right R-modules. Any (left) R-module M gives rise to a functor $(-\otimes_R M)\colon \operatorname{mod}(R^{\mathrm{op}}) \to \operatorname{Mod} \mathbb{Z}$, and we obtain in this way a functor

$$\Phi\colon \operatorname{Mod} R \to \operatorname{Mod}\operatorname{mod}(R^{\mathrm{op}}), \qquad \Phi(M) = (-\otimes_R M).$$

It is easy to see that this functor Φ is a full embedding[17]. The R-module M is algebraically compact if and only if $\Phi(M)$ is an injective object in $\operatorname{Mod}\operatorname{mod}(R^{\mathrm{op}})$. As a consequence, the restriction of Φ to the subcategory of all algebraically compact R-modules yields an equivalence with the category of injective objects in $\operatorname{Mod}\operatorname{mod}(R^{\mathrm{op}})$.

---

[17] In addition, it also commutes with direct sums, direct products and direct limits; the image consists of those additive functors which are right exact, and these are just the objects Q in $\operatorname{Mod}\operatorname{mod}(R^{\mathrm{op}})$ with $\operatorname{Ext}^1(F,Q) = 0$ for all finitely presented functor F, see for example [JL, Theorems B15 and B.16]. A sequence $0 \to M' \to M \to M'' \to 0$ in $\operatorname{Mod} R$ is pure exact if and only if its image under Φ is exact.

For any ring R, the R-modules which are indecomposable and algebraically compact are those we are interested in, thus we have to deal with the indecomposable injective objects in $\operatorname{Mod}\operatorname{mod}(R^{\mathrm{op}})$. The first observation to be stressed is the fact that such an object always has a local endomorphism ring, thus the endomorphism ring of an indecomposable algebraically compact R-module is local. Next, let us note that indecomposable injective objects in a Grothendieck category always are uniform; they are the injective envelopes of uniform objects. Thus, in order to get hold of some indecomposable algebraically compact R-module, it is sufficient to find an R-module M such that the functor $\Phi(M) = (- \otimes M)$ is uniform: an injective envelope of $\Phi(M)$ will be of the form $\Phi(\mu)\colon \Phi(M) \to \Phi(\overline{M})$, and $\mu\colon M \to \overline{M}$ is the so called pure-injective envelope[18]. Let us reformulate what it means that the functor $\Phi(M)$ is $\mathbb{N}$-uniform: we need modules N_i and maps $g_i\colon M \to N_i$, $h_i\colon N_{i+1} \to N_i$, for all $i \in \mathbb{N}$, with the following properties: first, $g_i = h_i \circ g_{i+1}$; second, no g_i is a pure monomorphism; and third, given any map $f\colon M \to N$ which is not a pure monomorphism, then there exists an index i and $f'\colon N_i \to N$ such that $f = f' \circ g_i$. (Let us show that these conditions imply that $\Phi(M)$ is $\mathbb{N}$-uniform; the converse can be shown in the same way. Let U_i be the kernel of the transformation $\Phi(g_i) = (- \otimes g_i)$. This U_i is a subobject of $\Phi(M) = (- \otimes M)$. Also, since g_i is not a pure monomorphism, we see that $\Phi(g_i)$ is not a monomorphism, thus $U_i \neq 0$. It follows from $g_i = h_i \circ g_{i+1}$ that $U_{i+1} \subseteq U_i$. Now assume there is given any subobject U of $\Phi(M)$. The injective envelope of $\Phi(M)/U$ is of the form $\Phi(N)$ for some module N, thus U is the kernel of some map $\Phi(M) \to \Phi(N)$. But since Φ is full, such a map is of the form $\Phi(f)$ for some $f\colon M \to N$. As we require, there exists some map $f_i\colon N_i \to N$ such that $f = f' \circ g_i$. But this means that $U_i \subseteq U$.) Actually, the third condition has to be checked only in very special cases, namely in case $f\colon M \to N$ is a monomorphism whose cokernel is indecomposable and of finite length, as well as in case f is an epimorphism with simple kernel. (Namely, assume the third condition holds true in these special cases, and let $f\colon M \to N$ be any map which is not a pure monomorphism. If f is not a monomorphism, then there is a simple submodule S of M which is contained in the kernel of f and f factors via the projection map $p\colon M \to M/S$, say $f = f' \circ p$. By assumption, $p = p' \circ g_i$ for some i, and therefore $f = f' \circ p = f' \circ p' \circ g_i$. On the other hand, if f is a monomorphism, then there is some finite length submodule N' of N such that the map $f\colon M \to N'$ does not split, since f is not a pure monomorphism. By assumption, we know that this $f\colon M \to N'$ factors via some g_i.)

The Ziegler spectrum. Denote by $\mathcal{Z}(R)$ the set of isomorphism classes of R-modules which are indecomposable and algebraically compact[19]. As Ziegler [Z]

[18] This means that μ is a pure monomorphism, $\overline{M}$ is algebraically compact, and μ is left minimal (any endomorphism ϕ of $\overline{M}$ with $\phi\mu = \mu$ is an automorphism).

[19] The number of isomorphism classes of indecomposable algebraically compact R-module is bounded by 2^λ, where $\lambda = \max(|R|, \aleph_0)$, see [JL],7.57.

has pointed out, the set $\mathcal{Z}(R)$ carries a natural topology which is useful for many considerations. There are several ways to define the Ziegler topology. Here we use the following approach: Given a class $\mathcal{X}$ of maps $f\colon X \to X'$ between finitely presented modules X, X', let $\mathcal{A}(\mathcal{X})$ be the indecomposable algebraically compact modules M such that $\operatorname{Hom}(f, M)$ is surjective for all $f \in \mathcal{X}$. As closed sets in $\mathcal{Z}(R)$, one takes the subsets of the form $\mathcal{A}(\mathcal{X})$. As one can show the closed sets are just the sets of isomorphism classes in $\mathcal{Z}(R)$ which belong to some definable subcategory; here, a subcategory $\mathcal{U}$ of $\operatorname{Mod} R$ is said to be *definable* provided it is closed under direct limits, products and pure submodules, or equivalently, if $\mathcal{U}$ is defined by the vanishing of a collection of functors $F\colon \operatorname{Mod} R \to \operatorname{Mod} \mathbb{Z}$ which commute with direct limits and products [CB4]. Given an indecomposable algebraically compact module M, we denote by $\operatorname{cl}(M) = \operatorname{cl}(\{M\})$ the Ziegler closure of the one-element set $\{M\}$. In general there do exist non-isomorphic indecomposable algebraically compact modules M, M' such that $\operatorname{cl}(M)$ and $\operatorname{cl}(M')$ coincide (thus the Ziegler spectrum is not necessarily a T_0-space). For example, if R is a simple von Neumann regular ring, then the only Ziegler closed sets are the empty set and $\mathcal{Z}(R)$ itself. We say that the modules M, M' are topologically equivalent provided $\operatorname{cl}(M) = \operatorname{cl}(M')$.

Elementary duality. The category $\operatorname{Mod} R$ of all R-modules where R is a ring is a Grothendieck category, the dual of this category is not. The dual of a full subcategory $\mathcal{U}$ of a module category is equivalent to a full subcategory of some other module category only in case there are severe restrictions on the size of the modules in $\mathcal{U}$. If we consider, as we do, full subcategories of module categories, the existence of a contravariant equivalence is very rare. The so called elementary duality which we are going to discuss does not concern maps, but only collections of objects in the Ziegler spectrum. It has been observed by Herzog [He] that there is a bijection D between the collection of closed sets of $\mathcal{Z}(R)$ and the collection of closed sets of $\mathcal{Z}(R^{\mathrm{op}})$ which respects finite unions and arbitrary intersections, in particular, it preserves and reflects inclusions, see also [P2,P6]. This duality is based on the duality between the categories of finitely presented functors on $\operatorname{mod} R$ and on $\operatorname{mod}(R^{\mathrm{op}})$ and it can be interpreted as well in terms of the so called positive primitive formulae in model theory. As Krause [K6] has pointed out, if R is a k-algebra, the elementary dual $D\mathcal{A}$ of any closed set $\mathcal{A}$ in $\mathcal{Z}(R)$ is obtained as the set of indecomposable direct summands of modules which belong to the definable subcategory generated by the modules $A^* = \operatorname{Hom}(A, k)$, with $A \in \mathcal{A}$.

Clearly, D induces a bijection between the equivalence classes with respect to topological equivalence. Unfortunately, even for a finite-dimensional k-algebra, where k is some field, a pointwise description of the elementary duality does not seem to be available: given an indecomposable algebraically compact R-module M, one may expect that all the modules N in $\mathcal{Z}(R^{\mathrm{op}})$ such that $D\operatorname{cl}(M) = \operatorname{cl}(N)$ are direct summands of the k-dual M^* of M, and then and one would like to have an effective procedure for obtaining such a direct summand. Of course, if M is finite-dimensional and indecomposable, then M^* has the same dimension, is indecomposable and is just the required module N. But if M is infinite-dimensional and indecomposable, then M^* may be decomposable. As Krause has shown [K3, K6], there is a class $\mathcal{R}(R)$ of indecomposable algebraically compact R-modules M, the so-called simple reflexive ones, such that one can define a duality $D\colon \mathcal{R}(R) \to \mathcal{R}(R^{\mathrm{op}})$ with the following properties: for any $M \in \mathcal{R}(R)$, the R^{op}-module DM is a direct summand of M^* and $D\,\mathrm{cl}(M) = \mathrm{cl}(DM)$. Of course, $DDM \simeq M$.

The denomination "elementary duality" refers to the fact that it is based on the elementary language of R-modules. It has to be stressed that the elementary duality does not preserve Σ-algebraic compactness. We will discuss below the Ziegler spectrum of a finite-dimensional algebra which is tame and hereditary; for these algebras, the Prüfer modules are Σ-algebraically compact whereas the adic modules are not, and the elementary dual of a Prüfer module is just an adic module.

Generic modules. Detailed examples of indecomposable algebraically compact modules will be presented in the next section. Here we want to mention only the most prominent class, namely the generic modules: these are those indecomposable modules which are of finite length when considered as modules over their endomorphism ring[20]. Of importance is the following observation of Krause [K4]: *The closure of any tube contains at least one generic module of infinite length.* One obtains such a generic module as follows [R9]: Take a ray $M_1 \to M_2 \to \cdots$ and its direct limit $P = \varinjlim M_i$ and form a countable product of copies of P; this will be a direct sum of copies of P and of copies of finitely many generic modules of infinite length. One conjectures that the closure of a tube contains precisely one generic module of infinite length[21]. Also, as Crawley-Boevey has shown, if R is a tame k-algebra, where k is an algebraically closed field, then *any generic module of infinite length is obtained in this way.* Namely, according to [CB1], such a module is of the form $M \otimes_{k[T]} k(T)$, where M is an R-$k[T]$-bimodule which is

[20] Some authors require in addition that the module itself is not of finite length. This may be a reasonable convention in case one deals with a tame algebra, but it is odd in general: The word "generic" refers to the fact that such a module G serves to parameterize an algebraic family of indecomposable modules, just in the same way, as in classical geometry generic points were used. To exclude the possibility for G to be of finite length would correspond to the requirement that only irreducible varieties of dimension at least 1 should have a generic point.

[21] In the general setting as considered in [R9], one deals with a module P with a locally nilpotent endomorphism ϕ such that ϕ has finite-dimensional kernel. Of course, the direct limit module for a ray in a tube has these properties, but there are other examples. A string algebra with a contracting $\mathbb{Z}$-word z provides the example of such a module $P = M(z)$ such that the infinite direct products of copies of P have two non-isomorphic generic modules as direct summands.

finitely generated and free as a $k[T]$-module and such that for almost all elements $\lambda \in k$, the modules $M \otimes_{k[T]} k[R]/(T-\lambda)^n$ form a tube.

If R is a finite-dimensional k-algebra which is connected, hereditary and tame, there is precisely one generic module of infinite length; details will be given below in the case where k is an algebraically closed field. There are other classes of algebras where all the generic modules are known, let us mention at least the tubular algebras, see for example [Le], and also the string algebras. For the relevance of maps between generic modules we refer to Bautista [Bt]. It has been shown in [R11] that for many string algebras one can find sequences $G_1, G_2, \ldots$ of generic modules such that for every i there do exist both proper monomorphisms $G_i \to G_{i+1}$ and proper epimorphisms $G_{i+1} \to G_i$.

4. Finite-dimensional algebras

The impetus for this collection of surveys came from a maybe surprising, but apparent need of using infinite-dimensional modules in order to understand the behaviour of finite-dimensional modules over a finite-dimensional algebra R. Much effort has been spent in order to define the representation type of such an algebra: this concerns the category of finite-dimensional R-modules, but the usual approaches involve infinite-dimensional R-modules, thus infinite length modules. Let us recall some of these developments.

Products of finite-dimensional modules. Of interest is Couchot's characterization [Ct] of the algebraically compact modules: *For a finite-dimensional k-algebra R, an R-module M is algebraically compact if and only if it is a direct summand of a product of finite length modules.* (Proof: On the one hand, the class of algebraically compact modules is closed under products and direct summands, for any ring R, and it includes, for any k-algebra R, all the finite-dimensional R-modules. Conversely, assume that R is a finite-dimensional k-algebra and take any R-module M. Consider the $R^{\rm op}$-module $M^* = \mathrm{Hom}(M,k)$. It is well-known that there exists a pure exact sequence $0 \to N' \to N \to M^* \to 0$ such that N is a direct sum of finite-dimensional $R^{\rm op}$-modules, and that the dual of a pure exact sequence is split exact; see for example [Fc2] 1.23 and 1.28. Thus M^{**} is isomorphic to a direct summand of N^*. But the canonical inclusion $M \to M^{**}$ is a pure monomorphism. If M is algebraically compact, then M is a direct summand of M^{**} and thus also of N^*. But since N is a direct sum of finite-dimensional $R^{\rm op}$-modules, N^* is the direct product of these modules.) Thus, in order to deal with all possible algebraically compact modules, no fancy constructions are needed: it is sufficient to form products and to take direct summands.

But we may interprete this result also differently: after all, as we will see, there do exist quite complicated algebraically compact modules. Thus we see that the process of forming products[22] of modules is not at all easy to control, see

[22] A general discussion of products in Grothendieck categories should be very worthwhile. The products we have considered here are always cartesian products,

also [HO]. Forming products of finite-dimensional modules is a very effective way in order to obtain new types of modules. In particular, we note the following well-known result: *Let* $(M_i)_i$ *be a set of finite-dimensional modules and assume that any indecomposable module is a direct summand of at most finitely many* M_i. *Then the module* $\prod M_i / \bigoplus M_i$ *has no finite-dimensional indecomposable direct summand.*

Representation types. We assume that R is a finite-dimensional algebra (or, more generally, an artin algebra). In case R is representation-finite, the structure of all the R-modules is known: **Theorem:** *If* R *is a representation-finite algebra, and* M *is any* R*-module, then* M *is a direct sum of indecomposable* R*-modules of finite length* [T,RT,A2]. Of course, such a direct sum is essentially unique, according to the Azumaya Theorem. On the other hand, *if* R *is not representation-finite, then there are always indecomposable* R*-modules of infinite length* [A3].

In [A1], Auslander has introduced the notion of a representation equivalence. A *representation equivalence* (or an *epivalence* [GR]) is a functor which is full, dense and reflects isomorphisms, and two categories $\mathcal{A}$ and $\mathcal{A}'$ were said to be *representation equivalent* (or to have equivalent representations) provided there is a sequence of representation equivalences $\mathcal{A} = \mathcal{A}_0 \to \mathcal{A}_1 \leftarrow \mathcal{A}_2 \to \cdots \mathcal{A}_n = \mathcal{A}'$. Actually, this equivalence relation is not very useful, for the following reason: If k is an algebraically closed field, then all the representation-infinite k-algebras R are representation-equivalent: the category $\operatorname{mod} R/\operatorname{rad}(\operatorname{mod} R)$ is the direct sum of copies of $\operatorname{mod} k$, the number of copies is just the cardinality of k (note that the number of isomorphism classes of indecomposables is equal to the cardinality of the field k).

The concept of a representation embedding as introduced by Crawley-Boevey in [CB2] is more appropriate and avoids such difficulties: a k-linear functor $F\colon \operatorname{Mod} S \to \operatorname{Mod} R$ is said to be a *representation embedding* provided it is exact, preserves direct sums and products, preserves indecomposability and non-isomorphy. Prest shows in [P4] that a representation embedding from $\operatorname{Mod} S$ to $\operatorname{Mod} R$ induces a homeomorphic embedding of $\mathcal{Z}(S)$ into $\mathcal{Z}(R)$.

On the basis of examples considered by Corner, by Brenner and Butler and others, there had been a vague feeling concerning a possible distinction between tame and wild algebras; a challenging conjecture was formulated by Donovan and Freislich at the Bonn workshop 1973 and proved by Drozd [Dd]. Let k be an

but one should be aware that a full subcategory $\mathcal{C}'$ of a Grothendieck category $\mathcal{C}$ which is closed under kernels, cokernels and direct limits (and thus a Grothendieck category on its own) does not have to be closed under products, a typical example is the subcategory $\mathcal{C}'$ of all abelian p-groups in the category of all abelian groups. In such a situation, the products in $\mathcal{C}'$ are subobjects of the products formed in $\mathcal{C}$ (in our example, the product in $\mathcal{C}'$ is the torsion subgroup of the cartesian product; note that for non-bounded p-groups the cartesian product is no longer a torsion group).

algebraically closed field. A k-algebra R is said to be *tame* provided for every dimension d, there are finitely many R-$k[T]$-bimodules M_i which are finitely generated free as $k[T]$-modules, such that almost all indecomposable R-modules of dimension d are isomorphic to modules of the form $M_i \otimes_{k[T]} N$, where N is an indecomposable finite length $k[T]$-module. Let us call R to be t-*domestic* provided t bimodules $M_1, \dots, M_t$, but not less, are needed altogether (for all d), and *non-domestic,* in case infinitely many bimodules M_i are needed. If we fix such a bimodule M_i and consider the set of all the R-modules $M_i \otimes_{k[T]} N$, where N runs through the indecomposable $k[T]$-modules say of dimension n, we obtain what may be called a (rational) one-parameter family of R-modules. Thus, an algebra is tame provided for every dimension d, almost all the indecomposable R-modules belong to a finite number of one-parameter families. In this way, one can reformulate the notion of tameness in terms of algebraic geometry. Since these bimodules M_i are free as $k[T]$-modules (and non-zero), they are infinite-dimensional over k, thus, as R-modules they also have infinite length. Only recently, Krause [K6,K7] gave the first characterization of tameness which only relies on the category of R-modules of finite length, without reference to an infinite length R-module, or the (external) algebraic geometrical structure.

It seems that an algebra R is tame (or representation-finite) if and only if any non-zero algebraically compact module has an indecomposable direct summand. If this is true, this would provide a very convenient and easy way for defining tameness, using only the notions of indecomposability and of algebraic compactness. Actually, taking into account Couchot's characterization, we may even avoid the notion of algebraically compactness, thus we arrive at the following reformulation: An artin algebra R should be tame if and only if any product of finite length modules is a discrete module.

The representation-finite algebras are quite well-understood (see [GR]): We recall that a representation-finite algebra with a faithful indecomposable representation is standard, so that one may use covering theory in order to recover all the indecomposables from a suitable representation-directed algebra, and there are very effective algorithms known in order to deal with the indecomposable representations of a representation-directed algebra. For algebras which are not representation-finite, no general theory is available at present: there does not yet exist a structure theory even for the 1-domestic algebras, the algebras nearest to the representation-finite ones.

Drozd's definition of wildness involves, as that of tameness, infinite length R-modules; here one uses an R-$k\langle X,Y\rangle$-bimodule M which is finitely generated free as $k\langle X,Y\rangle$-module.

Strictly wild and controlled wild algebras. Let us start with a strictly wild algebra R, here one requires that for any k-algebra S, there is a full and exact embedding $\operatorname{Mod} S \to \operatorname{Mod} R$ which sends finitely presented S-modules to finitely presented R-modules. Of course, as soon one knows one strictly wild algebra S_0, it is sufficient to find an embedding $\operatorname{Mod} S_0 \to \operatorname{Mod} R$ as required in order to know that also R is strictly wild. A typical example of a strictly wild algebra is the free k-algebra $k\langle X, Y\rangle$ in two generators X, Y. Also all the generalized Kronecker algebras $kK(n)$ with $n \geq 3$ are strictly wild; by definition, $kK(n)$ is the path algebra of the quiver $K(n)$ with two vertices, say a, b and n arrows $b \to a$

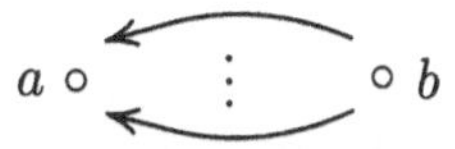

For example, for $n = 3$ one obtains an embedding as required $\operatorname{Mod} k\langle X, Y\rangle \to \operatorname{Mod} kK(3)$ by sending the $\langle X, Y\rangle$-module (M, X, Y) to the following representation of $K(3)$

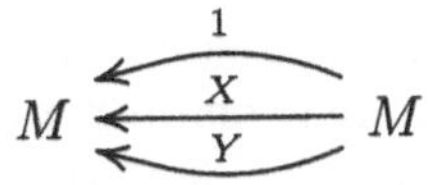

(here, M is a vector space, $X\colon M \to M$ denotes the multiplication by X, and similarly $Y\colon M \to M$ that by Y).

Why is it of interest to know that a k-algebra R is strictly wild? Of course, this implies that every k-algebra S occurs as the endomorphism ring of a suitable R-module, thus all module theoretical phenomena which can be read off from the endomorphism rings of a module occur for R-modules; in particular, this applies to all kinds of possible direct decompositions, since the direct decompositions of a module are encoded into the set of idempotents of its endomorphism ring. Strict wildness does not concern only individual modules, but also sets or even classes of R-modules. For example, for certain considerations it is good to have large sets of pairwise orthogonal modules at hand, a set $(M_i)_{i\in I}$ being called *pairwise orthogonal* provided $\operatorname{Hom}(M_i, M_j) = 0$ for all pairs of indices $i \neq j$, and this is the case for any strictly wild algebra.

It is well-known that there do exist k-algebras R which are not strictly wild, but which have the weaker property that any k-algebra S can be realized as a (nice) factor algebra of the endomorphism ring of an R-module. For example, consider the polynomial ring $k[X, Y, Z]$ in three variables and its factor algebra $R = k[X, Y, Z]/(X, Y, Z)^2$ modulo the square of the ideal generated by X, Y, Z. This is a local algebra, thus the only division ring which can be realized as an endomorphism ring of an R-module is k itself (the only module M with endomorphism ring being a division ring is the simple module k). Also, if $(M_i)_{i\in I}$ is a set of pairwise orthogonal R-modules, then the index set consists of at most one element! On the other hand, given any k-algebra S, there does exist an R-module M such that $S = \operatorname{End}(M)/J$, where J is an ideal of $\operatorname{End}(M)$, it is even a nicely defined ideal, namely the set of all endomorphisms of M with semisimple image. Such "wild" algebras were studied by Corner, Brenner and others. When Drozd formulated and proved his celebrated tame-and-wild theorem, he introduced a definition of wildness which deviated from the older intuitive notion: an algebra R is *wild* in the sense of Drozd provided there exists an R-$k\langle X, Y\rangle$-bimodule W which is finitely generated and projective as a $k\langle X, Y\rangle$-module such that the functor $F(-) = \big({}_RW \otimes_{k\langle X,Y\rangle} -\big)$ preserves indecomposability and non-isomorphy. Of course, in this definition, we may replace the (infinite-dimensional) algebra $k\langle X,Y\rangle$ by the (five dimensional) algebra $kK(3)$, considering a bimodule ${}_RW_{kK(3)}$ which is finitely generated and projective as a $kK(3)$-module and the functor

$$F(-) = \big({}_RW \otimes_{k\langle X,Y\rangle} -\big)\colon \operatorname{Mod} kK(3) \to \operatorname{Mod} R,$$

or at least its restriction to $\operatorname{mod} kK(3)$. Now, given a k-algebra S, there is a $kK(3)$-module N with $\operatorname{End}(N) = S$ and we may consider now the R-module $F(N) = W \otimes_{kK(3)} N$. Its endomorphism ring $\operatorname{End}(F(N))$ has S as a subring, but there is no reason that in this way S can be realized as the factor ring of an endomorphism ring. The problem we are dealing with is the following: given a R-$kK(3)$-bimodule W which is finitely generated and free as a $kK(3)$-module, such that the functor $F(-) = \big(W \otimes_{kK(3)} -\big)$ is faithful, then $F(-)$ is not necessarily full. In which way is it possible to control the subspaces

$$F(\operatorname{Hom}_{kK(3)}(N_1, N_2)) \subseteq \operatorname{Hom}_R(F(N_1), F(N_2))$$

We say that R is *controlled wild* provided the subspace $F(\operatorname{Hom}_{kK(3)}(N_1, N_2))$ of $\operatorname{Hom}_R(F(N_1), F(N_2))$ is complemented by the set $\operatorname{Hom}_R(F(N_1), F(N_2))_{\mathcal{U}}$ of homomorphisms $F(N_1) \to F(N_2)$ which factor through a prescribed additive subcategory $\mathcal{U}$ of R-modules (and $\mathcal{U}$ may be called the corresponding *control class*):

$$(*)\quad \operatorname{Hom}_R(F(N_1), F(N_2)) = F(\operatorname{Hom}_{kK(3)}(N_1, N_2)) \oplus \operatorname{Hom}_R(F(N_1), F(N_2))_{\mathcal{U}}.$$

For example, for the local algebra $R = k[X,Y,Z]/(X,Y,Z)^2$ the class $\mathcal{U}$ of all semisimple modules serves as such a control class. Recent investigations by Rosenthal and Han [Ha] support the conjecture that all finite-dimensional wild k-algebras (k is algebraically closed) are controlled wild. We can reformulate this concept[23] as follows: Let $\mathcal{V}$ be the class of R-modules which are images under F and consider the ideal $\langle\mathcal{U}\rangle \cap \mathcal{V}$ of $\mathcal{V}$. According to $(*)$, we see that the factor category $\mathcal{V}/\big(\langle\mathcal{U}\rangle \cap \mathcal{V}\big)$ is equivalent to the category $\operatorname{mod} k\langle X,Y\rangle$.

Tame is Wild. If we consider for a finite-dimensional k-algebra all modules and not only those of finite length, the difference between "tame" and "wild" vanishes. This is usually formulated as follows: tame algebras are Wild [R3,R12], here the small "t" in *tame* refers to tameness with respect to modules of finite length, the capital "W" in *Wild* refers to wildness with respect to arbitrary modules which

---

[23] This seems to be an appropriate setting for discussing the wildness of many categories; for example, we may take as $\mathcal{V}$ the category of all abelian p-groups and as $\mathcal{U}$ the subcategory of all bounded ones. Or, let $\mathcal{V}$ be the category of all abelian groups which are slender, and $\mathcal{U}$ the subcategory of all free abelian groups of finite rank.

are not necessarily of finite length. To be more precise, the quoted papers show that the Kronecker quiver $K(2)$

$$a \circ \leftleftarrows \circ\; b$$

is strictly Wild: they provide an explicit full exact embedding of the category of representations of $K(3)$ into the category of representations of $K(2)$, and, of course, $K(3)$ is a typical strictly wild quiver. Similar to the notion of a strictly wild algebra, one may call a k-algebra R *strictly tame* provided there is a full exact embedding of the category $\operatorname{Mod} kK(2)$ of all $kK(2)$-modules into the category $\operatorname{Mod} R$ of all R-modules. The precise formulation is: *All strictly tame algebras are Wild.* There is a strong belief that for all tame finite-dimensional k-algebras, where k is an algebraically closed field, the one-parameter families of indecomposables can be obtained using functors $\operatorname{Mod} kK(2) \to \operatorname{Mod} R$ which are quite well-behaved also with respect to infinite-dimensional modules. If this turns out to be true, then one will see that really all tame algebras are Wild. In this way, for the global behaviour of finite-dimensional algebras, the only relevant distinction seems to be that between finite representation type and infinite representation type.

The main observation behind the tame-is-Wild theorem concerns the existence of (many) infinite-dimensional $kK(2)$-modules with endomorphism ring k. Here is a recipe in case k is infinite: let $k(T)$ be the field of rational functions in one variable, let I be an infinite subset of k. Let U_b be the subspace of $k(T)$ generated by the elements $\frac{1}{T-\lambda}$ with $\lambda \in I$ and $U_a = U_b + k1$. The representation $U = U(I)$ we are interested in is

$$U_a \;\overset{1}{\underset{T\cdot}{\leftleftarrows}}\; U_b$$

This is a subrepresentation of the generic one $(k(T), k(T); 1, T\cdot)$, and it is not difficult to check that its endomorphism ring is k. Moreover, if I, J are disjoint infinite sets then $\operatorname{Hom}(U(I), U(J)) = 0$, but $\operatorname{Ext}^1(U(I), U(J)) \neq 0$. Using such representations $U(I)$, it is easy to construct many different full exact embeddings of strictly wild categories into $\operatorname{Mod} kK(2)$.

We have mentioned above that for any dualizing k-variety, the category $\operatorname{mod} \mathcal{C}$ has Auslander-Reiten sequences. As a consequence, one can consider the corresponding Auslander-Reiten quiver. It incorporates the basic concepts developed by Auslander and Reiten, in particular the notion of an irreducible map and that of the Auslander-Reiten translate τ ("dual of the transpose"), in order to provide a first overview over the category $\operatorname{mod} \mathcal{C}$. Here we will use this theory in the classical case where R is a finite-dimensional algebra over some field (or, more generally, an artin algebra) and denote by Γ_R its *Auslander-Reiten quiver.* This is a quiver with vertex set the isomorphism classes $[X]$ of the indecomposable R-modules of finite length, and with an arrow $[X] \to [Y]$, where X, Y are indecomposable, provided there exists an irreducible map $X \to Y$. In addition, this quiver is endowed with the partial translation τ, it has the following property: $\tau[Z]$ is defined if and only if Z is not projective, and then there is an arrow $[Y] \to [Z]$ if and only if there is an arrow $\tau[X] \to [Y]$. If R is connected and representation-finite, then Γ_R is connected. It has been conjectured that also the converse is true (this is known to be true in case k is algebraically closed). A component of Γ_R which does not contain any projective or injective module is said to be a *stable* component.

The tame hereditary algebras. One class of domestic algebras is very well-known, the tame hereditary k-algebras, where k is an algebraically closed field: these are just the path algebras of quivers of type $\widetilde{A}_n, \widetilde{D}_n, \widetilde{E}_6, \widetilde{E}_7, \widetilde{E}_8$. Let us recall the structure of the module category of such an algebra: There are two components of the Auslander-Reiten quiver which are not stable, one consists of the indecomposable projective modules and their τ^{-1}-translates (it is called the *preprojective* component $\mathcal{P}$), the other contains the injective modules and their τ-translates (the *preinjective* component $\mathcal{I}$). The remaining components are stable, all these components are "tubes", and the set of these components is indexed in a natural way by the projective line $\mathbb{P}_1 k$. The modules belonging to these stable components and their direct sums are said to be *regular*. The regular modules form an abelian subcategory $\mathcal{R}$, thus one may speak of simple regular representations (these are the regular representations which are simple objects when considered as objects in $\mathcal{R}$). Any regular module M is τ-periodic (this means that $\tau^p M \simeq M$ for some $p \geq 1$), the indecomposable modules in all but at most three of the tubes are *homogeneous* (this means that $\tau M \simeq M$); the remaining tubes are said to be *exceptional* and the minimal period for all the modules in a tube is said to be its *rank*. The global structure of the category $\operatorname{mod} R$ is as follows:

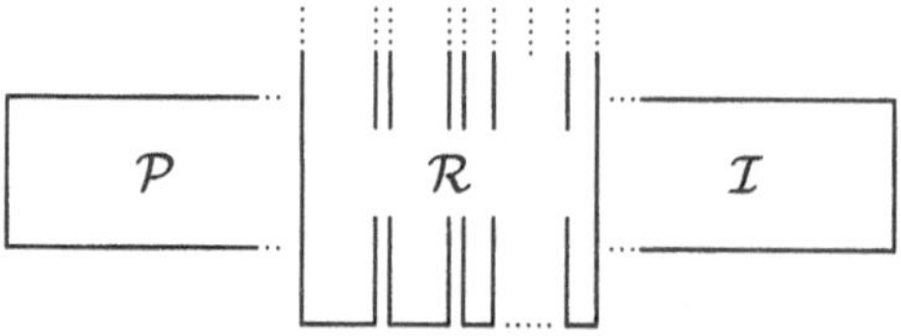

Globally, all the maps go from left to right: there are no maps from $\mathcal{R}$ to $\mathcal{P}$ and no maps from $\mathcal{I}$ to $\mathcal{P}$ or $\mathcal{R}$. (Actually, the subcategories $\mathcal{P}$ and $\mathcal{I}$ are also directed: the indecomposable modules in these subcategories can be arranged in such a way that there are no maps from right to left; but inside $\mathcal{R}$, there do exist cycles of maps.)

For a tame hereditary algebra R, the structure of the category $\operatorname{mod} \operatorname{mod} R$ is well-known, see [Gg]. Let us recall the shape of the indecomposable projective objects in $\operatorname{mod} \operatorname{mod} R$, these are the functors $\operatorname{Hom}(-, X)$ with X an indecomposable R-module of finite length. The top of $\operatorname{Hom}(-, X)$ is the simple functor $S_X = \operatorname{Hom}(-, X)/\operatorname{rad}(-, X)$, its socle is of the form $\bigoplus_i S_{P_i}$, where $\bigoplus_i P_i$ is a projective cover of $\operatorname{soc} X$. Of course, $\operatorname{Hom}(-, X)$ is of finite length if and only if X

is preprojective. If X is not preprojective, then the socle sequence $\operatorname{soc}_i \operatorname{Hom}(-,X)$ and the radical sequence $\operatorname{rad}_i \operatorname{Hom}(-,X)$ are controlled by the Auslander-Reiten structure of the category $\operatorname{mod} R$. It is instructive to specify suitable subobjects of $\operatorname{Hom}(-,X)$ in $\operatorname{Mod}\operatorname{mod} R$, namely the restriction to $\mathcal{P}$ and to $\mathcal{P}\vee\mathcal{R}$. The restriction $\operatorname{Hom}(-,X)|\mathcal{P}$ is just the union $\bigcup_i \operatorname{soc}_i \operatorname{Hom}(-,X)$, in particular this is an artinian functor. In case X is regular, $\operatorname{Hom}(-,X)/(\operatorname{Hom}(-,X)|\mathcal{P})$ is noetherian. Of special interest is the case where $X = E$ is simple regular. In this case the shape of $\operatorname{Hom}(-,E)$ (or the corresponding dual configuration) has been studied quite carefully in [R5] as the patterns which are relevant for tubular one-point extensions. If X is preinjective and $F = \operatorname{Hom}(-,X)$, let F'' be the restriction of F to $\mathcal{P}\vee\mathcal{R}$ and F'' the restriction to $\mathcal{P}$. Then $0 \subset F'' \subset F' \subset F$, the functor F'' is artinian, F/F' is noetherian, and F'/F'' is the direct sum of infinitely many functors (the summands correspond to the stable components, thus to the elements of $\mathbb{P}_1 k$), and none of these summands is artinian or noetherian.

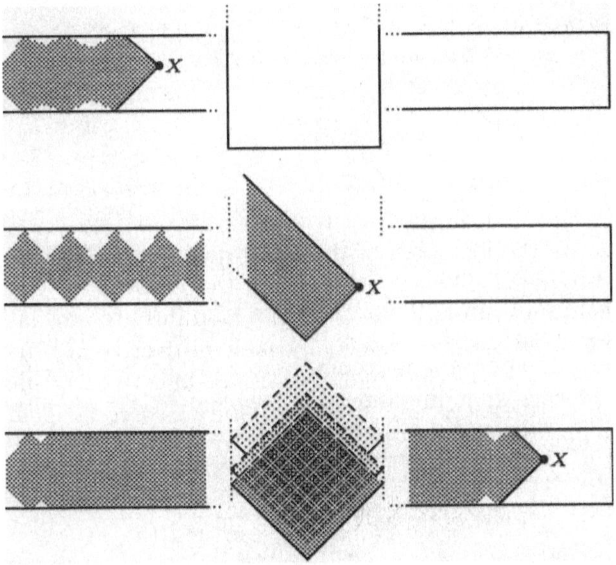

X preprojective

X regular

X preinjective

Recall that the functor category $\operatorname{mod}\operatorname{mod} R$ is a dualizing k-variety. In particular, this means that given any R-module X of finite length, the functor $D\operatorname{Hom}(X,-)$ is finitely presented. For example, starting with an indecomposable regular R-module X, this functor $D\operatorname{Hom}(X,-)$ can be displayed as follows:

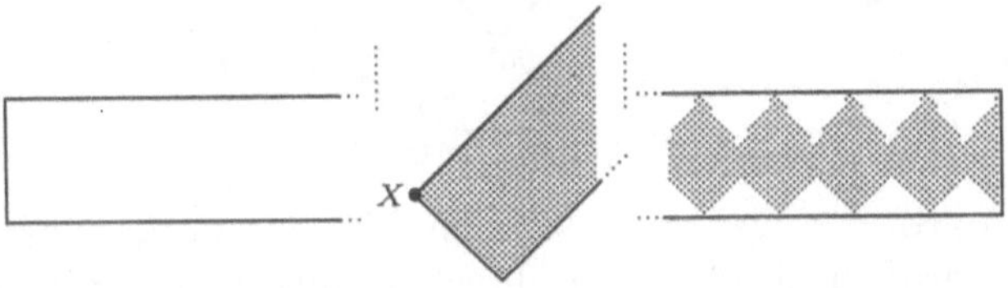

X regular

How do we obtain a finite presentation of $D\operatorname{Hom}(X,-)$? If $X = P(i)$ is an indecomposable projective R-module, then $D\operatorname{Hom}(P(i),-) \simeq \operatorname{Hom}(-,I(i))$, thus

$D\operatorname{Hom}(P(i), -)$ is an indecomposable projective functor. Otherwise we can use the Auslander-Reiten formula $D\operatorname{Hom}(-, X) \simeq \operatorname{Ext}^1(-, \tau X)$; we embed τX into an injective module I, the exact sequence $0 \to \tau X \to I \to I/\tau X \to 0$ yields the exact sequence $\operatorname{Hom}(-, I) \to \operatorname{Hom}(-, I/\tau X) \to \operatorname{Ext}^1(-, \tau X) \to 0$, thus a finite presentation of $\operatorname{Ext}^1(-, \tau X)$. Of course, since our ring R is hereditary, the module $I/\tau X$ is injective again. — Finally, let us single out also the shape of the projective-injective objects: recall that these are just the functors $\operatorname{Hom}(-, I(i)) \simeq D\operatorname{Hom}(P(i), -)$:

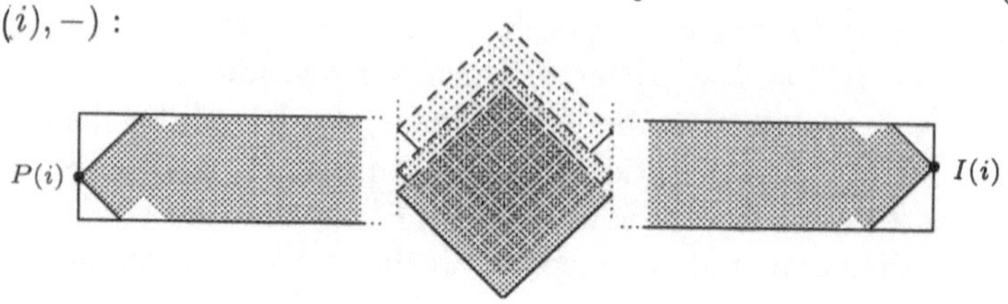

Let us deal with a specific example of a tame hereditary algebra. We consider the following quiver Q' of type $\widetilde{A}_5$

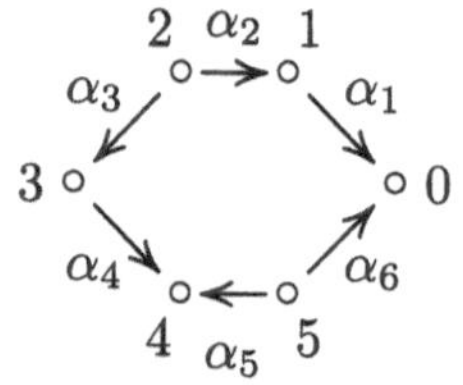

First of all, let us indicate, on the left, part of the preprojective component, and, on the right, part of the preinjective component. Here, the vertices are replaced by the corresponding dimension vectors[24], some of the modules are also labeled: Given a vertex i of Q', the corresponding indecomposable projective kQ'-module will be denoted by $P'(i)$; the corresponding indecomposable injective kQ'-module by $I'(i)$. Note that in both pictures, the solid horizontal lines have to be identified.

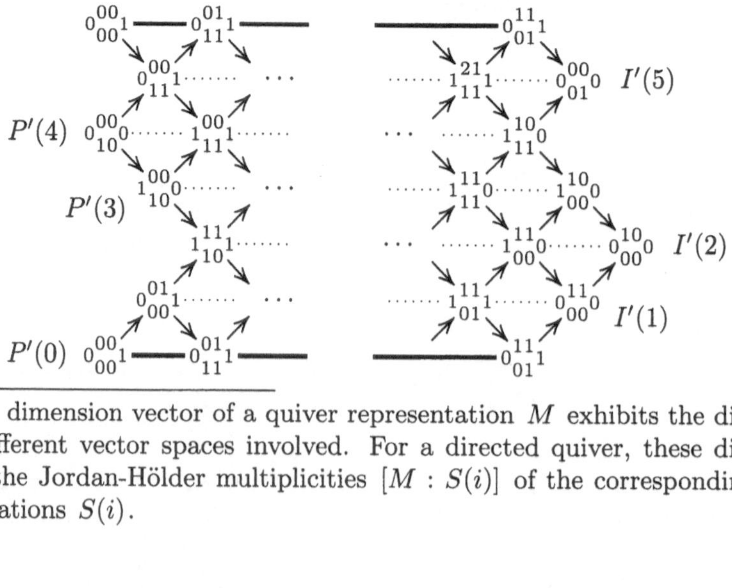

[24] The dimension vector of a quiver representation M exhibits the dimensions of the different vector spaces involved. For a directed quiver, these dimensions are just the Jordan-Hölder multiplicities $[M : S(i)]$ of the corresponding simple representations $S(i)$.

For this algebra, there are two exceptional tubes, both of rank 3. We are going to present also part of these tubes, again using dimension vectors and some labels.

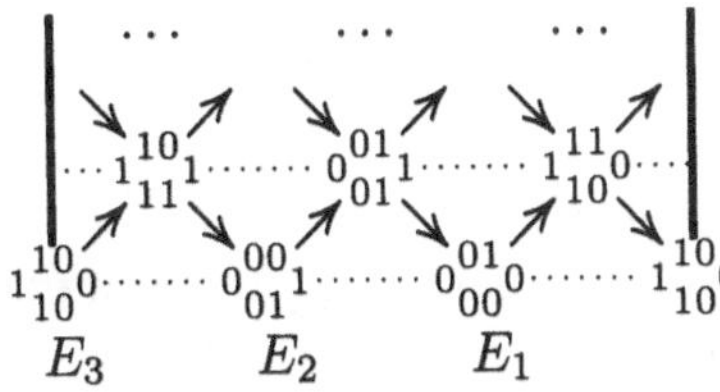

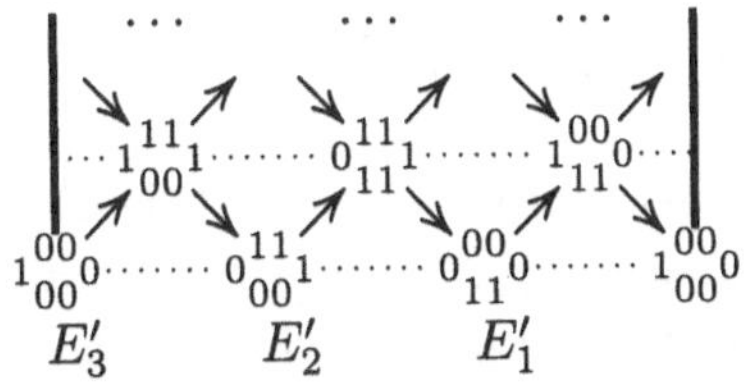

It is easy to list the indecomposable algebraically compact modules: Of course, all the indecomposable modules of finite length have to be mentioned. In addition, every simple regular module E gives rise to two indecomposable modules which are algebraically compact: the *Prüfer module* $E[\infty]$ and the *adic module* $\widehat{E}$. They are obtained as follows: Consider all the finite-dimensional indecomposable regular R-modules M with $\operatorname{Hom}(E, M) \neq 0$. These modules can be labeled in the form $M = E[s]$ and arranged as a so called *ray*:

$$E = E[1] \subset E[2] \subset \cdots \subset E[s] \subset \cdots$$

where all the inclusion maps $E[s] \subset E[s+1]$ are irreducible maps, and

$$E[\infty] = \bigcup E[s].$$

Similarly, consider all the finite-dimensional indecomposable regular R-modules M with $\operatorname{Hom}(M, E) \neq 0$. These modules can be labeled in the form $M = [s]E$ and arranged as a *coray*:

$$\cdots \to [s]E \to \cdots \to [2]E \to [1]E = E$$

where all the maps $E[s+1] \to E[s]$ are irreducible epimorphisms. The adic module $\widehat{E}$ is defined as the inverse limit

$$\widehat{E} = \lim_{\leftarrow} \, [s]E$$

and we denote by $\pi_s \colon \widehat{E} \to [s]E$ the canonical projection (the names Prüfer module and adic module are parallel to the use of the corresponding names for abelian groups, where one speaks of a Prüfer group and the p-adic integers; the p in "p-adic" specifies the simple top $\mathbb{Z}/\mathbb{Z}p$, in a similar way, we may call $\widehat{E}$ the E-adic module).

Let us indicate why *both modules* $E[\infty]$ *and* $\widehat{E}$ *are indecomposable.* It is well-known that for any $s \geq 1$, any non-zero map $E \to E[s]$ (and therefore also $E \to E[\infty]$) is the composition of an automorphism of E and the inclusion map. Thus, given a direct decomposition $E[\infty] = U \oplus U'$, we may assume that $E \subseteq U$

and $\operatorname{Hom}(E,U')=0$. Inductively, one sees that $E[s]\subseteq U$ for all s, thus $U'=0$. Similarly, we want to see that any non-zero map $\widehat{E}\to E$ is the composition of π_1 and an automorphism of E. Note that $\operatorname{Ext}^1(\tau^{-1}E,\widehat{E})\simeq\varprojlim\operatorname{Ext}^1(\tau^{-1}E,[s]E)$, and the maps $[s+1]E\to[s]E$ induce isomorphisms $\operatorname{Ext}^1(\tau^{-1}E,[s+1]E)\to\operatorname{Ext}^1(\tau^{-1}E,[s]E)\to\operatorname{Ext}^1(\tau^{-1}E,E)$. Now, we use the Auslander-Reiten formula which yields $\operatorname{Hom}(\widehat{E},E)\simeq D\operatorname{Ext}^1(\tau^{-1}E,\widehat{E})\simeq D\operatorname{Ext}^1(\tau^{-1}E,E)\simeq\operatorname{Hom}(E,E)$. This shows that any map $\widehat{E}\to E$ vanishes on the kernel of π_1. Using induction, it follows that any map $\widehat{E}\to[s]E$ vanishes on the kernel of π_s, and this implies that for any decomposition $\widehat{E}=U\oplus U'$, one of the summands has to be contained in the intersection of all these kernels, thus has to be zero.

Also, *both modules $E[\infty]$ and $\widehat{E}$ are algebraically compact.* This is trivial for $\widehat{E}$, since it is the inverse limit of finite-dimensional modules, and such an inverse limit is always algebraically compact. But it is also clear for the Prüfer module $E[\infty]$, since it is artinian when considered as a module over its endomorphism ring.

There is just one additional R-module which is indecomposable and algebraically compact, the *generic module G* of infinite length, see [R4]. There are several ways to construct G. Starting with a Prüfer module $E[\infty]$, note that there is an epimorphism $(\tau E)[\infty]\to E[\infty]$ whose kernel is simple regular. We obtain a sequence of maps

$$\cdots\to(\tau^nE)[\infty]\to\cdots\to(\tau E)[\infty]\to E[\infty]$$

and we may form the inverse limit $\varprojlim(\tau^nE)[\infty]$. This inverse limit is a direct sum of copies of G. Dually, starting with an adic module $\widehat{E}$, note that there is an embedding $\widehat{E}\subset(\tau^{-1}E)\widehat{\ }$ with simple regular cokernel. We obtain a sequence of inclusion maps

$$\widehat{E}\subset(\tau^{-1}E)\widehat{\ }\subset\cdots\subset(\tau^{-n}E)\widehat{\ }\subset\cdots$$

This time, we have to form the direct limit and obtain a module which again is the direct sum of copies of G.

We have constructed the infinite-dimensional indecomposable algebraically compact modules using limits and colimits (and direct summands) of regular modules. It is also possible to construct the adic modules as well as the generic module of infinite length as direct limits of preprojective modules, and the Prüfer modules as well as this generic module as direct summands of inverse limits of preinjective modules. We are going to insert the additional algebraically compact modules into our global picture of $\operatorname{mod}R$, the arrows indicate the directions of all the possible maps.

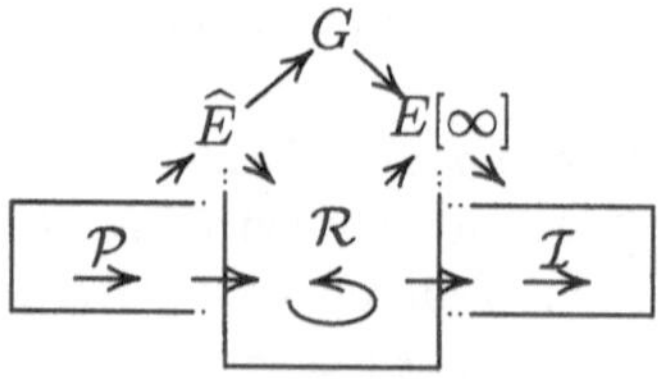

We recall that any algebraically compact R-module occurs as a direct summand of a product $X = \prod_{i\in I} X_i$ of finite-dimensional modules X_i. Since we only are interested in algebraically compact modules which are indecomposable, we may assume that we start with a collection of modules X_i where all these modules are preprojective, or regular, or preinjective. First, consider the case where all X_i are preinjective. Then X is a divisible module[25], thus it is a direct sum of indecomposable divisible modules. An indecomposable divisible module is either finite-dimensional and then preinjective, or else a Prüfer module or the generic module G of infinite length. Next, consider the case where all the modules X_i are regular. Then X will contain a submodule which is a direct sum of indecomposable modules which are finite-dimensional (and regular), or adic modules, and such that X/X' is a direct sum of copies of G. Usually X' will be a proper submodule of X, thus it will not be possible to write X as a direct sum of indecomposable modules. Finally, in case all the modules X_i are preprojective, then the only indecomposable summands of X are the finite-dimensional ones. Let us assume that all the modules X_i are indecomposable, let $P(1), P(2), \ldots$ be a complete list of all the indecomposable preprojective modules, one from each isomorphism class, and denote by $I(s)$ the set of indices i with X_i isomorphic to $P(s)$. Note that the product $\prod_{i\in I(s)} X_i$ can be written as a direct sum of copies of $P(s)$. The submodule $X' = \bigoplus_s \prod_{i\in I(s)} X_i$ of X is a direct sum of finite-dimensional indecomposable modules, and it is maximal with this property. Of course, in case $I(s)$ is non-empty for infinitely many s, then X' is a proper submodule of X. — It is of interest to compare the two extreme cases when dealing with preinjective or preprojective modules X_i. In the preinjective case, several new types of indecomposable direct summands occur, but X can be written as the direct sum of indecomposable modules. On the other hand, in the preprojective case, no new isomorphism classes of indecomposable direct summands do occur, but usually X cannot be written as a direct sum of indecomposable modules. — The case where all X_i are regular is intermediate and is similar to the well-known situation of analyzing reduced algebraically compact abelian groups [Fu1].

Having determined all the indecomposable algebraically compact modules, let us describe the Ziegler topology (see [Gr], [P5] and [R10]). A subset $\mathcal{X}$ of $\mathcal{Z}(R)$ is closed if and only if the following conditions are satisfied: First, if E is a simple regular R-module and if there are infinitely many finite length modules $X \in \mathcal{X}$ with $\mathrm{Hom}(E, X) \neq 0$, then $E[\infty]$ belongs to $\mathcal{X}$. Second, the dual condition, if E is a simple regular R-module and if there are infinitely many finite length modules $X \in \mathcal{X}$ with $\mathrm{Hom}(X, E) \neq 0$, then $\widehat{E}$ belongs to $\mathcal{X}$. And third, if there are infinitely many finite length modules in $\mathcal{X}$ or if there exists at least one module in $\mathcal{X}$ which is not of finite length, then the generic module of infinite length belongs to $\mathcal{X}$.

We see that the Ziegler closed subsets of $\mathcal{Z}(R)$ are related to the support

[25] For the notion of divisibility as well as the structure of products of preprojective modules, we refer to [R3], sections 5 und 2, respectively.

of the functors $\mathrm{Hom}(E, -)$ and $\mathrm{Hom}(-, E)$, where E is simple regular. These functors play an important role for many questions in the representation theory of tame hereditary algebras, for example for one-point extensions and coextensions.

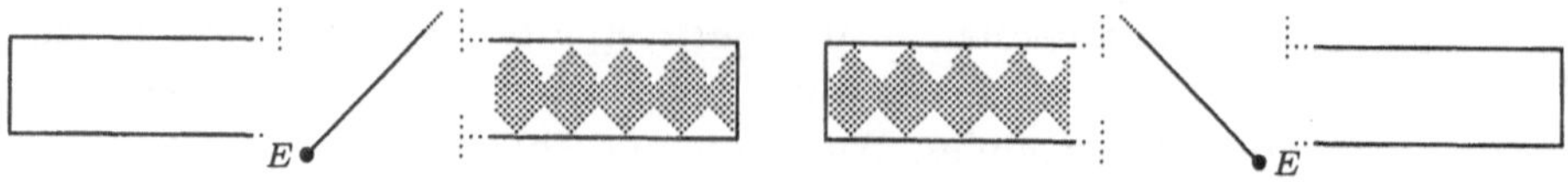

String algebras. We are going to describe the structure of some additional indecomposable algebraically compact modules. They are similar to the Prüfer modules and the adic modules, or to pairs of such modules. The algebras we will consider are string algebras, this is a class of tame algebras whose indecomposable modules can be constructed with bare hands. We are going to outline how to determine all the indecomposable algebraically compact modules for a string algebra R; in case R is domestic, we will give a complete description of these modules.

We recall that for a connected tame hereditary algebra[26] the indecomposable algebraically compact modules which are not of finite length are the Prüfer modules, the adic modules and one endofinite module of infinite length. These kinds of modules do also exist for any string algebra, more precisely: for any primitive cyclic word (or better: its equivalence class) there are Prüfer modules, adic modules and one endofinite module of infinite length, but there may be additional ones which are built up using Prüfer modules, adic modules and finite dimensional ones. For the almost periodic $\mathbb{N}$-words, one Prüfer module or one adic module is used, for the biperiodic $\mathbb{Z}$-words two such modules are used, and all three possible combinations occur. Of special interest seems to be the mixed case which involves at the same time a Prüfer module and an adic module.

Let us start with the case of R being domestic, even 1-domestic. First, let us mention an easy way for constructing some 1-domestic string algebras (following [R7]), this should help to illustrate some of the phenomena occurring for representation-infinite algebras. We start with a quiver Q' of type $\widetilde{A}_{n-1}$, thus there are n vertices, say labeled by the integers modulo n, and n arrows α_i, with $0 \le i < n$, such that $\{s(\alpha_i), t(\alpha_i)\} = \{i-1, i\}$. We assume that n is even and we assume that there is given also a fix point free involution $\omega\colon i \mapsto i'$ on the set Q'_0 of vertices of Q'. We form the algebra $R(Q', \omega) = kQ\langle \rho_0, \dots, \rho\rangle$, where the quiver Q is obtained from Q' by adding vertices a_ι for any ω-orbit ι of Q'_0 and arrows β_i with $\{s(\beta_i), t(\beta_i)\} = \{i, a_\iota\}$, for any $0 \le i < n$ with $i \in \iota$. The orientation of β_i is chosen so that the arrows α_i and β_i can be composed in order to form a path labeled ρ_i; thus, if $s(\alpha_i) = i$, let $t(\beta_i) = i$, and let $\rho_i = \alpha_i\beta_i$; similarly, if $t(\alpha_i) = i$, let $s(\beta_i) = i$, and let $\rho_i = \beta_i\alpha_i$. As an example, consider the quiver Q' of type $\widetilde{A}_5$ exhibited above and the map $\omega\colon Q'_0 \to Q'_0$ which exchanges 1 and 2; and 3 and 5; and 4 and 0. We obtain the following quiver Q, the dotted lines

[26] and similarly for a Dedekind ring.

indicate the relations:

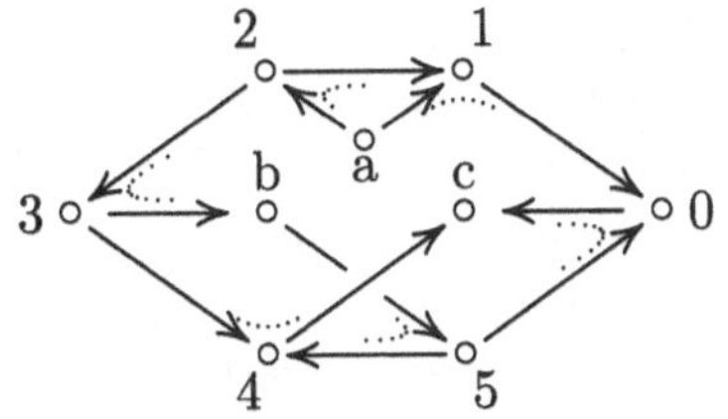

here, $a = a_{\{1,2\}}, b = a_{\{3,5\}}$ and $c = a_{\{0,4\}}$, and the six new arrows are those labeled β_i. This algebra $R(Q',\omega)$ is a string algebra [BuR], thus it is easy to list all the indecomposable modules of finite length. Since there is (up to equivalence) only one primitive cyclic word, namely $w = \alpha_1\alpha_2\alpha_3^{-1}\alpha_4^{-1}\alpha_5\alpha_6^{-1}$, we see that $R(Q',\omega)$ is 1-domestic. As an intuitive description of the new algebra $R(Q',\omega)$ we may say that we have added some "bridges" to the quiver Q' (connecting a vertex i with its image ωi), these bridges correspond to the orbits of ω.

Of course, the full subcategory of all kQ'-modules in $\operatorname{mod} R(Q',\omega)$ is known, and it consists of all representations M of Q for which all the vector spaces M_{a_ι} are zero.

Let us describe some interesting $R(Q',\omega)$-modules. Recall the following: A string algebra has two kinds of finite-dimensional indecomposable modules, the string modules and the band modules. In our case, all the band modules are actually kQ'-modules, thus let us concentrate on string modules. They are obtained by choosing a (finite) walk w in the quiver Q, this is a word $w = l_1l_2\dots l_n$, where we use as letters l_i the arrows and their formal inverses, subject to the requirement that never an arrow and its inverse are neighbors, and that consecutive arrows are composable. The kQ-module $M(w)$ is given by a k-space (here of dimension $n+1$, since w is supposed to have length n), and the word w describes the operation of the arrows of Q on this space. In order to obtain an $R(Q',\omega)$-module, we have to require in addition that w avoids the relations ρ_i; this means that these relations do not occur as a subword or as the inverse of a subword of w. We also may work with words using as letters the vertices of Q, say replacing the sequence $l_1l_2\dots l_n$ by the sequence $t(l_1)t(l_2)\dots t(l_n)s(l_n)$, provided this does not lead to confusion. This has the advantage that now the letters of the word w correspond bijectively to basis elements of $M(w)$.

In order to construct infinite-dimensional indecomposable modules, we will use $\mathbb{N}$-words $w = l_0l_2\dots l_n\dots$ or $\mathbb{Z}$-words $\dots l_{-1}l_0l_1l_2\dots$ (here $\mathbb{N}$ and $\mathbb{Z}$ refer to the sets of indices used). For example, for the algebra R exhibited above, there are precisely six $\mathbb{Z}$-words, namely the words

$$z(a) = {}^{\infty}(105432)\, a\, (123450)^{\infty},$$
$$z(b) = {}^{\infty}(210543)\, b\, (501234)^{\infty},$$
$$z(c) = {}^{\infty}(321054)\, c\, (012345)^{\infty},$$

and their inverses[27]. All the possible $\mathbb{N}$-words occur as subwords of these $\mathbb{Z}$-words. As for finite words, we can attach to every $\mathbb{Z}$-word or $\mathbb{N}$-word w a corresponding string module $M(w)$. In addition, we also may consider suitable completions of $M(w)$, see [R8], in order to obtain algebraically compact modules.

To be more precise, consider any $\mathbb{N}$-word $x = l_1 l_2 \dots$, where the letters l_i are arrows or inverses of arrows. The string module $M(x)$ is constructed as follows: take a (countably dimensional) vector space with basis $e_0, e_1, \dots$ and let R act according to the word x: in case the letter l_i is an arrow, say $l_1 = \alpha$, then define $\alpha(e_i) = e_{i-1}$, otherwise l_i is the inverse of an arrow, say $l_i = \alpha^{-1}$, then let $\alpha(e_{i-1}) = e_i$. In this way, we have defined the action of some arrows on some basis elements, the action of arrows on all other basis elements should be zero. Besides $M(x) = \bigoplus ke_i$, we also consider the product $\overline{M}(x) = \prod ke_i$ with a similarly defined action. It is well-known that the string modules $M(x)$ are indecomposable, and it is obvious that the modules $\overline{M}(x)$ are algebraically compact. Observe that at most one of the two modules can be both indecomposable and algebraically compact: The embedding $\iota\colon M(x) \to \overline{M}(x)$ is a pure embedding, thus, if $M(x)$ is algebraically compact, then ι splits and $\overline{M}(x)$ cannot be indecomposable.

Let us assume that x is almost periodic (this means that $x = x'w^\infty$, where x' and w are finite words). We claim that in this case, one of the two modules is both indecomposable and algebraically compact, and thus this is the module $C(x)$ we are interested in. An almost periodic $\mathbb{N}$-word x is either contracting or expanding, using the terminology of [R8]. There, it has been shown that for x contracting, the string module $M(x)$ is algebraically compact. But it is not difficult to see that for x expanding, the module $\overline{M}(x)$ is indecomposable[28].

Let us now consider $\mathbb{Z}$-words. In our example, the module $C(z(c))$ is constructed as follows: Again, we start with the string module $M(z(c)) = \bigoplus_{i\in\mathbb{Z}} ke_i$, where e_i, $i \in \mathbb{Z}$ is the defining basis following the letters of the $\mathbb{Z}$-word $z(c)$. Now, take the corresponding product module $C(z(c)) = \prod_{i\in\mathbb{Z}} ke_i$. This module $C(z(c))$ still has the simple module $S(c)$ as a submodule, and $C(z(c))/S(c)$ is the direct sum of two adic kQ'-modules, namely those corresponding to the simple regular modules E_2 and E_1. Second, consider the module $C(z(b))$, it is obtained from $M(z(b))$ by a partial completion ("on the left"), so that $C(z(b))$ has submodules $N' \subset N$, where N is the Prüfer kQ'-module with regular socle E_2',

---

[27] If u, v, w are finite words, we write vw^∞ instead of $vww\cdots$ and ${}^\infty uv$ instead of $\cdots uuv$. For quite obvious reasons we say that vw^∞ as well as the inverse of ${}^\infty uv$ are "almost periodic $\mathbb{N}$-words".

[28] For any j, we consider the subspace $C_j = \prod_{i\neq j} ke_i$ of $\overline{M}(x)$. Note that the intersection $\bigcap_j C_j$ is zero. We use the functorial filtration of the forgetful functor F as in [R1]. The word x defines a sequence of intervals $G_i \subset F_i$ of subfunctors of F. Now suppose there is given a direct decomposition $\overline{M}(x) = N \oplus N'$. Clearly, $(F_0/G_0)(\overline{M}(x))$ is one-dimensional, thus we may assume that $(F_0/G_0)(N') = 0$, and therefore $G_0(N') \subseteq C_0$. Inductively one shows that $G_i(N') \subseteq C_i$, but this implies that $N' = 0$, thus $\overline{M}(x)$ is indecomposable.

$N/N' = S(b)$ and $C(z(b))/N$ is an adic module with regular top E_3. And finally, consider $C(z(a)) = M(z(a))$. In this case $S(a)$ occurs as a factor module, the corresponding maximal submodule is the direct sum of two Prüfer modules with regular socles E_1', E_3'. Here are some schematic pictures:

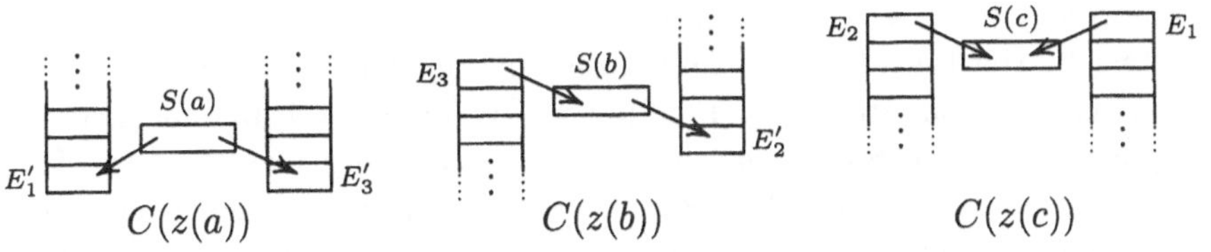

Similar pictures can be drawn in the case of an almost periodic $\mathbb{N}$-word x; to the left, we exhibit the case where x is contracting (there is a submodule which is a Prüfer module such that the factor module is indecomposable and of finite length), to the right, x is expanding (there is an indecomposable submodule of finite length such that the factor module is an adic module):

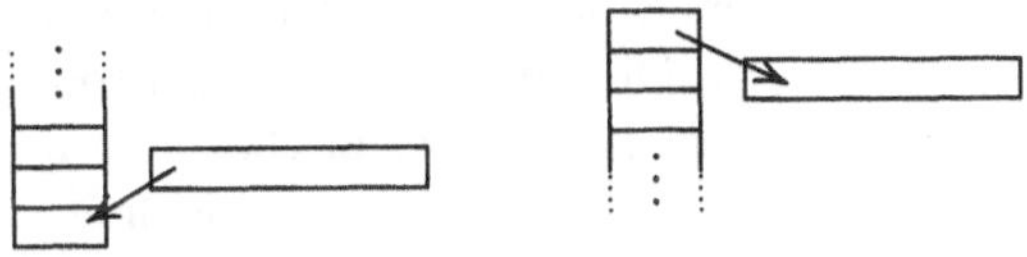

If R is a string algebra, then all the indecomposable algebraically compact modules can be determined and one can show that the only algebraically compact module which is superdecomposable is the zero module [R14]. For the proof, one determines R-modules M such that the functor $(-\otimes M)$ is $\mathbb{N}$-uniform; this implies that the pure injective envelope $\overline{M(z)}$ of $M(z)$ is indecomposable. *If* z *is a* $\mathbb{Z}$*-word which has no expanding end, then the functor* $(-\otimes M(z))$ *is* $\mathbb{N}$*-uniform.* For the proof, write $z = xy$, and further $x = \cdots x_2x_1x_0$, where the x_i are finite words with last letter being an arrow, and $y = y_0y_1y_2\cdots$, where the first letter of y_i is the inverse of an arrow. Observe that for all i, the two modules $N_i' = M(xy_0\cdots y_i)$ and $N_i'' = M(x_i\cdots x_0y)$ are factor modules of $M = M(z)$. Let $g_i\colon M \to N_i = N_i'\oplus N_i''$ be given by the canonical maps. One has to check that for any simple submodule S of M, the canonical projection $f\colon M \to M/S$ can be factored through one of the maps g_i, and also that any inclusion map $f\colon M \to N$ whose cokernel is finite-dimensional and indecomposable can be factored through some g_i. Similarly, one shows that *if* y *is a non-expanding* $\mathbb{N}$*-word, then the functor* $(-\otimes M(y))$ *is* $\mathbb{N}$*-uniform.*

5. Hammocks and quilts.

Hammocks. As mentioned above, the hammocks have been introduced in the realm of representation-finite algebras R by S. Brenner in order to obtain a combinatorial characterization of the translation quivers which occur as Auslander-Reiten quivers. The word "hammock" describes in a very intuitive way the shape of the representable functors $\operatorname{Hom}(-, I(S)) \simeq D\operatorname{Hom}(P(S), -)$, where S is a simple R-module, $I(S)$ its injective envelope, $P(S)$ its projective cover. One can attach to such a functor a translation quiver $\Gamma(S)$ whose vertices are equivalence classes of non-zero maps from $P(S)$ to indecomposable R-modules M, thus equivalence classes of composition factors of indecomposable R-modules which are isomorphic to S. These translation quivers $\Gamma(S)$ have a unique source (corresponding to the top of $P(S)$) and a unique sink (corresponding to the socle of $I(S)$); in addition, there is a function $\Gamma(S)_0 \to \mathbb{N}_1$, the hammock function, which plays a role: it counts the number of composition factors of the form S in suitable layers of the module in question; this function is additive on meshes. An axiomatic treatment of such translation quivers with hammock functions has been given in [RV], and there it has been shown that one obtains in this way precisely the Auslander-Reiten quivers of the categories of Ω-spaces[29], where Ω is a finite poset. Altogether the hammock philosophy uses three different approaches: a functorial one, dealing with functors which are both projective and injective, a combinatorial one, dealing with translation quivers and additive functions, and a linear one, dealing with subspace configurations in vector spaces.

An extension of these considerations to finite-dimensional algebras which are representation-infinite is needed. Of course, there is a strong interest to be able to extend all three approaches, but it seems clear that at present a combinatorial procedure can serve only as an auxiliary device: in dealing with representation-infinite algebras one cannot avoid to take the infinite radical $\operatorname{rad}^\omega$ into account and too little is known about the possibilities to handle it combinatorially. The main problem presently concerns the question under what conditions hammock functors $\operatorname{Hom}(-, I(S)) \simeq D\operatorname{Hom}(P(S), -)$ can be desribed in terms of S-space categories or related categories.

Note that these hammock functors are objects in a diamond category. To deal with objects in a diamond category has to be rated as a strong finiteness condition: such objects may be arbitrarily large, but the local structure of their subobject lattices is coined by their finite subfactors.

Tiles. Let us return to the algebras $R = R(Q', \omega)$ obtained from a quiver of type $\widetilde{A}_{n-1}$ by adding bridges. We consider in more detail whole parts of the category of all R-modules. Let a be a label of such a bridge, let $S(a)$ be the corresponding simple module. Let us consider all the words (finite words, as well as $\mathbb{N}$-words and $\mathbb{Z}$-words) w which contain the fixed letter a. For every such word,

---

[29] Given a poset Ω, an Ω-space $(V; V_s)_s$ is given by a k-space V and subspaces V_s of V indexed by the elements $s \in \Omega$, such that for $s \le s'$ one has $V_s \subseteq V_{s'}$. By definition, the dimension of $(V; V_s)_s$ is that of V. Given two Ω-spaces $(V; V_s)_s$ and $(W; W_s)_s$, a map $f : (V; V_s)_s \to (W; W_s)_s$ is given by a k-linear map $f : V \to W$ such that $f(V_s) \subseteq W_s$ for all $s \in \Omega$. The category of all Ω-spaces is an exact category. In case Ω is finite, the category of finite-dimensional Ω-spaces has Auslander-Reiten sequences.

there is given an indecomposable algebraically compact module $C(w)$ (for finite words, $C(w) = M(w)$, otherwise $C(w)$ may be a proper completion of $M(w)$). Given two such words w, w', write $w \leq w'$ provided there exists a homomorphism $f\colon C(w) \to C(w')$ with $f_a \neq 0$. We obtain a rectangle (a *tile*) of the following form (here, small elements with respect to the ordering are on the left, large ones on the right), and we denote it by $\mathcal{T}(a)$:

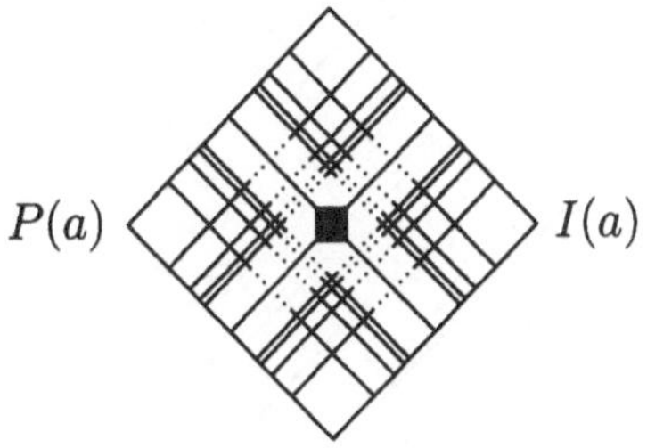

There is a smallest element, namely the word w with $C(w) = P(a)$, the projective cover of $S(a)$, and a largest element, the word w with $C(w) = I(a)$, the injective envelope of $S(a)$. The remaining two corners of the rectangle are given by suitable serial modules. The simple module $S(a)$ occurs in one of the four corners: if a is a sink, then $S(a) = P(a)$; if it is a source, then $S(a) = I(a)$; otherwise $S(a)$ is one of the other two corners. The center of this rectangle is the unique $\mathbb{Z}$-word which contains the letter a, on the two diagonals through the center, we have in addition just all the corresponding $\mathbb{N}$-words.

Given two posets Ω', Ω'', write $\Omega' \sqcup \Omega''$ for the disjoint union of these posets and $\Omega' \times \Omega''$ for the product. Instead of $\Omega' \sqcup \Omega'$ we also write $2\Omega'$. In addition, we also need the ordered sum $\Omega' \triangleleft \Omega''$, it is obtained from the disjoint union of Ω' and Ω'' by adding the relations $s' < s''$ for all $s' \in \Omega', s'' \in \Omega''$. When we consider subsets of $\mathbb{Z}$ as posets, then we use the natural ordering. Let $\Omega = \mathbb{N} \triangleleft (-\mathbb{N})$ and consider also its completion $\overline{\Omega} = \mathbb{N} \triangleleft \{*\} \triangleleft (-\mathbb{N})$:

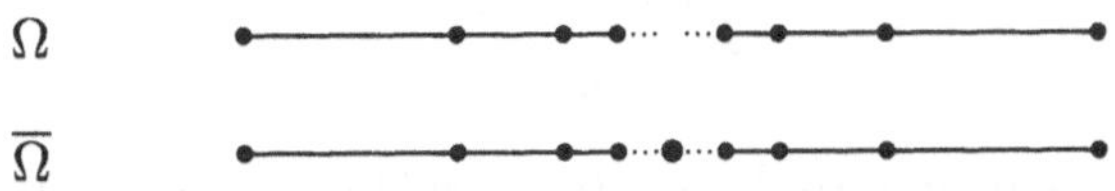

We may describe $\mathcal{T}(a)$ as the product $\overline{\Omega} \times \overline{\Omega}$ and, as Dräxler has pointed out, this product is the just Auslander-Reiten quiver $\Gamma_{2\Omega}$ of the category of all (2Ω)-spaces, thus

$$\mathcal{T}(a) = \overline{\Omega} \times \overline{\Omega} = \Gamma_{2\Omega}.$$

Indeed, the category which we consider when dealing with $\mathcal{T}(a)$ can be identified with the category of all (2Ω)-spaces: Let $I(a)$ be the ideal of $\operatorname{Mod} R$ of all maps f with $f_a = 0$. Then $(\operatorname{Mod} R)/I(a)$ is the category in question and it is equivalent to the category of (2Ω)-spaces. On the other hand, all the indecomposable (2Ω)-spaces are one-dimensional. The following fact should be stressed: Whereas some of the vertices of $\mathcal{T}(a)$ are infinite words, thus correspond to infinite-dimensional R-modules, the (2Ω)-spaces which are the vertices of $\Gamma_{2\Omega}$ are one-dimensional.

Hammock functors. Let us remove from $\mathcal{T}(a)$ for a while the modules of infinite length. What we obtain in this way is:

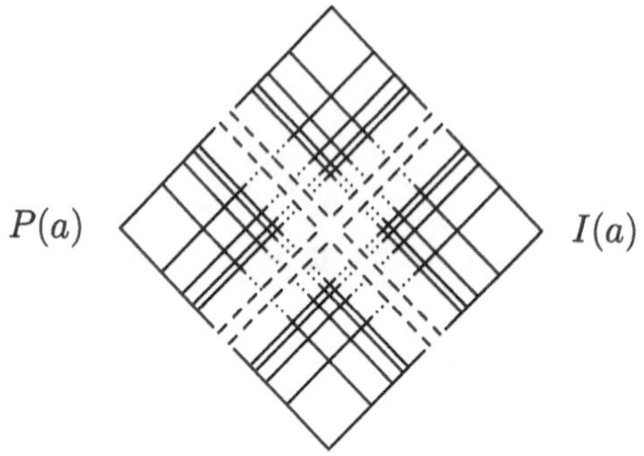

and this is just the product $\Omega \times \Omega$. On the other hand, this clearly describes the hammock functor $\mathrm{Hom}(-, I(S))$ which corresponds to the simple module $S = S(a)$.

In general, Schröer [Sc1] has described the structure of such hammock functors for all simple modules S over string algebras. The description is rather easy if one restricts the attention just to string modules. This part $\mathcal{H}(S)_0$ of the hammock functor can be visualized by a subset of the product $T \times T'$ of two totally ordered sets T, T' which are similar, but usually more complicated than $\mathbb{N} \triangleleft (-\mathbb{N})$.

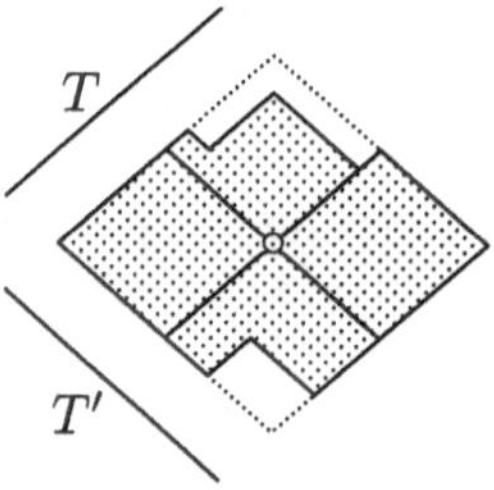

The chains T, T' are related to those introduced by Gelfand and Ponomarev ([GP], see also [R1]) in order to identify indecomposable modules by using functorial filtrations, but here we take into account only string modules. As before, the ordering $\leq$ goes from left to right; the unique minimal element is the projective cover of S, the unique maximal element the injective envelope of S. Note that the center is given by the module S itself. The position of S is important for describing the subset of $(T \times T')$ which corresponds to $\mathcal{H}(S)_0$, this subset is determined by the (finite!) set of words w such that $M(w)$ is serial, and is contained in the two quarters which contain elements which are incomparable with S. Let us stress that in general the structure of the chains T, T' is quite complicated, whereas it is easy to locate $\mathcal{H}(S)_0$ inside $T \times T'$.

How can we recover T and T' from the hammock $\mathcal{H}(S)$? Let us determine a subfactor as follows: Let $S = S(a)$, and suppose there are given arrows $x \xrightarrow{\alpha} a \xrightarrow{\alpha'} y$ such that $\alpha'\alpha = 0$. These arrows yield maps

$$I(x) \xleftarrow{I(\alpha)} I(a) \xleftarrow{I(\alpha')} I(y)$$

with zero composition. Let us denote by $M(\alpha)$ the kernel of $I(\alpha)$. The zero composition gives a map $I(y) \to M(\alpha)$, and we are interested in the cokernel $\mathrm{Hom}(-, M(\alpha))/\mathrm{Hom}(-, I(y))$. This is the functor which produces just one of the crossing lines through $S(a)$. Here is its support:

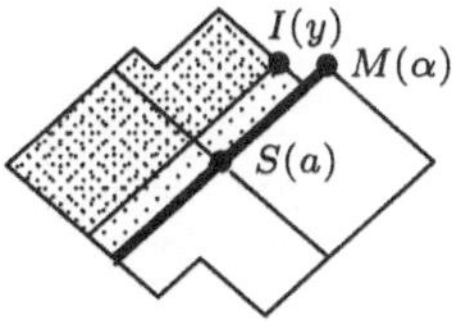

The support of $\mathrm{Hom}(-, M(\alpha))$ are the shaded parts, that of $\mathrm{Hom}(-, I(y))$ is shaded more heavily. As factor $F = \mathrm{Hom}(-, M(\alpha))/\mathrm{Hom}(-, I(y))$, there remains the bold line. This functor F is a serial functor. For a general discussion of serial subfactors of representable functors in the case of a string algebra we refer to [Sc2] and [PSc].

Quarters and Auslander-Reiten components. We return to the 1-domestic algebras $R = R(Q', \omega)$. As we have removed the modules of infinite length from a tile $\mathcal{T}(a)$, four connected parts (*quarters*) remain and these are actually parts of the usual Auslander-Reiten quiver Γ_R (since all the small rectangles occurring in $\mathcal{T}(S)$ are just usual meshes in Γ_R).

Let us consider the special case of the $\widetilde{A}_5$-quiver and the choice of ω as discussed above in greater detail. There are three tiles $\mathcal{T}(a), \mathcal{T}(b), \mathcal{T}(c)$ and they give rise to altogether 12 quarters. We label them as follows and indicate always the corresponding corner modules:

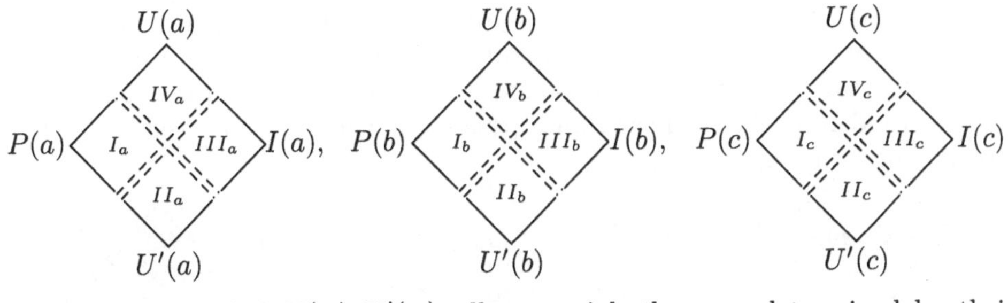

The modules labeled $U(-), U'(-)$ all are serial, they are determined by their composition factors: Here is the list of the factors:

$U(a)$	$U'(a)$	$U(b)$	$U'(b)$	$U(c)$	$U'(c)$
$a,2,3,4$	$a,1$	b	$3,b,5,0$	$5,4,c$	$2,1,0,c$

In order to describe Γ_R, we have to see how these quarters as well as the kQ'-modules which are string modules have to be fitted together. The four components of $\Gamma_{kQ'}$ which contain string modules have to be cut into rays or corays and then we have to rearrange these pieces. Let us consider in more detail our special example.

We cut the preprojective component in rays starting at the kQ'-projective modules $P'(0), P'(3), P'(4)$, such that the corresponding cokernels are of the form E_1, E_2, E_3. Similarly, we cut the preinjective component in corays ending at $I'(1)$, $I'(2), I'(5)$ such that the corresponding kernels are of the form E'_1, E'_2, E'_3. Here are these rays and corays:

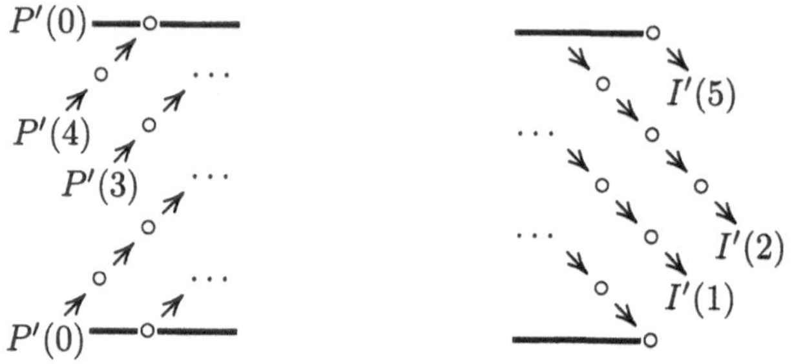

The two exceptional tubes also have to be cut; the one containing E_1, E_2, E_3 into corays, the other one into rays:

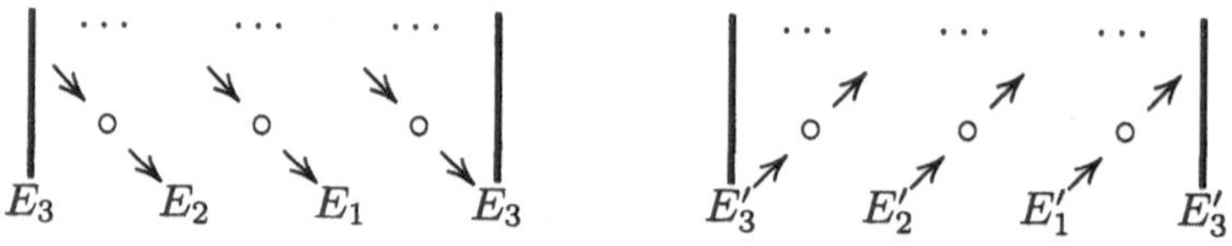

Altogether, we deal with the 12 quarters obtained from the three tiles $\mathcal{T}(a)$, $\mathcal{T}(b)$ and $\mathcal{T}(c)$, and in addition with 6 rays and 6 corays; if necessary, a fixed ray starting at a module M will be denoted by r_M, a coray ending in M by c_M. It is not difficult to see how these pieces fit together and that they yield two components $\mathcal{C}_1, \mathcal{C}_2$ of the Auslander-Reiten quiver Γ_R; the first of the following three pictures shows the structure of one of these components, the remaining two pictures still have to be put together in order to obtain the second component, this has to be done in the same way as one constructs the Riemann surface of the square root: one has to identify the bold solid lines (this is the coray $c_{E'_2}$) as well as the bold dashed lines (the coray $c_{I'(2)}$).

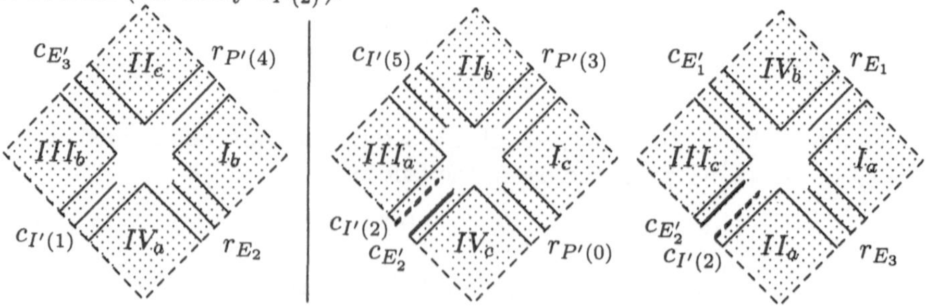

Topologically, both components are surfaces with a boundary; the left one is a punched plane, the right one is a twofold covering of a punched plane.

Note that the construction of non-stable components of a string algebra is easy, since we know all the modules lying on the boundary of these components [BuR]. Here is the boundary region of the first of the two components (the punched plane) exhibited above, it shows in which way rays, corays and quarters may be connected:

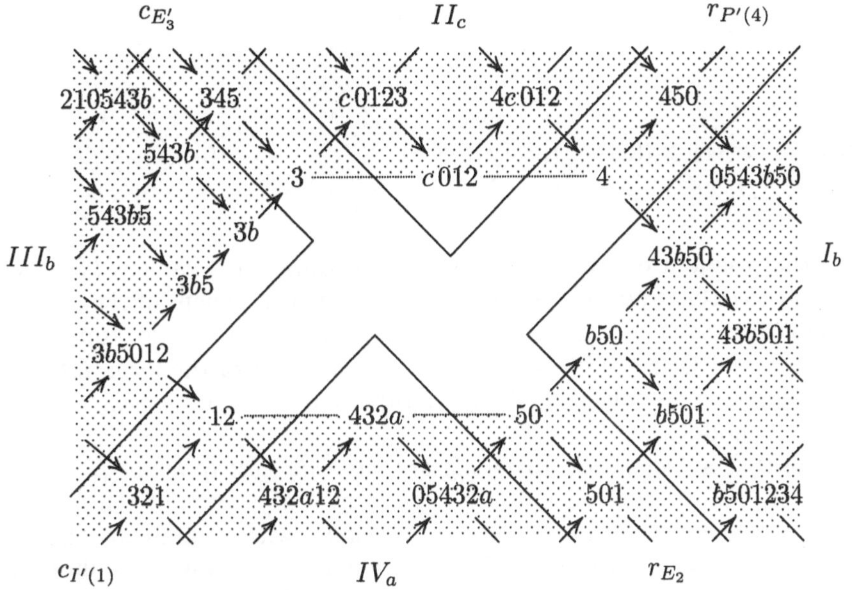

It seems to be of interest to compare this procedure of joining rays, corays and quarters with the well-understood process of ray insertions and coray insertions in tubes. For example, to deal with a ray insertion means to cut between two rays to insert there a new ray. Here we also cut between rays and we cut between corays, but we insert not rays or corays but quarters, for example the quarter I_b in between the two rays $r_{P'(4)}$ and r_{E_2} (but these are rays of quite different nature: the second is a ray coming from a tube, whereas the first one comes from a preprojective component), the quarter IV_a in between the ray r_{E_2} and the coray $c_{I'(1)}$ and so on. A clear recipe for this procedure of quarter insertion is not yet known.

In the example, the only stable components of Γ_R were tubes. A minor modification allows to produce also stable components of the form $\mathbb{Z}A_\infty^\infty$. Consider the following algebra

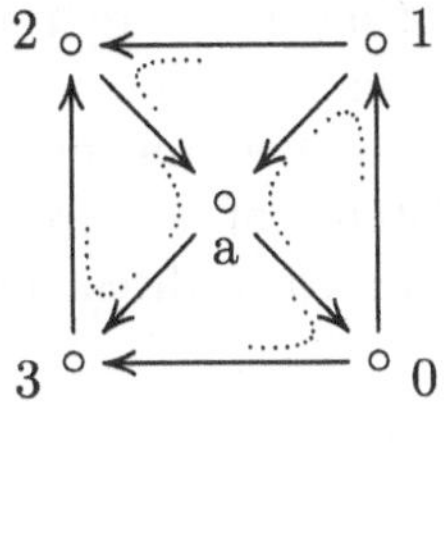

There are (up to inversion) four $\mathbb{Z}$-words:

$$z_1 = {}^{\infty}(2103)\, a\, (0123)^{\infty}, \quad z_2 = {}^{\infty}(1032)\, a\, (0123)^{\infty},$$
$$z_3 = {}^{\infty}(2103)\, a\, (1230)^{\infty}, \quad z_4 = {}^{\infty}(1032)\, a\, (1230)^{\infty}.$$

For any of these $\mathbb{Z}$-words z_i, we may consider all the (finite or infinite) subwords which contain the letter a, what we obtain in this way are again tiles. And, deleting the infinite-dimensional modules from the tiles, every tile is cut into four quarters:

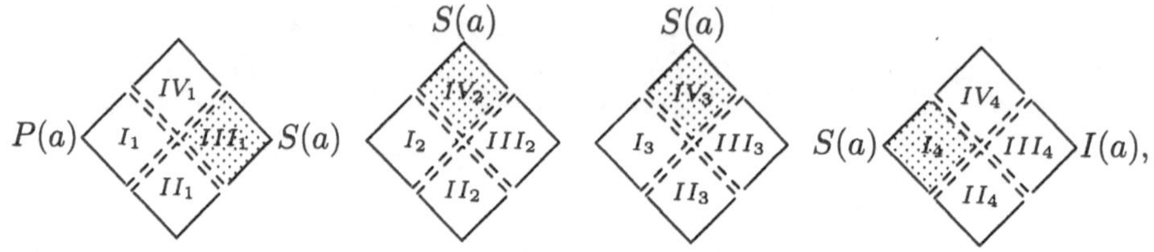

We see that the four quarters containing the module $S(a)$, namely III_1, IV_2, IV_3 and I_4, fit together and form an Auslander-Reiten component of the form $\mathbb{Z}A_\infty^\infty$. In general [BuR], stable components of a string algebra are either tubes or of the form $\mathbb{Z}A_\infty^\infty$, and Geiß [G] has shown that a component of the form $\mathbb{Z}A_\infty^\infty$ always contains a unique module of smallest length[30]. Of course, in our example $S(a)$ is such a Geiß module.

Note that a *1-domestic string algebra has only finitely many* $\mathbb{Z}$*-words,* thus only finitely many tiles and only finitely many components of the form $\mathbb{Z}A_\infty^\infty$. (Proof: write such a word in the form $(w^\infty)^{-1}uw^\infty$ with u minimal, where w is a fixed primitive cyclic word. Then there are only finitely many possibilities for u.) On the other hand, it is easy to construct examples of 2-domestic string algebras with infinitely many $\mathbb{Z}$-words, for example:

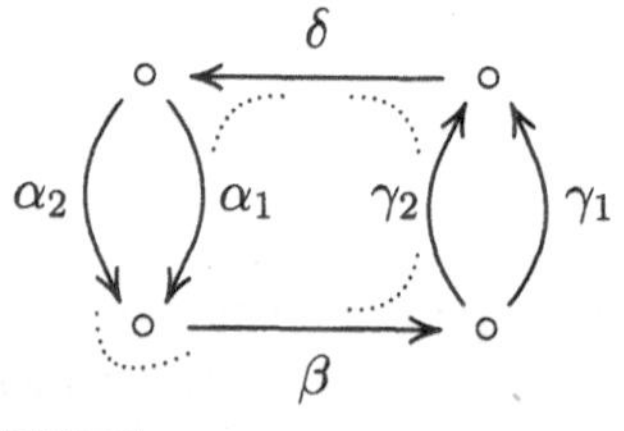

[30] A finite word w will be called a Geiß word provided there do exist $\mathbb{N}$-words x_1, x_2 which start with a direct letter and $\mathbb{N}$-words y_1, y_2 which start with an inverse letter such that $y_1^{-1}wx_1$ and $x_2^{-1}wy_2$ are $\mathbb{Z}$-words. If w is a Geiß word, then the string module $M(w)$ is contained in a component of the form $\mathbb{Z}A_\infty^\infty$ and all the other modules in this component have larger length. Also the converse is true: if $M(w)$ is a module of smallest length in a component of the form $\mathbb{Z}A_\infty^\infty$, then w is a Geiß word. Note that in case w contains both direct and inverse letters, then it is sufficient to require the existence of $\mathbb{N}$-words of the form $wx_1, w^{-1}y_1, w^{-1}x_2$ and wy_2, such that x_1, x_2 start with a direct letter and y_1, y_2 with an inverse letter.

Up to inversion, the $\mathbb{Z}$-words are of the form

$$z_n = {}^{\infty}\big(\alpha_1^{-1}\alpha_2\big)\,\delta\,\gamma_1\big(\gamma_2^{-1}\gamma_1\big)^n\beta\,\big(\alpha_1\alpha_2^{-1}\big)^{\infty}$$

for $n \geq 0$. Note that by adding bridges to the cycle $\gamma_2^{-1}\gamma_1$ we will obtain 2-domestic algebras which have even more $\mathbb{Z}$-words.

Quilts. Recall that we have removed the infinite-dimensional modules from the tiles $\mathcal{T}(a)$, $\mathcal{T}(b)$ and $\mathcal{T}(c)$ in order to obtain pieces of Γ_R. Of course, we may reinsert them. What we obtain in this way is a topological space which is connected and compact: a very nice compactification of the union $\mathcal{C}_1 \sqcup \mathcal{C}_2$.

Let us analyze this compactification process further: We start with a planar component of a string algebra, say either with a component of the form $\mathbb{Z}A_\infty^\infty$ or with a punched plane (similar considerations apply in the case of a finite covering of a punched plane). The usual rule for drawing components is that all meshes should have equal size (in the case of components with holes a slight squeezing may be necessary). However, the focus of such a visualization is the central part of the component and our aim is to understand also the relationship to other components. Thus, the first step is a very innocent one: to change the metric in such a way that the component fits into a finite region. We may think of a planar component as a square or a lozenge shape, standing on one of its corners:

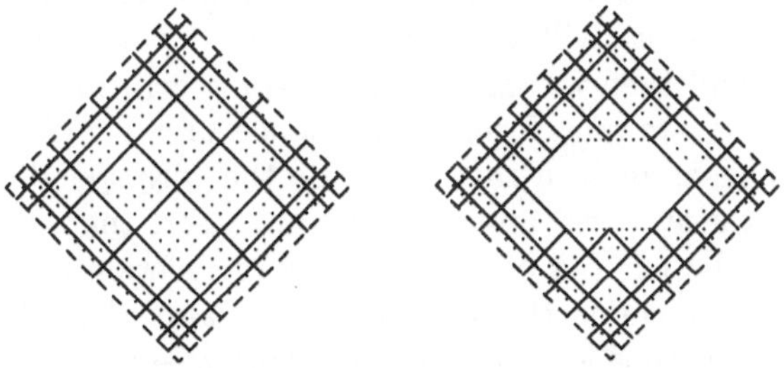

To the left, you see the case of a $\mathbb{Z}A_\infty^\infty$ component (the center is its Geiß module, the uniquely determined module of smallest length), to the right, the punched plane component considered before. This process of **shrinking** has mainly a psychological meaning[31], but it draws our attention to the endpoints of the rays and corays and to the four corners.

[31] A mathematical interpretation of such a metric could be as follows: fix a real number λ with $0 < \lambda < 1$ and define the length of an arrow $\alpha\colon [X] \to [Y]$ by $\sigma(\alpha) = \lambda^{|X|+|Y|}$. Then the length of a ray $[X_0] \to [X_1] \to [X_2] \to \cdots$ (and similarly that of a coray) will be bounded. When applied to a tube, such a metric realizes it as a punched disk, and it can be used for the following consideration: Call a cyclic walk *inessential* provided it is homotopic to walks of arbitrarily small length (with respect to this metric). Given a tame hereditary algebra R, all cyclic walks in the Auslander-Reiten quiver Γ_R are inessential.

Now comes the second step of **compactifying** these components by adding the boundary. We can formulate this purely combinatorially as follows: recall that the components of Γ_R we deal with have as vertices string modules $M(w)$, or, equivalently, just the strings w, and the arrows correspond to the operations on words of adding hooks or deleting cohooks (see [BuR]). Now, we also take into account $\mathbb{N}$-words (the small bullets on the four sides) and $\mathbb{Z}$-words (the black squares at the corners).

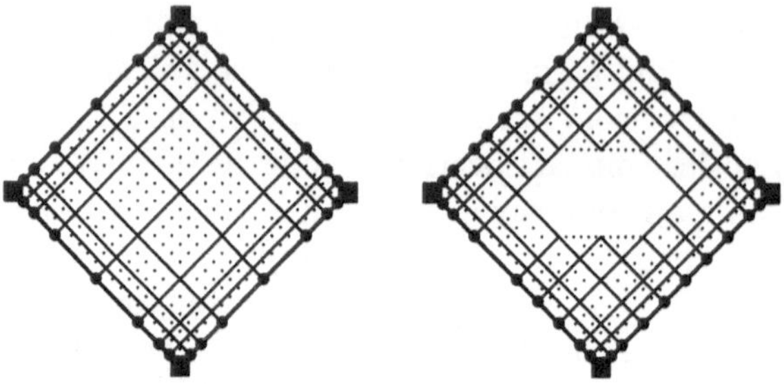

Recall that a countable sequence of arrows $w_0 \to w_1 \to w_2 \to \cdots$ is called a *ray*, provided $\tau w_{i+1} \neq w_{i-1}$, for all $i \geq 1$. Two such sequences are called *equivalent* provided they only differ by finitely many arrows. We may assume that we work with words w_i such that w_{i+1} is obtained from w_i by a change on the right hand side, for all i, and then x is constructed as follows: for every natural number n, almost all the words w_i will be of the form $w_i = x_n w_{in}$, where x_n is a fixed word of length n, and then the first n letters of x will just be x_n. In case no vertex w_i is injective, we may form the sequence $w_1' \to w_2' \to w_3' \to \cdots$ where $w_i' = \tau^{-1} w_{i-1}$. This is again a ray, let x' be the corresponding $\mathbb{N}$-word. Note that there are arrows $w_i \to w_i'$ and these arrows assert that w_i' is obtained from w_i by a change on the left (addition of a hook or deletion of a cohook). For all indices i, the same change on the left occurs, and also x' is obtained from x by this change on the left, thus we will draw an arrow $x \to x'$. For every component, we obtain in this way two sequences of consecutive $\mathbb{N}$-words x_i (where $i \in \mathbb{Z}$) and arrows $x_i \to x_{i+1}$. There is the dual procedure for obtaining an $\mathbb{N}$-word for any equivalence class of corays (the dual of a ray), which again is almost periodic, and also arrows between these $\mathbb{N}$-words. Again, we obtain two sequences of consecutive $\mathbb{N}$-words x_i (where $i \in \mathbb{Z}$) and arrows $x_i \to x_{i+1}$; altogether we obtain in this way the four boundary lines of our lozenge. Finally, note that these procedures combine and yield four biperiodic $\mathbb{Z}$-words, corresponding to the corners of the lozenge.

Here is the module theoretical interpretation: The vertices we have added are almost periodic $\mathbb{N}$-words x and biperiodic $\mathbb{Z}$-words z. In [R8], we have constructed corresponding indecomposable algebraically compact modules $C(x)$ and $C(z)$. Also, the new arrows $x_i \to x_{i+1}$ between $\mathbb{N}$-words indicate that x_{i+1} is obtained from x_i by adding a hook or deleting a cohook, and this corresponds to an irreducible map $C(x_i) \to C(x_{i+1})$.

In the case of a $\mathbb{Z}A_\infty^\infty$ component, the Geiß module may be considered as the origin, in case we need to index the modules. Note that any boundary line contains precisely one word which starts with a Geiß word, and it will be convenient to use the index 0 for this $\mathbb{N}$-word:

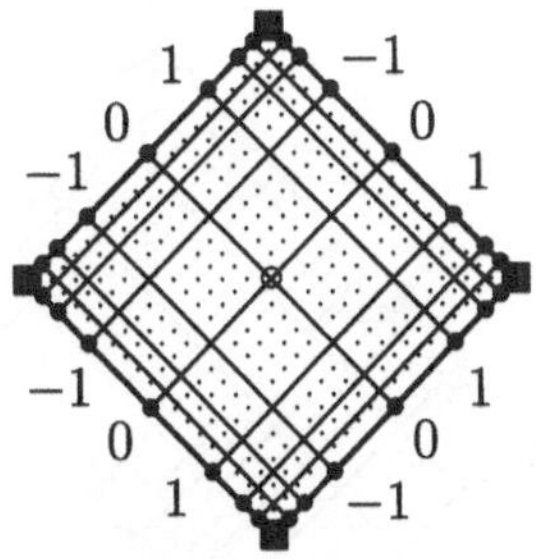

A similar process of shrinking and compactifying can be achieved in the case of components of the form $\mathbb{Z}D_\infty$. Such components do not occur for string algebras but they do for a related class of algebras, the so called clan algebras[32].

[32] Here are two typical such algebras:

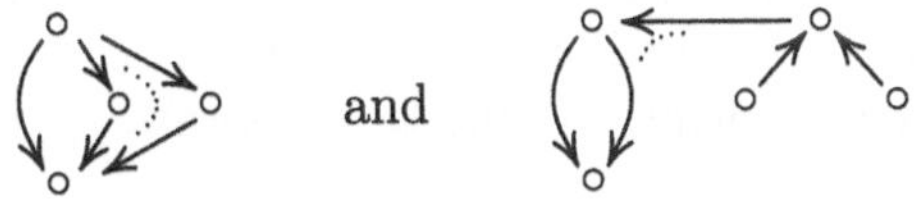

(the relation on the left is a commutativity relation). Let us exhibit a $\mathbb{Z}D_\infty$ component and its compactification: As for the $\mathbb{Z}A_\infty^\infty$ components, the boundary lines will be given by $\mathbb{N}$-words, the corners by $\mathbb{Z}$-words. However, at least two of these $\mathbb{Z}$-words yield decomposable modules, namely modules which are the direct sums of two indecomposable modules, thus we have inserted pairs $\vdots$ of bullets:

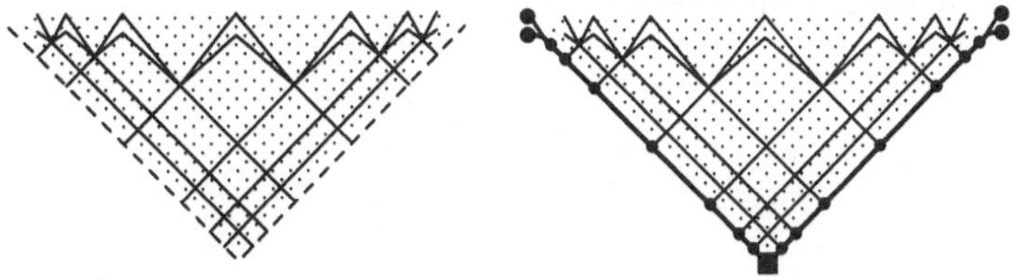

Many components of this form arise for the algebra above depicted left; for example the component containing the two simple modules which are neither projective nor injective is of this kind. In general, the third corner of the compactification (indicated by the black square ■) may also correspond to a decomposable module; in this case, it is more appropriate to replace also ■ by a pair of bullets $\vdots$. This happens for the algebra depicted right, namely for the component which contains the two simple injective modules; of course, this is a non-regular component, it is similar to a $\mathbb{Z}D_\infty$ component with a boundary part being missing — in the same way as our punched plane was similar to a $\mathbb{Z}A_\infty^\infty$ component.

Sewing. It may happen that the same boundary line (the same sequence $(x_i)_i$ of $\mathbb{N}$-words) is obtained twice: once by dealing with rays, the second time by dealing with corays. Then this boundary line can be considered as a **sewing** together of the two components (or of two different sides of one component):

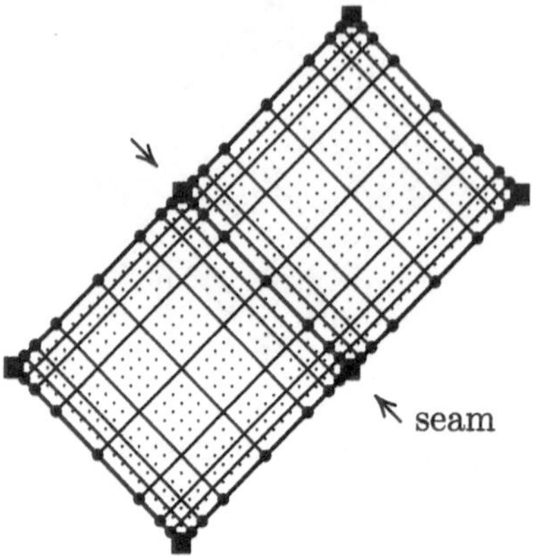

What we obtain in this way, by sewing together various components of a string algebra R may be called the *Auslander-Reiten quilt* of R. Let us stress that the sewing together is achieved by the infinite length modules which have been added.

Let us consider the following 2-domestic algebra:

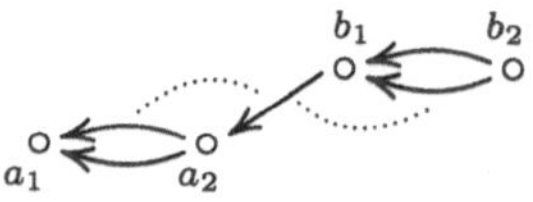

This algebra R has a an interesting Auslander-Reiten component, that containing the projective modules $P(b_1), P(b_2)$ and the injective modules $I(a_1), I(a_2)$. This component is a $\mathbb{Z}A_{\infty}^{\infty}$-component with a hole:

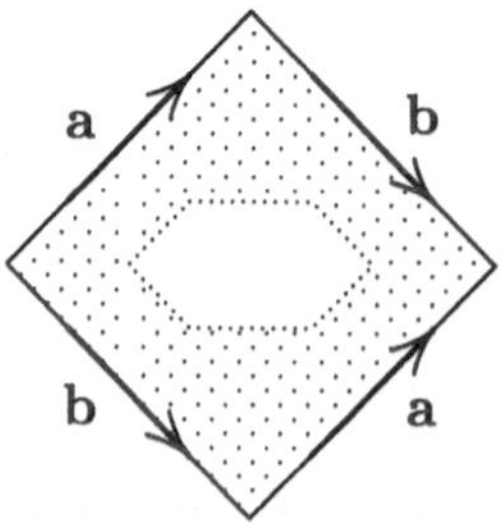

Using the sewing process, pairs of the sides of the square have to be ientified and we have indicated in which way: the two sides labeled **a** have to be identified, similarly, the two sides labeled **b** have to be identified, always in the direction of the arrows (but note that these arrows indicate the direction of the irreducible maps, thus no other identification would be possible). Clearly, in this way we obtain a torus with a hole. — We denote by K_a the Kronecker quiver with vertices a_1, a_2, by K_b that with vertices b_1, b_2. The preprojective component of K_a as well as a family of tubes of K_a-modules indexed by the affine line $\mathbb{A}^1$ remain components of Γ_R. Similarly, the preinjective component of K_b as well as a family of tubes of K_b-modules again indexed by the affine line $\mathbb{A}^1$ remain components of Γ_R. Let us exhibit the Auslander-Reiten quiver Γ_R. To the left, there are those components of the Kronecker quiver K_a with which remain components in Γ_R, to the right, those of K_b. In the middle, you see the torus[33]:

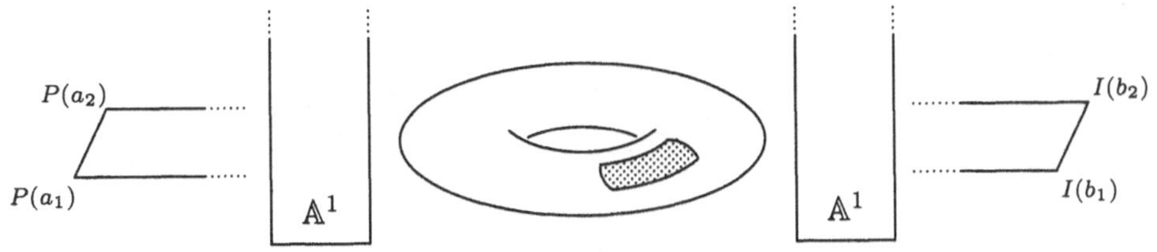

It is easy to see that such a torus and a similar Kleinian bottle occurs already for 1-domestic algebras: Let us present two typical examples, both being obtained from an $\widetilde{A}$-algebra by adding one bridge. Here is the first example: the algebra R', the punched plane with the labels **a**, **b** for identification and the torus:

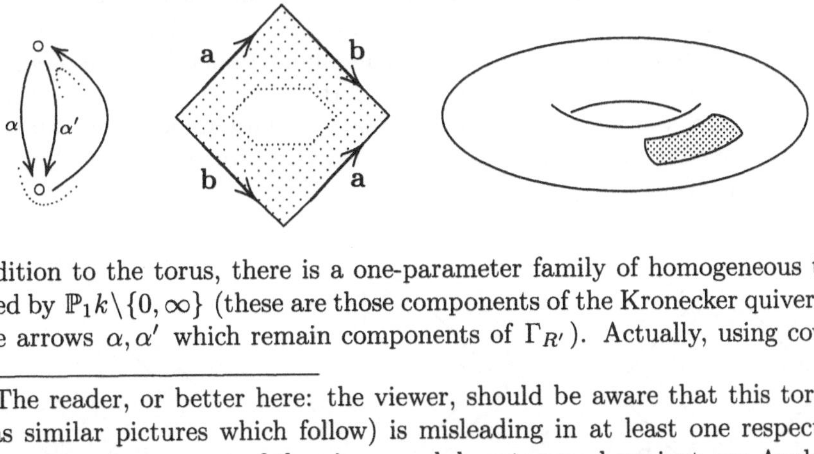

In addition to the torus, there is a one-parameter family of homogeneous tubes, indexed by $\mathbb{P}_1 k \setminus \{0, \infty\}$ (these are those components of the Kronecker quiver given by the arrows α, α' which remain components of $\Gamma_{R'}$). Actually, using covering

---

[33] The reader, or better here: the viewer, should be aware that this torus (as well as similar pictures which follow) is misleading in at least one respect: the torus concerns just a part of the given module category, here just one Auslander-Reiten component, but does not take into account other parts of the category. Indeed, the component $\mathcal{C}$ presented here contains the injective R-modules $I(a_1)$ and $I(a_2)$. Thus, for any non-zero R-module M whose support is not contained in K_b, there are non-zero homomorphisms f from M to modules in $\mathcal{C}$. In the same way, given a module M whose support is not contained in K_a, there are non-zero homomorphisms g from modules in $\mathcal{C}$ to M. But the picture exhibited here does not care about the nature of these maps, it does not even give a hint where such a map f arrives in the component or where a g will leave the component.

theory, one can reduce the study of the module category mod A' to the previous example mod A.

Next, consider the second example, in this case the sewed component is a Kleinian bottle with a hole. As additional components, there is a preprojective component as well as a one-parameter family of homogeneous tubes, but here indexed by the affine line $\mathbb{A}^1 = \mathbb{P}_1 k \setminus \{\infty\}$. Again, to the left, we present the quiver with relations, in the middle the punched plane with the labels **a**, **b** for the sewing process, to the right the Kleinian bottle which ones obtains in this way (see [HC]):

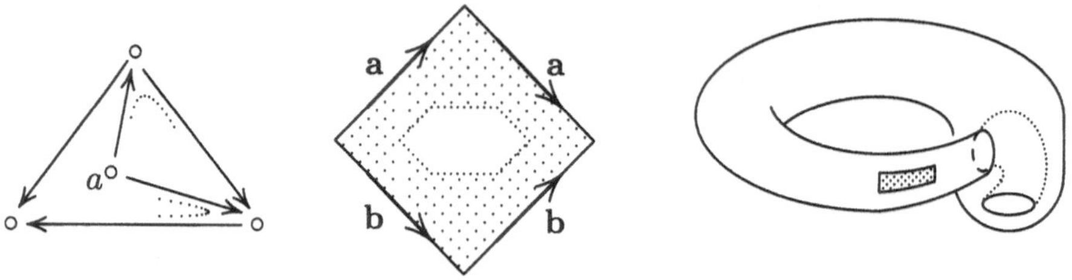

These quilts are convenient tools for the visualization of hammocks and for distinguishing the different behaviour of representable functors. The difference between the torus and the Kleinian bottle lies in the fact that the torus is orientable whereas the Kleinian bottle is not. To consider rays and corays on the torus amounts to consider foliations. What happens on the Kleinian bottle? Let us start with a ray, say in northeastern direction. We travel along the ray until we leave the component; thus we pass through an infinite dimensional module, say X, and return to the component, via first a coray then a ray, but now in southeastern direction. Again, we leave the component via an infinite-dimensional module, say Y, and return to a coray and a ray, in northeastern direction, parallel to the ray we have started with:

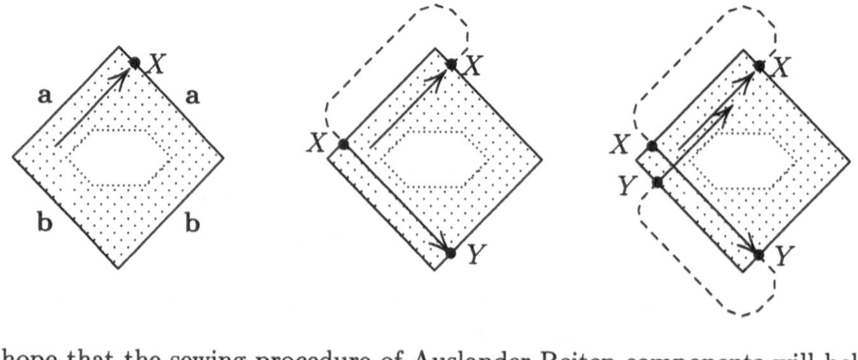

We hope that the sewing procedure of Auslander-Reiten components will help to get a better understanding of some hammocks, thus of some typical objects in diamond categories.

Let us reformulate the process of sewing together components and the way infinite length modules are used. There is given a ray of irreducible maps $X_0 \rightarrow$

$X_1 \to X_2 \to \cdots$, the direct limit $\varinjlim X_i$ being a string module for an $\mathbb{N}$-word, say the $\mathbb{N}$-word x. Second, there is given a coray of irreducible maps $\cdots \to X_2' \to X_1' \to X_0'$, and its inverse limit $\varprojlim X_i'$ is the product module corresponding to the same $\mathbb{N}$-word x; in particular, there is a canonical embedding $\iota\colon \varinjlim X_i \to \varprojlim X_i'$. Altogether, we deal with the following configuration of modules:

$$X_0 \to X_1 \to X_2 \to \cdots \quad \to \varinjlim X_i \xrightarrow{\iota} \varprojlim X_i' \to \quad \cdots \to X_2' \to X_1' \to X_0'.$$

Now, the direct limit module $\varinjlim X_i$ is always indecomposable, whereas the inverse limit module $\varprojlim X_i'$ always is algebraically compact, and as we have mentioned, one of the two modules (and only one) will be both indecomposable and algebraically compact. This is the module to be selected.

As the Auslander-Reiten quiver itself, also the Auslander-Reiten quilt is a purely combinatorial object which yields generators and relations for describing an additive category.

Unfortunately, the sewing procedure using rays and corays of irreducible maps can be applied only in very special cases, since usually there are not enough such rays and corays available. This is already the case for most of the tame hereditary algebras: the rays in the tubes produce all the Prüfer modules, their corays give rise to the adic modules; but for the adic modules, we also would need rays in the preprojective component, and for the Prüfer modules, we would need corays in the preinjective component.

Note that the Kronecker quiver $K(2)$ has enough rays and corays of irreducible maps, but the sewing of the preprojective component via the adic modules with the tubes, and dually the sewing of the tubes via the Prüfer modules with the preinjective component gives a rather involved and unintelligible picture.

The algebras R of type $\widetilde{D}_n, \widetilde{E}_6, \widetilde{E}_7, \widetilde{E}_8$ have no rays in the preprojective component, and no corays in the preinjective component. Of course, in all these cases, the category $\operatorname{mod} R$ contains as a full subcategory the category $\operatorname{mod} K(2)$, thus we could use the preprojective rays and the preinjective corays in this subcategory, but this adds new difficulties. On the other hand, we should stress that the dissertation of Geigle [Gg] has to be named as first investigation dealing with the process of sewing of components, and he discusses precisely this complicated case: the tame hereditary algebras.

In the case of a tame hereditary algebra, the missing rays (and similarly, the missing corays) concern indecomposable modules which belong to one component. In general, we will have to deal with sequences $X_0 \to X_1 \to \cdots$ or $\cdots \to X_1 \to X_0$ where all the modules X_i may belong to pairwise different components. Usually, it should be hopeless to keep track of all sequences needed. But there are special cases of algebras where one obtains a quite satisfactory description of the module category.

As our last example, let us consider the 3-domestic algebra R with the following quiver and relations:

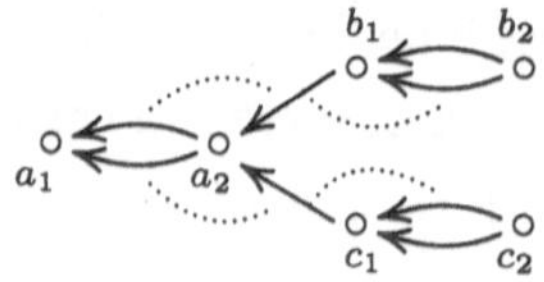

Here is a first attempt to visualize the global structure of the module category:

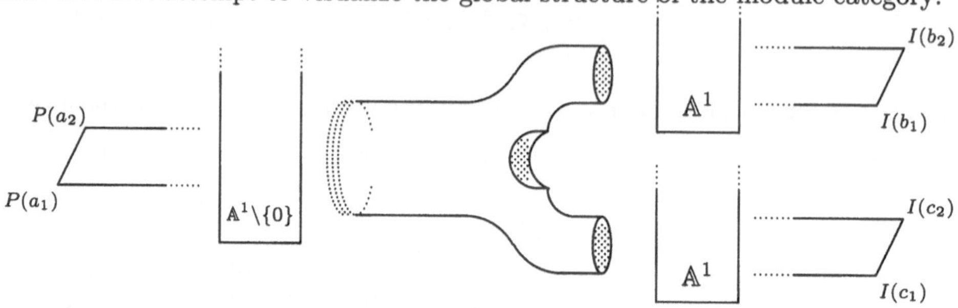

This picture exhibits all the Auslander-Reiten components, but taking into account the sewing procedure for $\mathbb{Z}A_\infty^\infty$ components as outlined above. To the left, there are those components of the Kronecker quiver K_a with vertices a_1, a_2 which remain components in Γ_R: there is the preprojective component of K_a as well as a family of tubes indexed by the punched affine line $\mathbb{A}^1 \setminus \{0\}$. Let K_b, K_c be the Kronecker quivers with vertices b_1, b_2 or c_1, c_2, respectively. To the right of the picture, there are those components of K_b and K_c, which remain components in Γ_R: the preinjective components as well as tubular families indexed by the affine line $\mathbb{A}^1$.

Of interest are the remaining components and there are countably many. There is one additional non-regular component $\mathcal{C}$, it contains the projective modules $P(b_1), P(b_2), P(c_1), P(c_2)$, and the injective modules $I_1 = I(a_1), I_2 = I(a_2)$. Its shape is obtained from the Riemann surface of the square root by cutting a central hole (in the same way as one of the components of the 1-domestic algebras discussed above). For later use, decompose $I_2/\operatorname{soc} I_2 = E(b) \oplus E(c)$, such that the support of $E(b)$ is K_b, that of $E(c)$ is K_c. For $i \geq 1$, we denote by M_i the preinjective K_a-module of dimension $2i-1$. It is easy to see that this is a Geiß module, thus we obtain countably many $\mathbb{Z}A_\infty^\infty$ components $\mathcal{C}_i$. The components $\mathcal{C}$ and $\mathcal{C}_i$ are all the remaining ones, they can be sewed together to a big connected quilt. What one obtains in this way are the "Chinese baby trousers" seen in the middle. Here is another view of the trousers, this time using three cuts:

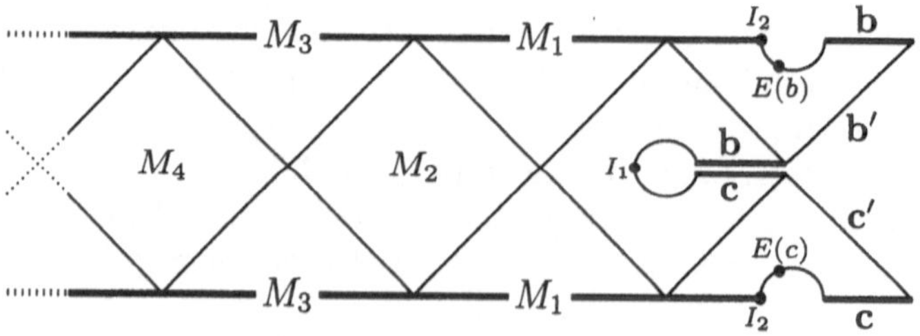

one cut yields the solid line through the modules M_i with odd index i, the other two cuts are labelled **b** and **c**. The position of the Geiß modules indicate the $\mathbb{Z}A_\infty^\infty$ components $\mathcal{C}_i$. On the right we see the non-regular component $\mathcal{C}$, there are three pieces which are sewed along two seams; we also provide the position of the injective modules I_1, I_2 and of their factor modules $E(b), E(c)$.

It remains to consider the boundary lines labelled $\mathbf{b}'$ and $\mathbf{c}'$ (the openings for the legs in the trousers). There are indeed infinite-dimensional indecomposable algebraically compact modules which live on these lines and as usually they are approached by rays of irreducible maps. The modules for the seam $\mathbf{c}'$ are given by $\mathbb{N}$-words and the unique biperiodic $\mathbb{Z}$-word ${}^\infty(a_1a_2)(c_1c_2)^\infty$ with support on K_a and K_c; those for the seam $\mathbf{b}'$ are similarly given by $\mathbb{N}$-words and the unique biperiodic $\mathbb{Z}$-word ${}^\infty(a_1a_2)(b_1b_2)^\infty$ with support on K_a and K_b. Since the support of the boundary line $\mathbf{b}'$ is related to K_b, we are tended to direct the corresponding leg of the trousers towards the K_b-components, and similarly, we have directed the leg $\mathbf{c}'$ towards the K_c-components, but as we will see this is questionable.

The modules occurring on the boundary lines $\mathbf{b}'$ and $\mathbf{c}'$ are approached by rays of irreducucible maps, but, in contrast to previous situations, not by corays of irreducible maps. For example, consider $x = {}^\infty(a_1a_2)(c_1c_2)^n$, this is (the inverse of) an $\mathbb{N}$-word lying on the boundary line $\mathbf{c}'$. As we have mentioned, the R-module $C(x) = M(x)$ is approached by a ray, namely it is the union of the sequence

$$M((c_1c_2)^n) \subset M(a_1a_2(c_1c_2)^n) \subset M((a_1a_2)^2(c_1c_2)^n) \subset \cdots$$

of inclusions and these inclusions are irreducible maps. On the other hand, we may consider the modules $N_i = M((a_2a_1)^i a_2(c_1c_2)^n)$ and the canonical surjective maps

$$\cdots \to N_2 \to N_1 \to N_0.$$

However, these modules belong to pairwise different Auslander-Reiten components: the module N_0 belongs to $\mathcal{C}$, the module N_i to $\mathcal{C}_i$. But note that the maps can be controlled quite well: they are jumps from one component to a neighboring one and are compositions of a ray and a coray of irreducible maps; the corresponding arrows all point in northeastern direction. The following picture indicates a ray r approaching the module $M(x)$ as well as the modules N_i and the maps $N_{i+1} \to N_i$.

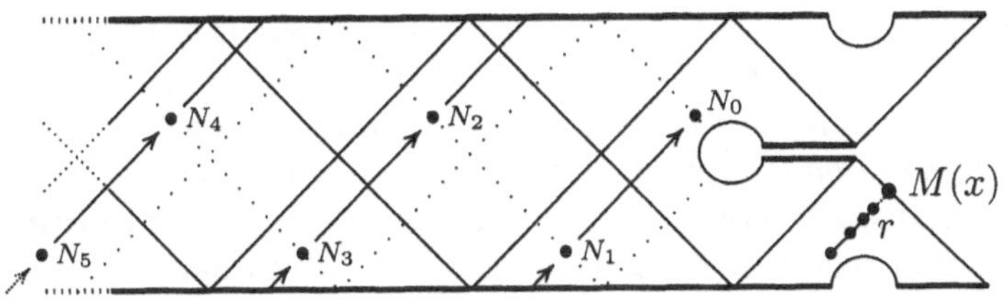

In the same way, we may approach any module on the boundary line **b**′ by a sequence of modules in the various components C_i, this time using sequences of arrows which point in southeastern direction.

The sequences of modules needed approach the waist of the trousers, thus we sew a second time, now joining the legs of the trousers with the waist:

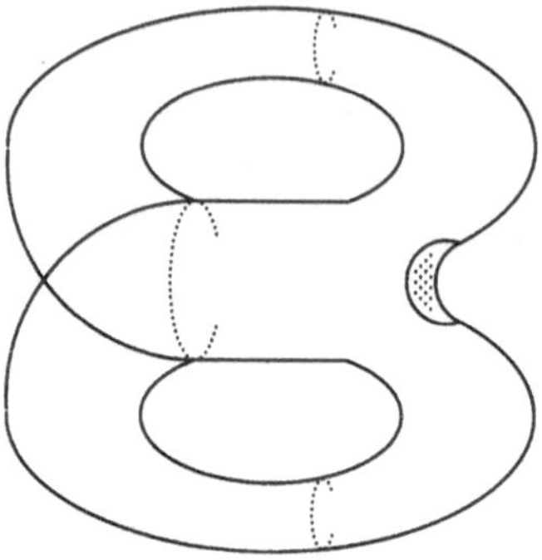

Since we join both legs to the waist, the waist becomes a branching locus. We stress that also the second sewing procedure is achieved by infinite length modules which we have added.

Conclusion. When dealing with finite length modules, there are two essentially different ways in which infinite length modules or similar objects in related abelian categories play a decisive role: on the one hand, as objects which can be used in order to describe infinite families of R-modules (two typical ways: a Prüfer group incorporates all indecomposable p-groups as all its proper non-zero subgroups; Crawley-Boevey introduced the generic modules for tame algebras in order to parameterize the one-parameter families), on the other hand in order to describe in a module-theoretical language the behaviour of functors on $\operatorname{mod} R$. But these two ways turn out to be just two sides of one and the same coin, it is a challenging demand to descibe the correspondences.

Epilogue. A clear misunderstanding.

There are by now several books available which deal with questions in the representation theory of finite dimensional algebras, starting with Curtis and Reiner: *Representation Theory of Finite Groups and Associative Algebras (1962)* and its successor *Methods of Representation Theory I* (1981) and *II* (1987); then the volume 73 of the Encyclopaedia of Mathematical Sciences by Gabriel and Roiter, with the title *Representations of Finite-Dimensional Algebras* (1992) and finally the book *Representation Theory of Artin Algebras* (1995) by Auslander, Reiten and Smalø. All these treatises restrict their attention to **finite-dimensional** representations and take it for granted that a title which does not mention this specification will not be considered as misleading. We do not object that the books are confined to a specific class of representations with nice properties, namely the finite-dimensional ones, but we wonder why the authors do not see the necessity to mention this in the title. Actually, the orientation on these titles had the effect that some mathematicians now seem to distinguish between the *representation theory* and the *module theory* of finite-dimensional algebras, meaning that the first one concerns finite-dimensional representations, in contrast to the second. But, of course, this does not correspond to the usual use of the word representation theory, say when dealing with representations of Lie algebras or algebraic groups, where the study of infinite-dimensional representations is considered as an essential part of the theory. There seems to be the attitude that dealing with finite-dimensional algebras (or also with finite groups), the most natural representations are the finite-dimensional ones, but there should be real doubts! The belief that the natural representations of a finite-dimensional algebra are just the finite-dimensional representations has to be rated as very naive. In the same vein, the natural setting for considering finite Galois groups should be the realm of finite fields only.

The examples discussed up to now were motivated from within representation theory. Here, let us draw the attention to an application, the possible use of Kronecker modules in order to deal with operators on a vector space. The Kronecker modules are the representations of the Kronecker quiver $K(2)$, the corresponding path algebra $kK(2)$ is four dimensional, thus really a very small algebra. Looking at the algebra, one may be inclined to consider its indecomposable projective representations (their dimension is 1 and 3) or its indecomposable injective representations (again dimension 1 and 3) as the most natural ones. But is this justified? The Kronecker algebra $kK(2)$ is one of the tame hereditary algebras, thus the structure of the category $\operatorname{mod} kK(2)$ is as displayed above: besides the preprojective and the preinjective modules, there are the indecomposable regular modules, they belong to tubes, and these tubes are indexed by the projective line $\mathbb{P}_1(k)$. If we delete one of these tubes, the remaining regular representations form a category which is equivalent to the category $\operatorname{fin} k[T]$ of all finite-dimensional $k[T]$-modules, where $k[T]$ is the polynomial ring in one variable. As usual, the additional tube is indexed by the symbol ∞. Of course, $k[T]$-modules are nothing else than pairs (V, β), where V is a vector space over k and $\beta\colon V \to V$ a linear transformation (where one writes $T \cdot v$ instead of $\beta(v)$, for $v \in V$), or, as it is also called, a linear operator on V. The embedding

$$\iota\colon \operatorname{Mod} k[T] \longrightarrow \operatorname{Mod} kK(2)$$

is achieved by sending the pair (V, β) to the Kronecker module

$$V \underset{\beta}{\overset{1}{\leftleftarrows}} V$$

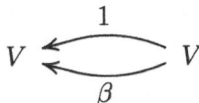

The image under this functor ι are all the representations of $K(2)$ for which the upper map α is an identity map; up to a categorical equivalence, these are all the representations of $K(2)$ for which α is an isomorphism. Of course, for this subcategory one tends to rate the module $\iota(k[T])$ as being a very natural representation. The indecomposable finite-dimensional $k[T]$-modules are (in case k is algebraically closed) of the form $k[T]/(T-\lambda)^n$, and we may assume that $\iota(k[T]/(T-\lambda)^n)$ belongs to the tube with index λ, thus the labeling of the tubes reflects the eigenvalue behaviour of the operator. The label ∞ refers to situations where the map α is not invertible. Of particular interest is the module $G = \iota(k(T))$, the only infinite-dimensional generic representation of $K(2)$. We may consider the one-dimensional projective representation $P(2)$ of $K(2)$ as a submodule of G (up to automorphisms of G there is just one such embedding), and $G/P(2)$ is just the direct sum of all possible Prüfer modules, each one occurring with multiplicity one.

What are typical operators? The vector spaces dealt with in functional analysis are usually function spaces, the operators are differential or integral operators. Let us look at a very elementary example: consider a space F of functions closed under differentiation, and let $\beta = \frac{d}{dT}$. If we take for F the space of all polynomial functions, the module we obtain is a very interesting one, it is the Prüfer module corresponding to the eigenvalue 0, provided we work with a field k of characteristic zero. For an example concerning the use of infinite-dimensional Kronecker modules in studying perturbations of differential operators we refer to [AB].

A more detailed look at the development of the representation theory of finite-dimensional algebras reveals an interest in arbitrary representations from its beginning, at least in some papers. For example, when dealing with serial finite-dimensional algebras, Nakayama and later again Eisenbud and Griffith were very proud to be able to decompose all the modules, not just the finite-dimensional ones, as a direct sum of indecomposables. There has been a parallel development in the representation theory of finite groups: until recently, the main emphasis was lying on finite-dimensional representations. New approaches which are based on the use of infinite-dimensional representations are outlined in the paper by Benson [Be].

References

[Ar] D. M. Arnold: Finite Rank Torsion Free Abelian Groups and Rings. Springer LNM 931 (1982).

[A1] M. Auslander: Representation Dimension of Artin Algebras. Queen Mary College Mathematics Notes. London (1971).

[A2] M. Auslander: Representation theory of artin algebras II. Comm. Algebra 2 (1974), 269-310.

[A3] M. Auslander: Large modules over artin algebras. In: Algebra, Topology and Category Theory. Academic Press New York (1976), 1-17.

[AR1] M. Auslander, I. Reiten: Stable equivalence of dualizing R-varieties. Advances Math. 12 (1974), 306-366.

[AR2] M. Auslander, I. Reiten: Uniserial functors. In: Representation Theory II. Springer LNM 832 (1980), 1-47.

[AB] N. Aronszajn, R.D. Brown: Finite dimensional perturbations of spectral problems and variational method for eigenvalue problems. I, Studia Mathematica 36 (1970), 1-76.

[Ba] H. Bass: Algebraic K-Theory. Benjamin. New York (1968).

[Bt] R. Bautista: On some tame and discrete families of modules. This volume.

[Be] D.J. Benson: Infinite dimensional modules for finite groups. This volume.

[BrR] S. Brenner, C.M. Ringel: Pathological modules over tame rings. J. London Math. Soc. 14 (1976), 207-215.

[BuR] M.C.R. Butler, C.M. Ringel: Auslander–Reiten sequences with few middle terms, with applications to string algebras. Comm. Algebra 15 (1987), 145-179.

[CD] R. Camps, W. Dicks: On semilocal rings. Israel J. Math.81 (1993), 203-211.

[Cz] J.H. Cozzens: Homological properties of the ring of differential operators. Bull. Amer. Math. Soc.76 (1970), 66-75

[Cn] P.M. Cohn: Algebra 3. Wiley & Sons. Chichester 1991.

[Ct] F. Couchot: Sous-modules purs et modules de type cofini. Séminaire d'Algèbre P. Dubreil. Springer LNM 641 (1978), 198-208.

[CB1] W. Crawley-Boevey: Tame algebras and generic modules. Proc. London Math. Soc. 63 (1991), 241-265.

[CB2] W. Crawley-Boevey: Modules of finite length over their endomorphism rings. In: Representations of algebras and related topics. London Math. Soc. LNS 68 (1992), 127-184.

[CB3] W. Crawley-Boevey: Locally finitely presented additive categories. Comm. Algebra 22 (1994), 1641-1674.

[CB4] W. Crawley-Boevey: Infinite-dimensional modules in the representation theory of finite-dimensional algebras. Trondheim lectures (1996). In: Algebras and Modules I. CMS Conference Proceedings 23. Providence (1998), 29-54.

[CJ] P. Crawley, B. Jónsson: Refinements for infinite direct decompositions of algebraic systems. Pacific J. Math. 14 (1964), 797-855.

[Dl] P. Dräxler: Representation directed diamons. In preparation.

[Dd] Yu. A. Drozd: Tame and wild matrix problems. Representations and quadratic forms. Inst. Math., Acad. Sciences. Ukrainian SSR, Kiev 1979, 39-74. Amer. Math. Soc. Transl. 128 (1986), 31-55.

[E] P.C. Eklof: Modules with strange decomposition properties. This volume.

[Fc1] A. Facchini: Krull-Schmidt fails for serial modules. Trans. Amer. Math. Soc.348 (1996), 4561-4575.

[Fc2] A. Facchini: Module Theory. Birkhäuser. Basel (1998).

[Fc3] A. Facchini: Failure of the Krull-Schmidt theorem for artinian modules and serial modules. This volume.

[FHLV] A. Facchini, D. Herbera, L.S. Levy, P. Vámos: Krull-Schmidt fails for artinian modules. Proc. Amer. Math. Soc. 123 (1995), 3587-3592.

[Fa1] C. Faith: Algebra: Rings, Modules, Categories I. Springer. Berlin 1973.

[Fa2] C. Faith: Algebra II. Ring Theory. Springer. Berlin 1976.

[Fu1] L. Fuchs: Infinite abelian groups I. Academic Press 1970.

[Fu2] L. Fuchs: Infinite abelian groups II. Academic Press 1973.

[GO] P. Gabriel, U. Oberst: Spetralkategorien und reguläre Ringe im von–Neumannschen Sinn. Math. Z. 92 (1966), 389-395.

[GR] P. Gabriel, A.V. Roiter: Representations of finite-dimensional algebras. Encycl. Math. Sciences 73. Springer. Berlin 1992.

[GP] I.M. Gelfand, V.A. Ponomarev: Indecomposable representations of the Lorentz group. Russ. Math. Surv. 23 (1968), 95-112.

[Gg] W. Geigle: The Krull-Gabriel dimension of the representation theory of a tame hereditary algebra and applications to the structure of exact sequences. Manuscr. Math. 54 (1985), 83-106.

[Gr] G. Geisler: Zur Modelltheorie von Moduln. Dissertation Freiburg/Brg. (1994), 276 pp.

[Gß] Chr. Geiß: On components of type $\mathbb{Z}A_\infty^\infty$ for string algebras. Comm. Alg. 26 (1998), 749-758.

[Gö] R. Göbel: Some combinatorial principles for solving algebraic problems. This volume.

[GJ] L. Gruson, C.U. Jensen: Dimensions cohomologiques reliées aux foncteurs $\lim\limits_{\leftarrow}^{(i)}$. Séminaire d'Algébre P. Dubreil et M.-P. Malliavin. Springer LNM 867 (1981), 234-294.

[Ha] Han Y.: Controlled wild algebras. Diss. Bielefeld (1999).

[He] I. Herzog: Elementary duality of modules. Trans. Amer. Math. Soc. 340 (1993), 37-69.
[HC] D. Hilbert, S. Cohn-Vossen: Anschauliche Geometrie. Wiss. Buchgesellschaft. Darmstadt (1973).
[H] B. Huisgen-Zimmermann: Purity, algebraic compactness, direct sum decompositions, and representation type. This volume.
[HO] B. Huisgen-Zimmermann, F. Okoh: Direct products of modules and the pure semisimplicity conjecture. Comm. Algebra (To appear).
[JL] C.U. Jensen, H. Lenzing: Model Theoretic Algebra. Gordon and Breach, 1989.
[Kc] V. Kac: Root systems, representations of quivers and invariant theory. In: Invariant Theory. Springer LNM 996 (1983), 74-108.
[Ka] I. Kaplansky: Infinite Abelian Groups. Ann Arbor (1954). Revised edition. Ann Arbor (1969).
[K1] H. Krause: Constucting large modules over artin algebras. J. Algebra 187 (1997), 413-421.
[K2] H. Krause: The spectrum of a locally coherent category. J. Pure Appl. Algebra 114 (1997), 259-271.
[K3] H. Krause: An axiomatic description of a duality for modules. Advances Math. 130 (1997), 280-286.
[K4] H. Krause: Generic modules over artin algebras. Proc. London Math. Soc. 76 (1998), 276-306.
[K5] H. Krause: Stable equivalence preserves representation type. Comment. Math. Helv. 72 (1997), 266-284.
[K6] H. Krause: The spectrum of a module category. Habilitation Schrift. Bielefeld 1998. SFB-Preprint 98-004. Memoirs of the Amer. Math. Soc. (To appear).
[K7] H. Krause: Finite versus infinite dimensional representations — a new definition of tameness. This volume.
[La] T.-Y. Lam: Modules with isomorphic multiples an rings with isomorphic matrix rings. L'Enseignement Math. Monogr. 35. Genève 1999.
[Le] H. Lenzing: Generic modules over a tubular algebra. Advances in Algebra and Model Theory. Gordon-Breach. London (1997), 375-385.
[PY] K.I. Pimenov, A.V. Yakovlev: Artinian modules over a matrix ring. This volume.
[P1] M. Prest: Model Theory and Modules. London Math. Soc. LNS 130: Cambridge University Press. (1988)
[P2] M. Prest: Remarks on elementary duality. Ann. Pure Appl. Logic. 62 (1993), 183-205.
[P3] M. Prest: A note concerning the existence of many indecomposable infinite-dimensional pure-injective modules over some finite-dimensional algebras. (1995) Preprint.
[P4] M. Prest: Representation embeddings and the Ziegler spectrum. J. Pure Appl. Algebra 113 (1996), 315-323.
[P5] M. Prest: Ziegler spectra of tame hereditary algebras. J. Algebra 207 (1998), 146-164
[P6] M. Prest: Topological and geometrical aspects of the Ziegler spectrum. This volume.
[PSc] M. Prest, J. Schröer: Uniserial functors and the infinite radical of a module category. In preparation.
[R1] C.M. Ringel: The indecomposable representations of the dihedral 2–groups. Math. Ann. 214 (1975), 19-34.
[R2] C.M. Ringel: Representations of k-species and bimodules. J. Algebra 41 (1976), 269-302.
[R3] C.M. Ringel: Infinite dimensional representations of finite dimensional hereditary algebras. Symposia Math. 23 (1979), 321-412.
[R4] C.M. Ringel: The spectrum of a finite dimensional algebra. Marcel Dekker Lecture Notes Pure Appl. Math 51 (1979), 535-597.
[R5] C.M. Ringel: Tame algebras. In: Representation Theory I. Springer LNM 831 (1980), 104-136.
[R6] C.M. Ringel: Tame algebras and integral quadratic forms. Springer LNM 1099 (1984).
[R7] C.M. Ringel: The use of infinite dimensional modules in the representation theory of finite dimensional algebras. Lecture Leeds (1994).
[R8] C.M. Ringel: Some algebraically compact modules I. In: Abelian Groups and Modules (ed. A. Facchini and C. Menini). Kluwer (1995), 419-439.
[R9] C.M. Ringel: A construction of endofinite modules. Advances in Algebra and Model Theory. Gordon-Breach. London (1997), 387-399

[R10] C.M. Ringel: The Ziegler spectrum of a tame hereditary algebra. Coll. Math. 76 (1998), 105-115.
[R11] C.M. Ringel: On generic modules for string algebras. (To appear).
[R12] C.M. Ringel: Tame algebras are Wild. Algebra Coll. 6 (1999), 473-480.
[R13] C.M. Ringel: Krull-Remak-Schmidt fails for artinian modules over local rings. (To appear).
[R14] C.M. Ringel: Some algebraically compact modules II. In preparation.
[RT] C.M. Ringel, H. Tachikawa: QF–3 rings. J. Reine Angew. Math. 272 (1975), 49-72.
[RV] C.M. Ringel, D. Vossieck: Hammocks. Proc. London Math. Soc. 54 (1987), 216-246.
[Sc1] J. Schröer: Hammocks for string algebras. Diss. Bielefeld (1998). SFB-Ergänzungsreihe 97-010.
[Sc2] J. Schröer: On the Krull-Gabriel dimension of an algebra. Math. Z. (To appear).
[Sc3] J. Schröer: On the infinite radical of a module category. Proc. London Math. Soc. (To appear).
[Sc4] J. Schröer: The Krull-Gabriel dimension of an algebra — open problems and conjectures. This volume.
[Sk] A. Skowroński: Module categories over tame algebras. In: Representation Theory of Algebras and Related Topics. CMS Conference Proceedings 19. Providence (1996), 281-313.
[Sw] R.G. Swan: Vector bundles and projective modules. Trans. Amer. Math. Soc. 105 (1962), 264-277.
[T] Tachikawa: Quasi-Frobenius Rings and Generalizations. Springer LNM 351 (1973).
[Wa] R.B. Warfield Jr.: A Krull-Schmidt theorem for infinite sums of modules. Proc. Amer. Math. Soc.22 (1969), 460-465.
[Wi] R. Wisbauer: Grundlagen der Modul- und Ringtheorie. R. Fischer. München 1988.
[Z] M. Ziegler: Model theory of modules. Ann. Pure Appl. Logic. 26(2) (1984), 149-213.
[ZZ] B. Zimmermann-Huisgen, W. Zimmermann: Algebraically compact rings and modules. Math. Z. 161 (1978), 81-93.

Fakultät für Mathematik, Universität Bielefeld,
POBox 100 131,
D-33 501 Bielefeld
Germany

E-mail address: ringel@mathematik.uni-bielefeld.de

Trends in Mathematics, © 2000 Birkhäuser Verlag Basel/Switzerland

MODULES WITH STRANGE DECOMPOSITION PROPERTIES

PAUL C. EKLOF

It is well-known that modules with pathological properties can be obtained by constructing modules with "nearly prescribed" endomorphism rings. (See, for example, [5], [2], [3], or [10].) Here we shall investigate that method using a rather weak notion of nearly prescribing the endomorphism ring, namely one that just requires that a certain ring be algebraically closed in the endomorphism ring of the module. As we shall see, in certain cases this method will suffice to construct pathological modules where it seems difficult, if not impossible, for set-theoretic reasons, to prescribe the endomorphism ring more precisely.

We will focus on modules M which are pathological in the sense that for some $r \geq 2$, $M^m \cong M^n$ if and only if $m \equiv n \pmod r$. Any class, $\mathcal{C}$, of modules (closed under finite direct sums and direct summands) which contains such a pathological module M for $r = 2$ lacks a satisfactory classification theorem in the sense that there are negative answers to the following two *Kaplansky test problems* (cf. [22, pp. 12f]):

(I) Are $A, B \in \mathcal{C}$ isomorphic whenever A is isomorphic to a direct summand of B and B is isomorphic to a direct summand of A?

(II) Are $A, B \in \mathcal{C}$ isomorphic whenever A^2 is isomorphic to B^2?

In fact, one obtains negative answers to both problems by letting $A = M$ ($\cong M^3$) and $B = M^2$ ($\cong M^4$).

Throughout R is a commutative ring with 1 and A is a unital R-algebra. An A-module M can then also be regarded as an R-module; we shall be interested in conditions on the endomorphism ring of M as R-module. We will always use λ to denote an infinite cardinal and κ an uncountable cardinal.

This work is an extension of joint work with Shelah [16].

1. Pathological modules and algebraically closed subrings

If M is a (right) A-module such that the action of A is faithful, then A is naturally a subring of $\mathrm{End}_R(M)$. We say that A is *algebraically closed* in $\mathrm{End}_R(M)$ when every finite set of ring equations with parameters from A which is satisfied in $\mathrm{End}_R(M)$ is also satisfied in A; that is, every set of finitely many polynomial equations (where the polynomials, in several variables, have coefficients from A) which has a solution in $\mathrm{End}_R(M)$ has one also in A. Say that A is *weakly algebraically closed* in $\mathrm{End}_R(M)$ when every finite set of ring equations with parameters from $\{0,1\}$ which is satisfied in $\mathrm{End}_R(M)$ is also satisfied in A.

For example, A is algebraically closed in $\mathrm{End}_R(M)$ if $A = \mathrm{End}_R(M)$ or if there is a ring homomorphism $\beta : \mathrm{End}_R(M) \to A$ which is the identity on A. In the latter case, if I is the kernel of β, then $\mathrm{End}_R(M) = A \oplus I$ as R-modules and we say that $\mathrm{End}_R(M)$ is a *split extension* of A by I. But, as we shall see, it may be possible to prove that A is (weakly) algebraically closed in $\mathrm{End}_R(M)$ even when we cannot prove that $\mathrm{End}_R(M)$ is a split extension of A.

Given an R-module M, we will say that an R-algebra A is (weakly) algebraically closed in $\mathrm{End}_R(M)$ if A is isomorphic to a subalgebra of $\mathrm{End}_R(M)$ which is (weakly) algebraically closed in $\mathrm{End}_R(M)$. We will be particularly interested in the case when A is the following ring.

Definition 1. [The Leavitt-Cohn-Corner Ring] *Let $V_{r+1} = V_{1,r+1}(R)$ be the R-algebra which is freely generated by symbols x_{1i} and y_{i1} $(i = 1, \ldots, r+1)$ subject to the relations*

$$y_{j1}x_{1i} = \begin{cases} 1 & \text{if } i = j \\ 0 & \text{otherwise} \end{cases}$$

and

$$\sum_{i=1}^{r+1} x_{1i}y_{i1} = 1.$$

Thus, if

$$X = \begin{bmatrix} x_{11} & x_{12} & \ldots & \ldots & x_{1,r+1} \end{bmatrix} \text{ and } Y = \begin{bmatrix} y_{11} \\ y_{21} \\ \cdot \\ \cdot \\ y_{r+1,1} \end{bmatrix}$$

are matrices over V_{r+1}, then $XY = I_1$ and $YX = I_{r+1}$. Corner [8] has shown that if $V_{r+1}^m \cong V_{r+1}^n$ as (left) V_{r+1}-modules, then $m \equiv n \pmod r$, and also that V_{r+1} is free as an R-module. (See also [4] and [21].)

Lemma 2. *Let M be an R-module. If $V_{r+1}(R)$ is weakly algebraically closed in $\mathrm{End}_R(M)$, then $M^m \cong M^n$ if and only if $m \equiv n \pmod r$.*

PROOF. The matrices X and Y, which may be regarded as having entries in $\mathrm{End}_R(M)$, define inverse isomorphisms from M to M^{r+1} and from M^{r+1} to M, respectively. Now suppose that $M^m \cong M^n$. Then there are matrices U and W of sizes $m \times n$ and $n \times m$ respectively, with entries from $\mathrm{End}_R(M)$, satisfying $UW = 1_m$ and $WU = 1_n$. The latter matrix equations may be regarded as a finite set of ring equations in $\mathrm{End}_R(M)$ with parameters from $\{0,1\}$. Since V_{r+1} is weakly algebraically closed in $\mathrm{End}_R(M)$, these equations are satisfied in V_{r+1} and hence $V_{r+1}^m \cong V_{r+1}^n$, so $m \equiv n \pmod r$. □

We could have weakened further the definition of weakly algebraically closed, and still obtained Lemma 2, since only quadratic polynomials are used in the proof. However, in the application in the last section we will in fact be using M such that $V_{r+1}(R)$ is algebraically closed in $\mathrm{End}_R(M)$.

Corner [5] proved that every countable reduced torsion-free ring A equals $\mathrm{End}_{\mathbb{Z}}(M)$ for some countable torsion-free $\mathbb{Z}$-module M. Corner [7] also proved an analogous result for p-primary groups, but in this case the endomorphism ring is a split extension of the given ring A. (For example, A can be the p-adic completion of $V_{r+1}(\mathbb{Z})$.)

We shall consider in detail another example involving split extensions. A (torsion-free) abelian group is called *separable* if every finite subset is contained in a free direct summand of the group. Dugas and Göbel [12] proved that for certain rings A, among them $V_{r+1}(\mathbb{Z})$ and $\mathbb{Z}$, there is a separable abelian group M such that $\mathrm{End}_{\mathbb{Z}}(M) = A \oplus I$, where I is the ideal of endomorphisms whose range is finitely-generated. If we take $A = \mathbb{Z}$, we obtain a separable group M such that in any direct decomposition $M = M_1 \oplus M_2$, either M_1 or M_2 is finitely-generated; we say that M is "essentially indecomposable". For $A = V_{r+1}(\mathbb{Z})$, Lemma 2 applies, and in particular we obtain negative solutions to the Kaplansky Test Problems for the class of separable abelian groups.

The Dugas-Göbel result is proved by means of a prediction principle which is a theorem of ZFC, called Shelah's Black Box. For more on the Black Box, which has many forms, see [10], [15, Chap. XIII] or [18].

Separable groups are $\aleph_1$-free, that is, every countable subgroup is free. More generally, an abelian group M is called *κ-free* if every subgroup of cardinality $< \kappa$ is free. For any ring A whose underlying abelian group is free, there is an $\aleph_1$-free group M such that $\mathrm{End}_{\mathbb{Z}}(M) = A$. (See [10]; for a construction where M is of cardinality $\aleph_1$ when A is countable, see [20].) Note that, in general, M will not be separable; indeed A may have few, if any, non-trivial idempotents.

If we try to extend these realization results for separable and $\aleph_1$-free abelian groups to analogs for larger cardinals, we run into set-theoretic obstacles. Let n be an integer ≥ 1. A $\mathbb{Z}$-module M is called *$\aleph_n$-separable* (respectively, *strongly $\aleph_n$-free*) if every subset of cardinality $< \aleph_n$ is contained in a free direct summand of M (resp., contained in a free submodule H which is a direct summand of every submodule of M which contains H and has cardinality $< \aleph_n$).

Clearly, every $\aleph_n$-separable $\mathbb{Z}$-module is strongly $\aleph_n$-free. The weak Continuum Hypothesis ($2^{\aleph_0} < 2^{\aleph_1}$) implies that there are strongly $\aleph_1$-free abelian groups of cardinality $\aleph_1$ which are not $\aleph_1$-separable; on the other hand, it is consistent with $2^{\aleph_0} = 2^{\aleph_1}$ that every strongly $\aleph_1$-free abelian group of cardinality $\aleph_1$ is $\aleph_1$-separable. (See [13].)

The following theorem holds when A is $\mathbb{Z}$ or $V_{r+1}(\mathbb{Z})$. The assumption $\diamondsuit_{\omega_1}$ in part (b) is a prediction principle, due to R. Jensen, which is a consequence of

Gödel's Axiom of Constructibility, but is not provable from ZFC. (See, for example, [15, Chap. VI].)

Theorem 3. *Let A be a countable ring whose additive group is free.*

(a) [11] *(* $2^{\aleph_0} < 2^{\aleph_1}$ *) There is a strongly* $\aleph_1$*-free* $\mathbb{Z}$*-module M of cardinality* $\aleph_1$ *such that* $\mathrm{End}_{\mathbb{Z}}(M) = A$.

(b) [27] *(*$\diamondsuit_{\omega_1}$*) There is an* $\aleph_1$*-separable* $\mathbb{Z}$*-module M of cardinality* $\aleph_1$ *such that* $\mathrm{End}_{\mathbb{Z}}(M) = A \oplus I$ *where I is the ideal of endomorphisms with a countable image.* □

In particular, for $A = \mathbb{Z}$, we obtain from (a) an indecomposable strongly $\aleph_1$-free abelian group M, and from (b) an $\aleph_1$-separable M such that whenever $M = M_1 \oplus M_2$, either M_1 or M_2 is countable. For $A = V_3(\mathbb{Z})$, we obtain negative solutions to Kaplansky's Test Problems for these classes, as before.

These (near-)prescriptions of the endomorphism ring cannot be carried out in ZFC, since it is consistent with ZFC that *every* strongly $\aleph_1$-free abelian group of cardinality $\aleph_1$ is $\aleph_1$-separable and of the form $F \oplus \bigoplus_{\nu<\omega_1} M_\nu$ where F is free of uncountable rank and each M_ν is non-free (cf. [13], [24], or [14]). However, the method of the next sections will allow us to prove in section 4 the following:

Theorem 4. *Let* $n \geq 1$*. For* $n \geq 2$*, assume that* $2^{\aleph_{n-2}} \leq \aleph_n$*. Let A be a countable ring whose additive group is free. Then there is an* $\aleph_n$*-separable* $\mathbb{Z}$*-module M of cardinality* $\aleph_n$ *such that A is algebraically closed in* $\mathrm{End}_{\mathbb{Z}}(M)$.

In particular, for $n = 1$ we obtain a theorem of ZFC (proved in [16]). The extension to arbitrary $n \geq 2$ involves an interesting induction. We do not know if the result can be proved without the hypothesis on cardinal exponentiation.

The method of weak realization—that is, realizing the ring A only as an algebraically closed subring of the endomorphism ring—does not work for the construction of pathological modules using the ring of [5, Lemma 5.3] (i.e., modules without indecomposable summands) or for the construction of indecomposables (i.e., where the ring has no non-trivial idempotents). It can however be applied to the following ring:

Definition 5. [The Corner-Crawley Ring] *Let* $C = C(R)$ *be the R-algebra of all* 2×2 *matrices of the form*

$$\begin{bmatrix} f_{11}(t) & (1-t^2)f_{12}(t) \\ f_{21}(t) & b + (1-t^2)f_{22}(t) \end{bmatrix}$$

where $b \in R$ *and* $f_{ij}(t) \in R[t]$.

Then if

$$x = \begin{bmatrix} t & 1-t^2 \\ 0 & 0 \end{bmatrix} \text{ and } y = \begin{bmatrix} t & 0 \\ 1 & 0 \end{bmatrix},$$

we have $xyx = x$, $yxy = y$. Let $\alpha = xy$, $\beta = yx$; then α and β are idempotents which are equivalent but such that $1-\alpha$ and $1-\beta$ are not equivalent. Also, C is free as R-module. (See [9] and [28].)

Similarly to the proof of Lemma 2 it can be proved that if C is algebraically closed in $\mathrm{End}_R(M)$, then $M \cong M\alpha \oplus M(1-\alpha) \cong M\beta \oplus M(1-\beta)$ and $M\alpha \cong M\beta$ but $M(1-\alpha) \not\cong M(1-\beta)$. (Weak algebraic closure of C in $\mathrm{End}_R(M)$ is not enough here.) If $C = \mathrm{End}_R(M)$, then the summands $M\alpha$, $M\beta$, $M(1-\alpha)$, and $M(1-\beta)$ are indecomposable (cf. [28, pp. 297f]). Absent that condition, or a finiteness hypothesis, the failure of cancellation can sometimes be achieved by "cheap" means (e.g. $R^{(\omega)} \oplus R \cong R^{(\omega)} \oplus R^2$ but $R \not\cong R^2$). But we will use this ring in section 4 to give a more interesting cancellation failure.

We conclude this section by mentioning an important case where similar ideas are applied to the case of a non-commutative ring R. In fact, Shelah [25, p. 118] has shown that pathological modules occur whenever they can, that is, whenever the ring R is not pure semisimple. More precisely he proved the following result. (His proof in [25] assumes $V = L$, but he claims a proof in ZFC, using the Black Box.)

Theorem 6. *Suppose R is not left pure semisimple. Let K be the subring of R generated by 1, and let A be any ring which is free as K-module. Then there is an $R-A$ bimodule M which is a faithful A-module and has the property that any finite set of ring equations with parameters from A which is solvable in* $\mathrm{End}_R(M)$ *is solvable in $D \otimes_K A$ for some field D (which is a K-algebra).*

If we take $A = V_{1,r+1}(K)$, then $D \otimes_K A = V_{1,r+1}(D)$ and the conclusion of Lemma 2 can be obtained as before.

2. Set-theoretic tools

The *cofinality* of a limit ordinal δ is the minimal cardinality of an unbounded (that is, cofinal) subset of δ, and will be denoted $\mathrm{cof}(\delta)$. For example, the cofinality of the cardinal $\aleph_\omega$ is $\aleph_0$ $(= \omega)$. A cardinal is called *regular* if and only if it is equal to its own cofinality. Every successor cardinal, that is one of the form $\aleph_{\alpha+1}$, is regular. If $\lambda = \aleph_\alpha$, we will denote $\aleph_{\alpha+1}$ by λ^+.

A subset C of a regular uncountable cardinal κ is a *club* (or closed unbounded set) iff $\sup C = \kappa$ (i.e., C is unbounded in κ) and for every subset X of C of cardinality $< \kappa$, $\sup X \in C$. A subset E of κ is called *stationary* (in κ) iff for all clubs C, $E \cap C \neq \emptyset$. The intersection of two clubs is easily shown to be a club, so, in particular, clubs are stationary. (See, for example, [15, Chap. II].)

Proposition 7. *Suppose that κ is a regular cardinal $\geq |R|$ and M is a faithful A-module such that $M = \bigcup_{\alpha<\kappa} M_\alpha$ is the union of a continuous chain of submodules of cardinality $< \kappa$. Suppose also that there is a stationary subset E of κ such that for every $\alpha \in E$ and every $\beta > \alpha$, A acts faithfully on M_β/M_α and is (weakly) algebraically closed in* $\mathrm{End}_R(M_\beta/M_\alpha)$. *Then A is (weakly) algebraically closed in* $\mathrm{End}_R(M)$.

PROOF. For any $\sigma \in \mathrm{End}_R(M)$, the set C_σ of all $\alpha \in \kappa$ such that $\sigma[M_\alpha] \subseteq M_\alpha$ is a club subset of κ. Indeed, it is easy to see that the continuity of the chain of submodules implies that C_σ is closed; and C_σ is unbounded because given $\beta < \kappa$ we can choose a strictly increasing sequence

$$\beta < \alpha_1 < \alpha_2 < \ldots < \alpha_n < ..$$

such that $\sigma[M_{\alpha_n}] \subseteq M_{\alpha_{n+1}}$ and then $\sup\{\alpha_n : n \in \omega\}$ will be in C_σ.

For any $\sigma_1, \ldots, \sigma_n$ in $\mathrm{End}_R(M)$, $C_{\sigma_1} \cap \ldots \cap C_{\sigma_n}$ is a club so $E \cap C_{\sigma_1} \cap \ldots \cap C_{\sigma_n}$ is non-empty and we can choose $\alpha < \beta$ in $C_{\sigma_1} \cap \ldots \cap C_{\sigma_n}$ so that also $\alpha \in E$. Then each σ_i induces an endomorphism, $\bar{\sigma}_i$, of M_β/M_α. Clearly, if $\sigma_i + \sigma_j = \sigma_k$ (resp. $\sigma_i\sigma_j = \sigma_k$), then $\bar{\sigma}_i + \bar{\sigma}_j = \bar{\sigma}_k$ (resp. $\bar{\sigma}_i\bar{\sigma}_j = \bar{\sigma}_k$) and if $\sigma_i = a \in A$ (regarded as an element of $\mathrm{End}_R(M)$), then $\bar{\sigma}_i = a$. Hence if $\sigma_1, \ldots, \sigma_m$ satisfy some ring equations over A, then $\bar{\sigma}_1, \ldots, \bar{\sigma}_m$ satisfy those equations in $\mathrm{End}_R(M_\beta/M_\alpha)$. The conclusion follows since A is (weakly) algebraically closed in $\mathrm{End}_R(M_\beta/M_\alpha)$. □

Let κ be a regular uncountable cardinal and E a stationary subset of κ consisting of limit ordinals. A *ladder system (on E)* is an indexed family $\{\eta_\delta : \delta \in E\}$ of functions $\eta_\delta : \mathrm{cof}(\delta) \to \delta$ which are strictly increasing and which have image cofinal in δ and consisting only of successor ordinals. A *tree-like* ladder system has the additional property that for every $\delta, \gamma \in E$ and every $\mu \in \mathrm{cof}(\delta)$, $\nu \in \mathrm{cof}(\gamma)$, $\eta_\delta(\mu) = \eta_\gamma(\nu)$ implies that $\mu = \nu$ and $\eta_\delta \upharpoonright \mu = \eta_\gamma \upharpoonright \mu$.

Trivially, ladder systems exist on any stationary subset of any regular κ (consisting of limit ordinals). Tree-like ladder systems exist on any stationary set of limit ordinals in $\aleph_1$ without any special set-theoretic hypotheses; the proof is a variation of the proof of Lemma 8. (See also [15, pp. 369, 386].) So from now on we shall assume that ρ is a regular infinite cardinal and consider subsets of ρ^{++} whose elements are limit ordinals of cofinality ρ^+. From the assumption $2^\rho \leq \rho^{++}$ we will conclude that there is a tree-like ladder system on a stationary set of this form. (I thank Matt Foreman for this proof of a result of Shelah regarding $I[\lambda]$.)

Let E' be a subset of ρ^{++} consisting of limit ordinals; E' is said to be an *approachable* set if there is a sequence $\langle S_\nu : \nu < \rho^{++} \rangle$ of subsets of ρ^{++} each of cardinality $\leq \rho^+$ such that for every $\delta \in E'$, S_δ is a cofinal subset of δ of order type $\mathrm{cof}(\delta)$ and for every $\alpha < \delta$, $S_\delta \cap \alpha = S_\nu$ for some $\nu < \delta$.

Lemma 8. *If there is a stationary approachable subset E' of ρ^{++} consisting of limit ordinals of cofinality ρ^+, then there is a tree-like ladder system on a stationary set E (of the form $E' \cap C$ for some club C).*

PROOF. Let $\langle S_\nu : \nu < \rho^{++} \rangle$ be as in the definition. Without loss of generality we can assume that $\nu_1 \neq \nu_2$ implies $S_{\nu_1} \neq S_{\nu_2}$. Choose a one-one function $\theta : \{S_\nu : \nu < \rho^{++}\} \to Succ(\rho^{++})$ such that for all ν, $\theta(S_\nu) > \sup S_\nu$ and such that whenever S_μ is an initial segment of S_ν, then $\theta(S_\mu) \leq \theta(S_\nu)$. Such a function can be defined by transfinite induction, making sure to define $\theta(S_\mu)$ at the same time that $\theta(S_\nu)$ is defined for all μ such that S_μ is an initial segment of S_ν. Then $C = \{\alpha \in \rho^{++} : \theta(S_\nu) < \alpha \text{ for all } \nu < \alpha\}$ is a club in ρ^{++}. So $E' \cap C$ is stationary:

call it E. For each $\delta \in E$ and $\beta \in \rho^+$, define $\eta_\delta(\beta) = \theta(S_\delta \restriction \beta)$ where $S_\delta \restriction \beta$ denotes the initial segment of S_δ of order type β. Then $\eta_\delta(\beta) < \delta$ since $\delta \in C$; moreover, $\eta_\delta : \rho^+ \to \delta$ is strictly increasing with range cofinal in δ by properties of θ. The family $\{\eta_\delta : \delta \in E\}$ is tree-like since θ is one-one. □

Theorem 9. *If $2^\rho \leq \rho^{++}$, then there is a tree-like ladder system on a stationary subset of ρ^{++} consisting of limit ordinals of cofinality ρ^+*

PROOF. By Lemma 8 it suffices to prove that there is a stationary approachable subset E' of ρ^{++} consisting of ordinals of cofinality ρ^+. If Y is the set of elements of ρ^{++} of cofinality different from ρ^+, then by the assumption that $2^\rho \leq \rho^{++}$, there is an enumeration $\langle S_\nu : \nu \in Y\rangle$ of all the subsets of ρ^{++} which have cardinality $\leq \rho$. For δ of cofinality ρ^+, let S_δ be a subset of δ of order type ρ^+ which is cofinal in δ and has the property that for all $\alpha < \delta$, $S_\delta \cap \alpha = S_\nu$ for some $\nu \in Y \cap \delta$ —if there is such a subset; otherwise, let $S_\delta = \emptyset$. It suffices to prove that $E' = \{\delta \in \rho^{++} : \operatorname{cof}(\delta) = \rho^+ \text{ and } S_\delta \neq \emptyset\}$ is stationary in ρ^{++}, that is, for all clubs D, $E' \cap D \neq \emptyset$.

Given a club D, we will choose an increasing sequence $\langle \beta_\mu : \mu < \rho^+\rangle$ of elements of D as follows. Let $\beta_0 \in D$ be arbitrary. Suppose that β_μ has been defined for $\mu < \tau$; then $\{\beta_\mu : \mu < \tau\} = S_\nu$ for some $\nu \in Y$; choose $\beta_\tau \in D$ such that $\beta_\tau > \max(\{\nu\} \cup \{\beta_\mu : \mu < \tau\})$. This defines the sequence; let $\delta = \sup\{\beta_\mu : \mu < \rho^+\}$. Then $\delta \in D$ because D is closed and $\delta \in E'$, as witnessed by the set $\{\beta_\mu : \mu < \rho^+\}$.
□

The conclusion of Theorem 9 (for $\rho = \aleph_0$) is consistent with $2^{\aleph_0} > \aleph_2$ (because it follows from $\square_{\omega_1}$ which is consistent with $2^{\aleph_0} > \aleph_2$: cf. [1]); but it seems to be open whether the conclusion follows from ZFC alone.

3. A GENERAL CONSTRUCTION

In what follows, all modules are A-modules. "Free" will mean free as A-module; we say that a free module has rank λ (not necessarily uniquely determined) if it has a basis of cardinality λ. We will say that M is *κ-separable* if every subset of M of cardinality $< \kappa$ is contained in a direct summand M' of M such that M' is free. (In case A is as in Theorem 4, $R = \mathbb{Z}$, and $\kappa = \aleph_n$, this agrees with the definition of $\aleph_n$-separable given previously.)

We are going to describe some machinery for constructing κ-separable modules; the machinery will have a "feedback" capability that will allow us, by iteration, to prove Theorem 4 in the next section.

As an *ad hoc* definition, a *λ-template* is a pair $H \supseteq B$ of free modules each of rank λ such that B is the union of a continuous chain $\bigcup_{\nu<\lambda} B_\nu$ of free submodules of rank λ and for all $\nu < \lambda$, $B_{\nu+1}/B_\nu$ and H/B_ν are free of rank λ. The following theorem constructs a λ^+-separable module starting with a λ-template. (The template gives the isomorphism type of a "large"—that is, stationary—number of quotients $M_{\alpha+1}/M_\alpha$ in the constructed chain.)

Theorem 10. *Let λ be a regular infinite cardinal $\geq |A|$ and let $\kappa = \lambda^+$. Suppose there is a tree-like ladder system $\{\eta_\delta : \delta \in E\}$ on some stationary subset E of κ consisting of ordinals of cofinality λ. Given a λ-template $H \supseteq B$, there is a module $M = \bigcup_{\beta<\kappa} M_\beta$ which is the union of a continuous chain of free submodules M_β of rank λ. Moreover, for all $\alpha \notin E$, M_α is a direct summand of M and M_β/M_α is free for all $\beta > \alpha$; and for all $\alpha \in E$, $M_{\alpha+1}/M_\alpha$ is isomorphic to H/B.*

PROOF. For each $\nu \in \lambda$, fix a basis $\{b_{\nu,i} + B_\nu : i \in \lambda\}$ of $B_{\nu+1}/B_\nu$ (as A-module). Also, fix a set of representatives $\{h_i : i \in \lambda\}$ for H/B with $h_0 = 0$; thus each coset $h + B$ equals $h_i + B$ for a unique $i \in \lambda$.

Inductively define free modules M_β $(\beta < \kappa)$ as follows: $M_0 = 0$; if β is a limit ordinal, $M_\beta = \bigcup_{\alpha<\beta} M_\alpha$; if $\beta = \alpha + 1$ where $\alpha \notin E$, let

$$M_\beta = M_\alpha \oplus \bigoplus_{i\in\lambda} x_{\alpha,i}A.$$

(So the $x_{\alpha,i}$ are A-linearly independent over M_α.) If $\beta = \delta + 1$ where $\delta \in E$, define an embedding $\iota_\delta : B \to M_\delta$ by sending the basis element $b_{\nu,i}$ to $x_{\eta_\delta(\nu),i}$. Essentially $M_{\delta+1}$ is the pushout of

$$\begin{array}{ccc} M_\delta & & \\ \uparrow \iota_\delta & & \\ B & \hookrightarrow & H \end{array}$$

but we will be more explicit in order to avoid the necessity of identifying isomorphic copies. Let $y_{\delta,0} = 0$ and let $\{y_{\delta,i} : i \in \lambda \setminus \{0\}\}$ be a new set of distinct elements (not in M_δ). Then define $M_{\delta+1}$ to be $\{y_{\delta,i} + u : u \in M_\delta,\ i \in \lambda\}$ where the operations on $M_{\delta+1}$ extend those on M_δ and are otherwise determined by the rules

$$\begin{array}{lll} y_{\delta,i} + y_{\delta,j} = y_{\delta,k} + \iota_\delta(b) & \text{if} & h_i + h_j = h_k + b \\ y_{\delta,i}a = y_{\delta,k} + \iota_\delta(b) & \text{if} & h_i a = h_k + b \end{array}$$

where $b \in B$ and $a \in A$. Then there is an embedding $\theta_\delta : H \to M_{\delta+1}$ extending ι_δ which takes $h_i + b$ to $y_{\delta,i} + i_\delta(b)$ and induces an isomorphism of H/B with $M_{\delta+1}/M_\delta$:

$$\begin{array}{ccc} M_\delta & \hookrightarrow & M_{\delta+1} \\ \uparrow \iota_\delta & & \uparrow \theta_\delta \\ B & \hookrightarrow & H \end{array}$$

This completes the inductive definition of the M_β. Let $M = \bigcup_{\beta<\kappa} M_\beta$.

Note that it follows from the construction that every element of M has a unique representation in the form

$$\sum_{j=1}^{s} y_{\delta_j,\nu_j} + \sum_{\ell=1}^{t} x_{\alpha_\ell,i_\ell} a_\ell$$

where $\delta_1 < \delta_2 < ... < \delta_s$ are elements of E, $\nu_j \in \lambda \setminus \{0\}$, $\alpha_\ell \in \kappa \setminus E$, $i_\ell \in \lambda$, $a_\ell \in A$, and the pairs (α_ℓ, i_ℓ) $(\ell = 1, ..., t)$ are distinct.

We claim that for all $\gamma \notin E$,

(✠) M_γ is a direct summand of M

and

(✠✠) for all $\beta > \gamma$, M_β/M_γ is free.

Notice that (✠✠) implies that every M_β is free since $0 \notin E$ and $M_0 = 0$. To prove (✠), it suffices to consider the case when γ is a successor, since $M_{\gamma+1} = M_\gamma \oplus F$ for some free F when $\gamma \notin E$. For all $\alpha < \kappa$ we will define a projection π_α of M onto $M_{\alpha+1}$ (that is, $\pi_\alpha \restriction M_{\alpha+1}$ is the identity). For every $\tau < \lambda$ there is a projection $\rho_\tau : H \to B_\tau$ since H/B_τ is free. For each $\delta \in E$ with $\delta > \alpha$, let $\tau_\delta = \sup\{\sigma + 1 : \eta_\delta(\sigma) \leq \alpha\}$; then let π_α act like ρ_{τ_δ} on the isomorphic copy, $\theta_\delta[H]$, of H. More precisely, for each element z of $\operatorname{ran}\theta_\delta$, define $\pi_\alpha(z)$ to be $\theta_\delta(\rho_{\tau_\delta}(\theta_\delta^{-1}(z)))$; if $\nu \notin \bigcup\{\operatorname{ran}(\eta_\delta) : \delta \in E\}$ and $\nu > \alpha$, define $\pi_\alpha(x_{\nu,i}) = 0$. Extend to an arbitrary element of M by additivity; this will define a homomorphism on M provided that π_α is well-defined. It is easy to see, using the unique representation of elements, that the question of well-definition reduces to showing that the definition of $\pi_\alpha(x_{\beta,i})$ for $x_{\beta,i} \in \operatorname{ran}\theta_\delta$ is independent of δ; here is where we use the tree-like property of the ladder system. If $\beta \leq \alpha$, then $\pi_\alpha(x_{\beta,i}) = x_{\beta,i}$. If $\beta > \alpha$ and $\beta = \eta_\delta(\nu) = \eta_\gamma(\nu')$, then $\nu = \nu'$ and $\eta_\delta(\mu) = \eta_\gamma(\mu)$ for all $\mu \leq \nu$; hence $\tau_\delta = \tau_\gamma$ and $\tau_\delta < \nu$. Therefore $\pi_\alpha(x_{\beta,i})$ is well-defined because $\theta_\delta(\rho_{\tau_\delta}(\theta_\delta^{-1}(x_{\beta,i}))) = \theta_\delta(\rho_{\tau_\delta}(x_{\nu,i})) = \theta_\gamma(\rho_{\tau_\delta}(x_{\nu,i}))$ since $\theta_\delta \restriction B_{\tau_\delta} = \theta_\gamma \restriction B_{\tau_\delta}$ because $\eta_\delta \restriction \tau_\delta = \eta_\gamma \restriction \tau_\delta$.

For (✠✠), we prove by induction on β that M_β/M_α is free for all $\alpha < \beta$ with $\alpha \notin E$. If β is a limit ordinal, choose a continuous increasing sequence $\langle \xi_\nu \rangle_{\nu \in \operatorname{cof}(\beta)}$ cofinal in β; then $M_\beta/M_\alpha = \bigcup_{\nu<\operatorname{cof}(\beta)} M_{\xi_\nu}/M_\alpha$ is free because $M_{\xi_{\nu+1}}/M_{\xi_\nu}$ is free for all ν since either ξ_ν is a successor or $\operatorname{cof}(\xi_\nu) < \operatorname{cof}(\beta) \leq \lambda$ so $\xi_\nu \notin E$. If β is a successor ordinal we will do the case $\beta = \delta + 1$ where $\delta \in E$; the other cases are similar, but easier. We will inductively define $S_\nu \subseteq M_{\delta+1}$ so that the cosets of the elements of

$$\mathcal{S} = \bigcup_{\nu \in \lambda \cup \{-1\}} S_\nu \cup \{x_{\xi,i} : \xi \in \delta \setminus (E \cup \bigcup\{\operatorname{ran}(\eta_\mu) : \mu \in E \cap (\delta+1)\}), i \in \lambda\}$$

form an A-basis of $M_{\delta+1}/M_\alpha$. Let S_{-1} be the image under θ_δ of a basis of H. Fix a bijection $\psi : \lambda \to \{\sigma \in E : \alpha < \sigma < \delta\}$; also, for convenience, let $\psi(-1) = \delta$. Suppose that for some $\nu < \lambda$, S_μ has been defined for $\mu < \nu$ so that modulo M_α, $\bigcup_{\mu<\nu} S_\mu$ is A-linearly independent and generates $\bigcup\{\theta_{\psi(\mu)}[H] : -1 \leq \mu < \nu\}$. Let $\gamma = \psi(\nu)$ and let

$$\zeta = \inf\{\xi < \lambda : \eta_\gamma(\xi) > \alpha \text{ and } \eta_\gamma(\xi) \neq \eta_{\psi(\mu)}(\xi) \text{ for any } \mu < \nu\}.$$

Notice that, since the ladder system is tree-like, $\theta_\gamma[B_\zeta]$ is contained in the A-submodule generated by $\bigcup_{\mu\leq\nu} S_\mu \cup M_\alpha$. Since H/B_ζ is free, we can write $H = B_\zeta \oplus C_\zeta$ for some free module C_ζ ($= \ker(\rho_\zeta)$); let S_ν be the image under θ_γ of a basis of C_ζ. This completes the inductive construction. One can then verify that $\{b + M_{\alpha+1} : b \in \mathcal{S}\}$ is an A-basis of $M_{\delta+1}/M_\alpha$. □

We use $A^{(\lambda)}$ to denote the direct sum of λ copies of A.

Theorem 11. *Suppose M is as described in the statement of Theorem 10. Suppose also that A is algebraically closed in* $\mathrm{End}_R((H/B)\oplus A^{(\lambda)})$. *Then A is algebraically closed in* $\mathrm{End}_R(M)$ *and in* $\mathrm{End}_R(M\oplus A^{(\kappa)})$.

PROOF. Let M' denote either M or $M\oplus\bigoplus_{\nu<\kappa}A_\nu$ where $A_\nu\cong A$; let M'_α denote, respectively, M_α or $M_\alpha\oplus\bigoplus_{\nu<\alpha}A_\nu$. For $\alpha\in E$, M'_β/M'_α is isomorphic to $H/B\oplus A^{(\mu)}$ for some cardinal $\mu\leq\lambda$. So we can apply Proposition 7. □

The following theorem takes a chain of length κ such as that which is the output of Theorem 10 and produces a κ-template (cf. [15, Thm XII.1.4]).

Theorem 12. *Let κ be a regular cardinal $\geq\aleph_0$. Suppose that $M=\bigcup_{\beta<\kappa}M_\beta$ is the union of a continuous chain of free submodules M_β of rank $<\kappa$. Then there is a κ-template $H\supseteq B=\bigcup_{\beta<\kappa}B_\beta$ such that $H/B\cong M$.*

PROOF. It suffices to prove that there is a short exact sequence

$$0\to B\hookrightarrow H\xrightarrow{\varphi}M\to 0$$

where $H\supseteq B$ are free modules such that $H=\bigoplus_{\beta<\kappa}F_\beta$, $B=\bigoplus_{\beta<\kappa}K_\beta$ and for all $\beta<\kappa$, F_β and K_β are free of rank κ and there is a commutative diagram

$$\begin{array}{ccccccccc}
0 & \to & \bigoplus_{\nu<\beta}K_\nu & \hookrightarrow & \bigoplus_{\nu<\beta}F_\nu & \xrightarrow{\varphi_\beta} & M_\beta & \to & 0\\
 & & \downarrow & & \downarrow & & \downarrow^{1_A} & & \\
0 & \to & B & \hookrightarrow & H & \xrightarrow{\varphi} & M & \to & 0
\end{array}$$

such that the rows are exact and the vertical maps are inclusions. For then we will define $B_\beta=\oplus_{\nu<\beta}K_\nu$. Notice that $H/B_\beta\cong(\oplus_{\nu<\beta}F_\nu/B_\beta)\oplus\bigoplus_{\nu\geq\beta}F_\nu\cong M_\beta\oplus\bigoplus_{\nu\geq\beta}F_\nu$ is free because M_β is free.

The modules are defined by induction on β. Choose

$$0\to K_0\to F_0\xrightarrow{\varphi_0}M_1\to 0$$

with $K_0=0$ and φ_0 an isomorphism. Suppose F_ν, K_ν and φ_ν have been defined for all $\nu<\beta$ so that φ_γ is an extension of φ_ν if $\nu<\gamma$. If β is a limit ordinal it is clear how to define φ_β, so suppose $\beta=\gamma+1$ for some γ. Choose F'_γ such that $\psi_\gamma:F'_\gamma\to M_{\gamma+1}$ is an isomorphism. Let $F_\gamma=F'_\gamma\oplus L_\gamma$ where L_γ is free of rank κ. Let $\varphi_\beta:\bigoplus_{\nu<\beta}F_\nu\to M_{\gamma+1}$ be such that $\varphi_\beta\restriction\bigoplus_{\nu<\gamma}F_\nu=\varphi_\gamma$, $\varphi_\beta\restriction F_\gamma=\psi_\gamma$, and $\varphi_\beta\restriction L_\gamma$ is zero. Let $\{m_i:i\in I\}$ be an A-basis for M_γ. Choose $\{y_i:i\in I\}\subseteq\bigoplus_{\nu<\gamma}F_\nu$ so that for all i, $\varphi_\gamma(y_i)=m_i$. Let K'_γ be the submodule of $\bigoplus_{\nu<\beta}F_\nu$ generated by $\{y_i-\psi_\gamma^{-1}(m_i):i\in I\}$. It is easy to check that this set is a basis of K'_γ and that if $K_\gamma=K'_\gamma\oplus L_\gamma$, then $\bigoplus_{\nu<\beta}K_\nu$ is the kernel of φ_β. This completes the inductive step in the construction. Finally, define φ to be the union of the φ_ν's. □

4. $\aleph_n$-SEPARABLE ABELIAN GROUPS

In this section R is the ring $\mathbb{Z}$ (or any countable p.i.d.) We will write $\mathrm{End}(M)$ instead of $\mathrm{End}_{\mathbb{Z}}(M)$ and sometimes refer to M as a "group", i.e., abelian group. We will prove Theorem 4. To get the inductive construction started we use the following (from [19] or [16, Lemma 1.2]).

Lemma 13. *If A is a countable ring whose underlying group is free, there is an A-module L such that $\mathrm{End}(L) = A$ and L is the union, $\bigcup_{m\in\omega} L_m$, of an increasing chain of free A-modules. Moreover, $\mathrm{Hom}(L, \mathbb{Z}) = 0$.* □

We will also need the following:

Lemma 14. *Let A be a countable ring whose underlying group is free. Suppose that L is a group such that $\mathrm{End}(L) \cong A$ and $\mathrm{Hom}(L, \mathbb{Z}) = 0$. Then A is algebraically closed in $\mathrm{End}(L \oplus A^{(\lambda)})$ for any λ.*

PROOF. Restriction to L defines a homomorphism of $\mathrm{End}(L \oplus A^{(\lambda)})$ onto $\mathrm{End}(L) \cong A$ because $\mathrm{Hom}(L, \mathbb{Z}) = 0$ and hence $\mathrm{Hom}(L, F) = 0$ for any free group F. □

Proof of Theorem 4. Let L be as in Lemma 13. Apply Theorem 12 to obtain an ω-template $H \supseteq B = \cup_{m<\omega} B_m$ with $H/B \cong L$. Applying Theorems 10, we obtain an $\aleph_1$-separable M_1 such that, by Theorem 11 and Lemma 14, A is algebraically closed in $\mathrm{End}(M_1)$ and also in $\mathrm{End}(M_1 \oplus A^{(\aleph_1)})$. Applying Theorem 12, we obtain an $\aleph_1$-template to which we apply Theorem 10 to obtain an $\aleph_2$-separable M_2. (The hypothesis of Theorem 10 about the existence of a tree-like ladder system is satisfied because of the assumption that $2^{\aleph_0} \leq \aleph_2$ and Theorem 9.) By Theorem 11, A is algebraically closed in $\mathrm{End}(M_2)$ and in $\mathrm{End}(M_2 \oplus A^{(\aleph_2)})$. The proof continues in this way for arbitrary $n \in \omega$. At first sight it may seem that because of the iteration we need the hypothesis that $2^{\aleph_{k-2}} \leq \aleph_k$ for $2 \leq k \leq n$ (instead of just for $k = n$). However, the hypothesis in Theorem 10 about the existence of the *tree-like* ladder system is essential only to insure that M_α is a direct summand of M when $\alpha \notin E$ and this property is not needed in constructing the new template using Theorem 12: a slight modification of the proof allows one to verify (✠✠) even when the ladder system is not tree-like. (For the same reason, no assumption on cardinal arithmetic is needed for a weaker version of Theorem 4 where M is only required to be strongly $\aleph_n$-free instead of $\aleph_n$-separable.) □

Applying Theorem 4 to the rings in Definitions 1 and 5, we obtain:

Corollary 15. *Under the hypotheses of Theorem 4:*

(i) the Kaplansky test problems have negative solutions for the class of $\aleph_n$-separable groups of cardinality $\aleph_n$;

(ii) there are $\aleph_n$-separable groups M_0, M_1, and M_2 each of cardinality $\aleph_n$ such that $M_0 \oplus M_1 \cong M_0 \oplus M_2$ but $M_1 \not\cong M_2$. □

In contrast to the result in (ii), it is independent of ZFC whether $M_0 \oplus M_1 \cong M_0 \oplus M_2$ implies $M_1 \cong M_2$ when M_1 and M_2 are $\aleph_1$-separable of cardinality $\aleph_1$ and M_0 is free. (See [14].)

To construct κ-separable groups as in Theorem 4 when κ is a regular cardinal larger than $\aleph_\omega$, additional set-theoretic hypotheses and/or different methods will be needed (cf. [23] or [15, pp. 190f]).

It is an open question whether one can construct (in ZFC or with the cardinal arithmetic hypotheses of Theorem 4) $\aleph_n$-separable groups M such that $\mathrm{End}(M)$ is a split extension of A by some ideal I.

Abelian group theorists in proving theorems about the realization of rings as endomorphism rings have sought to construct rigid systems of such groups, that is, families of groups realizing the ring with only the zero homomorphism (or only the "essential" homomorphisms) between different members of the family. It is not clear how to construct such rigid families in this setting, since already for cardinality $\aleph_1$ it is consistent with ZFC, as we have noted, that every $\aleph_1$-separable group of cardinality $\aleph_1$ has an uncountable free summand, and hence any two have $2^{\aleph_1}$ homomorphisms between them.

Added in proof: Shelah has informed me of a recent result of his regarding the existence of tree-like ladder systems ([26, Thm 10.2]) that allows one to prove, without any assumptions on cardinal arithmetic, a somewhat weaker version of Theorem 4, namely, for any A as there and any $m \geq 2$ there exists $n \in \{m, m+1\}$ such that there is an $\aleph_n$-separable $\mathbb{Z}$-module M of cardinality $\aleph_n$ such that A is algebraically closed in $\mathrm{End}_{\mathbb{Z}}(M)$.

References

[1] J. E. Baumgartner, M. Foreman and O. Spinas, *The spectrum of the Γ-invariant of a bilinear space*, J. Algebra **189** (1997), 406–418.

[2] S. Brenner, *Some modules with nearly prescribed endomorphism rings*, J. Algebra **23** (1972), 250–262.

[3] S. Brenner and C. M. Ringel, *Pathological modules over tame rings*, J. London Math. Soc. (2) **14** (1976), 207–215.

[4] P. M. Cohn, *Some remarks on the invariant basis property*, Topology **5** (1966), 215–228.

[5] A. L. S. Corner, *Every countable reduced torsion-free ring is an endomorphism ring*, Proc. London Math. Soc. **13** (1963), 687–710.

[6] A. L. S. Corner, *On a conjecture of Pierce concerning direct decompositions of Abelian groups*, in Proceedings of the Colloquium on Abelian Groups, Tihany, Budapest (1964), 43–48.

[7] A. L. S. Corner, *On endomorphism rings of primary Abelian groups*, Quart. J. Math. Oxford **20** (1969), 277–296.

[8] A. L. S. Corner, *Additive categories and a theorem of W. G. Leavitt*, Bull. Amer. Math. Soc. **75** (1969), 78–82.

[9] A. L. S. Corner and P. Crawley, *An abelian p-group without the isomorphic refinement property*, Bull Amer. Math. Soc. **74** (1968), 743–746.

[10] A. L. S. Corner and R. Göbel, *Prescribing endomorphism algebras, a unified treatment*, Proc. London Math. Soc (3) **50** (1985), 447–479.
[11] M. Dugas and R. Göbel, *Every cotorsion-free ring is an endomorphism ring*, Proc. London Math. Soc. (3) **45** (1982), 319–336.
[12] M. Dugas and R. Göbel, *Endomorphism rings of separable torsion-free abelian groups,* Houston J. Math. **11** (1985), 471–483.
[13] P. C. Eklof, *The structure of ω_1-separable groups*, Trans. Amer. Math. Soc. **279** (1983), 497–523.
[14] P. C. Eklof, *Set theory and structure theorems*, in **Abelian Group Theory**, Lec. Notes in Math. No. 1006 (1983), Springer-Verlag, 275-284.
[15] P. C. Eklof and A. H. Mekler, **Almost Free Modules**, North-Holland (1990).
[16] P. C. Eklof and S. Shelah, The Kaplansky test problems for $\aleph_1$-separable groups, Proc. AMS **126** (1998), 1901–1907.
[17] L. Fuchs, **Infinite Abelian Groups**, vol. I, Academic Press (1970).
[18] R. Göbel, *Some combinatorial principles for solving algebraic problems*, this volume.
[19] R. Göbel and B. Goldsmith, *The Kaplansky test problems - an approach via radicals*, J. Pure and Appl. Algebra **99** (1995), 331–344.
[20] R. Göbel and S. Shelah, *Indecomposable almost free modules - the local case*, Canad. J. Math. **50** (1998), 719–738.
[21] W. G. Leavitt, *The module type of a ring*, Trans. Amer. Math. Soc **103** (1962), 113–130.
[22] I. Kaplansky, **Infinite abelian groups**, rev. ed., Univ. of Michigan Press (1969).
[23] M. Magidor and S. Shelah, *When does almost free imply free? (For groups, transversals, etc.)*, J. Amer. Math. Soc **7** (1994), 769–830
[24] A. H. Mekler, *The structure of groups that are almost the direct sum of countable abelian groups*, Trans. Amer. Math. Soc. **303** (1987), 145–160.
[25] S. Shelah, *Kaplansky test problem for R-modules*, Israel J. Math. **74** (1991), 91–127.
[26] S. Shelah, *On what I do not understand (and have something to say): part I*, preprint.
[27] B. Thomé, *$\aleph_1$-separable groups and Kaplansky's test problems*, Forum Math. **2** (1990), 203–212.
[28] R. B. Warfield, Jr., *Countably generated modules over commutative Artinian rings*, Pacific J. Math. **60** (1975), 289–302.

MATHEMATICS DEPARTMENT, IRVINE, CA 92697-3875, USA
E-mail address: `peklof@math.uci.edu`

Trends in Mathematics, © 2000 Birkhäuser Verlag Basel/Switzerland

FAILURE OF THE KRULL-SCHMIDT THEOREM FOR ARTINIAN MODULES AND SERIAL MODULES

ALBERTO FACCHINI

The purpose in writing this note has been three-fold. First, we wanted to present the solution of a problem posed by Wolfgang Krull in 1932 [K]. Krull asked whether what is now called the "Krull-Schmidt Theorem" holds for artinian modules. A negative answer was published only in 1995 by Herbera, Levy, Vámos and the author [FHLV]. Second, we wanted to present the answer to a question posed by Warfield in 1975 [W2], namely, whether the Krull-Schmidt Theorem holds for serial modules. The author published a negative answer in 1996 [F1]. The solution to Warfield's problem shows an interesting behavior. Briefly, the Krull-Schmidt Theorem holds for some classes of modules and not for others. When it does hold, any indecomposable decomposition is uniquely determined up to a permutation. For serial modules the Krull-Schmidt Theorem does not hold, but any indecomposable decomposition is uniquely determined up to *two* permutations. Third, we wanted to present the structure of the semigroup $\mathcal{S}_\oplus(\mathcal{P}\text{-Mod}\,R)$ of isomorphism classes of finitely generated projective modules over a semilocal ring R [FH1]. Both artinian modules and serial modules of finite Goldie dimension have semilocal endomorphism ring.

Complete proofs can be found in part in the monograph [F2] and in part in the paper [FH1]. We shall consider unital right modules over an associative ring R with identity $1_R \neq 0_R$. If k is a commutative ring, a *module-finite k-algebra* is a k-algebra R which is finitely generated as a k-module. For any ring R we denote the Jacobson radical of R by $J(R)$.

1. Krull-Schmidt fails for artinian modules

We all know the classical Krull-Schmidt Theorem, which was published by Krull [K] in 1932:

Theorem 1.1. (The Krull-Schmidt Theorem) *Let M be an R-module of finite length and $M = M_1 \oplus M_2 \oplus \cdots \oplus M_n = N_1 \oplus N_2 \oplus \cdots \oplus N_t$ be two direct sum decompositions of M into indecomposable direct summands M_i and N_j. Then $n = t$ and there is a permutation σ of $\{1, 2, \ldots, n\}$ such that $M_i \cong N_{\sigma(i)}$ for every $i = 1, 2, \ldots, n$.*

Partially supported by Consiglio Nazionale delle Ricerche and Ministero dell'Università e della Ricerca Scientifica e Tecnologica (Italy).

In the same paper in which Krull published this result, he asked a question, that is, he posed a problem: he asked whether Theorem 1.1 holds more generally for any artinian right module. Note that every artinian right module decomposes as a finite direct sum of indecomposable modules. Warfield showed in 1969 [W1, Proposition 5] that the answer is "yes" when the ring R is either right noetherian or commutative. The negative answer to Krull's question for an arbitrary noncommutative ring R appeared in [FHLV] in 1995. In that paper we constructed the following examples:

Example 1.2. (Non-uniqueness of the number of indecomposable summands, [FHLV, Example 1.6]) *Let $n \geq 2$ be an integer. There exists an artinian module M that has a direct sum decomposition $M = M_{1,i} \oplus M_{2,i} \oplus \cdots \oplus M_{i,i}$ into i indecomposable direct summands $M_{1,i}, M_{2,i}, \ldots, M_{i,i}$ for every $i = 2, 3, \ldots, n$.*

Example 1.3. (Simple failure of Krull-Schmidt for artinian modules, [FHLV, Example 1.7]) *There exist four indecomposable, pairwise non-isomorphic, artinian modules M_1, M_2, M_3, M_4 such that $M_1 \oplus M_2 \cong M_3 \oplus M_4$.*

Sketch of the construction: Let k be any commutative semilocal noetherian ring and let S be any module-finite k-algebra. It is possible to prove that any such module-finite k-algebra S can be realized as the endomorphism ring of a suitable cyclic artinian module M_R over a suitable ring R. There exist module-finite k-algebras S over commutative semilocal noetherian rings k in which 1_S has pathological decompositions (for instance, in which $1_S = e_{1,i} + e_{2,i} + \cdots + e_{i,i}$ can be written as the sum of i orthogonal primitive idempotents $e_{1,i}, e_{2,i}, \ldots, e_{i,i}$ for every $i = 2, 3, \ldots, n$). Realizing these k-algebras S as endomorphism rings of artinian modules M_R, we see that there exist artinian modules M_R with the required pathological decompositions into indecomposable direct summands (for instance, with a direct sum decomposition $M = M_{1,i} \oplus M_{2,i} \oplus \cdots \oplus M_{i,i}$ into i indecomposable direct summands $M_{1,i}, M_{2,i}, \ldots, M_{i,i}$ for every $i = 2, 3, \ldots, n$).

In this context there are some further nice results and examples due to Yakovlev and Pimenov [Y, PY].

2. Krull-Schmidt fails for serial modules

Problems similar to the question asked by Krull were posed for a number of classes of modules. Recall that an R-module M is *uniserial* if its lattice of submodules $\mathcal{L}(M)$ is linearly ordered under inclusion, that is, if for any submodules A and B of M either $A \subseteq B$ or $B \subseteq A$. A module is *serial* if it is a direct sum of uniserial submodules. Hence a serial module of finite Goldie dimension is a direct sum of finitely many uniserial modules. A ring R is *serial* if both R_R and ${}_RR$ are serial modules.

For example, every right vector space over a division ring is a serial module. Every finite abelian group is a serial module over the ring $\mathbf{Z}$ of integers. As simple modules are uniserial, semisimple modules are serial modules. In particular, semisimple artinian rings are serial rings. Another example of serial rings is given by rings of $n \times n$ upper triangular matrices with entries in a field. Direct products of two serial rings and homomorphic images of serial rings are serial rings.

Warfield [W2] published in 1975 a beautiful paper in which he described the structure of serial rings and proved that

Theorem 2.1. (Warfield [W2]) *Every finitely presented module M over a serial ring is a finite direct sum of uniserial modules, i.e., is serial.*

In that paper Warfield asked whether such a finite decomposition of a finitely presented module M into uniserial summands is essentially unique, i.e., he asked whether the Krull-Schmidt Theorem holds for finitely presented modules over serial rings. In order to present the negative answer to Warfield's question, which was given by the author in [F1], we need some further definitions.

If A and B are two modules, we say that A and B *belong to the same monogeny class*, and write $[A]_m = [B]_m$, if there exist a monomorphism $A \to B$ and a monomorphism $B \to A$. Similarly, we say that A and B *belong to the same epigeny class*, and write $[A]_e = [B]_e$, if there exist an epimorphism $A \to B$ and an epimorphism $B \to A$. Clearly, belonging to the same monogeny class and belonging to the same epigeny class are two equivalence relations. The reason why monogeny classes and epigeny classes appear in the study of uniserial modules is given by the following result.

Proposition 2.2. [F1, Proposition 1.6] *Let A and B be two uniserial modules over a ring R. Then $A \cong B$ if and only if $[A]_m = [B]_m$ and $[A]_e = [B]_e$.*

Also in this case the technical starting point is the study of the endomorphism ring of uniserial modules:

Theorem 2.3. [F1] *Let A_R be a uniserial module over an arbitrary ring R and let $E = \mathrm{End}(A_R)$ be its endomorphism ring. Let I be the subset of E whose elements are all the endomorphisms of A_R that are not injective, and let K be the subset of E whose elements are all the endomorphisms of A_R that are not surjective. Then I and K are two two-sided completely prime ideals of E, and every proper right ideal of E and every proper left ideal of E is contained either in I or in K. Moreover, exactly one of the following two conditions hold:*

(a) *Either the ideals I and K are comparable, so that E is a local ring and $I \cup K$ is its maximal ideal, or*
(b) *I and K are not comparable, $J(E) = I \cap K$, and $E/J(E)$ is canonically isomorphic to the direct product of the two division rings E/I and E/K.*

Our main result is the following weak form of the Krull-Schmidt Theorem, which holds for uniserial modules.

Theorem 2.4. (Weak Krull-Schmidt Theorem for uniserial modules, [F1, Theorem 1.9]) *Let $A_1, \dots, A_n, B_1, \dots, B_t$ be non-zero uniserial modules over an arbitrary ring. Then the direct sums $A_1 \oplus \cdots \oplus A_n$ and $B_1 \oplus \cdots \oplus B_t$ are isomorphic if and only if $n = t$ and there are two permutations σ, τ of $\{1, 2, \dots, n\}$ such that $[A_i]_m = [B_{\sigma(i)}]_m$ and $[A_i]_e = [B_{\tau(i)}]_e$ for every $i = 1, 2, \dots, n$.*

The negative answer to Warfield's question is given by the following example.

Example 2.5. [F1, Example 2.1] *If $n \geq 2$ is an integer, then there exist n^2 pairwise non-isomorphic finitely presented uniserial modules $U_{i,j}$ $(i, j = 1, 2, \ldots, n)$ over a suitable serial ring R satisfying the following properties:*

(a) *for every $i, j, k, \ell = 1, 2, \ldots, n$, $[U_{i,j}]_m = [U_{k,\ell}]_m$ if and only if $i = k$;*

(b) *for every $i, j, k, \ell = 1, 2, \ldots, n$, $[U_{i,j}]_e = [U_{k,\ell}]_e$ if and only if $j = \ell$.*

(Note that if we put every uniserial module $U_{i,j}$ as (i, j)-entry in a square $n \times n$ matrix, then the $U_{i,j}$'s on the same row are in the same monogeny class, and the $U_{i,j}$'s on the same column are in the same epigeny class.)

Hence, by Theorem 2.4,

$$U_{1,1} \oplus U_{2,2} \oplus \cdots \oplus U_{n,n} \cong U_{\sigma(1),\tau(1)} \oplus U_{\sigma(2),\tau(2)} \oplus \cdots \oplus U_{\sigma(n),\tau(n)}$$

for every pair of permutations σ, τ of $\{1, 2, \ldots, n\}$. ∎

Some remarks on Theorem 2.4:

• In some sense, Theorem 2.4 is the best possible theorem. Given *any two permutations* σ, τ of $\{1, 2, \ldots, n\}$, there is always a serial module with two direct sum decompositions such that the direct summands corresponding with respect to the first permutation σ are in the same monogeny class, and the direct summands corresponding with respect to the second permutation τ are in the same epigeny class. This is shown by Example 2.5.

• We do not get a better theorem even if we restrict our attention to the class of finitely generated uniserial modules, or to the class of finitely presented uniserial modules, or to the class of finitely presented uniserial modules over a serial ring, which is the case Warfield was interested in. This also is shown by Example 2.5.

• There are some partial generalizations of Theorem 2.4 to the case of infinite direct sums of uniserial modules due to Nguyen Viet Dung and the author [DF1].

• Also there is a generalization of Theorem 2.4 to the case of *biuniform* modules, that is, modules with Goldie dimension and dual Goldie dimension both equal to one [F2, Chapter 9]. We shall come back to this class of modules later.

Open Problem. We do not know whether every direct summand of a serial module is serial. We do not know either whether every direct summand of a serial module of finite Goldie dimension is serial. For some partial results see [DF2].

3. Semilocal rings and modules with semilocal endomorphism rings

3.1. **Semilocal rings.** Let us turn now to consider one of the main properties that artinian modules and finite direct sums of uniserial modules have in common. Both artinian modules and finite direct sums of uniserial modules have semilocal endomorphism rings. A ring R is *semilocal* if $R/J(R)$ is semisimple artinian. Equivalently, a ring R is semilocal if and only if $R/J(R)$ is a right artinian ring, if and only if $R/J(R)$ is a left artinian ring.

Think of being semilocal as a finiteness condition on the ring R. For instance, a semilocal ring has only finitely many simple modules up to isomorphism. Also, every set of orthogonal idempotents in a semilocal ring is finite. Here are some examples of semilocal rings:

(1) A commutative ring is semilocal if and only if it has finitely many maximal ideals.

(2) Every right (or left) artinian ring is semilocal.

(3) Every local ring is semilocal.

(4) If R is a semilocal ring, the ring $\mathbf{M}_n(R)$ of $n \times n$ matrices with entries in R is semilocal.

(5) The direct product $R_1 \times R_2$ of two semilocal rings R_1 and R_2 is semilocal.

(6) Every homomorphic image of a semilocal ring is semilocal.

(7) Every module-finite algebra over a commutative semilocal ring is semilocal.

(8) If R is a semilocal ring and e is a non-zero idempotent of R, then eRe is a semilocal ring.

3.2. **Dual Goldie dimension.** The notion of Goldie dimension is a notion that concerns modular lattices with 0 and 1 [GP]. Let $(L, \vee, \wedge)$ be a modular lattice with a smallest element 0 and a greatest element 1. A finite subset $\{\, a_i \mid i \in I \,\}$ of $L \setminus \{0\}$ is said to be *join-independent* if $a_i \wedge (\bigvee_{j \neq i} a_j) = 0$ for every $i \in I$. The *Goldie dimension* $\dim(L)$ of the lattice L is the least upper bound of the set of the cardinalities of the join-independent finite subsets of $L \setminus \{0\}$. Thus either $\dim(L)$ is a non-negative integer (the greatest element in the set of the cardinalities of the join-independent finite subsets of $L \setminus \{0\}$, when this set is bounded) or $\dim(L) = \infty$ (when the set of the cardinalities of the join-independent finite subsets of $L \setminus \{0\}$ is unbounded). The *Goldie dimension* $\dim(A_R)$ of a module A_R is the Goldie dimension of its lattice of submodules $\mathcal{L}(A_R)$. Now the dual (=opposite) L^{op} of any modular lattice L is a modular lattice. Therefore it is always possible to define the *dual Goldie dimension* $\operatorname{codim}(A_R)$ of a module A_R as the Goldie dimension of the modular lattice $\mathcal{L}(A_R)^{\mathrm{op}}$. Thus

$$\operatorname{codim}(A_R) = \dim(\mathcal{L}(A_R)^{\mathrm{op}}).$$

The next result shows exactly in which sense being semilocal is a finiteness condition on a ring. For a proof see [SV, Corollary 1.14] or [F2, Proposition 2.43].

Proposition 3.1. *A ring R is semilocal if and only if R_R has finite dual Goldie dimension, if and only if ${}_RR$ has finite dual Goldie dimension. Moreover, if these equivalent conditions hold, then* $\operatorname{codim}(R_R) = \operatorname{codim}({}_RR) =$ *"Goldie dimension of the semisimple R-module $R/J(R)$".*

3.3. **Modules with semilocal endomorphism rings.** Next consider modules with semilocal endomorphism rings. Having semilocal endomorphism ring is also a finiteness condition on a module. For instance, if a module A_R is a direct sum of infinitely many non-zero submodules, then its endomorphism ring cannot be semilocal. In the next three results we collect some properties of modules with semilocal endomorphism ring. The first theorem follows from results of Bass and Evans. It says that modules with semilocal endomorphism ring cancel from direct sums. For a proof see [L, Theorem 20.11] or [F2, Corollary 4.6].

Theorem 3.2. *Let A_R, B_R, C_R be modules over an arbitrary ring R. Suppose that* $\mathrm{End}(A_R)$ *is a semilocal ring. Then $A_R \oplus B_R \cong A_R \oplus C_R$ implies $B_R \cong C_R$.*

The following two propositions were proved in [FHLV, Proposition 2.1] (cf [F2, Propositions 4.8 and 4.9]).

Proposition 3.3. (n-th root property) *Let R be a ring and let A_R, B_R be R-modules. If A_R has a semilocal endomorphism ring, n is a positive integer and $A^n \cong B^n$, then $A \cong B$.*

For the next proposition recall that every semilocal ring S has finite dual Goldie dimension $\mathrm{codim}(S)$ (Proposition 3.1).

Proposition 3.4. *Let A_R be a module with semilocal endomorphism ring. Set* $n = \mathrm{codim}(\mathrm{End}(A_R))$. *Then A_R has at most 2^n isomorphism classes of direct summands.*

If a module has only finitely many isomorphism classes of direct summands, then it has also only finitely many direct sum decompositions up to isomorphism. Therefore, not only are all direct sum decompositions of a module with semilocal endomorphism ring finite, but also there are just a finite number of such decompositions.

Here are some examples of classes of modules with semilocal endomorphism ring:

(1) artinian modules [CD];
(2) noetherian right modules of finite dual Goldie dimension;
(3) noetherian right modules over a semilocal ring;
(4) serial modules of finite Goldie dimension;
(5) biuniform modules, that is, modules A_R with

$$\dim(A_R) = \mathrm{codim}(A_R) = 1;$$

(6) modules A_R with finite Goldie dimension and such that every injective endomorphism of A_R is bijective;
(7) modules A_R with finite dual Goldie dimension and such that every surjective endomorphism of A_R is bijective;
(8) modules with finite Goldie dimension and finite dual Goldie dimension.

Apart from the first example (artinian modules), most of the other examples were discovered by Herbera and Shamsuddin [HS]. In that paper further examples of modules with semilocal endomorphism ring can be found.

3.4. **From modules with semilocal endomorphism ring to finitely generated projective modules over a semilocal ring.** As we have already said, both artinian modules and serial modules of finite Goldie dimension have semilocal endomorphism ring, and the Krull-Schmidt Theorem fails for these modules. The problem we shall consider now is the following: *Study the direct sum decompositions of a module A_R with semilocal endomorphism ring. In particular, when does the Krull-Schmidt Theorem hold for A_R?*

Let R be any ring and $\mathcal{P}$-Mod R be the full subcategory of Mod-R whose objects are the finitely generated projective right R-modules. For any right R-module A_R, let $\operatorname{add}(A_R)$ denote the full subcategory of Mod-R whose objects are the direct summands of finite direct sums A_R^n of copies of A_R. Let $E = \operatorname{End}(A_R)$. Then the categories $\operatorname{add}(A_R)$ and $\mathcal{P}$-Mod E are equivalent. The equivalence is given by the functors $\operatorname{Hom}_R(A_R, -)\colon \operatorname{add}(A_R) \to \mathcal{P}\text{-Mod}\, E$ and $- \otimes_E A\colon \mathcal{P}\text{-Mod}\, E \to \operatorname{add}(A_R)$. Instead of considering the direct sum decompositions of a module A_R with semilocal endomorphism ring, we shall consider, more generally, the direct sum decompositions in the category $\operatorname{add}(A_R)$. Via the category equivalence $\operatorname{add}(A_R) \cong \mathcal{P}\text{-Mod}\, E$, our problem becomes: *Study the direct sum decompositions of finitely generated projective modules over a semilocal ring. In particular, when does the Krull-Schmidt Theorem hold for finitely generated projective modules over a semilocal ring?*

4. The semigroup $\mathcal{S}_\oplus(\mathcal{P}\text{-Mod}\, R)$ for a semilocal ring R

We need some easy concepts from semigroup theory. All semigroups we consider in this note are commutative additive semigroups with zero. If M is a semigroup and x, y are elements in M, define $x \le y$ if there exists $z \in M$ such that $x+z = y$. The relation $\le$ is reflexive and transitive, i.e., it is a pre-order. Moreover, $\le$ is invariant under translation (that is, for any $t \in M$, $x \le y$ implies $x+t \le y+t$). The pre-order $\le$ is called the *algebraic pre-order* of M. An *order-unit* u of a semigroup M is an element $u \in M$ such that for any $x \in M$ there exists $n \in \mathbf{N}$ with $x \le nu$. If M is a semigroup and u is an order-unit of M, we shall say that the pair (M, u) is a semigroup with order-unit. Semigroups with order-unit are the objects of a category **Swou**, in which the morphisms $f\colon (M, u) \to (M', u')$ are the semigroup homomorphisms such that $f(0_M) = 0_{M'}$ and $f(u) = u'$.

For example, consider the semigroup $\mathbf{N}^n$, where the operation on the set $\mathbf{N}^n$ of all n-tuples of non-negative integers is defined by $(a_1, \dots, a_n) + (b_1, \dots, b_n) = (a_1 + b_1, \dots, a_n + b_n)$ for every $(a_1, \dots, a_n), (b_1, \dots, b_n) \in \mathbf{N}^n$. The algebraic pre-order on $\mathbf{N}^n$ is the componentwise order defined by $(a_1, \dots, a_n) \le (b_1, \dots, b_n)$ if and only if $a_i \le b_i$ for every $i = 1, \dots, n$. An element $(u_1, \dots, u_n)$ of $\mathbf{N}^n$ is an order-unit in $\mathbf{N}^n$ if and only if $u_i \ne 0$ for every $i = 1, \dots, n$.

As a second example, let R be any ring. As in the previous section, let $\mathcal{P}$-Mod R denote the full subcategory of Mod-R whose objects are all finitely generated projective right R-modules, and for any R-module $A_R \in \mathcal{P}\text{-Mod}\, R$ let $\langle A_R \rangle$ denote the isomorphism class of A_R. The set

$$\mathcal{S}_\oplus(\mathcal{P}\text{-Mod}\, R) = \{\, \langle A_R \rangle \mid A_R \in \mathcal{P}\text{-Mod}\, R \,\}$$

is an additive semigroup with respect to the operation defined by

$$\langle A_R\rangle + \langle B_R\rangle = \langle A_R \oplus B_R\rangle$$

for every $A_R, B_R \in \mathcal{P}\text{-Mod}\,R$. In this semigroup, $\langle R_R\rangle$ is an order-unit. If R, S are rings, a ring homomorphism $\varphi\colon R \to S$ induces a semigroup homomorphism $\mathcal{S}_\oplus(\varphi)\colon \mathcal{S}_\oplus(\mathcal{P}\text{-Mod}\,R) \to \mathcal{S}_\oplus(\mathcal{P}\text{-Mod}\,S)$ defined by $\mathcal{S}_\oplus(\varphi)(\langle A_R\rangle) = \langle A\otimes_R S\rangle$ for every $A_R \in \mathcal{P}\text{-Mod}\,R$. This semigroup homomorphism maps the order-unit $\langle R_R\rangle$ of $\mathcal{S}_\oplus(\mathcal{P}\text{-Mod}\,R)$ to the order unit $\langle S_S\rangle$ of $\mathcal{S}_\oplus(\mathcal{P}\text{-Mod}\,S)$. Thus there is a functor $\mathcal{S}_\oplus$ from the category of associative rings with identity to the category **Swou**.

For a semilocal ring R there are only finitely many finitely generated indecomposable projective R-modules up to isomorphism [FS, Theorem 9], and every finitely generated projective R-module is a direct sum of indecomposables (because every finitely generated projective R-module $\neq 0$ has finite dual Goldie dimension $\neq 0$). Therefore for a semilocal ring R, the Krull-Schmidt Theorem holds for finitely generated projective right R-modules if and only if there exist elements $a_1, \dots, a_t \in \mathcal{S}_\oplus(\mathcal{P}\text{-Mod}\,R)$ such that every element of $\mathcal{S}_\oplus(\mathcal{P}\text{-Mod}\,R)$ can be written in a unique way as a linear combination of $a_1, \dots, a_t$ with non-negative integral coefficients, that is, if and only if $\mathcal{S}_\oplus(\mathcal{P}\text{-Mod}\,R) \cong \mathbf{N}^t$ for some $t \geq 1$. In order to describe the structure of $\mathcal{S}_\oplus(\mathcal{P}\text{-Mod}\,R)$ for an arbitrary semilocal ring R we need a further easy notion of semigroup theory: the notion of full affine semigroup, a type of semigroup studied for instance by Hochster in [H].

Lemma 4.1. *Let M be a subsemigroup of $\mathbf{N}^n$. The following conditions are equivalent:*

(a) *There exists a subgroup G of $\mathbf{Z}^n$ such that $M = G \cap \mathbf{N}^n$.*
(b) *If $\langle M\rangle$ is the subgroup of $\mathbf{Z}^n$ generated by M, then $\langle M\rangle \cap \mathbf{N}^n \subseteq M$.*
(c) *If $a, b \in M$ and $a \leq b$ in $\mathbf{N}^n$ with the componentwise order, then there exists $c \in M$ such that $a + c = b$.*
(d) *If $a, b \in M$, the equation $a + x = b$ has a solution in M if and only if it has a solution in $\mathbf{N}^n$.*

A subsemigroup M of $\mathbf{N}^n$ is said to be a *full affine subsemigroup of* $\mathbf{N}^n$ if the equivalent conditions of Lemma 4.1 are satisfied. A *full affine semigroup* is any semigroup M that is isomorphic to a full affine subsemigroup of $\mathbf{N}^n$ for some n.

For instance, let R be any semilocal ring. If we apply the functor $\mathcal{S}_\oplus$ to the canonical projection $\pi\colon R \to R/J(R)$ of R onto the semisimple artinian ring $R/J(R)$, we obtain a semigroup homomorphism $\mathcal{S}_\oplus(\pi)\colon \mathcal{S}_\oplus(\mathcal{P}\text{-Mod}\,R) \to \mathcal{S}_\oplus(\mathcal{P}\text{-Mod}\,R/J(R))$ defined by $\langle A_R\rangle \mapsto \langle A_R/A_RJ(R)\rangle$ for every $A_R \in \mathcal{P}\text{-Mod}\,R$. This mapping $\mathcal{S}_\oplus(\pi)$ is an embedding of semigroups with order-unit. Now the ring $R/J(R)$ is semisimple artinian, so that every $R/J(R)$-module is semisimple. If n is the number of isomorphism classes of simple $R/J(R)$-modules, then $\mathcal{S}_\oplus(\mathcal{P}\text{-Mod}\,R/J(R))$ is canonically isomorphic to $\mathbf{N}^n$ via an isomorphism

$$h\colon \mathcal{S}_\oplus(\mathcal{P}\text{-Mod}\,R/J(R)) \longrightarrow \mathbf{N}^n$$

that maps the free canonical set of generators $\{\langle N_R\rangle \mid N_R$ a simple R-module$\}$ of $\mathcal{S}_\oplus(\mathcal{P}\text{-Mod}\,R/J(R))$ to the free canonical set of generators $\{e_1,\dots,e_n\}$ of $\mathbf{N}^n$. The composite mapping

$$h\circ\mathcal{S}_\oplus(\pi)\colon \mathcal{S}_\oplus(\mathcal{P}\text{-Mod}\,R)\longrightarrow \mathbf{N}^n$$

gives an isomorphism between $\mathcal{S}_\oplus(\mathcal{P}\text{-Mod}\,R)$ and the full affine subsemigroup $h(\mathcal{S}_\oplus(\mathcal{P}\text{-Mod}\,R))$ of $\mathbf{N}^n$ (it is a full affine subsemigroup of $\mathbf{N}^n$ because it is the intersection of $\mathbf{N}^n$ with the image of the abelian group $K_0(R)$ in $K_0(R/J(R))\cong\mathbf{Z}^n$). Therefore $\mathcal{S}_\oplus(\mathcal{P}\text{-Mod}\,R)$ is a full affine semigroup for every semilocal ring R.

Theorem 4.2. [FH1] *Let k be a field and let M be a full affine semigroup, so that for some n there exists an injective semigroup homomorphism $T\colon M\to\mathbf{N}^n$ that is an isomorphism between M and the full affine subsemigroup $T(M)$ of $\mathbf{N}^n$. Assume that $u\in M$ is such that $T(u)$ is an order-unit in $\mathbf{N}^n$, i.e., $T(u)=(d_1,\dots,d_n)$ with $d_1,\dots,d_n\neq 0$. Then there exists a semilocal right and left hereditary k-algebra R such that the diagram of commutative semigroups and semigroup homomorphisms*

$$\begin{array}{ccc}\mathcal{S}_\oplus(\mathcal{P}\text{-Mod}\,R) & \overset{\mathcal{S}_\oplus(\pi)}{\longrightarrow} & \mathcal{S}_\oplus(\mathcal{P}\text{-Mod}\,R/J(R))\\ \downarrow & & \downarrow h\\ M & \overset{T}{\longrightarrow} & \mathbf{N}^n\end{array}$$

commutes, where $\pi\colon R\to R/J(R)$ denotes the canonical projection, the vertical arrows are isomorphisms, the vertical arrow on the left sends $\langle R_R\rangle$ to u, and h maps the free canonical set of generators $\{\langle N_R\rangle \mid N_R$ a simple R-module$\}$ of $\mathcal{S}_\oplus(\mathcal{P}\text{-Mod}\,R/J(R))$ to the free canonical set of generators $\{e_1,\dots,e_n\}$ of $\mathbf{N}^n$.

Note that the commutative diagram in the statement of Theorem 4.2 is a commutative diagram in the category **Swou**, because its arrows are homomorphisms of semigroups with order-unit between the semigroups with order-unit $(\mathcal{S}_\oplus(\mathcal{P}\text{-Mod}\,R),\langle R_R\rangle)$, $(\mathcal{S}_\oplus(\mathcal{P}\text{-Mod}\,R/J(R)),\langle R/J(R)\rangle)$, (M,u) and $(\mathbf{N}^n, T(u))$.

Corollary 4.3. *The following conditions are equivalent for a semigroup M:*

(a) *M is a full affine semigroup;*
(b) *$M\cong\mathcal{S}_\oplus(\mathcal{P}\text{-Mod}\,R)$ for some semilocal ring R;*
(c) *$M\cong\mathcal{S}_\oplus(\mathcal{P}\text{-Mod}\,R)$ for some semilocal right and left hereditary ring R.*

Let us go back to the Krull-Schmidt Theorem for finitely generated projective modules over a semilocal ring. Let R be a semilocal ring. We have already remarked that the Krull-Schmidt Theorem holds for finitely generated projective right R-modules if and only if there exist $a_1,\dots,a_t\in\mathcal{S}_\oplus(\mathcal{P}\text{-Mod}\,R)$ such that every element of $\mathcal{S}_\oplus(\mathcal{P}\text{-Mod}\,R)$ can be written in a unique way as a linear combination of $a_1,\dots,a_t$ with non-negative integral coefficients, that is, if and only if $\mathcal{S}_\oplus(\mathcal{P}\text{-Mod}\,R)\cong\mathbf{N}^t$ for some $t\geq 1$. Suppose that these equivalent conditions hold. For every $i=1,\dots,t$ let P_i be a finitely generated projective right R-module with $\langle P_i\rangle=a_i$, so that $\{P_1,\dots,P_t\}$ is a complete set of representatives of the isomorphism classes of indecomposable finitely generated projective right R-modules. If we fix an index $i=1,\dots,t$, the module P_i is a direct summand of a free module R^k for some $k \geq 1$, so that $P_i \oplus Q \cong R^k$ for some projective module Q. Thus $Q \cong P_1^{m_1} \oplus \cdots \oplus P_t^{m_t}$ and $R_R \cong P_1^{n_1} \oplus \cdots \oplus P_t^{n_t}$ for suitable $m_1, \ldots, m_t, n_1, \ldots, n_t \in \mathbf{N}$. So $P_i \oplus Q \cong R^k$ implies that $m_j = kn_j$ for $j \neq i$, and $1 + m_i = kn_i$. In particular, $n_i \geq 1$. This shows that every P_i is isomorphic to a direct summand of R_R, so that every indecomposable finitely generated projective right R-module is isomorphic to eR for some idempotent $e \in R$. Therefore:

Theorem 4.4. *The following conditions are equivalent for a semilocal ring R:*

(a) *The Krull-Schmidt Theorem holds for finitely generated projective R-modules, that is, if P is a finitely generated projective R-module and $P = Q_1 \oplus \cdots \oplus Q_m = Q_1' \oplus \cdots \oplus Q_{m'}'$ are two direct sum decompositions of P into indecomposable direct summands Q_i and Q_j', then $m = m'$ and there is a permutation σ of $\{1, 2, \ldots, n\}$ such that $Q_i \cong Q_{\sigma(i)}'$ for every $i = 1, 2, \ldots, m$.*

(b) $\mathcal{S}_\oplus(\mathcal{P}\text{-Mod}\, R) \cong \mathbf{N}^t$ *for some $t \geq 1$.*

(c) *There exists a direct sum decomposition $R_R = I_1 \oplus \cdots \oplus I_m$ such that if $\{I_1, \ldots, I_t\}$ is a complete set of representatives of the isomorphism classes of $I_1, \ldots, I_m$, then every finitely generated projective right R-module is isomorphic to a direct sum of copies of $I_1, \ldots, I_t$ in a unique way.*

For instance, every semiperfect ring satisfies condition (c) of Theorem 4.4 [F2, Theorem 3.10]. But there are semilocal rings that satisfy the equivalent conditions of Theorem 4.4 and are not semiperfect. For example, let R be a commutative semilocal integral domain that is not local. Then every finitely generated projective R-module is free [B, §5, n. 3], so that R satisfies the equivalent conditions of Theorem 4.4, but R is not semiperfect.

Finally, note that the condition "there exists a direct sum decomposition $R_R = I_1 \oplus \cdots \oplus I_m$ such that every finitely generated projective right R-module is isomorphic to a direct sum of the I_j's" is strictly weaker than condition (c). For instance, the subsemigroup M of $\mathbf{N}^2$ generated by $A = \{(2,0), (1,1), (0,2)\}$ is a full affine subsemigroup of $\mathbf{N}^2$, and A is the least set of generators of M [FH2, Example 2.3]. By Theorem 4.2 there exists a semilocal right and left hereditary ring R such that $\mathcal{S}_\oplus(\mathcal{P}\text{-Mod}\, R) \cong M$ via an isomorphism that sends $\langle R_R \rangle$ to $(3,3)$. Since $(3,3) = (2,0) + (1,1) + (0,2)$, there is a direct sum decomposition $R_R = I_1 \oplus I_2 \oplus I_3$, where I_1, I_2, I_3 are indecomposable finitely generated projective right ideals corresponding to the elements $(2,0), (1,1), (0,2)$ of M respectively. The set $\{I_1, I_2, I_3\}$ is a complete set of representatives of the indecomposable finitely generated projective right R-modules up to isomorphism (so that every finitely generated projective right R-module is isomorphic to a direct sum of the I_j's), but Krull-Schmidt fails because $I_1 \oplus I_3 \cong I_2 \oplus I_2$.

References

[B] N. Bourbaki, "Algèbre commutative", Chapitre 2, Hermann, Paris, 1961

[CD] R. Camps and W. Dicks, *On semilocal rings*, Israel J. Math. **81** (1993), 203–211.

[DF1] N. V. Dung and A. Facchini, *Weak Krull-Schmidt for infinite direct sums of uniserial modules,* J. Algebra **193** (1997), 102–121.

[DF2] N. V. Dung and A. Facchini, *Direct summands of serial modules*, J. Pure Appl. Algebra **133** (1998), 93–106.

[F1] A. Facchini, *Krull-Schmidt fails for serial modules*, Trans. Amer. Math. Soc. **348** (1996), 4561–4575.

[F2] A. Facchini, "Module theory. Endomorphism rings and direct sum decompositions in some classes of modules", Progress in Math. 167, Birkhäuser Verlag, Basel, 1998.

[FH1] A. Facchini and D. Herbera, *K_0 of a semilocal ring*, preprint, 1998.

[FH2] A. Facchini and D. Herbera, *Projective modules over semilocal rings*, preprint, 1998.

[FHLV] A. Facchini, D. Herbera, L. S. Levy and P. Vámos, *Krull-Schmidt fails for artinian modules*, Proc. Amer. Math. Soc. **123** (1995), 3587–3592.

[FS] K. R. Fuller and W. A. Shutters, *Projective modules over non-commutative semilocal rings*, Tôhoku Math. J. **27** (1975), 303–311.

[GP] P. Grzeszczuk and E. R. Puczyłowski, *On Goldie and dual Goldie dimensions*, J. Pure Appl. Algebra **31** (1984), 47–54.

[HS] D. Herbera and A. Shamsuddin, *Modules with semi-local endomorphism ring*, Proc. Amer. Math. Soc. **123** (1995), 3593–3600.

[H] M. Hochster, *Rings of invariants of tori, Cohen-Macaulay rings generated by monomials, and polytopes,* Ann. of Math. **96** (1972), 318–337.

[K] W. Krull, *Matrizen, Moduln und verallgemeinerte Abelsche Gruppen im Bereich der ganzen algebraischen Zahlen,* Heidelberger Akademie der Wissenschaften **2** (1932), 13–38.

[L] T. Y. Lam, "A First Course in Noncommutative Rings", Graduate Texts in Math. 131, Springer-Verlag, New York, 1991.

[PY] K. I. Pimenov and A. V. Yakovlev, *Artinian modules over a matrix ring*, in this volume.

[SV] B. Sarath and K. Varadarajan, *Dual Goldie Dimension - II*, Comm. Algebra **7** (1979), 1885–1899.

[W1] R. B. Warfield, Jr., *A Krull-Schmidt theorem for infinite sums of modules*, Proc. Amer. Math. Soc. **22** (1969), 460–465.

[W2] R. B. Warfield, Jr., *Serial rings and finitely presented modules*, J. Algebra **37** (1975), 187–222.

[Y] A. V. Yakovlev, *On the direct decomposition of artinian modules*, preprint, 1997.

Dipartimento di Matematica e Informatica, Università di Udine, 33100 Udine, Italy

Trends in Mathematics, © 2000 Birkhäuser Verlag Basel/Switzerland

ARTINIAN MODULES OVER A MATRIX RING

K.I. PIMENOV, A.V. YAKOVLEV

Let $\mathbb{Z}$, $\mathbb{Q}$ be the ring of integers and the field of rational numbers. Denote by Λ the matrix ring

$$\begin{pmatrix} \mathbb{Q} & 0 \\ \mathbb{Q} & \mathbb{Z} \end{pmatrix} = \left\{ \begin{pmatrix} \alpha & 0 \\ \beta & m \end{pmatrix} \middle| \alpha, \beta \in \mathbb{Q},\ m \in \mathbb{Z} \right\}.$$

The ring Λ is rather famous: it is the most known example of ring which is Noetherian on the right but is not Noetherian on the left. We shall show in this paper that it is not the only "anomaly" of the ring Λ. Namely, we prove that the Krull-Schmidt theorem fails for Artinian modules over Λ and that there exist cyclic Artinian Λ-modules which are not Noetherian. It was known that both phenomena are impossible over commutative or Noetherian rings and the fact that they can be realized for modules over such a small ring as the ring Λ is rather surprising.

1. Notations and definitions

All Λ-modules will be left Λ-modules. Set

$$e_1 = \begin{pmatrix} 1 & 0 \\ 0 & 0 \end{pmatrix}, \quad e_2 = \begin{pmatrix} 0 & 0 \\ 0 & 1 \end{pmatrix}, \quad u = \begin{pmatrix} 0 & 0 \\ 1 & 0 \end{pmatrix}.$$

Let X be an abelian group; make it into a Λ-module by the following rule:

$$\begin{pmatrix} \alpha & 0 \\ \beta & m \end{pmatrix} x = mx \quad \text{for any } \alpha, \beta \in \mathbb{Q},\ m \in \mathbb{Z},\ x \in X.$$

We shall call such Λ-modules *trivial.* In other words, a Λ-module X is trivial if $e_2X = X$. We call a Λ-module *reduced* if it does not contain any nonzero direct summand which is trivial or isomorphic to the Λ-module Λe_1.

Let us give another characterisation of reduced Λ-modules. Let X be a Λ-module. Let $V = e_1X$ and let V_0 be the set of all elements $v \in V$ such that $uv = 0$. The module X is reduced if and only if V is the rational hull of V_0 and $X = V + uV$. Indeed, V is a module over the field $e_1\Lambda e_1 = \mathbb{Q}$, i.e., it is a divisible torsion free abelian group. If V is not the rational hull of V_0, then we can find an element $z \in V$ such that $u(\alpha z) \neq 0$ for all $\alpha \in \mathbb{Q}$, $\alpha \neq 0$; then the element $z = e_1 z$ generates a direct summand of X isomorphic to Λe_1. Further, remark that $uV = ue_1X = u(e_1 + e_2)X = uX = e_2uX \subseteq e_2X$. If $X \neq V + uV$, then $uV \neq e_2X$, and, since the group uV is divisible, there exists a nonzero subgroup $X_1 \in X$, such that $e_2X = uV \oplus X_1$. It is clear that X_1 is a trivial direct summand

This work was supported by RFFI.

of X. Conversely, if X contains a direct summand which is a nonzero trivial module then $V + uV \neq X$, and if X contains a direct summand isomorphic to Λe_1 then V is not the rational hull of V_0.

Denote by $\mathfrak{M}$ the category of reduced Λ-modules and by $\mathfrak{M}'$ the category of Artinian reduced Λ-modules. Further, let $\mathfrak{A}$ be the category of torsion free abelian groups and $\mathfrak{A}'$ be its full subcategory consisting of all torsion free abelian groups A of finite rank such that $pA = A$ for almost all (i.e., for all but a finite number) prime integers p.

2. The functor F

Define a functor $F : \mathfrak{A} \to \mathfrak{M}$. Let A be a group from the category $\mathfrak{A}$. Denote by $E(A)$ the rational hull of A and by π_A the canonical epimorphism $E(A) \to E(A)/A$. Make the group of columns $F(A) = \{(a, \bar{a})^{\mathrm{T}} \mid a \in E(A), \bar{a} \in E(A)/A\}$ into a left Λ-module in the following way:

$$\begin{pmatrix} \alpha & 0 \\ \beta & m \end{pmatrix} \begin{pmatrix} a \\ \bar{a} \end{pmatrix} = \begin{pmatrix} \alpha a \\ m\bar{a} + \pi_A(\beta a) \end{pmatrix} \quad \text{for any } \begin{pmatrix} a \\ \bar{a} \end{pmatrix} \in F(A),\ \alpha, \beta \in \mathbb{Q},\ m \in \mathbb{Z}.$$

Let now B be another group from the category $\mathfrak{A}$ and let $\xi : A \to B$ be a homomorphism of groups. The group A, B are torsion free, therefore, the homomorphism ξ can be uniquely extended till a homomorphism $\xi_0 : E(A) \to E(B)$ of rational hulls. Denote by $\bar{\xi}$ the homomorphism $E(A)/A \to E(B)/B$ induced by ξ_0. Then the mapping $F(\xi) : F(A) \to F(B)$ defined by the formula

$$F(\xi)((a, \bar{a})^{\mathrm{T}}) = (\xi_0(a), \bar{\xi}(\bar{a}))^{\mathrm{T}} \quad \text{for any } a \in E(A),\ \bar{a} \in E(A)/A$$

is a homomorphism of Λ-modules.

It is clear that the correspondence $A \mapsto F(A)$, $\xi \mapsto F(\xi)$ is a functor from $\mathfrak{A}$ to $\mathfrak{M}$.

Theorem 1. *The functor F is a natural equivalence of the category of torsion free abelian groups $\mathfrak{A}$ onto the category of reduced Λ-modules $\mathfrak{M}$. The restriction of F to the category $\mathfrak{A}'$ is a natural equivalence of $\mathfrak{A}'$ onto the category of Artinian reduced Λ-modules $\mathfrak{M}'$.*

Proof. Let A, B be torsion free abelian groups. We shall prove that the homomorphism $F : \mathrm{Hom}(A, B) \to \mathrm{Hom}_\Lambda(F(A), F(B))$ is an isomorphism. Let $\xi \in \mathrm{Hom}(A, B)$ and $F(\xi) = 0$; it follows that the extension ξ_0 of ξ to the homomorphism of rational hulls is the zero homomorphism; but this is possible only if $\xi = 0$. Hence, the homomorphism F is injective. Let now Ξ be any Λ-homomorphism of $F(A)$ in $F(B)$. Ξ is a group homomorphism of the group of columns $F(A) = \{(a, \bar{a})^{\mathrm{T}} \mid a \in E(A), \bar{a} \in E(A)/A\}$ in the group of columns $F(B) = \{(b, \bar{b})^{\mathrm{T}} \mid b \in E(B), \bar{b} \in E(B)/B\}$; therefore, the action of Ξ can be described by a matrix:

$$\Xi(\begin{pmatrix} a \\ \bar{a} \end{pmatrix}) = \begin{pmatrix} \xi_0 & \mu \\ \tau & \bar{\xi} \end{pmatrix} \begin{pmatrix} a \\ \bar{a} \end{pmatrix},$$

where $\xi_0 : E(A) \to E(B)$, $\mu : E(A)/A \to E(B)$, $\tau : E(A) \to E(B)/B$, $\bar{\xi} : E(A)/A \to E(B)/B$ are homomorphisms of groups. The group $E(A)/A$ is a torsion group, and the group $E(B)$ is torsion free; hence, $\mu = 0$. Since Ξ is a homomorphism of Λ-modules, we have for every $a \in E(A)$:

$$\begin{pmatrix} \xi_0 a \\ 0 \end{pmatrix} = e_1 \begin{pmatrix} \xi_0 & 0 \\ \tau & \bar{\xi} \end{pmatrix} \begin{pmatrix} a \\ 0 \end{pmatrix} = \begin{pmatrix} \xi_0 & 0 \\ \tau & \bar{\xi} \end{pmatrix} e_1 \begin{pmatrix} a \\ 0 \end{pmatrix} = \begin{pmatrix} \xi_0 & 0 \\ \tau & \bar{\xi} \end{pmatrix} \begin{pmatrix} a \\ 0 \end{pmatrix} = \begin{pmatrix} \xi_0 a \\ \tau a \end{pmatrix},$$

$$\begin{pmatrix} 0 \\ \pi(\xi_0 a) \end{pmatrix} = u \begin{pmatrix} \xi_0 & 0 \\ \tau & \bar{\xi} \end{pmatrix} \begin{pmatrix} a \\ 0 \end{pmatrix} = \begin{pmatrix} \xi_0 & 0 \\ \tau & \bar{\xi} \end{pmatrix} u \begin{pmatrix} a \\ 0 \end{pmatrix} = \begin{pmatrix} \xi_0 & 0 \\ \tau & \bar{\xi} \end{pmatrix} \begin{pmatrix} 0 \\ \pi(a) \end{pmatrix} = \begin{pmatrix} 0 \\ \bar{\xi}\pi(a) \end{pmatrix}.$$

Therefore, $\tau a = 0$, $\pi_B(\xi_0 a) = \bar{\xi}\pi_A(a)$ for all $a \in E(A)$. Hence, $\tau = 0$ and $\xi_0 a \in \operatorname{Ker} \pi_B = B$ for every $a \in \operatorname{Ker} \pi_A = A$. We see that the homomorphism ξ_0 is an extension of a homomorphism $\xi : A \to B$. Since π_A is an epimorphism, it follows from the equality $\pi_B(\xi_0 a) = \bar{\xi}\pi_A(a)$ that the homomorphism $\bar{\xi}$ is induced by the homomorphism ξ_0. Therefore,

$$\Xi(\begin{pmatrix} a \\ \bar{a} \end{pmatrix}) = \begin{pmatrix} \xi_0 & 0 \\ 0 & \bar{\xi} \end{pmatrix} \begin{pmatrix} a \\ \bar{a} \end{pmatrix} = \begin{pmatrix} \xi_0 a \\ \bar{\xi}\bar{a} \end{pmatrix} = F(\xi)(\begin{pmatrix} a \\ \bar{a} \end{pmatrix}),$$

and $\Xi = F(\xi)$. Thus, we have proved that the homomorphism F is surjective.

Show that $F(A)$ is a reduced Λ-module. Otherwise there is an idempotent endomorphism d' of the module $F(A)$ the image of which is trivial or isomorphic to Λe_1. As we have already proved, there is an idempotent endomorphism d of the group A such that $d' = F(d)$. Let $B = dA$; then $F(B) = d'F(A)$ is trivial or isomorphic to Λe_1. But it is obvious that for every torsion free abelian group $B \neq 0$ the Λ-module $F(B)$ is not trivial and is not isomorphic to Λe_1.

Let now X be any reduced Λ-module. Then $e_1 X$ is a module over $e_1 \Lambda e_1 = \mathbb{Q}$; in particular, $e_1 X$ is a torsion free divisible group. Let $A = \{a \in e_1 X \mid ua = 0\}$. The group A is torsion free and $e_1 X$ is the rational hull of A; it is evident that the Λ-module $F(A) = e_1 X + u e_1 X$ is isomorphic to X.

The module $F(A)$ is Artinian if and only if $E(A)$ is a finite-dimensional vector space over $\mathbb{Q}$ and $E(A)/A$ is an Artinian abelian group. But this means exactly that $pA = A$ for all but a finite number of prime integers p, i.e., that A is a group from the category $\mathfrak{A}'$.

Theorem 1 is completely proved.

3. Cyclic Artinian Λ-modules

We use here the functor F for the construction of cyclic Artinian Λ-modules which are not Noetherian Λ-modules. Let P be any finite nonempty set of prime integers, and let A be the subgroup of the additive group of rational numbers $\mathbb{Q}^+$ consisting of all fractions the denominators of which are divisible only by primes out of P. Then $E(A) = \mathbb{Q}$ and the group $E(A)/A$ is Artinian; therefore, A belongs to the category $\mathfrak{A}'$ and the Λ-module $F(A)$ is Artinian.

It is easy to see that the Λ-module $F(A)$ is cyclic. Indeed, $F(A)$ is the group of columns $(a, \bar{b})^{\mathrm{T}}$, where $a \in \mathbb{Q}, \bar{b} \in \mathbb{Q}/A$, and the ring Λ acts on $F(A)$ in the following way:

$$\begin{pmatrix} \alpha & 0 \\ \beta & m \end{pmatrix} \begin{pmatrix} a \\ \bar{b} \end{pmatrix} = \begin{pmatrix} \alpha a \\ m\bar{b} + \pi(\beta a) \end{pmatrix} \quad \text{for any } \begin{pmatrix} a \\ \bar{b} \end{pmatrix} \in \tilde{A}, \ \alpha, \beta \in \mathbb{Q}, \ m \in \mathbb{Z}$$

(here π denotes the canonical epimorphism $\mathbb{Q} \to \mathbb{Q}/A$). The column $(1,0)^{\mathrm{T}}$ generates $F(A)$, because for any $a \in \mathbb{Q}$, $\bar{b} \in \mathbb{Q}/A$ there exists an element $b \in \mathbb{Q}$ such that $\pi(b) = \bar{b}$ and

$$\begin{pmatrix} a & 0 \\ b & 0 \end{pmatrix} \begin{pmatrix} 1 \\ 0 \end{pmatrix} = \begin{pmatrix} a \\ \bar{b} \end{pmatrix}.$$

It is obvious that the Artinian cyclic module $F(A)$ is not Noetherian because the abelian group $\mathbb{Q}/A$ is not Noetherian. If the set P consists only of one prime integer, then the module $F(A)$ is uniserial.

4. Failure of the Krull-Schmidt theorem for Artinian Λ-modules

Let A be a torsion free abelian group of finite rank which belongs to the category $\mathfrak{A}'$ and for which the Krull-Schmidt theorem fails. Since F is a natural equivalence of categories, the Krull-Scmidt theorem fails for the Artinian Λ-module $F(A)$ from the category $\mathfrak{M}'$. But many examples of torsion free abelian groups of finite rank for which the Krull-Schmidt theorem fails are known (see, for example, [1], §90). Though many of such groups are not contained in the category $\mathfrak{A}'$, almost all constructions can be modified so that we obtain groups which do not satisfy the Krull-Schmidt theorem and belong to $\mathfrak{A}'$. Therefore, we obtain many examples of Artinian Λ-modules for which the Krull-Schmidt theorem fails.

We can obtain more precise information about direct decompositions of Artinian Λ-modules. Let $T(X)$ be the torsion subgroup of a reduced Λ-module X considered as an abelian group. Then $X/T(X)$ is a linear space over $\mathbb{Q}$; we call *rank* of X and denote $\operatorname{rk} X$ the dimension of this space. It is easy to see that the rank of a reduced Λ-module coincides with the minimal number of generators of this module. Besides, the rank of the module $F(A)$ coincides with the rank of the abelian group A.

The following results are immediate corollaries of Theorem 1 and the theorems of Blagoveshchenskaya-Yakovlev and Blagoveshchenskaya [2].

Theorem 2. *Let $n = r_1 + \cdots + r_s = l_1 + \cdots + l_t$ $(1 < s, t < n)$ be two partitions of a positive integer n into sums of positive integers, and let s_0 and t_0 be the numbers of the summands r_i (respectively, l_j) which are equal to 1. There exists an Artinian Λ-module X of rank n for which there are two decompositions $X = Y_1 \oplus \cdots \oplus Y_s = Z_1 \oplus \cdots \oplus Z_t$ into direct sums of indecomposable modules, such that $\operatorname{rk} Y_i = r_i$, $\operatorname{rk} Z_j = l_j$ for every i, j, if and only if the following conditions are fulfilled: (i) $r_i \le n - t_0$ for all i, $1 \le i \le s$, $l_j \le n - s_0$ for all j, $1 \le j \le t$; (ii) if $r_i = n - t_0$ for an index i, then only one of the integers l_j differs from 1, and, similarly, if $l_j = n - s_0$ for an index j, then only one of the integers r_i differs from 1.*

Theorem 3. *Let* $1 < r_1 < r_2 < \cdots < r_s < n$*. There exists a* Λ*-module* X *of rank* n *which has a direct decomposition into the sum of* r_1 *indecomposable* Λ*-modules, a direct decomposition into the sum of* r_2 *indecomposable* Λ*-modules, ..., a direct decomposition into the sum of* r_s *indecomposable* Λ*-modules if and only if* $r_s < n - n/2r_1$.

The existence of Artinian modules for which the Krull-Schmidt theorem fails was proved by A.Facchini, D.Herbera, L.S.Levy, P.Vamos [3].

References

[1] L. Fuchs, *Infinite abelian groups*, vol.2, Academic Press (1973).

[2] E.A. Blagoveshchenskaya, A.V.Yakovlev, *Direct decomposition of torsion-free groups of finite rank*, in Russian, Algebra & Analyses **1** (1989), 111–129.

[3] A. Facchini, D. Herbera, L.S. Levy, P. Vamos *Krull-Schmidt fails for Artinian modules*, Proc. Amer. Math. Soc. **123** (1995), 3587–3592.

St.Petersburg State University, Bibliotechnaya pl.2, Stary Peterhof, St.Petersburg 198904 RUSSIA

E-mail address: yakovlev@yak.pdmi.ras.ru

Trends in Mathematics, © 2000 Birkhäuser Verlag Basel/Switzerland

SOME COMBINATORIAL PRINCIPLES FOR SOLVING ALGEBRAIC PROBLEMS

RÜDIGER GÖBEL

ABSTRACT. This is an extension of my talk at the Bielefeld conference in September 1998, which offers various infinite combinatorial principles proved by model theorists in the last two decades. These theorems are either based on ordinary set theory, like Shelah's Black Box or the Shelah Elevator or need additional set theoretic axioms like CH or GCH or more which hold in Gödel's universe. They are designed for applications in different areas of mathematics, mainly for proving non-structure theorems closely related to tame or wild representation type. In any case we will give examples of recent work in algebra in order to illustrate how these methods can be useful to algebraists in solving problems related with infinite structures.

INTRODUCTION

The aim of this paper is an introduction into the use of combinatorial results for proving theorems in algebra, mainly in module theory. We will proceed as follows.

In the first part (section 1) we want to discuss an often applied principle, Shelah's Black Box; see also Corner, Göbel [6].

We will state the black box for use in torsion-free abelian groups, indicate its model theoretic nature, which will make it possible for the reader to transfer the details to his/her own mathematical problems in order to decide whether this principle is of any use here. For this reason we will also sketch the main steps of the proof, to convince that the only ingredients are simple, but clever counting arguments. I will concentrate on the crucial steps only and will leave out the uninteresting book-keeping. This also helps to rewrite and modify the black box easily, to adopt mathematical needs for particular applications, which normally will be the case!

Secondly (in section 2), as an example taken from the talk of the meeting at Bielefeld university, I will illustrate the use of the black box in showing the existence of arbitrarily large torsion-free abelian groups with endomorphism ring $\mathbb{Z}$, a result shown first in a weaker form by Shelah [75] in 1974 by different techniques.

Key words and phrases. combinatorial principles, predicting symmetry groups, realizing rings as endomorphism rings of self-splitting modules.

This work is supported by the project No. G-0545-173.06/97 of the German-Israeli Foundation for Scientific Research & Development.

Again I will only concentrate on the interesting steps and leave out boring and obvious arguments which do not contribute to the understanding of the use of the black box.

In the third part (section 3) we will offer several different combinatorial principles, partly based on ordinary set-theory (ZFC), however some-times using additional axioms of set-theory like CH, GCH or even stronger $\diamondsuit$ which all hold in Gödel's universe. I would like to take the point of view that these principles are helpful tools to understand algebraic problems and their solutions - under additional set theoretic condition which are consistent with ordinary set theory the arguments are often simpler and the algebra becomes more transparent. These offered principles were used to settle open problems in algebra over the last years.

Hence, in the fourth section, we will underline their importance by a discussion of results obtained by them. This section will also include results not yet published and obtained in a German-Israel project of GIF. They also include the existence of non-classical splitters (stones), the solution of a problem of E. Dror Farjoun in homotopy theory, a different way of looking at Shelah's solution of the Kurosh problem, a theorem realizing any group as the outer automorphism group of a metabelian group (see Kourovka Notebook). Moreover we want to recall the solution of a problem of Hall's using black-box arguments and results related to this theorem.

1. Shelah's Black Box

Let λ be an infinite cardinal such that $\lambda = \lambda^{\aleph_0}$. If κ is any infinite cardinal, then $\kappa^{\aleph_0}$ is a candidate for λ. The cardinal condition ensures that the set of all countable subsets of λ has size λ as well.

There are many good reasons why we want to build algebraic structures, in this case a torsion-free abelian group, on a tree $T = {}^{\omega>}\lambda$ as its underlying set of 'supports'. The most obvious reason is the additional geometric structure, an easy way to find countable, almost disjoint subsets from branches. Let T be the set of all finite sequences $\tau = \lambda_0{}^\wedge \cdots {}^\wedge\lambda_{n-1}$ in λ, hence

$$T = \{\tau : n \longrightarrow \lambda : n \in \omega\}.$$

Recall that τ above is a finite branch of length $\lg\tau = n$ and similarly we define the set $\operatorname{Br} T$ of all infinite branches of length ω, which is

$$\operatorname{Br} T = {}^{\omega}\lambda = \{v : \omega \longrightarrow \lambda\}.$$

This set has cardinality λ by assumption on λ. Finite and infinite branches v have a canonical *support* which is the subset

$$[v] = \{v \restriction n \in T : n < \lg(v)\}$$

of T. Note that $[v], [w]$ are almost disjoint if v, w are distinct branches. Trees also have a natural ordering by extensions, say

$$\tau < \nu \Longleftrightarrow \tau \subseteq \nu \Longleftrightarrow [\tau] \subseteq [\nu] \text{ and } \nu \restriction [\tau] = \tau \text{ for any } \tau, \nu \in T.$$

We transport these supports to an abelian group, taking

$$B = \bigoplus_{\tau \in T} \tau \mathbb{Z}$$

to be the free abelian group generated by $T \subset B$ as our base group. Let p be a fixed prime and let $\widehat{B}$ denote its p-adic completion. Any element $g \in \widehat{B}$ can be expressed as a countable sum $g = \sum_{n \in \omega} g_n \tau_n$ for some $g_n \in \widehat{\mathbb{Z}}$ and $g_n \in p^m \widehat{\mathbb{Z}}$ for almost all $n \in \omega$. We denote by $[g] = \{\tau_n : g_n \neq 0, n \in \omega\}$ the *support* of g. If v is an infinite branch, then we also denote by $v = \sum_{n \in \omega} p^n v \upharpoonright n \in \widehat{B}$ and call this group element a branch element, which obviously has the same support as the branch v, namely $[v]$. These branch elements are useful tools to recognize elements of the group G under construction, which will satisfy $\operatorname{End} G = \mathbb{Z}$. We will work between B and the completion $\widehat{B}$:

$$B \subseteq_* G \subseteq_* \widehat{B} \tag{1.1}$$

where $\subseteq_*$ denotes pure subgroups. In order to control endomorphisms we need Shelah's black box adapted to (1.1). For different problems it is (1.1) which must be changed accordingly.

The black box needs two easy preliminary definitions, the norm and the notion of a trap which comes from countable elementary submodels:

As before let ${}^{\omega>}\omega$ be the countable tree of finite sequences in ω and let $f : {}^{\omega>}\omega \longrightarrow T$ denote a tree embedding. Let $\varphi : B \longrightarrow \widehat{B}$ denote a partial homomorphism with countable domain $\operatorname{dom} \varphi \subseteq B$ a subgroup of B and suppose that $[\operatorname{dom} \varphi]$ is a countable subtree of T, where $[X] = \bigcup\{[x] : x \in X\}$ is the natural extension of supports to subsets X of $\widehat{B}$. We will require

$$\operatorname{Im} f \subseteq [\operatorname{dom} \varphi] \subseteq T \tag{f,φ}$$

and call (f, φ) a trap.

Next we define the norm of a subset X of $\widehat{B}$. If X is a subset of T, then $||X|| = \sup \operatorname{ord}([X])$ where $\operatorname{ord}(...)$ denotes the ordinals in λ involved in defining the subset ... of T. Note that the norm is defined if X is countable, because the cofinality $\operatorname{cf}(\lambda)$ of λ is greater than $\aleph_0$. The following theorem is due to Shelah [78], see also the appendix of [6].

Black Box 1.1. *Let $\lambda = \lambda^{\aleph_0}$ be an infinite cardinal and $T = {}^{\omega>}\lambda$ be a tree which is the basis of a free abelian group $B = \bigoplus_{\tau \in T} \tau \mathbb{Z}$. Then there exists an ordinal λ^* of cardinality λ and a list of traps*

$$(\varphi_\alpha, f_\alpha) \quad \alpha \in \lambda^*$$

with the following properties:

(a) $||\operatorname{dom}(\varphi_\alpha)||$ *is a limit ordinal with* $||v|| = ||\operatorname{dom}(\varphi_\alpha)||$ *for all* $v \in \operatorname{Br}(\operatorname{Im} f_\alpha)$

(b) *If* $\beta < \alpha \in \lambda$ *then* $||\operatorname{dom}(\varphi_\beta)|| \leq ||\operatorname{dom}(\varphi_\alpha)||$ *and* $\operatorname{Br}(\operatorname{Im} f_\beta) \cap \operatorname{Br}(\operatorname{Im} f_\alpha) = \emptyset$

(c) *If* $\beta + 2^{\aleph_0} \leq \alpha$, *then* $\operatorname{Br}(\operatorname{dom}(\varphi_\beta)) \cap \operatorname{Br}(\operatorname{Im} f_\alpha) = \emptyset$ *and the* **prediction**

(d) *If X is a countable subset of B and $\varphi : B \longrightarrow \widehat{B}$ is a homomorphism, then there exists an ordinal $\alpha \in \lambda^*$ such that $X \subseteq \operatorname{dom}(\varphi_\alpha)$ and $\varphi \upharpoonright \operatorname{dom}(\varphi_\alpha) = \varphi_\alpha$.*

A discussion of the conditions (a) – (d) of the black box: In the following proof we will see that the only relevant feature of B used is the fact that B is freely generated in some algebraic sense over the independent set T, e.g. B could be a function field in commuting variables from T over $\mathbb{Q}$ or a free group in some variety of groups or a polynomial ring over $\mathbb{Z}$ in λ commuting variables from the set T. Also note that $\widehat{B}$ could be any reasonable extension of B which has enough elements depending on B by systems of equations, e.g. by polynomial equations. It helps that the homomorphisms $\varphi : B \longrightarrow \widehat{B}$ above extend uniquely to $\widehat{B} \longrightarrow \widehat{B}$, but this is not needed either.

Any object constructed by the black box will have cardinality $\lambda^{\aleph_0} = \lambda$, hence $|G|$ is at least $2^{\aleph_0}$ the size of the continuum. The cardinal gap $< 2^{\aleph_0}$ can only be filled by different techniques, in many cases this needs 'more algebra', see for example Corner [4], Göbel, May [31], Göbel, Shelah [40] or Göbel, Paras [34].

The abelian group G under construction will be the union of an ascending, continuous chain $G = \bigcup_{\alpha \in \lambda^*} G_\alpha$ where

$$G_{\alpha+1} = \langle G_\alpha, g_\alpha \rangle_*$$

and the element g_α is in charge of killing possibly unwanted homomorphisms φ extending φ_α from the list of traps. In order to have enough room for finding g_α we need the trap condition $(f_\alpha, \varphi_\alpha)$. However we also must ensure that g_α will not destroy the purpose of any 'earlier' g_β for $\beta < \alpha$. This would clearly follow if we are able to choose these elements linear independently and this is made possible by condition (b) and (c). The prediction (d) ensures that no homomorphism is left out and sneaks in secretly.

We now come to a sketch of the essential steps in showing the black box. These steps are based only on counting all possible cases that could occur, some back and forth arguments in order to have nicer substructures. The interesting feature is the way a partial homomorphism can be coded into a branch.

The set of countable subtrees of T has cardinality λ by assumption on λ. And similarly the set $\mathfrak{C}$ of all homomorphisms from subgroups of B generated by these countable subtrees into $\widehat{B}$ has size $|\mathfrak{C}| = \lambda$. This set $\mathfrak{C}$, *mutatis mutandis,* satisfies the prediction (d), however the missing tree embeddings $f : {}^{\omega>}\omega \longrightarrow T$ must be found such that some set of (φ, f)'s for certain $\varphi \in \mathfrak{C}$ will constitute the list of traps. To this end we use induction on the length of the tree embedding f which is

$$f^n = f \restriction {}^{n\geq}\omega : {}^{n\geq}\omega \longrightarrow T.$$

We assume that all the desired pairs (φ^n, f^n) with $\varphi^n \in \mathfrak{C}$ and f^n as above are constructed. The set of these 'partial traps' has cardinality λ and we label them as

$$(\varphi^n_\alpha, f^n_\alpha) \quad \alpha \in \lambda.$$

In order to ensure the 'disjointness condition' (b) for infinite branches we need more room and choose a

data bank with the 'gene-function' $g : \lambda \times \lambda \times \lambda \longrightarrow \lambda$

which is a bijection, increasing continuously in each component. Now we copy all the information on $(\varphi^n, f^n) := (\varphi^n_\alpha, f^n_\alpha)$ for any fixed α into the next node of length $n+1$ such that each new node also uses ordinals strictly larger than all its predecessors. This way we will ensure that (b) holds and the norm of each branch converges to the same limit ordinal (cofinal to ω). If $\sigma \in {}^{n+1}\omega \setminus {}^n\omega$ is such a new node, then $\sigma : n+1 \longrightarrow \omega$ and we put

$$f^{n+1}(\sigma) = f^n(\sigma \restriction n)^\wedge g(\sigma(n), \alpha, ||\mathrm{dom}\,\varphi^n||) \in {}^{n+1}\lambda.$$

Obviously f^n extends uniquely to f^{n+1} and choose any $\varphi^{n+1} \in \mathfrak{C}$ such that φ^{n+1} extends φ^n and $\mathrm{Im}\, f^{n+1} \subseteq [\mathrm{dom}\,\varphi^{n+1}]$ (moreover we could assume that $\mathrm{Im}\,\varphi^n \subseteq \widehat{\mathrm{dom}\,\varphi^{n+1}}$ which at the end would give us that the homomorphisms on each trap map their domain into the completion). Hence (φ^{n+1}, f^{n+1}) is well-defined and induction gives us an ascending chain of such partial traps. We take the union of this chain and find this way an arbitrary member of the desired set $\mathfrak{T}$ of all traps. If the two traps $(\varphi, f), (\varphi', f')$ in $\mathfrak{T}$ have an infinite branch $v \in \mathrm{Br}\,(\mathrm{Im}\, f) \cap \mathrm{Br}\,(\mathrm{Im}\, f')$ in common, then at level $n+1$, using the gene function, we can read off (φ^n, f^n) and (φ'^n, f'^n) which must be the same, hence $(\varphi, f) = (\varphi', f')$ and (b) follow.

It is easy to see that for any fixed trap (φ, f) there are only $< 2^{\aleph_0}$ other traps (φ', f') with $\mathrm{Br}\,(\mathrm{dom}\,(\varphi)) \cap \mathrm{Br}\,(\mathrm{Im}\, f') \neq \emptyset$, see [6]. From this it is also easy to introduce an enumeration of $\mathfrak{T}$ such that (c) holds. The black box follows.

2. A class of abelian groups with endomorphism ring $\mathbb{Z}$

We want to apply the black box to show that there is an abelian group G of cardinality λ with $\mathrm{End}\, G = \mathbb{Z}$ as promised in the last section. The group will be the union $G = \bigcup_{\alpha \in \lambda^*} G_\alpha$ of an ascending, continuous chain of cotorsion-free groups G_α such that

$$B \subseteq_* G_\alpha \subseteq_* \widehat{B}.$$

A group G is cotorsion-free if and only if it is torsion-free, reduced and has no isomorphic copy of the p-adic integers for any prime p as a subgroup (or equivalently as a summand). This is equivalent to saying that the only cotorsion subgroup of G is 0 or equivalently all pure-injective subgroups of G are 0. Equivalent characterization for cotorsion-free groups can be found in [47, 9]. Recall that cotorsion subgroups are the epimorphic images of pure injective groups, hence cotorsion-freeness is also necessary for any (large enough) indecomposable group G. Moreover we could show that G in preparation is slender and $\aleph_1$-free. A group is $\aleph_1$-free if its countable subgroups are free and a group is slender if and only if it is cotorsion-free and does not have an isomorphic copy of the Baer-Specker group $\mathbb{Z}^{\aleph_0}$ as a subgroup - by a celebrated theorem of Nunke's, see [26]

We begin with $G_0 = B$ and by continuity we only need to consider the inductive step at any $\alpha \in \lambda^*$. We want to find $g_\alpha \in \widehat{B}$ such that

$$G_{\alpha+1} = \langle G_\alpha, g_\alpha \rangle_*$$

and the new element g_α must fulfill several tasks.

First of all we require that this new element is a 'branch-like' element of $\widehat{B}$: Recall that any branch $v \in \mathrm{Br}(\mathrm{Im}\, f_\alpha)$ gives rise to a branch-element, which we also denoted by v. Any sum $b + v$ with

$$b \in \widehat{\mathrm{dom}\, \varphi_\alpha} \text{ and } ||b|| < ||v||$$

is called branch-like element. The point is that a branch-like element has a support which at the top looks like a branch from $\mathrm{Br}(\mathrm{Im}\, f_\alpha)$. This particular choice of generators for G allows us a better description of the action of endomorphisms of G. We interrupt the precise construction of G and state a result which gives more insight into endomorphisms which are not multiplication by an integer (we say $\varphi \notin \mathbb{Z}$). Note that we will identify any homomorphism $B \longrightarrow \widehat{B}$ with its unique extension $\widehat{B} \longrightarrow \widehat{B}$.

Proposition 2.1. *Let G be the group constructed so far and let $\varphi \in \mathrm{End}\, G \setminus \mathbb{Z}$. Then there is an $x \in \widehat{B}$ such that*

$$x\varphi \notin \langle G, x \rangle_*.$$

We postpone the proof, which is algebraically the crucial step, to the end of this section. However we want to note that also the additional properties of G mentioned above follow by variations of this proposition.

We now continue the construction of G. By the proposition it is clear that the first of the following two tasks of the branch-like element g_α is a rephrasing that φ_α is not scalar multiplication by an integer. We summarize the two tasks depending on $(\varphi_\alpha, f_\alpha)$ from the black box in order to control the endomorphisms of G:

(I) First we consider the following 'bad case for α:
If there is an $x \in \widehat{\mathrm{dom}\, \varphi_\alpha}$ such that $||x|| < ||\mathrm{dom}\, \varphi_\alpha||$ and

$$x\varphi_\alpha \notin \langle G_\alpha, x \rangle_*$$

then choose a branch $v \in \mathrm{Br}(\mathrm{Im}\, f_\alpha)$ and put $g_\alpha = v$ or $g_\alpha = x + v$; the choice of g_α depends on the requirement that

$$g_\alpha \varphi_\alpha \notin \langle G_\alpha, g_\alpha \rangle_* = G_{\alpha+1}.$$

If α is not bad, *then choose any branch-like g_α taking care of (II).*

(II) *If $\beta < \alpha$ was bad, that is $g_\beta \varphi_\beta \notin G_{\beta+1}$ then we also want $g_\beta \varphi_\beta \notin G_{\alpha+1}$.*

We pause again in order to see that the two tasks suffice to show that $\mathrm{End}\, G = \mathbb{Z}$: Let $\varphi \in \mathrm{End}\, G \setminus \mathbb{Z}$. By the proposition there is an element $x \in \widehat{B}$ such that $x\varphi \notin G$ and by the black box we can find an $\alpha \in \lambda^*$ such that φ extends φ_α, $x \in \widehat{\mathrm{dom}\, \varphi_\alpha}$ and $||x|| < ||\mathrm{dom}\, \varphi_\alpha||$. Hence α is a bad case and g_α in the construction must satisfy (I) and (II). It follows by task (I) that $g_\alpha \varphi_\alpha = g_\alpha \varphi \notin G_{\alpha+1}$. By task

(II) we also have $g_\alpha\varphi \notin G_\gamma$ for any later ordinal $\alpha < \gamma \in \lambda^*$, hence $g_\alpha\varphi \notin G$ and φ was not an endomorphism of G, a contradiction.

Hence it remains to show that the two tasks (I) and (II) are possible for g_α: The work on condition (I) is put into the proposition, hence (II) must be verified. When defining $g_\alpha = g_{\alpha,v} = x + v$ we have a free choice of branches $v \in \mathrm{Br}\,(\mathrm{Im}\, f_\alpha)$ which we now use. If (II) is violated for some $\beta = \beta_v$, then

(2.1) $$g_\beta\varphi_\beta \in \langle G_\alpha, g_{\alpha,v}\rangle_*, \text{ hence } p^{k_v} g_{\beta_v}\varphi_{\beta_v} - g_{\alpha,v}a_v \in G_\alpha.$$

A support argument and (c) from the black box show that

$$\beta_v < \alpha < \beta_v + 2^{\aleph_0}$$

and if β_0 is the least ordinal satisfying this inequality, then

$$\beta_0 < \beta_v < \beta_0 + 2^{\aleph_0}.$$

There are two distinct branches $v, w \in \mathrm{Br}\,(\mathrm{Im}\, f_\alpha)$ such that $\beta_v = \beta_w$. Suppose $k_v \geq k_w$ and $k = k_v - k_w$. Subtracting the according expressions (2.1) we get

$$g_{\alpha,v}a_v - p^k g_{\alpha,w}a_w \in G_\alpha$$

and by an easy support argument it can be seen that this is only possible for $v = w$, a contradiction. Hence (II) can be arranged for $g_\alpha = x + v$ and some $v \in \mathrm{Br}\,(\mathrm{Im}\, f_\alpha)$.

It remains to show the proposition: First we use the assumption on φ to show that there exists a countable subset

(2.2) $C \subseteq T$ such that $D = \langle C\rangle$ satisfies $\widehat{D}(\varphi - a) \nsubseteq G$ for all $a \in \mathbb{Z}$

Choose a 'constant branch' $v : \omega \longrightarrow \{\eta\}$ at some $\eta < \lambda$ and let C' be a countable subset of T such that the branch-element v belongs to $\widehat{D'}$ where $D' = \langle C'\rangle$. If (2.2) fails for $D = D'$, there is $a \in \mathbb{Z}$ with $\widehat{D'}(\varphi - a) \subseteq G$. But $\varphi \notin \mathbb{Z}$ and there is $x \in B$ such that $x(\varphi - a) \notin G$. Enlarge C' to C such that $x \in D = \langle C\rangle$. If (2.2) fails again, then $\widehat{D}(\varphi - a') \subseteq G$ for some $a' \in \mathbb{Z}$, hence $v(a - a') \in G$ which, by supports, forces $a = a'$. We derive the contradiction $x(\varphi - a) \in G$ and (2.2) follows.

Condition (2.2) can be extended to (2.3). We may assume that

(2.3) D in (2.2) also satisfies $\widehat{D}(p^n\varphi - a) \nsubseteq G$ for all $n > 0$, $a \in \mathbb{Z} \setminus p\mathbb{Z}$

Suppose (2.3) fails for $\psi = p^n\varphi - a$. Now we choose elements $\sigma_m \in T$ which constitute an 'anti-branch': $\lg(\sigma_m) = m$ and any two σ_m's are incomparable in T. Moreover we require

$$\sup_{m \leq k} \|\sigma_m\psi\| < \|\sigma_{k+1}\| \text{ and let } h = \sum_{m\in\omega} p^{m+n}\sigma_m.$$

Then $h\psi \equiv \sum_{m\leq k}\sigma_m\psi \bmod p^{(k+1)n}\widehat{B}$, hence $h\psi \restriction \sigma_k \equiv -p^{kn}a \bmod p^{(k+1)n}\mathbb{Z}$ which cannot be 0 because $a \notin p\mathbb{Z}$. Hence $||h\psi|| = \sup||\sigma_k||$ and the element $h\psi$ cannot be in G by support. Now we enlarge D such that all the σ_m's belong to D and this failure for (a, p^n) is impossible for the enlarged D. Similarly we deal with the other (at most countably many) potential failures of (2.3) and correct D, hence (2.3) holds.

We are now ready to find the desired element $x \in \widehat{B}$. First we choose a new constant branch w with $||D||, ||D\varphi|| < ||w||$. If the branch-element $x = w$ satisfies the proposition, then the proof is finished. Otherwise $x\varphi \in \langle G, x\rangle_*$ and there are $n \in \omega, a \in \mathbb{Z}$ such that

$$p^n w\varphi - wa \in G.$$

If $n = 0$ we can apply (2.2) directly, but if $n > 0$, we may assume that $a \in \mathbb{Z}\setminus p\mathbb{Z}$ and (2.3) applies as well. There is a $z \in \widehat{D}$ such that $z(p^n\varphi - a) \notin G$. Now it is easy to check by support arguments that $x = z + w$ meets the requirements of the proposition.

3. Prediction Principles

In the last section we used the black box for showing the existence of many (indecomposable) torsion-free abelian groups with prescribed endomorphism ring $\mathbb{Z}$. Other applications of the black box are discussed in section 4. The special feature of these constructions is that inductively we concentrate on one task only, e.g. on killing endomorphisms or outer automorphisms etc. However often algebraic problems demand to carry out several tasks at the same time while constructing e.g. a 'complete module' with prescribed endomorphism ring: We must solve systems of equations (a task φ^1) **and** kill endomorphisms (a task φ^2). The black box needs some substantial - but again not complicated - changes and becomes a *stationary black box* which allows us to carry out many tasks φ^i while constructing the algebraic object. Very recent applications concerning cotilted modules and splitters can be found in Göbel, Shelah [41] and Göbel, Trlifaj [46], which are also discussed in section 4. Here we want to provide the stationary black box. An indication of the model theoretic proof of this prediction principle can be found in Shelah [78], and a proof for pedestrians, based on counting as in section 1, is published in Franzen, Göbel [25]. The statement of the stationary black box needs the old notion of a stationary set. If κ is a regular cardinal, that is $\operatorname{cf}\kappa = \kappa$, e.g. if κ is the successor λ^+ of any cardinal λ, then $C \subseteq \kappa$ is a cub if C is closed and unbounded in the ordering of the well-ordered set κ. More generally $S \subseteq \kappa$ is a stationary subset of κ if $S \cap C \neq \emptyset$ for any cub C of κ. As in section 1 we want to formulate the new combinatorial principle in terms of torsion-free abelian groups. It then will be obvious to rewrite the stationary black box in terms of other algebraic structures. If $T = {}^{\omega>}\lambda$ is the tree over the cardinal λ, then first we refine the notion of a norm. Let $\kappa = \operatorname{cf}\lambda$ be the cofinality of λ which is a regular cardinal and assuming $\lambda = \lambda^{\aleph_0}$ we have $\kappa > \aleph_0$ from König's lemma. Then we fix any continuously strictly increasing, unbounded function $||..|| : \kappa \longrightarrow \lambda$. If $g \in \widehat{B}$, then let

$$||g|| = \min\{\nu < \kappa : [g] \subseteq {}^{\omega>}||\nu||\}.$$

As in section 1 this norm function can be extended to arbitrary countable subsets of $\widehat{B}$, see (1.1).

$$B \subseteq_* G \subseteq_* \widehat{B} \tag{3.1}$$

Also let I be an indexing set of cardinality $\leq \kappa$ of the different tasks labeled by $(i \in I)$ and carried out often enough while constructing G. In many cases we will have $|I| = 2$. Any such task i will need a 'personal' stationary subset of κ on which the construction only deals λ-times with task i. Hence we choose a disjoint union $S = \coprod_{i \in I} S_i$ of stationary subsets S_i of the set S of all ordinals $\alpha < \kappa$ which are cofinal to ω. This decomposition exists by Solovay's theorem, see Jech [57]. Next the old trap $(\varphi_\alpha, f_\alpha)$ with the partial homomorphism φ_α and domain $P_\alpha = \operatorname{dom}\varphi_\alpha$ must be replaced according to a new task carried out on a 'canonical summand' P_α and its completion $\widehat{P_\alpha}$. The task i is nothing sophisticated, usually it allows us to add particular elements from $\widehat{P_\alpha}$ to G in a particular way prescribed by i for instance. It is only crucial that this happens often enough - on a specially chosen stationary set S_i with $||P_\alpha|| \in S_i$. Again, in order to have uniform notation as before, we denote this specific procedure i dealing with elements from $\widehat{P_\alpha}$ by φ^i. The particular task i will be carried out on P_α if and only if $||P_\alpha|| \in S_i$. So φ_α could be a system of equations of some type i with coefficients in P_α waiting for solutions in $\widehat{P_\alpha}$. We will then say that *φ_α is a task $i \in I$ taking place at $\widehat{P_\alpha}$*.

Note that the black box in section 1 can be rephrased as predicting partial homomorphisms φ. In this sense we want to predict tasks φ^i carried out on countable subgroups. This restriction to countable subgroups (e.g. considering countable set of equations) can be removed at the expense of restrictions on cardinals λ, see e.g. [46]. We are ready to formulate the

Stationary Black Box 3.1. *Let $\lambda = \lambda^{\aleph_0}$ be an infinite cardinal and $T, B, \widehat{B}$ as in section 1 , I be the set of tasks φ^i with $|I| \leq \kappa = \operatorname{cf}\lambda$, $S = \coprod_{i \in I} S_i$ and traps of the form $(\varphi_\alpha, f_\alpha, P_\alpha)$ with φ_α a task to be carried out at P_α defined as above. Then there is an ordinal λ^* with $||\lambda^*|| = \lambda$ and a sequence of traps, $((\varphi_\alpha, f_\alpha, P_\alpha) \mid \alpha \in \lambda^*)$, such that*

- *(i)* $\forall \beta < \alpha \in \lambda^* : ||P_\beta|| \leq ||P_\alpha||$,
- *(ii)* $\forall \alpha \neq \beta \in \lambda^* : \operatorname{Br}(\operatorname{Im} f_\alpha) \cap \operatorname{Br}(\operatorname{Im} f_\beta) = \emptyset$,
- *(iii)* $\forall \alpha, \beta \in \lambda^* : \beta + 2^{\aleph_0} \leq \alpha \Rightarrow \operatorname{Br}(\operatorname{Im} f_\alpha) \cap \operatorname{Br}([P_\beta]) = \emptyset$,
- *(iv)* $\forall \alpha \in \lambda^* : ||P_\alpha|| \in S_i$ *if and only if* $(\varphi_\alpha, f_\alpha, P_\alpha)$ *is a trap with* φ_α *a task i for* P_α,
- *(v) for all countable subsets $X \subseteq \hat{B}$, all tasks φ^i and any $i \in I$ there exists an ordinal $\alpha \in \lambda^*$ such that $||P_\alpha|| \in S_i \,\&\, X \subseteq \hat{P}_\alpha \,\&\, ||X|| < ||P_\alpha||$ and φ_α is the task φ^i if restricted to X.*

These black box prediction principles are inspired by a result due to R. Jensen [58] that is a much stronger combinatorial lemma which holds in Gödel's universe $V = L$. For the sake of completeness we add this principle $\diamondsuit_\lambda$ for regular, not weakly compact cardinals λ which can be found at many places, in Eklof, Mekler [23, p.139] or in Kanamori [59]. For the $\square_\kappa$-principle we also refer to [23, p.156].

Let $G = \bigcup_{\alpha<\lambda} G_\alpha$ be a λ-filtration of the set G which is the union of a continuously, increasing chain of subsets G_α of cardinality $< \lambda$. If S is a stationary subset of λ, then $\diamondsuit_\lambda(S)$ is the following statement:

There are Jensen-functions

$$h_\alpha : G_\alpha \longrightarrow G_\alpha \quad (\alpha \in S)$$

such that for any $h : G \longrightarrow G$, the set $\{\alpha \in S : h \restriction G_\alpha = h_\alpha\}$ is stationary in λ.

Then $\diamondsuit_\lambda$ is just $\diamondsuit_\lambda(\lambda)$. In many applications this principle is needed for $S = \{\alpha < \lambda : \operatorname{cf} \alpha = \omega\}$, also in the cases discussed in section 4. While the black box principles predict maps with domain of size λ on only countable subsets, Jensen ($\diamondsuit_\lambda$) predicts such functions on any subsets of cardinality $< \lambda$. This helps to improve the algebra; see applications in section 4. Often we are able to invoke almost freeness of the objects G when passing from Shelah's black box to Jensen's $\diamondsuit$. However note that these result need not be true in any model of ZFC. Compare the hard work in showing the existence of almost-free, non-free abelian groups for particular cardinals by Magidor, Shelah [64] and also the existence of $\aleph_1$-free abelian groups of cardinality $\aleph_1$ with prescribed countable, free endomorphism ring in Göbel, Shelah [40]. This follows relatively easy with the help of $\diamondsuit$, as a theorem of $ZFC + V = L$', see [9]. It is also interesting to note that often GCH implies diamond principles:

An example of such a useful combinatorial lemma (applied in [39]) is the following

Theorem 3.2. (Shelah [80]) *Suppose $\lambda = 2^\nu = \mu^+$ and for some regular cardinal $\kappa < \mu$ either*

(i) $\mu^\kappa = \mu$ *or*
(ii) μ *is singular with* $\operatorname{cf} \mu \neq \kappa$ *and* $|\delta|^\kappa < \mu$ *for all* $\delta < \mu$

Then $\diamondsuit_\lambda(S_\kappa)$ holds, where $S_\kappa = \{\alpha < \lambda : \operatorname{cf} \alpha = \kappa\}$.

In particular we have a

Corollary 3.3. *Assume GCH and that μ is some regular cardinal. Then $\diamondsuit_\mu S_\lambda$ holds for any regular cardinal $\lambda < \mu$.*

A useful combinatorial principle $\Phi_{\omega_1}(\omega_1)$ holds under weak CH, that is $2^{\aleph_0} < 2^{\aleph_1}$. More generally weak diamond $\Phi_\lambda(E)$ for $E \subseteq \lambda$ is the following statement:

Let $G = \bigcup_{\alpha<\lambda} G_\alpha$ be a λ-filtration of the set G and

$$P_\nu : \mathcal{P}(G_\nu) \to 2 = \{0, 1\} \ (\nu \in E)$$

a given partition of subsets of G_ν*, then* $\Phi_\lambda(E)$ *provides a prediction function* $\varphi : E \to 2$ *such that for all*

$$X \subseteq G \text{ the set } \{\nu \in E : \; P_\nu(X \cap G_\nu) = \varphi(\nu)\}$$

is stationary in λ*. In this case* E *is called non-small.*

Hence ω_1 is non-small if we assume CH. See Eklof, Mekler [23, pp.175-178], Shelah [80, p.376, Theorem 32] and Gregory [48] for predictions under GCH and weaker.

We also would like to refer to a combinatorial principle used in [33] and [44, 45] which generalizes a result of Ringel [68] and Brenner, Ringel [2]. The statement of this 'Shelah elevator' in categorical terms for applications in representation theory over infinite dimensional modules is given in [33]. For proving a theorem in group theory (like the solution of the Kurosh problem on the existence of Jonsson groups) the following recent principle from Hart, Laflamme and Shelah [51] could be used to simplify some of the arguments in Shelah [77]. We will use this in answering a problem in homotopy theory from [37] and discussed in section 4. We will close this section explaining this new important prediction principle from [51], see also [81].

Model theorists like to phrase their theorems in terms of games, see also Hodges [56]. The desired structure G from [51] is the union of a directed system of size λ^+ of substructures of cardinality $< \lambda$ chosen by two players I and II in a game (by certain rules) with λ^+ moves from a *standard* λ^+*-uniform subset*

$$\mathbb{P} \text{ of } \mathcal{P} = \lambda \times \mathcal{P}_{<\lambda}(\lambda^+) \text{ with } \mathcal{P}_{<\lambda}(\lambda^+) = \{X \subseteq \lambda^+ : |X| < \lambda\}.$$

Player II has a winning strategy by cheating with the help of a combinatorial axiom $\mathbb{D}_\lambda$. And $\mathbb{D}_\lambda$ holds if either $\lambda = \aleph_0$ or λ is strongly inaccessible or if $\diamondsuit_\lambda$ applies. The work for an algebraist afterwards is to show that a standard λ^+-uniform subset $\mathbb{P}$ of $\mathcal{P}$ exists which meets the requirements such that the final structure G will have the desired algebraic properties we want. In case of [77, 37] this will need cancellation theory, or just free products of groups with amalgamations, see [63] for instance. We now must explain the undefined notions and leave the lengthy proof in [51] that player II has a winning strategy to the more ambitious reader.

We begin at the end and define the axiom $\mathbb{D}_\lambda$ for a regular cardinal λ. If $\alpha < \lambda$, then $\mathcal{P}(\alpha)$ denotes the power set on the set α.

$\mathbb{D}_\lambda$ *asserts that there are sets* $\mathfrak{A}_\alpha \subseteq \mathcal{P}(\alpha)$*, with* $|\mathfrak{A}_\alpha| < \lambda$ *such that for all* $X \subseteq \lambda$ *the set* $\{\alpha \in \lambda : X \cap \alpha \in \mathfrak{A}_\alpha\}$ *is stationary in* λ*.*

The standard λ^+-uniform set

Next we define a *standard* λ^+*-uniform subset* of approximations $\mathbb{P}$ (of G) in $\mathcal{P}$. Elements in $\mathcal{P}$ are pairs $p = (\alpha, u)$ where $u = \operatorname{dom} p$, and at the end G is the union of a special choice of subsets $u \subseteq \lambda^+$ and $G \subseteq \lambda^+$ has size λ^+.

Definition 3.4. *If* $\mathbb{P} = (\mathbb{P}, <)$ *is a partially ordered set and a subset of* $\mathcal{P}$*, then* $\mathbb{P}$ *is a standard* λ^+*-uniform subset if the following holds.*

(i) *Compatibility of the orders: If* $p \le q$ *then* $\operatorname{dom} p \subseteq \operatorname{dom} q$

(ii) *If* $p, q, r \in \mathbb{P}$ *with* $p, q \le r$ *then there is* $r' \in \mathbb{P}$ *so that* $p, q \le r' \le r$ *and* $\operatorname{dom} r' = \operatorname{dom} p \cup \operatorname{dom} q$.

(iii) *If* p_α $(\alpha < \delta < \lambda)$ *is an increasing sequence in* $\mathbb{P}$ *then it has a least upper bound* $q \in \mathbb{P}$ *with* $\operatorname{dom} q = \bigcup_{\alpha<\delta} \operatorname{dom} p_\alpha$*; we say that* $q = \bigcup_{\alpha<\delta} p_\alpha$.

(iv) *If* $p \in \mathbb{P}$ *and* $\alpha < \lambda^+$ *then there is* $q \in \mathbb{P}$ *such that* $q \le p$ *and* $\operatorname{dom} q = \operatorname{dom} p \cap \alpha$ *and there is a unique maximal such* q *for which we write* $q = p \restriction \alpha$.

(v) *Continuity I: If* δ *is a limit then* $p \restriction \delta = \bigcup_{\alpha<\delta} p \restriction \alpha$.

(vi) *Continuity II: If* p_i $i < \delta < \lambda$ *is an increasing sequence in* $\mathbb{P}$ *and* $\alpha < \lambda$ *then*

$$(\bigcup_{i<\delta} p_i) \restriction \alpha = \bigcup_{i<\delta} (p_i \restriction \alpha).$$

(vii) *Indiscernibility: If* $p = (\alpha, u) \in \mathbb{P}$ *and* $\gamma : u \longrightarrow u' \subseteq \lambda^+$ *is an order-isomorphism onto* u'*, then* $\gamma(p) =: (\alpha, \gamma(u)) \in \mathbb{P}$ *such that* $\le$ *is compatible in the sense that* $q \le p$ *implies* $\gamma(q) \le \gamma(p)$.

(viii) *Amalgamation property: For every* $p, q \in \mathbb{P}$ *and* $\alpha < \lambda^+$ *with* $p \restriction \alpha \le q$ *and*

$$\operatorname{dom} p \cap \operatorname{dom} q = \operatorname{dom} p \cap \alpha$$

we have $r \in \mathbb{P}$ *such that* $p, q \le r$.

Next we must define the 'playing cards', the rules of the match and say what means 'winning the game'.

The playing cards:

Let $\mathbb{P}_\alpha = \{p \in \mathbb{P} : \operatorname{dom} p \subseteq \alpha\}$ for any ordinal $\alpha < \lambda^+$. A subset $G \subseteq \mathbb{P}_\alpha$ is an *admissible ideal of* $\mathbb{P}_\alpha$ if G is a maximal subset which is λ-directed [there exist upper bounds for subsets of size $< \lambda$ in G] and which is closed downwards.

If G is an admissible ideal of $\mathbb{P}_\alpha$, then $\mathbb{P}/G$ denotes the restriction of $\mathbb{P}$ to $\{p \in \mathbb{P} : p \restriction \alpha \in G\}$.

Let G be an admissible ideal of $\mathbb{P}_\alpha$ and $\alpha < \beta < \lambda^+$. An (α, β)*-density system for* G is a function

$$D : \mathfrak{P} := \{(u, v) : u, v \in \mathcal{P}_{<\lambda}(\lambda^+) \text{ and } u \subseteq v\} \longrightarrow \mathcal{P}(\mathbb{P})$$

such that the following holds:

(i) If $(u, v) \in \mathfrak{P}$ then $D(u, v) \subseteq \{p \in \mathbb{P}/G : \operatorname{dom}(p) \subseteq v \cup \beta\}$ is a dense and upward-closed subset, or more explicitly:
If $p \in D(u, v), q \in \mathbb{P}/G, p \le q, \operatorname{dom} q \subseteq v$ then $q \in D(u, v)$, and if $p \in \mathbb{P}/G$ and $\operatorname{dom} p \subseteq v$ there is $p \le q \in D(u, v)$.

(ii) If $(u, v), (u', v') \in \mathfrak{P}$ and $u \cap \beta = u' \cap \beta$, $v \cap \beta = v' \cap \beta$ and there is an order-isomorphism from v onto v' which maps u onto u', then for any ordinal γ we have

$$(\gamma, v) \in D(u, v) \Longleftrightarrow (\gamma, v') \in D(u', v').$$

An admissible ideal G' of $\mathbb{P}_{\alpha'}$ for some $\alpha' < \lambda^+$ *meets* the (α,β)-density system D for G if $\alpha < \alpha'$, $G \subseteq G'$ and for each $u \in \mathcal{P}_{<\lambda}(\alpha')$ there is a $u \subseteq v \in \mathcal{P}_{<\lambda}(\alpha')$ such that $D(u,v) \cap G' \neq \emptyset$.

The game

Let $\mathbb{P}$ be a λ^+-uniform partially ordered set as above, then players I and II will have λ^+ moves and player I moves first. At the αth step player II will choose an increasing sequence of ordinals $\zeta_\beta < \lambda^+$ $(\beta < \alpha)$ and along with this admissible ideals G_β on $\mathbb{P}_{\zeta_\beta}$.

Now player I will choose an element $g_\alpha \in \mathbb{P}/G_{<\alpha}$ [where $G_{<\alpha}$ is the supremum of the G_β's which is again an admissible ideal of $\mathbb{P}_{\sup \zeta_\beta}$. Moreover he will choose $|I_\alpha| \leq \lambda$ density systems D_i^α $(i \in I_\alpha)$ over $G_{<\alpha}$.

At the next step player II will pick again an ordinal ζ_α and an admissible ideal in $\mathbb{P}_{\zeta_\alpha}$

Player II wins the game if the sequences ζ_α and G_α are increasing, we have $g_\alpha \in G_\alpha$ and G_β for all $\beta \geq \alpha$ meets D_i^α for all $i \in I_\alpha$. After the game we take the union $\mathcal{G}$ of the ascending chain of admissible sets G_α which approximate the union G of this directed system $\mathcal{G}$ Then G is the desired structure which has the properties 'forced' by player I who picks the algebric properties in form of the dense sets $D(u,v)$. The 'meet' condition ensures that this has the right effect on G. So it is a problem in group theory, for instance, to formulate desired properties like 'being simple' or 'complete' in terms of dense sets, as carried out in [37].

4. Some Applications of Prediction Principles to Problems in Algebra

We will order the selection of results which are chosen only by my own view after the combinatorial principles in use.

A. Results based on the black box

An early version of the black box was used in Dugas, Göbel [10] to show the following

Theorem 4.1. *A ring R is the endomorphism ring of some cotorsion-free abelian group if and only if R^+ is cotorsion-free. The group realizing R can be chosen arbitrarily large.*

Recall that an abelian group is cotorsion-free if and only if it has no proper pure-injective subgroups. This is also necessary for the endomorphism ring if the group is cotorsion-free, and if the ring has no non-trivial central idempotents and the group is larger than the continuum $2^{\aleph_0}$ then conversely the group must be cotorsion-free as well. The theorem answered a problem by Corner from 1961 and it also motivated Saharon Shelah [78, 79] to develop the combinatorial principle which helped to decrease the size of the underlying cotorsion-free group, which we like to call Shelah's black box since Corner, Göbel [6]. In [6] various 'realization theorems' for rings as endomorphism rings (often modulo some well-known ideal of 'inessential endomorphisms') are derived for certain classes of abelian groups, like torsion-free groups, p-groups or mixed abelian groups. More recently, in Dugas, Göbel [18] we showed that a ring R with free additive groups R^+ is the endomorphism ring of some Butler group with only two types. This shows (strengthening the cotorsion-free result above) that all pathologies discussed by Eklof [22] in this volume will occur for cotorsion-free groups, even for these special Butler groups - and this despite the fact that a little more restriction on the class of Butler groups makes them well-behaving with nice characterization, see Files, Göbel [24] and Dugas, Göbel [19].

The results in [18] concerning Butler groups depend on an investigation of free R-modules with distinguished (partially ordered) submodules. These results can be found in [32, 33, 20] which is connected to earlier work [68] and [2], and more recent papers [44, 45] establish links to the Kronecker modules in representation theory.

A different direction of research based on arguments from section 3 and the black box in section 2 (and [6]) was taken up by Dugas, Mader, Vinsonhaler [21]. They derived the existence of arbitrarily large E-rings, which answered a problem raised by Pierce and others some time ago. Recall that a ring R is an E-ring if the canonical mapping

$$R \longrightarrow \mathrm{End}_{\mathbb{Z}} R \ (r \longrightarrow id_R \cdot r)$$

is an isomorphism. Small E-rings like $\mathbb{Z}$ up to cardinality $2^{\aleph_0}$ were known for a long time.

In Dugas, Göbel [12] the black box was used to construct locally finite p-groups with prescribed countable outer automorphism group. If the outer automorphism group is trivial, this is to say that all automorphisms are inner, the group is called complete. Hence there are complete, locally finite p-groups, answering a question by Philip Hall 1967. Note that we know for a long time by theorems of Gaschütz [27, 88], Zalesskii [89] and Menegazzo, Stonehewer [65] that such groups had to be uncountable. In [13, 14] we used the black box for generalization of the results on E-rings [21] in order to construct nilpotent groups of class 2 with prescribed automorphism groups modulo the stabilizer. And in a recent paper [35] jointly with Paras we are luckily able to show (after many years) that any group is the outer automorphism group of some metabelian group, see also [36, 34]. The algebraic part needs Magnus representations of free metabelian groups.

Another application leads to automorphism groups of groups of fields and endomorphism monoids of fields. In particular, given any field and group, then the automorphism group of an extension field can be prescribed as this group, and if the group is infinite then the extension field is given in such a way that the fixed field is the given field, see [15]. In another paper we discuss the minimal sizes of such field extension, a problem which surprisingly depends on the choice whether we consider endomorphism monoids or automorphism groups of fields; we would like to refer the reader to [16, 17]. If the given field are the complex numbers and the group is 1, then this answers a question by C. Jensen.

B. Results based on the stationary black box

Following [73], a pair $(\mathfrak{F}, \mathfrak{C})$ of maximal classes $\mathfrak{F}, \mathfrak{C}$ of R-modules is a cotorsion theory if $\operatorname{Ext}(\mathfrak{F}, \mathfrak{C}) = 0$ in an obvious sense. Then $\mathfrak{C}$ is the cotorsion part and $\mathfrak{F}$ is the torsion-free part of this theory $(\mathfrak{F}, \mathfrak{C})$. Harrison and many other algebraists investigated the classical case (torsion-free, cotorsion), see Fuchs [26]. The classical cotorsion theory is cogenerated by $\mathbb{Q}$:

$$\operatorname{Ext}(C, \mathbb{Q}) = 0 \text{ if and only if } C \in \mathfrak{C}.$$

Cotorsion theories cogenerated by subgroups of $\mathbb{Q}$ are called *rational cotorsion theories*; they are well-studied in [73]. However the existence of enough projectives and injectives (a question raised in [73]) could be answered only when investigating 'splitters' in [41]. Splitters were introduced in Schultz [74]. They also come up under different names; see [61, 70, 69]. Splitters G are defined by

$$\operatorname{Ext}_R(G, G) = \operatorname{Ext}_{\mathbb{Z}}(G, G) = 0$$

where R is the nucleus of the torsion-free group $G \neq 0$ which is the largest subring $R = \operatorname{nuc} G$ of $\mathbb{Q}$ such that G is an R-module. Surprisingly we have a

Theorem 4.2. ([42] and assume CH) *If $R = \operatorname{nuc} G$ is the nucleus and G is an $\aleph_1$-free splitter of cardinality $\aleph_1$ then G is a free R-module.*

Under the negation of CH the implication of Theorem 4.2 is undecidable, see [43] which corrects a statement from [42].

The other 'classical splitters' are torsion-free pure-injective R-modules, so all of them are classified by cardinal invariants. In [41] we used the stationary black box to answer Schultz's [74, Problem 4] in the negative by providing a list of splitters in ZFC which are neither free over their nuclei nor cotorsion. Our examples also show that there is no hope of classifying splitters because any prescribed R-algebra A which is free as a R-module, R a proper subring of $\mathbb{Q}$, is an endomorphism algebra $\operatorname{End} G \cong A$ of some splitter G with nucleus R. Hence contrary to the classical splitters all kind of nasty decompositions may occur, Kaplansky's test problems are violated etc., see Eklof [22] or Corner, Göbel [6] for strange decompositions. We need two stationary subsets of λ for the black box in order to complete G in some sense to make it a splitter and on the other hand to control its endomorphism ring. The idea of finding injective and projective covers indicated above from [41] stimulated recent work which lead to the solution of the flat cover conjecture.

Related aspects can be found in [46].

C. Results using Jensen-functions and $\diamondsuit$ or Φ_κ

As mentioned earlier, black box results often can be simplified and strengthened towards freeness if we are willing to accept additional set theoretic axioms besides ZFC. A predecessor of Theorem 4.1 is the same result [9] under $V = L$ with an additional property of the group G realizing the ring R (with R^+ free) as endomorphism ring: $|G|$ can be any regular, not weakly compact cardinal $\lambda > |R|$

and G is strongly λ-free, in particular any subgroup of G of cardinality $< \lambda$ is free. Similar freeness condition in connection with realizing rings were also derived in a number of papers, we mention only one of them [30]. So even assuming this strong freeness for the class of groups under consideration we are faced with the pathological decompositions as in [22]. In his recent PhD-thesis Strüngmann [84] also proved that E-rings of cardinality $|R| = \lambda$ can be constructed with this additional freeness property which in this setting reads as follows: Any subring of cardinality $< \lambda$ is a subring of a polynomial subring $\mathbb{Z}[X] \subseteq R$ in a set X of $< \lambda$ commuting variables. In any of the mentioned results in this paragraph we also have a particular case which needs only a weak statement of CH, the assertion $2^{\aleph_0} < 2^{\aleph_1}$ which is equivalent to the weak diamond Φ, see [7] and section 3. This principle implies such freeness results for groups, modules or rings of cardinality $\aleph_1$.

D. Results based on the last match of section 3

Shelah [77], applying small cancellation theory, answered the Kurosh problem by showing the existence of a Jonsson group, say of cardinality $\aleph_1$, with no proper uncountable subgroups. His arguments could also be carried out by proving first a combinatorial theorem like the last match in section 3, and then preparing the necessary group theory, which would uniform and simplify the proof, see [66]. Using free products with amalgamations (only), as explained in [63], a problem of E. Dror Farjoun on homotopy theory can be answered by the last match, see [37]. In terms of group theory the main result in [37] reads as follows.

Theorem 4.3. (Assume GCH.) *Any finite simple group H is the subgroup of a group G with the following properties.*

(i) G has cardinality λ^+.
(ii) Any monomorphism of G is an inner automorphism of G, in particular G is complete.
(iii) G is simple (has no non-trivial normal subgroups).
(iv) Any two subgroups of G isomorphic to H are conjugate.
(v) G is generated by copies of H.
(vi) Any element of G which commutes with all elements of a copy of H in G will be trivially 1, i.e. $\mathfrak{c}_G H' = 1$ for any copy H' of H.

It answering a question raised by one of the referees we note a 'monster' group G with the properties as in the theorem can be constructed such that all finite simple groups appear as subgroups of G. The proof is by inspection of the arguments in [37] - just vary H during the inductive construction.

E. Open Problems having a good chance to be solved using the discussed combinatorial principles

Following a suggestion of the referee I would like to finish this paper giving a list of carefully selected classical problems from (mostly) non-commutative group theory which may fall by using the discussed principles. There are obvious other

- say home made - problems which could also belong to this list but which are excluded. The reader will easily find them comparing results known for cardinals $\geq 2^{\aleph_0}$ with those known for structures of cardinality $< 2^{\aleph_0}$. In any case it would be nice to fill the obvious gaps concerning the strength of two related kinds of theorems.

(i) **Warren May's conjecture on group rings:** Let G be a p-group and F a field of characteristic p. Then May's conjecture says that the group ring FG determines the isomorphism class of G. It should be possible to construct two 'nasty' groups two contradict this statement. The conjecture is known to be true only for 'well-behaving' groups, for instance for totally projective groups. For details see Karpilovsky's book [60] or Hill's survey in [55].

(ii) **Two problems by Philip Hall:** Is there a non-trivial group which is isomorphic with every proper extension of itself by itself. This problem is related to the (commutative) recently solved splitter problem discussed above, details can be seen in J. Lennox and S. Stonehewers's book [62]

(iii) **Moreover Hall asked:** Must a non-trivial group, which is isomorphic to each of its non-trivial normal subgroups be either free of infinite rank, simple or infinite cyclic? Surely we guess 'NO' and suggest to hunt for a counter example and use a construction based on the above principles.

(iv) We mentioned from [35, 34] that any group is the outer automorphism group of some (torsion-free) metabelian group. Can we assume that the metabelian groups used in the theorem have additonal nice properties towards freeness? I would think that this needs a change of both the combinatorial principle as well as the group theory used in [35, 34].

(v) **A problem raised by A. I. Mal'cev:** Describe the automorphism group of a free solvable group. By a diamond-construction it should be possible to find 'wild' (at least) almost free solvable groups. This could tell us more about their free relatives.

(vi) Similar to the case of abelian ($p-$)groups it would be interesting to realize near rings as rings of endomorphisms of non-commutative groups modulo the ideal generated by the inner automorphisms.

(vii) **(L. Bokut')** Can the group ring of a torsion-free group contain non-trivial zero-divisors?

(viii) **(B. I. Plotkin)** Is every locally-nilpotent group epimorphic image of a torsion-free locally-nilpotent group?

(ix) **(R. Phillips)** Can we embed any torsion group into a simple torsion group?

References

[1] **T. Becker, L. Fuchs, S. Shelah**, Whitehead modules over domains, Forum Mathematicum **1** (1989), 53–68.

[2] **S. Brenner, C. M. Ringel**, Pathological modules over tame rings, J. London Math. Soc. (2) **14** (1976), 207 – 215.

[3] **C. Casacuberta, J. L. Rodríguez, J.-Y. Tai,** Localization of abelian Eilenberg–Mac Lane spaces of finite type, preprint (1997).

[4] **A.L.S. Corner**, Every countable reduced torsion–free ring is an endomorphism ring, Proc. London Math. Soc. (3) **13** (1963), 687–710.

[5] **A.L.S. Corner**, Additive categories and a theorem of W. G. Leavitt, Bull. Amer. Math. Soc. **75** (1969), 78–82.

[6] **A.L.S. Corner, R. Göbel**, Prescribing endomorphism algebras - A unified treatment, Proceed. London Math. Soc. (3) **50** (1985), 471 – 483.

[7] **K. Devlin, S. Shelah**, A weak version of $\diamondsuit$ which follows from $2^{\aleph_0} < 2^{\aleph_1}$, Israel J. Math. **29** (1978), 239–247.

[8] **S. E. Dickson**, A torsion theory for abelian categories, Trans. Amer. Math. Soc. **121** (1966), 223–235.

[9] **M. Dugas, R. Göbel** Every cotorsion-free ring is an endomorphism ring, Proceed. London Math. Soc. **45** (1982), 319–336.

[10] **M. Dugas, R. Göbel** Every cotorsion–free algebra is an endomorphism algebra, Math. Zeitschr. **181** (1982), 451–470.

[11] **M. Dugas, R. Göbel**, On radicals and products, Pacific J. Math. **18** (1985), 70–104.

[12] **M. Dugas, R. Göbel**, Solution of Philip Hall's problem on the existence of complete locally finite p-groups and results on E.C. groups, J. of Algebra **159** (1993), 115 - 138.

[13] **M. Dugas, R. Göbel**, Automorphisms of torsion-free nilpotent groups of class two, Trans. Amer. Math. Soc. **332** (1992), 633-646.

[14] **M. Dugas, R. Göbel**, Torsion-free nilpotent groups and E-modules, Arch. Math. **54** (1990), 340-351.

[15] **M. Dugas, R. Göbel**, All infinite groups are Galois groups over any field, Trans. Amer. Math. Soc. **304** (1987), 355-384.

[16] **M. Dugas, R. Göbel** Automorphism groups of fields, manuscripta mathematica **85** (1994), 227 – 242.

[17] **M. Dugas and R. Göbel**, Automorphism groups of fields II, Commun. in Algebra, **25** (1997), 3777–3785.

[18] **M. Dugas, R. Göbel**, Endomorphism rings of B_2-groups of infinite rank, Israel Journal Math. **101** (1997), 141 – 156.

[19] **M. Dugas, R. Göbel**, Classification of modules with two distinguished pure submodules and bounded quotients, Results in Math. **30** (1996), 264–275.

[20] **M. Dugas, R. Göbel, W. May**, Free modules with two distinguished submodules, Commun. Algebra **25** (1997), 3473 – 3481.

[21] **M. Dugas, A. Mader, and C. Vinsonhaler**, Large E-rings exist, J. Algebra 108 (1987), 88–101.

[22] **P. Eklof**, Modules with strange decomposition properties, this volume, Birkhäuser Verlag Basel 1999

[23] **P. Eklof, A. Mekler** Almost free modules, Set-theoretic methods, North-Holland, Amsterdam 1990

[24] **S. Files, R. Göbel**, Gauß's theorem for two submodules, Mathematische Zeitschrift **228** (1998), 511 – 536.

[25] **B. Franzen, R. Göbel**, Prescribing endomorphism algebras. The cotorsion-free case. Rend. Sem. Mat. Padova **80** (1989), 215 – 241.

[26] **L. Fuchs**, Infinite abelian groups - Volume 1,2 Academic Press, New York (1970, 1972)

[27] **W. Gaschütz**, Nicht abelsche p-Gruppen besitzen äussere p-Automorphismen, J. Algebra **4** (1966), 1-2.

[28] **R. Göbel**, Cotorsion-free abelian groups with only small cotorsion images, Austral. Math. Soc. (Ser. A) **50** (1991), 243–247.

[29] **R. Göbel**, New aspects for two classical theorems on torsion splitting, Comm. Algebra **15** (1987), 2473–2495.

[30] **R. Göbel, B. Goldsmith**, On almost–free modules over complete discrete valuation rings, Rendiconti Padova **86** (1991), 75 – 87.

[31] **R. Göbel, W. May**, Independence in completions and endomorphism algebras, Forum Mathematicum **1** (1989), 215 – 226.

[32] **R.Göbel, W. May**, Four submodules suffice for realizing algebras over commutative rings, J. Pure Appl. Algebra **65** (1990), 29-43.

[33] **R.Göbel, W. May**, Endomorphism algebras of peak I-spaces over posets of infinite prinjective type, Trans. Amer. Math. Soc. **349** (1997), 3535-3567.

[34] **R, Göbel, A. Paras**, Outer automorphism groups of countable metabelian groups, pp. 309–317, in Proceedings of the Dublin Conference on Abelian Groups, Birkhäuser Verlag, Basel (1999)

[35] **R. Göbel, A. Paras**, Outer automorphism groups of metabelian groups, to appear in Journal of Pure and Applied Algebra 2000

[36] **R, Göbel, A. Paras**, Automorphisms of metabelian groups with trivial center, Illinois J. Math. **42** (1998), 333–346.

[37] **R. Göbel, J. L. Rodríguez, S. Shelah**, Infinite localizations of finite simple groups, submitted to Journal reine und angew. Mathematik 1999

[38] **R. Göbel, S.Shelah**, On the existence of rigid $\aleph_1$-free abelian groups of cardinality $\aleph_1$, pp. 227–237 in Proceedings of the Padova Conference on Abelian Groups and Modules, Kluwer, London 1995

[39] **R. Göbel, S. Shelah**, G.C.H. implies the existence of many rigid almost free abelian groups, in Abelian Groups and Modules, pp. 253 – 271 in Proceedings of the international conference at Colorado Springs 1995, Lecture Notes in Pure and Appl. Math. **182** Marcel Dekker, New York 1996

[40] **R. Göbel, S. Shelah**, Almost free indecomposable modules, the local case, Canadian Journal **50** (1998), 719–738.

[41] **R. Göbel, S. Shelah**, Cotorsion theories and splitters, to appear Trans. Amer. Math. Soc. (June, 2000)

[42] **R. Göbel, S. Shelah**, Almost free splitters, Colloquium Mathematicum **81** (1999), 193–221

[43] **R. Göbel, S. Shelah**, Almost free splitters under negation of CH, new results and a correction, to be submitted to Colloquium Math. 2000

[44] **R. Göbel, D. Simson**, Embeddings of Kronecker modules into the category of prinjective modules and the endomorphism ring problem, Colloquium Math. **75** (1998), 213-244.

[45] **R. Göbel, D. Simson**, Rigid families and endomorphism algebras of Kronecker modules, Israel Journal of Math **110** (1999), 293 – 315.

[46] **R. Göbel, J. Trlifaj**, Cotilting and a hierarchy of almost cotorsion groups, Journal of Algebra **224** (2000), 110 – 122.

[47] **R. Göbel, B. Wald, P. Westphal**, Groups of integer valued functions, Springer Lecture Notes in Math. **874** (1981), 161 – 178.

[48] **J. Gregory**, Higher Souslin trees and the generalized continuum hypothesis, Journ. Symb. Logic **41** (1976), 663-667.

[49] **P. A. Griffith**, A solution of the splitting mixed problem of Baer, Trans. American Math. Soc. **139** (1969), 261–269.

[50] **P. Griffith**, $\aleph_n$–free abelian groups, Quart. J. Math. (2) **23** (1972), 417–425.

[51] **B. Hart, C. Laflamme, S. Shelah**, Models with second order properties V. A general principle, Annals Pure Appl. Logic, Annals of Pure and Appl. Logic **64** (1993) 169 –194.

[52] **G. Higman**, Almost free groups, Proc. London Math. Soc. **1** (1951), 184–190.

[53] **G. Higman**, Some countably free groups, pp. 129–150 in Proceedings "Group Theory", Singapore 1991.

[54] **J. Hausen**, Automorphismen gesättigte Klassen abzählbarer abelscher Gruppen, Studies on Abelian Groups, Springer, Berlin (1968), 147–181.

[55] **P. Hill**, Modular group algebras and simply presented groups, to appear in 'Proceddings of the Dublin Conference on Abelian Groups', (eds. P. Eklof, R. Göbel), Birkhäuser Verlag, Basel 1999

[56] **W. Hodges**, Building models by games, Stud. Texts **2**, Cambr. Univ. Press 1985
[57] **T. Jech**, Set Theory, Academic Press, New York (1978)
[58] **R. Jensen**, The fine structure of the constructible hierarchy, Ann. Math. Logic **4** (1972), 229 – 308.
[59] **A. Kanamori**, The higher infinite, Perspectives in Math. Logic, Springer, Berlin, Heidelberg, New York 1994
[60] **G. Karpilovsky**, Commutative group algebras, Lecture Notes Pure and Appl. Math., **78**, Marcel Dekker, New York 1983
[61] **O. Kerner**, Elementary stones, Comm. Algebra, **22** (1994), 1797 – 1806.
[62] **J. Lennox, S. Stonehewer**, Subnormal subgroups of groups, Oxford Mathem. Monographs 1987
[63] **R. C. Lyndon, P. E. Schupp**, Combinatorial Group Theory, Springer Ergebnisberichte, Vol. **89** Berlin, Heidelberg, New York 1977
[64] **M. Magidor, S. Shelah**, When does almost free imply free? (for groups, transversals, etc.), Journ. Amer. Math. Soc. **7** (4) (1994), 769–830.
[65] **F. Menegazzo and S. Stonehewer**, On the automorphism groups of a nilpotent p-group, J. London Math. Soc. (2) **3** (1985), 272-276.
[66] **N.N.**, this should be carried out, unpublished
[67] **M. Prest**, Model theory and modules, London Math. Soc. L.N. **130**, Cambridge University Press 1988
[68] **C. M. Ringel**, Infinite-dimensional representations of finite dimensional hereditary algebras, Symposia Math. **23** (1979), 321-412.
[69] **C. M. Ringel**, Bricks in hereditary length categories, Resultate der Mathematik **6** (1983), 64 – 70.
[70] **C. M. Ringel**, The braid group action on the set of exceptional sequences of a hereditary artin algebra, pp. 339 – 352 in *Abelian group theory and related topics*, Contemporary Math. **171** American Math. Soc., Providence, R.I. 1994
[71] **P. Rothmaler** , Purity in model theory, pp. 445 – 469 in *Advances in Algebra and Model Theory*, Series Algebra, Logic and Applications, Vol. 9, Gordon and Breach, Amsterdam 1997.
[72] **A. N. Rudakov**, Helices and vector bundles, London Math. Soc. LNM **148**
[73] **L. Salce**, Cotorsion theories for abelian groups, Symposia Math. **23** (1979), 11–32.
[74] **P. Schultz**, Self-splitting groups, Preprint series of the University of Western Australia at Perth (1980)
[75] **S. Shelah** Infinite abelian groups, Whitehead problem and some constructions, Israel Journal Math. **18** (1974), 243 – 256.
[76] **S. Shelah** On uncountable abelian groups, Israel Journal Math. **32** (1979), 311 –330.
[77] **S. Shelah**, On a problem of Kurosh, Jonsson groups, and applications, S. I. Adian, W. W. Boone, G. Higman, eds. Word Problems II. North-Holland Publ. Co. (1980), 373–394.
[78] **S. Shelah**, A combinatorial theorem and endomorphism rings of abelian groups II, pp. 37 – 86, in *Abelian groups and modules*, CISM Courses and Lectures, **287**, Springer, Wien 1984
[79] **S. Shelah**, A combinatorial principle and endomorphism rings. I: On p–groups, Israel J. Math. **49** (1984), 239–257.
[80] **S. Shelah**, On successors of singular cardinals, Logic Colloquium '78, **97** (1978) of Stud. Logic Foundations Math., 357-380, North Holland, Amsterdam-New York.
[81] **S. Shelah**, Non Structure Theory, Oxford University Press (2000) in preparation
[82] **S. Shelah, Z. Spasojevic**, A forcing axiom on strongly inaccessible cardinals making all uniformizations for all ladder systems lying on a fixed stationary set and applications for abelian groups, manuscript No. 587.
[83] **M. Sollert** Unabhängigkeitsaussagen in der Algebra - illustriert an der Theorie der abelschen Gruppen, Staatsexamensarbeit, Essen (1981)

[84] **L. Strüngmann**, Almost-Free $E(R)$-algebras over countable domains, Ph. D. Thesis, Essen 1998

[85] **L. Unger**, Schur modules over wild, finite dimensional path algebras with three non isomorphic simple modules, Journal Pure Appl. Algebra **64** (1990), 205 – 222.

[86] **T. Wakamatsu**, On modules with trivial self extension, Journal of Algebra **114** (1988), 106–114.

[87] **R. Warfield**, Purity and algebraic compactness for modules, Pacific J. Math. **28** (1969), 699 – 719.

[88] **U. H. Webb**, An elementary proof of Gaschütz's theorem, Arch. Math. **35** (1980), 23-26.

[89] **A. E. Zalesskii**, A nilpotent p-group has an outer automorphism, Dokl. Akad. Nauk SSSR **196** (1971), 751-754; Soviet Math. Dokl. **12** (1971), 227-230.

[90] **M. Ziegler**, Model theory of modules, Ann. Pure Appl. Logic **26** (1984), 149 – 213.

FACHBEREICH 6, MATHEMATIK UND INFORMATIK, UNIVERSITÄT ESSEN, 45117 ESSEN, GERMANY

E-mail address: `R.Goebel@Uni-Essen.De`

Trends in Mathematics, © 2000 Birkhäuser Verlag Basel/Switzerland

DIMENSION THEORY OF NOETHERIAN RINGS

T H LENAGAN

INTRODUCTION

This article is a report of three lectures given on "Dimension Theory of Noetherian Rings", at the meeting "Infinite Length Modules", held in Bielefeld, from 7th to 11th September 1998. I would like to thank the organisers for the opportunity to present these lectures, and for the invitation to write up the lectures for this volume. I was asked to present an introduction to the uses of dimension theory in noetherian rings for an audience of algebraists who were not specialists in the area of noetherian rings. In view of this, the material presented here is not original work, nor is it necessarily the most important work in this area. Rather, the results and examples were chosen to illustrate typical uses of dimension theory. Since most of the audience consisted of experts in the Representation theory of Artin rings and finite dimensional algebras, I decided to concentrate on two dimension functions, Krull dimension and Gelfand-Kirillov dimension, which are generalisations of the notions of artinian and finite dimensional, respectively.

1. NOETHERIAN RINGS AND MODULES

We start with a reminder of some of the basic definitions and results in the area of noetherian ring theory. A good reference for most of the notions introduced is the book "Noncommutative Noetherian Rings", by J C McConnell and J C Robson, [11].

Definition A ring R is **right noetherian** if the following three equivalent conditions hold.

1. Each right ideal of R is finitely generated as a right module over R.
2. For each ascending chain of right ideals

$$I_1 \subseteq I_2 \subseteq \dots I_n \subseteq \dots \subseteq R$$

there exists an integer N such that $I_N = I_{N+i}$, for each $i \geq 0$.

3. Each nonempty set of right ideals of R contains a maximal element.

The ring R is **noetherian** if R is both left and right noetherian.

There is a similar definition for the idea of a noetherian module, and it is easy to check that a finitely generated right module over a right noetherian ring is a right noetherian module.

In commutative ring theory, the notion of a prime ideal is one of the key ideas. In the noncommutative theory, it is important, but has to be defined in terms of ideals rather than elements, since otherwise there would be too few prime ideals in an arbitrary ring to have much influence on the structure of rings.

An ideal P of R is a **prime** ideal of R if, whenever A, B are ideals of R with $AB \subseteq P$ then either $A \subseteq P$ or $B \subseteq P$.

A ring R is a **prime ring** if 0 is a prime ideal.

Example The ring of $n \times n$ matrices, for $n > 1$, over a field or commutative integral domain, is a prime ring that is not a domain.

The cornerstone of the whole theory of noetherian rings is Goldie's theorem which describes the structure of noetherian prime rings in terms of the existence of rings of fractions, and links the noetherian theory to artinian theory via the Artin-Wedderburn theorem. An artinian ring that is prime is semisimple artinian. The key point of departure from the theory of semisimple artinian rings is the realization that although it is not true that, in a noetherian prime ring, every right ideal has a complement direct summand; nevertheless, introducing the notion of an essential extension is enough to deal with this deficiency.

A submodule N of a module M is said to be an **essential submodule** of M if and only if $N \cap L$ is nonzero whenever L is a nonzero submodule of M; in this situation, M is also said to be an **essential extension** of N.

Specializing to the ring R, we obtain the notion of an essential right ideal: a right ideal E of R is an **essential right ideal** of R if and only if $E \cap I \neq 0$, whenever I is a nonzero right ideal of R.

Goldie's Theorem A prime noetherian ring R has a ring of fractions $Q = Q(R)$ which is a simple artinian ring. Consequently, invoking the Artin-Wedderburn Theorem, R has a ring of fractions that is a full matrix ring over a division ring.

The key point of Goldie's Theorem is that a right ideal E of R is essential if and only if E contains a non-zerodivisor c of R. Nonzero divisors are often called **regular** elements.

Exercise Show that every nonzero two-sided ideal of a prime noetherian ring is essential as a right ideal (and hence contains a regular element).

A module M for which every nonzero submodule is essential is called a **uniform** module. Thus, in a uniform module, the intersection of any two nonzero submodules is nonzero. It is easy to check that any noetherian module M contains a uniform submodule, and in fact, there is a finite direct sum of uniform submodules $U_1 \oplus \cdots \oplus U_n$ which is an essential submodule of M. The number of summands n in such a direct sum of uniform submodules is an invariant of the module called the **uniform dimension** of M, or, alternatively, the **Goldie rank** of M, denoted by $\text{u.dim}(M)$. If R is a prime noetherian ring with simple artinian quotient ring $Q = M_n(D)$, for some division ring D, then $\text{u.dim}(R_R) = \text{u.dim}(Q_Q) = n$.

Goldie's Theorem admits an extension to the case of semiprime noetherian rings. A ring R is **semiprime** if it has no nonzero nilpotent ideals. If R is a noetherian semiprime ring then there are a finite number of minimal prime ideals, $P_1, \ldots, P_n$, such that $P_1 \cap \cdots \cap P_n = 0$.

Goldie's Theorem for semiprime rings A semiprime noetherian ring R has a ring of fractions $Q = Q(R)$ which is a semisimple artinian ring.

Example The **Weyl Algebra** is the algebra A generated over a field k by two indeterminates x, y satisfying the commutation relation $xy - yx = 1$.

The Weyl Algebra is a noetherian domain. If the field k has characteristic zero then the Weyl Algebra is a simple ring; that is, the only two-sided ideals are 0 and R. However, the Weyl Algebra is not a division ring.

In characteristic zero, the Weyl Algebra has no nonzero finite dimensional representations. Otherwise, we would have matrices A and B satisfying the relation $AB - BA = I$, and a trace argument gives a contradiction (the left hand side has trace zero, the right hand side has trace n if we are in $n \times n$ matrices).

Exercise Prove that the first Weyl algebra, A, over a field of characteristic zero, is simple (consider $[x, -], [y, -]$, to see that the ideal generated by any nonzero element contains a unit).

Here is an interesting result about modules over noetherian nonartinian rings that has no counterpart in the artinian theory, and which has far-reaching consequences in the noetherian theory.

Eisenbud-Robson Let M be a module of finite length over a simple noetherian ring R that is not artinian. Then M is cyclic.

Sketch of Proof We argue by induction on the length of the module. If M has length one then M is simple, and certainly cyclic. Suppose that M has length n and that modules of length less than n over R are cyclic. Choose a simple submodule B of M, and any $b \in B$. Then $B = bR$, and $M/bR = \bar{a}R$ for some $a \in M$, by the inductive hypothesis. Thus, $M = aR + bR$, with bR simple. Now, $aR \cong R/\mathrm{Ann}(a)$, so $\mathrm{Ann}(a) \neq 0$, since R is not artinian. Choose $0 \neq x$ with $ax = 0$.

Now,

$$bRx = 0 \Rightarrow bRxR = 0 \Rightarrow bR = 0 \Rightarrow b = 0,$$

a contradiction; so $brx \neq 0$, for some r.

Then

$$(a + br)R \supseteq (a + br)xR = brxR = bR$$

and

$$a = (a + br) - br \in (a + br)R,$$

and so $M = (a + br)R$. □

2. Dimension functions

In commutative noetherian ring theory, one of the most powerful dimensions is the **Krull dimension** which measures the longest possible chain of prime ideals in the ring. In the case of Krull dimension zero, the ring is artinian, and if the ring is a finitely generated algebra over a field k then it is finite dimensional as a k vector space, see, for example, [11, 13.10.3].

We will try to find dimension functions for noncommutative rings that will replace the commutative Krull dimension. The obvious generalization is the **classical Krull dimension**, which again just measures chains of prime ideals. This works when there are lots of two-sided ideals (eg. for polynomial identity rings, cf. [11, 6.4.8 and Chapter 13]), but would give every simple ring dimension zero, and this is not much use!

We will ask for dimensions defined for noncommutative noetherian rings and modules, which correspond to the commutative setting in dimension zero, and are sufficiently delicate to enable progress to be made. First, what kinds of properties do we want?

A **dimension function**, d, defined on noetherian modules over noetherian rings, assigns a value $d(M_R)$ to each noetherian R-module M. Assume that $d(M)$ is either -1 (when $M = 0$) or in $\mathbb{R}^{\geq 0} \cup \infty$, or an ordinal number.

Let us try to explore the structure of noetherian rings by using a putative dimension function d, and see what desireable properties we would like to hold for the function d. We will use the following general strategy as a model for our attack. In a noetherian ring, the nil radical N is nilpotent, the ring R/N is semiprime (that is has no nilpotent ideals) and R/N has a semisimple artinian ring of quotients, by Goldie's Theorem. In a general problem, one hopes to be able to factor out N and then solve the problem in the resulting semiprime ring R by exploiting the existence of a semisimple artinian quotient ring. We will try to follow a similar strategy using d. Right ideals with dimension less than the dimension of R will be considered small; we factor out such right ideals and are then in a ring with all right ideals having the same dimension as that of the ring. We hope to be able to say something about the structure of such rings (for example, that they have rings of fractions). However, for all this to work, we need d to have certain desireable properties.

The following are some desireable properties that we might hope that the function d satisfies.

Suppose that $d(R) = \alpha$. Look at the sum I of all right ideals E of R with $d(E) < \alpha$. We would like $d(I) < \alpha$, and that each nonzero right ideal A of R/I has $d(A) = \alpha$. This will hold, provided that d is **exact** in the sense given below.

Now assume we have factored out this ideal I so that each nonzero right ideal A of R has $d(A) = d(R) = \alpha$. Can we prove that R has a ring of quotients? Let c be an element of R. Suppose that $d(R/cR) < \alpha$. If $xc = 0$ then we have R/cR

maps onto xR and so $d(R/xR) < \alpha$, forcing $xR = 0$, and so $x = 0$, and the left annihilator of c is zero. With a suitable symmetry hypothesis we would also know that the right annihilator is zero, so that c is a regular element. Thus, we will have that $d(R/cR) < \alpha$ implies that c is regular. The torsion property defined below is sufficient to give the converse to this. We then have a way of passing between "large" right ideals I (ie. $d(R/I) < d(R)$) and regular elements. The strategy of proof of Goldie's Theorem can then be used to produce a ring of fractions.

A dimension function d is **exact** if whenever

$$0 \to A \to B \to C \to 0$$

is a short exact sequence of modules then

$$d(B) = \max\{d(A), d(C)\}.$$

Exactness ensures that the dimension of submodules and quotient modules are no bigger than the original modules, and enables arguments to be constructed by passing from a module to suitable submodules or quotients.

A dimension function d is **finitely partitive** if for each finitely generated module M there is an integer $n = n(M)$ such that for every chain

$$M \supseteq M_1 \supseteq M_2 \supseteq \ldots \supseteq M_m$$

with $d(M_i/M_{i+1}) = d(M)$, one has $m \leq n$.

Partitivity enables us to use artinian type techniques in proofs.

A dimension function d has the **torsion property** if whenever c is a regular element of R then $d(R/cR) + 1 \leq d(R)$.

A dimension function d **separates primes** if whenever $P \subsetneqq Q$ are prime ideals of R then

$$d(R/Q) + 1 \leq d(R/P).$$

Prime separation follows from the torsion property. One use of these properties is to establish isomorphisms. For example, suppose that R and S are noetherian domains and we have an epimorphism θ from R onto S which we suspect should be an isomorphism. If θ is not an isomorphism then $d(S) = d(R/\ker(\theta)) < d(R)$, and conversely; so if we can show R and S have the same dimension, then the isomorphism is established.

A dimension function d is **symmetric** if whenever A and B are noetherian rings in the class of rings under consideration and M is an A-B–bimodule that is finitely generated on each side. Then

$$d({}_AM) = d(M_B).$$

Many proofs in noncommutative rings require one to pass from left to right and vice-versa. Symmetry retains control of dimensions in this situation. In fact, in this discussion, we only need a weak symmetry condition on subfactors of ${}_RR_R$.

Suppose that d is an exact, symmetric dimension function on a class noetherian rings and assume the torsion property. Let $d(R) = \alpha$. By exactness, there is a unique maximal ideal A with $d(A) < d(R)$, and in R/A every nonzero right ideal has dimension equal to α. Such a ring is said to be **α-homogeneous**. By symmetry, each nonzero left ideal has dimension equal to α.

A further technical condition is needed to prove the existence of a ring of quotients. A dimension function d is **ideal invariant** if whenever M is a noetherian module and T is a two-sided ideal of R then $d(M \otimes T) \leq d(M)$; in particular, if I is a right ideal of R, then ideal invariance implies that $d(T/IT) \leq d(R/I)$.

Theorem Suppose that d is an exact, symmetric, ideal invariant dimension function on a class noetherian rings, and that R is α-homogeneous. Assume the torsion property. Then R has an artinian ring of quotients.

Sketch of Proof By Small's Theorem, [11, Corollary 4.1.4], it is enough to show that an element c that is a regular element modulo the nilpotent radical N of R is also a regular element in R. Let c be such an element.

By exactness, $d(R/N) = d(R) = \alpha$, and by the torsion property, $d(R/cR+N) < \alpha$. We aim to show that $d(R/cR) < \alpha$. If $N^m = 0$ then $R/cR = R/(cR + N^m)$, so it is enough to show that $d(R/(cR + N^s) < \alpha$, for each s. We prove this by induction, the case $s = 1$ being given above. Thus, assume that $d(R/(cR + N^{s-1})) < \alpha$. By exactness, it is enough to show that $d((cR+N)/(cR+N^s) < \alpha$. However, $(cR+N)/(cR+N^s) \cong N/(N \cap (cR+N^s))$ is a homomorphic image of $N/(cN+N^s) = N/(cR+N^{s-1})N$ and this latter module has Krull dimension less than α, by the inductive hypothesis and the fact that N is ideal invariant. Thus, $d((cR+N)/(cR+N^s)) < \alpha$, and $d(R/(cR+N^s)) < \alpha$, as required.

Now suppose that $xc = 0$. Then xR is a homomorphic image of R/cR; and so $d(xR) \leq d(R/cR) < \alpha$. Hence, $x = 0$, since R is homogeneous.

By a symmetric argument, $cx = 0 \Rightarrow x = 0$; so c is a regular element. Hence, R has an artinian ring of quotients, by Small's Theorem. □

3. Krull dimension

The following are suitable references for the basic material on noncommutative Krull dimension:
R Rentschler and P Gabriel, *Sur la dimension des anneaux et ensembles ordonnés*, [14];
R Gordon and J C Robson, *Krull dimension*, [5];
J C McConnell and J C Robson, *Noncommutative noetherian rings*, [11].

Krull dimension is a measure of how far a module deviates from being artinian. The definition is inductive, starting with artinian modules as the base case.

Krull dimension of a noetherian module.

$\mathrm{Kdim}(0) := -1(\text{or } -\infty)$

$\operatorname{Kdim}(M) := 0$ if and only if M is artinian.

$\operatorname{Kdim}(M) := 1$ if and only if in any chain

$$M \supseteq M_1 \supseteq M_2 \supseteq \dots$$

almost all factors have $\operatorname{Kdim}(M_i/M_{i+1}) < 1$, we then continue inductively for all ordinals.

In general, we define $\operatorname{Kdim}(M) = \alpha$ if and only if $\operatorname{Kdim}(M) \not< \alpha$ but in any chain

$$M \supseteq M_1 \supseteq M_2 \supseteq \dots$$

all but finitely many factors have Krull dimension less than α.

An alternative way to look at this definition is via **quotient categories**: Let $\mathbb{A}$ be the category of noetherian R-modules, and $\mathbb{C}$ the subcategory of all modules with Krull dimension less than α. Let T be the natural map $\mathbb{A} \to \mathbb{A}/\mathbb{C}$. Then $\operatorname{Kdim}(M) = \alpha$ if and only if $T(A)$ is an artinian object of $\mathbb{A}/\mathbb{C}$. For a discussion of this approach, see [6].

Example Let G be a finitely generated abelian group (that is, a $\mathbb{Z}$-module), say,

$$G \cong \text{Finite} \oplus \{\mathbb{Z}^n\}.$$

In this case, the **torsionfree rank**, $\operatorname{Rank}(G)$, is n. Note that $\operatorname{Kdim}(G) = 0$ iff G artinian iff G is finite iff $\operatorname{Rank}(G) = 0$.

If $H \leq G$ then

$$\operatorname{Rank}(G) = \operatorname{Rank}(H) + \operatorname{Rank}(G/H);$$

and so if

$$G \geq G_1 \geq G_2 \geq \dots$$

is a descending chain of subgroups of G then at most n factors G_i/G_{i+1} can be infinite, so all but finitely many factors are finite; that is, have Krull dimension less than 1. Hence, $\operatorname{Kdim}(G) \leq 1$. □

Elementary properties of Krull dimension

- $\operatorname{Kdim}(R) := \operatorname{Kdim}(R_R)$
- Every noetherian module has a Krull dimension
- If $N \subseteq M$ then

$$\operatorname{Kdim}(M) = \max\{\operatorname{Kdim}(N), \operatorname{Kdim}(M/N)\}$$

- If c is a regular element of R then

$$\operatorname{Kdim}(R/cR) + 1 \leq \operatorname{Kdim}(R)$$

- $\operatorname{Kdim}(R[x]) = \operatorname{Kdim}(R) + 1$

The proofs of these basic properties can be found in [11, Chapter 6]. In the next example, we need a generalization of the last property in the list to the case of skew polynomial rings: if δ is a derivation of the ring R then $\mathrm{Kdim}(R) \leq \mathrm{Kdim}(R[x;\delta]) \leq \mathrm{Kdim}(R)+1$, while if R is artinian then $\mathrm{Kdim}(R[x;\delta]) = 1$. [11, 6.5.4]. In general, the behaviour of Krull dimension under extensions of rings is very difficult to calculate; see the paper by Vladimir Bavula, [1], in these proceedings for more information about this.

Example Let $A = k[x,y]$, with $xy - yx = 1$ be the first Weyl algebra over a field of characteristic zero. Then $\mathrm{Kdim}(A) = 1$.

Sketch of proof. The Weyl algebra is not artinian; so $\mathrm{Kdim}(A) \geq 1$. The sets $k[x]\backslash 0$ and $k[y]\backslash 0$ are Ore sets in A, and so the localizations $B = k(x)[y]$ and $C = k(y)[x]$ can be constructed. Both of these rings are skew polynomial algebras, for example, $k(x)[y] = k(x)[y;\delta]$, where $\delta(f) = df/dx$, for $f \in k(x)$. Set $R := B \oplus C$. Note that R is a flat extension of A, since B and C are both localizations of A. Since Krull dimension is defined on lattices of right ideals, and since right ideals in a localization are extensions of right ideals in the original ring, it is obvious that $\mathrm{Kdim}(A) \geq \max\{\mathrm{Kdim}(B), \mathrm{Kdim}(C)\}$. It is easy to establish that $\mathrm{Kdim}(R) = \max\{\mathrm{Kdim}(B), \mathrm{Kdim}(C)\} = 1$.

Consider $A \hookrightarrow R$ by $a \mapsto a \oplus a$. We will show that R is a faithfully flat extension of A. Since this then gives an injection of the lattice of right ideals of A into the lattice of right ideals of R, we deduce that $\mathrm{Kdim}(A) \leq \mathrm{Kdim}(R) = 1$. In order to establish faithful flatness, it is enough to show that if I is a right ideal of A with $IR = R$ then $I = A$, [11, 7.2.3].

Suppose that I is a right ideal of A such that $IR = R$. Then $IB = B$, and $IC = C$, and so there exist $0 \neq f(x) \in I$ and $0 \neq g(y) \in I$, and it follows that A/I is finite dimensional over k, a contradiction unless $I = A$.

Hence, R is a faithfully flat extension of A and so $\mathrm{Kdim}(A) \leq \mathrm{Kdim}(R) = 1$. □

A result that applies to right ideals of the Weyl algebra can now be proved.

Theorem If R is a simple noetherian ring, and $\mathrm{Kdim}(R) = 1$, then each right ideal of R can be generated by 2 elements.

Sketch of proof. Assume R is a domain, to simplify the argument. Let I be a nonzero right ideal of R, and choose $0 \neq c \in I$. Then

$$\mathrm{Kdim}(I/cR) \leq \mathrm{Kdim}(R/cR) < \mathrm{Kdim}(R) = 1,$$

so that either $\mathrm{Kdim}(I/cR) = -1$ and $I = cR$, or $\mathrm{Kdim}(I/cR) = 0$.

In the latter case, I/cR is an artinian module over a simple nonartinian ring, and so is cylic, generated by the coset $b + cR$, say. Hence, $I = bR + cR$. □

Corollary Every right ideal of the Weyl algebra can be generated by 2 elements.

In fact, this is usually known as a $1\frac{1}{2}$**-generator** result, since there is a free choice of the first generator, among the regular elements in the right ideal.

If M is a noetherian module with the property that $\mathrm{Kdim}(M/N) < \mathrm{Kdim}(M)$, for each nonzero submodule N of M then M is said to be a **critical module**. Critical modules are generalizations of simple modules.

It is not difficult to establish that every noetherian module contains a critical submodule.

If M is a noetherian module with $\mathrm{Kdim}(M) = \alpha$, then there is an integer n, called the **α-length** of M such that in any chain,

$$M \supseteq M_1 \supseteq M_2 \supseteq \dots$$

at most n factors can have $\mathrm{Kdim}(M_i/M_{i+1}) = \alpha$.

A general strategy for proving results in noetherian rings, using Krull dimension is to use artinian type methods, arguing on the α-length of a module, and then to replace simple modules by critical modules.

Example Levitski's Theorem states that a nil right ideal in a noetherian ring is nilpotent. Here is a proof of this fundamental result, using an artinian perspective and Krull dimension techniques.

Sketch of proof. Note that if I is a nil right ideal that is not nilpotent and N is a nilpotent ideal, then $(I + N)/N$ is a nonzero nil right ideal of R/N. By using this observation, we may first reduce to the case that R has no nilpotent ideals, and then we need to show that any nil right ideal is equal to 0. Suppose not, and choose a nonzero nil right ideal I with $\alpha = \mathrm{Kdim}(I)$ as small as possible. Since I contains a critical right ideal which will also be nil and have Krull dimension α, by the minimality of α, we may assume that I itself is critical. If $I^2 = 0$, then $(RI)^2 = 0$, and we have a nonzero nilpotent ideal, a contradiction. Thus, $I^2 \neq 0$ and we may choose $0 \neq c \in I$, with $cI \neq 0$.

Consider the homomorphism

$$\theta : I \to I, \qquad i \mapsto ci.$$

Note that $\ker(\theta) \neq 0$, since $c^n = 0$, for some n.

Thus,

$$\mathrm{Kdim}(cI) = \mathrm{Kdim}(I/\ker(\theta)) < \alpha,$$

since I is critical; and so $cI = 0$, by the minimality of α, a contradiction. □

A famous result of Stafford starts with a generalization to arbitrary noetherian modules, using Krull dimension, of the Eisenbud-Robson result for modules of finite length given earlier.

In order to prove Stafford's result, we use the idea of **noetherian induction**. Suppose one wants to prove that all modules in some class $\mathcal{C}$, which is closed under taking factors, have a certain property P. Let M be a module in $\mathcal{C}$. If M does not have property P, then the set $\mathcal{S}$ of submodules N of M such that M/N does not have property P is not empty, since $0 \in \mathcal{S}$. By the noetherian property, we may choose a maximal member $N \in \mathcal{S}$. Then M/N does not have property P, but M/A has property P for all submodules A of M which properly contain N. Thus, M/N does not have property P, but all of its proper factors do have property P. In using noetherian induction, we replace the original M by M/N, so that we may assume that M does not have the desired property P, but each of its proper factors does have property P. We then aim to produce a contradiction.

Lemma Let M be a nonzero noetherian module over a simple noetherian ring R, and suppose that $\mathrm{Kdim}(M) < \mathrm{Kdim}(R)$. Then, there exists $d \in M$ such that $\mathrm{Kdim}(M/dR) < \mathrm{Kdim}(M)$.

Sketch Proof By noetherian induction, we may assume that the result holds in all proper factors of M. Let B be any critical submodule of M and choose any $0 \neq b \in B$. Then $0 \neq bR$ is also critical, and by noetherian induction, the result holds in $\bar{M} = M/bR$. Thus, there is an element $a \in M$ such that $\mathrm{Kdim}(\bar{M}/\bar{a}R) < \mathrm{Kdim}(\bar{M}) \leq \mathrm{Kdim}(M)$. It follows that

$$\mathrm{Kdim}(M/aR + bR) < \mathrm{Kdim}(M)$$

with bR critical.

Suppose that we can find a submodule dR of $aR + bR$ such that $\mathrm{Kdim}(aR + bR/dR) < \mathrm{Kdim}(M)$. Then $\mathrm{Kdim}(M/dR) < \mathrm{Kdim}(M)$, since $\mathrm{Kdim}(M/aR + bR) < \mathrm{Kdim}(M)$. Thus, we may assume that $M = aR + bR$.

Note that

$$\begin{aligned} \mathrm{Kdim}(R/\mathrm{Ann}(a)) &= \mathrm{Kdim}(aR) \\ &\leq \mathrm{Kdim}(M) < \mathrm{Kdim}(R); \end{aligned}$$

and so $\mathrm{Ann}(a) \neq 0$. Choose $0 \neq x$ with $ax = 0$.

As in the Eisenbud-Robson argument,

$$bRx = 0 \Rightarrow bRxR = 0 \Rightarrow bR = 0 \Rightarrow b = 0,$$

a contradiction; so $brx \neq 0$, for some r.

Set $d = a + br$. Then $a = d - br \in dR + bR$ and $b \in dR + bR$, so that $M = aR + bR \subseteq dR + bR$ and it follows that $M = dR + bR$. The element $dx = (a + br)x = brx \neq 0$ is in $dR \cap bR$, and so this submodule is nonzero. Thus, $\mathrm{Kdim}(bR/(dR \cap bR)) < \mathrm{Kdim}(bR) \leq \mathrm{Kdim}(M)$, since bR is critical. Hence, $\mathrm{Kdim}(M/dR) = \mathrm{Kdim}(dR + bR/dR) = \mathrm{Kdim}(bR/(dR \cap bR)) < \mathrm{Kdim}(M)$, as required.

Theorem (Stafford) Let R be a simple noetherian ring with $\mathrm{Kdim}(R) = n$. Then every right ideal of R can be generated by $n + 1$ elements.

Sketch of proof. First, we show that it is enough to establish the result for essential right ideals. Suppose that I is a right ideal that is not essential. Then choose a nonzero right ideal J maximal such that $I \cap J = 0$. It is easy to check that $E := I \oplus J$ is essential right ideal, and any generating set for E projects onto a generating set for I.

Thus, we assume that I is an essential right ideal. Choose a regular element $c \in I$ and set $M = I/cR$. Note that

$$\mathrm{Kdim}(M) \leq \mathrm{Kdim}(R/cR) < \mathrm{Kdim}(R).$$

Now apply the previous result to M reducing Krull dimension by at least one each time we factor out an element. □

Notice that in the Eisenbud-Robson Theorem, and Stafford's result above, the generator $(a + br)$ arises in a "stable" way from $aR + bR$. By developing these ideas much further, Stafford obtains noncommutative versions of Serre's theorem on free summands, and Bass cancellation:

SERRE'S THEOREM If M is a 'big' module, then M has a free summand.

BASS CANCELLATION If M is a 'big' module then

$$M \oplus R \cong N \oplus R \Rightarrow M \cong N.$$

In each case, 'big' is specified in terms of Krull dimension. Exact statements of these results can be found in [11, 11.7.13].

4. Gelfand-Kirillov dimension

Suitable references for Gelfand-Kirillov dimension are:

W Borho and H-P Kraft, *Über die Gelfand-Kirillov Dimension*, [2];
G R Krause and T H Lenagan, *Growth of algebras and Gelfand-Kirillov dimension*, [7];
J C McConnell and J C Robson, *Noncommutative noetherian rings*, [11, Chapter 8];
V A Ufnarovskij, *Combinatorial and asymptotic methods in algebra*, [15].

Let A be a finitely generated k-algebra and let V be a finite dimensional k-subspace that contains 1 and a set of algebra generators of A.

Then

$$k = V^0 \subseteq V \subseteq V^2 \subseteq V^3 \subseteq \ldots \subseteq \bigcup V^n = A.$$

The **growth function of A (relative to V)** is the function F given by

$$F(n) := \dim(V^n).$$

It is sometimes easier to calculate the first difference function $f(n)$, where

$$\begin{aligned} f(n) &:= F(n) - F(n-1) \\ &= \dim(V^n) - \dim(V^{n-1}) \\ &= \dim(V^n/V^{n-1}). \end{aligned}$$

(i) The algebra A is said to have **polynomially bounded growth** if there is a constant d such that $F(n) \leq n^d$, for $n \gg 0$.

(ii) The algebra A is said to have **exponential growth** if there is a constant $c > 1$ such that $F(n) \geq c^n$, for $n \gg 0$.

(iii) Otherwise, A is said to have **intermediate growth**.

Example $R = k[x_1, \ldots, x_d]$ a commutative **polynomial algebra**.

$$f(n) = \#\text{ distinct } x_1^{a_1} x_2^{a_2} \ldots x_d^{a_d}$$

with $\sum a_i = n$.

Thus

$$f(n) = \binom{n+d-1}{d-1},$$

which is a polynomial of degree $d-1$, and so

$$F(n) = f(0) + f(1) + \ldots + f(n) = \binom{n+d}{d}$$

is a polynomial of degree d, and the polynomial algebra has **polynomial growth**! □

Exercise Prove the statement above.

Example The free algebra $R = k\{x, y\}$

$f(n)$ is the number of words of length n, so

$$f(n) = 2^n.$$

The free algebra has **exponential growth**. □

Example Let $\mathfrak{g}$ be the infinite dimensional Lie algebra with basis $e_1, e_2, \ldots$ and relations

$$[e_i, e_j] = (i-j)e_{i+j}.$$

Let $\mathcal{U}$ be the universal enveloping algebra of $\mathfrak{g}$; so $\mathcal{U} = k[e_1, e_2, \ldots]$ with relations

$$e_i e_j - e_j e_i = (i-j)e_{i+j}.$$

Then, $\mathcal{U} = k[e_1, e_2]$, and, in fact, $\mathcal{U}$ is finitely presented, [15].

To calculate $f(n)$ we need to find

$$p(n) = \#\text{ partitions of } n$$

and

$$p(n) \sim \frac{1}{4n\sqrt{3}} e^{\pi\sqrt{\frac{2n}{3}}}$$

so the growth is **intermediate**. □

(The methods we present here will not deal with algebras of intermediate growth. However, Petrogradsky, [12, 13], has recently introduced a family of functions designed to measure different types of subexponential growth.)

The distinction between subexponential and exponential growth is important, as indicated by the following structure result for domains with subexponential growth.

Proposition (Jategaonkar) If R is a domain with subexponential growth then R has a division ring of fractions.

Sketch of Proof. The crucial point concerning the existence of the division ring of quotients is to see that left fractions can be re-written as right fractions; this is the content of the Ore condition for the existence of a ring of fractions.

Given $a, c \in R$, we need to be able to find $b, d \in R$ such that $c^{-1}a = bd^{-1}$, or, $ad = cb$ $(\neq 0)$; ie.

$$aR \cap cR \neq 0.$$

Now, $aR \cap cR = 0$ implies $k\{a, c\}$ is a free algebra, and so $k\{a, c\}$ has exponential growth and hence, so does R, a contradiction. □

Exercise Prove $aR \cap cR = 0$ implies $k\{a, c\}$ is a free algebra.

In case (i), we define the **Gelfand-Kirillov dimension** of the algebra R by

$$\mathrm{GKdim}(R) = \overline{\lim}\left\{\frac{\log(\dim(V^n))}{\log(n)}\right\}$$

where V is a finite dimensional subspace of R that contains 1 and a set of algebra generators of R.

Exercise Prove that the definition is independent of the choice of V, using the fact that if W is another such subspace then there are integers s, t with $V \subseteq W^s \subseteq V^t$.

An alternative definition of **Gelfand-Kirillov dimension** is

$$\mathrm{GKdim}(R) := \inf\{\gamma \in \mathbb{R} \mid \dim(V^n) \leq n^\gamma, n \gg 0\}$$

(Again, this is independent of V.)

Exercise Prove that the two definitions are equivalent.

In cases (ii) and (iii), (exponential and intermediate growth),

$$\mathrm{GKdim}(R) = \infty.$$

Example $R = k[x_1, \dots, x_d]$ a commutative **polynomial algebra**.

$$f(n) = \#\text{ distinct } x_1^{a_1}x_2^{a_2}\dots x_d^{a_d}$$

with $\sum a_i = n$.

Set $V := k + kx_1 + kx_2 + kx_n$. Then

$$\dim(V^n) = F(n) = f(0) + f(1) + \ldots + f(n) = \binom{n+d}{d}$$

is a polynomial of degree d, and

$$\mathrm{GKdim}(k[x_1,\ldots,x_d]) = d. \quad \square$$

Example The Weyl algebra $R = k[x, y]$, with relation $xy - yx = 1$.

Let V be spanned by $1, x, y$. Start writing out the basis for R given by the powers of V.

$$1$$

$$x, \quad y$$

$$x^2, \quad xy, \quad yx, \quad y^2$$

$$x^3,\ x^2y,\ xyx,\ xy^2,\ yx^2,\ yxy,\ y^2x,\ y^3$$

$$\vdots$$

Notice that at any stage where there is a monomial with a y occuring before an x then the defining relation $xy - yx = 1$ can be used to show that the monomial is dependent on earlier monomials, and so does not contribute to the dimension count.

Thus the growth is exactly the same as the commutative polynomial ring $k[x, y]$; and so the Weyl Algebra has Gelfand-Kirillov dimension 2. $\square$

The Bergman Gap

$$\mathrm{GKdim}(R) = 0$$

iff $\quad V^n = V^{n+1} = \ldots$, for $n \gg 0$,
iff $\quad R$ is a finite dimensional algebra.

Otherwise, $\dim V^n \geq n$, for all n, and so

$$\mathrm{GKdim}(R) \geq 1.$$

Borho and Kraft showed that any real number in the interval $[2, \infty)$ occurs as a Gelfand-Kirillov dimension,

and

Bergman showed that there is **no** algebra R such that $1 < \mathrm{GKdim}(R) < 2$.

Example Let A be the algebra generated by x and y subject to the relations

$$yx^m yx^n y = 0,$$

for all $m, n \geq 0$; while

$$yx^n y = 0,$$

whenever $n \geq 1$ is a nonsquare.

Exercise Show that the number of nonzero monomials of length n is $\mathcal{O}(n^{\frac{3}{2}})$; so that $\mathrm{GKdim}(A) = \frac{5}{2}$.

Let A be a k-algebra, and let M be a finitely generated right A-module. Let M_0 be a finite dimensional vector space such that $M = M_0A$. We define the **Gelfand-Kirillov dimension** of M by

$$\mathrm{GKdim}(M) := \inf\{\gamma \in \mathbb{R} \mid \dim(M_0V^n) \leq n^\gamma, n \gg 0\}$$

Proposition Let c be a regular element of a finitely generated k-algebra A. Then

$$\mathrm{GKdim}(A/cA) + 1 \leq \mathrm{GKdim}(A).$$

Proof Set $I = cA$. Let V generate A, and $1, c \in V$. Then $\bar{V}$ generates $\bar{A} := A/I$. Let D_n be a complement of $I \cap V^n$ in V^n. Then, $\dim(\bar{V}^n) = \dim(D_n)$. However,

$$D_n \cap cA \subseteq D_n \cap (I \cap V^n) = 0,$$

and it follows that

$$D_n + cD_n + c^2D_n + \ldots + c^nD_n \ (*)$$

is a direct sum of k-subspaces of A contained in V^{2n}.

Suppose that $\mathrm{GKdim}(A) = \beta$, and let $\gamma > \beta$; so $\dim(V^n) \leq n^\gamma$, for $n \gg 0$. Then

$$n \times \dim(\bar{V}^n) \leq (2n)^\gamma,$$

for $n \gg 0$, since

$$D_n \oplus cD_n \oplus c^2D_n \oplus \ldots \oplus c^nD_n \subseteq V^{2n};$$

and so

$$\dim(\bar{V}^n) \leq 2^\gamma n^{\gamma-1},$$

for $n \gg 0$. The result then follows by using the definition of Gelfand-Kirillov dimension. □

Exercise Prove $(*)$ is a direct sum.

Proposition Let A and B be finitely generated K-algebras, and let M be an A-B–bimodule that is finitely generated on each side. Then

$$\mathrm{GKdim}({}_AM) = \mathrm{GKdim}(M_B).$$

If, in addition, ${}_AM$ is faithful as an A-module, then

$$\mathrm{GKdim}({}_AM) = \mathrm{GKdim}(A).$$

Sketch of proof. Let F and V be finite dimensional subspaces of M and B, respectively, with $\cup FV^n = M$. Since ${}_AM$ is finitely generated, there exists a finite dimensional subspace $G \supseteq F$ such that $AG = M$. Obviously $FV^n \subseteq GV^n$ for all integers $n \geq 0$. Now GV is a finite dimensional subspace of M; so, since $AG = M \supseteq GV$, there exists a finite dimensional subspace W of A such that $WG \supseteq GV$. Then $FV^n \subseteq GV^n \subseteq W^nG$ for any n, and the claim follows from the definition of the Gelfand-Kirillov dimension for modules. □

Proposition Let A be a k-algebra, let M be a right A-module, and let ${}_AN_A$ be a bimodule, finitely generated on each side. Then

$$\mathrm{GKdim}(M \otimes_A N) \leq \mathrm{GKdim}(M).$$

Exercise Prove this

Gelfand-Kirillov dimension is known to be an integer for many classes of affine noetherian algebras. In fact, there are no known examples of such algebras with finite Gelfand-Kirillov dimension where the Gelfand-Kirillov dimension is not an integer.

Also, Gelfand-Kirillov dimension is known to be finitely partitive for many classes of affine noetherian algebras. In fact, there are no known examples of such algebras which are not finitely partitive.

Gelfand-Kirillov dimension is also known to be exact for many classes of affine noetherian algebras, and again, there are no known examples of such algebras which are not exact.

However, there are examples, due to Bergman of affine PI algebras which are not exact.

Stephenson and Zhang have shown that a connected graded noetherian k-algebra cannot have exponential growth. It is an open problem whether intermediate growth is possible

The situation for noetherian rings or noetherian algebras with finite Gelfand-Kirillov dimension is summarized by the following table.

	GK	Kdim
Exact		$\surd$
Partitive		$\surd$
Symmetric	$\surd$	
Torsion property	$\surd$	$\surd$
Separates primes	$\surd$	$\surd$
Ideal invariant	$\surd$	

Filling in any of these gaps would be a welcome development, and (if positive) would lead to substantial progress.

For example, Jategaonkar proved symmetry of Krull dimension for fully bounded noetherian rings, and used this to prove Jacobson's conjecture for fullly bounded noetherian rings:

$$\bigcap_{i=1}^{\infty} J^n = 0.$$

(A noetherian ring R is **fully bounded noetherian** if in each prime factor ring of R every nonzero one-sided ideal contains a nonzero two-sided ideal. This guarantees that a fully bounded noetherian ring has many two-sided ideals; in fact, the Krull dimension and the classical Krull dimension coincide for fully bounded noetherian rings. The main examples of fully bounded noetherian rings are the noetherian polynomial identity rings.)

I showed that the artinian condition ($\mathrm{Kdim}(M) = 0$) is symmetric for noetherian bimodules and used this to prove Jacobson's conjecture for rings with Krull dimension one. A major question in the area is whether Krull dimension is symmetric on noetherian bimodules.

Incidentally, the symmetry of the artinian condition for noetherian bimodules has a nice consequence in representation theory: a finite length module which is a bimodule and artinian or noetherian as a module over its endomorphism ring, is endofinite, [3, Prop 4.5].

Substantial progress can often be made if a link can be found between one dimension function and another, since then desireable properties of both can be exploited.

In another paper in this volume, Vladimir Bavula, [1], will explore some of the connections between Gelfand-Kirillov dimension and Krull dimension, so I won't go into these.

However, connections between growth and homological conditions have also been particularly useful. Gabber used such a connection to establish catenarity for enveloping algebras of finite dimensional solvable Lie algebras. Goodearl and I have recently put Gabber's result into an abstract form which has enabled catenarity to be established for many classes of algebras.

The ring R is said to be **catenary** if, for any two prime ideals $P < Q$ of R, all saturated chains of prime ideals between P and Q have the same length.

Matsumura, [9, p31], says

"Practically all important (commutative) noetherian rings arising in applications are catenary. The first example of a non-catenary ring is due to Nagata 1956", see the back of Nagata's book on local rings, [10], for the example,

and on page 123

"The theorem that quotients of (commutative) Cohen-Macaulay rings are catenary is of great importance in dimension theory".

The abstract form of Gabber's Theorem proved by Goodearl and myself says informally:

Let R be an affine, noetherian, algebra over a field, with finite Gelfand-Kirillov dimension, good homological properties and a link between the Gelfand-Kirillov dimension and the homological properties. If R has lots of normal elements, then R is catenary.

In order to give a precise statement, we need to introduce some homological notions.

The algebra R is said to be **Auslander-Gorenstein** provided

(a) the injective dimension of R (as both a right and a left R-module) is finite, and

(b) for any integers $0 \leq i < j$ and any finitely generated (right or left) R-module M, we have

$$\mathrm{Ext}^i_R(N, R) = 0$$

for all R-submodules N of $\mathrm{Ext}^j_R(M, R)$.

The **grade** of a finitely generated R-module M is defined to be

$$j(M) := \inf\{j \geq 0 \mid \mathrm{Ext}^j_R(M, R) \neq 0\}.$$

The algebra R is called **Cohen-Macaulay** (or **CM**) if

$$j(M) + \mathrm{GKdim}(M) = \mathrm{GKdim}(R)$$

for all finitely generated R-modules M.

An element $x \in R$ is said to be a **normal** element of R if

$$xR = Rx.$$

If $P < Q$ are prime ideals of the noetherian k-algebra R with $\mathrm{height}(Q/P) = 1$, then

$$\mathrm{GKdim}(R/P) \geq \mathrm{GKdim}(R/Q) + 1.$$

Let $P < Q$ be prime ideals of a ring R such that $\mathrm{height}(Q/P) = 1$. We say that Q and P are **normally separated** if there exists an element $x \in Q \setminus P$ that is normal modulo P.

Theorem [4] Let R be Auslander-Gorenstein and Cohen-Macaulay with $\mathrm{GKdim}(R)$ finite, and let $P < Q$ be prime ideals of R with $\mathrm{height}(Q/P) = 1$. If Q and P are normally separated, then

$$\mathrm{GKdim}(R/P) = \mathrm{GKdim}(R/Q) + 1.$$

Let R be an affine, noetherian, Auslander-Gorenstein, Cohen-Macaulay algebra over a field, with finite Gelfand-Kirillov dimension. If $\mathrm{Spec}(R)$ is normally separated, then R is catenary.

STOP PRESS

In a recent paper, Yekutieli and Zhang, [16], have shown that results such as the previous one hold for rings with an Auslander Dualizing Complex, and they give criteria for the existence of Auslander Dualizing complexes which show that they occur quite frequently.

Letzter and Lorenz, [8], have just announced a proof that group algebras of polycylic-by-finite groups are catenary. Roseblade obtained this, up to a finite extension part, long ago, and recent developments in the understanding of the prime ideals of noetherian rings have now enabled the last step to be made. The proof uses Krull dimension techniques

REFERENCES

[1] V V Bavula, *Krull, Gelfand-Kirillov, Filter, Faithful and Schur dimensions*, this proceedings, 149-166.

[2] W Borho and H-P Kraft, *Über die Gelfand-Kirillov Dimension*, Math Annalen 220 (1976), 1-24.

[3] W Crawley-Boevey, *Modules of finite length over their endomorphism rings*, Representations of algebras and related topics (Kyoto, 1990), 127-184, London Math. Soc. Lecture Note Ser., 168, Cambridge Univ. Press, Cambridge, 1992.

[4] K R Goodearl and T H Lenagan, *Catenarity in quantum algebras*, J. Pure Appl. Algebra 111 (1996), 123-142.

[5] R Gordon and J C Robson, *Krull dimension*, Mem Amer Math Soc 133, Amer Math Soc, Providence 1973.

[6] R Gordon and J C Robson, *The Gabriel dimension of a module*, J Algebra 29 (1974), 459-473.

[7] G R Krause and T H Lenagan, *Growth of algebras and Gelfand-Kirillov dimension*, Pitman, London, 1985.
Revised edition: Graduate Studies in Mathematics 22, American Mathematical Society, Providence RI, 2000.

[8] E S Letzter and M Lorenz, *Polycyclic-by-finite group algebras are catenary*, Math Res Lett 6 (1999), 183-194.

[9] H Matsumura, *Commutative ring theory*, Translated from the Japanese by M. Reid. Second edition. Cambridge Studies in Advanced Mathematics, 8. Cambridge University Press, Cambridge-New York, 1989.

[10] M Nagata, *Local rings*, Interscience Tracts in Pure and Applied Mathematics, No. 13, Wiley, New York-London 1962.

[11] J C McConnell and J C Robson, *Noncommutative noetherian rings*, Wiley, Chichester, 1987.

[12] V M Petrogradsky, *Some types of intermediate growth in Lie algebras*, Russian Math. Surveys 48 (1993), 181-182.

[13] V M Petrogradsky, *Intermediate growth in Lie algebras and their enveloping algebras*, J Algebra 179 (1996), 459-482.

[14] R Rentschler and P Gabriel, *Sur la dimension des anneaux et ensembles ordonnés*, C R Acad Sci Paris, 265, (1967) 712-715.

[15] V A Ufnarovskij, *Combinatorial and asymptotic methods in algebra*, Encyclopaedia of Mathematical Sciences, Vol 57, 1990.

[16] A Yekutieli and J J Zhang, *Rings with Auslander dualizing complexes*, J of Algebra 213 (1999), 1-51.

DEPARTMENT OF MATHEMATICS AND STATISTICS, KING'S BUILDINGS, MAYFIELD ROAD, EDINBURGH EH9 3JZ

Trends in Mathematics, © 2000 Birkhäuser Verlag Basel/Switzerland

KRULL, GELFAND-KIRILLOV, FILTER, FAITHFUL AND SCHUR DIMENSIONS

VLADIMIR BAVULA

The aim of the paper is to explain relations between 5 mentioned dimensions.

In Section 1 a connection is established between Krull, Gelfand-Kirillov and Filter dimension of simple affine algebras, and an analog of Bernstein's inequality is proved for nonzero finitely generated modules over simple affine algebras. These results are applied to compute the Krull dimension of the ring of differential operators on a smooth affine variety.

In Section 2 the results of Section 1 are generalized for affine algebras (not necessarily simple). For a new dimension, the faithful dimension, is to be introduced.

In Section 3 we consider the Schur dimension and its relations with holonomic modules over the ring of differential operators on a smooth affine variety. In particular, we prove that every holonomic module M which has multiplicity $e(M) = 1$ over the n'th Weyl algebra A_n has Schur dimension $\mathrm{sd}(M) = 1$.

In Section 4 we give a review of (recent) results on the Krull dimension of skew Laurent extensions and generalized Weyl algebras with left Noetherian coefficients.

This paper can be considered as an addition to and continuation of the paper of Tom Lenagan in the present volume.

1. Krull, Gelfand-Kirillov and Filter Dimensions of Simple Affine Algebras

Let K be a field, module means left module, $\otimes = \otimes_K$. Let A be a simple affine infinite dimensional algebra and let M be a nonzero finitely generated A-module; then $\dim_K M = \infty$ (since the linear map $A \to \mathrm{Hom}_K(M, M)$, $a \mapsto (m \mapsto am)$, $m \in M$, is injective). Then the Gelfand-Kirillov dimension $\mathrm{GK}(M) \geq 1$.

QUESTION. *How to find (estimate) the number*

$$i_A := \inf\{\mathrm{GK}(M),\ M \textit{ is a nonzero finitely generated } A-\textit{module}\}?$$

The (left) filter dimension gives a good approximation of i_A (Theorem 1.2). To find such an approximation for simple affine infinite dimensional algebras was the main motivation of introducing the filter dimension, [Bav1].

Let A_n be the n'th **Weyl algebra**, that is the K-algebra with $2n$ generators $X_1, \dots, X_n, \partial_1, \dots, \partial_n$ and relations

$$\partial_i X_j - X_j \partial_i = \delta_{ij}, \text{ the Kronecker delta, } X_i X_j - X_j X_i = \partial_i \partial_j - \partial_j \partial_i = 0,$$

$i, j = 1, \dots, n$. If char $K = 0$, then A_n is a simple Noetherian algebra canonically isomorphic to the ring of differential operators $K[X_1, \dots, X_n, \partial/\partial X_1, \dots, \partial/\partial X_n]$ with polynomial coefficients ($X_i \leftrightarrow X_i, \partial_i \leftrightarrow \partial_i/\partial X_i$, $i = 1, \dots, n$).

Let K.dim and GK be the *(left) Krull* (in the sense of Rentschler and Gabriel, [RG]) and *Gelfand-Kirillov* dimension respectively.

Theorem 1.1. (the Bernstein inequality, [Be]). *Let A_n be the n'th Weyl algebra over a field of characteristic zero. Then* $\mathrm{GK}(M) \geq n$ *for every nonzero finitely generated A_n-module M.* ∎

For each function f from the set of natural numbers $\mathbf{N} = \{0, 1, \dots\}$ to $\mathbf{R}'$, where $\mathbf{R}' = \{r \in \mathbf{R}, r \geq 1\}$ we can associate a number $\gamma(f) \in \mathbf{R} \cup \{\infty\}$, a *degree* of f,

$$\gamma(f) = \inf\{r \in \mathbf{R} : f(n) \leq n^r \text{ for sufficiently large} \, n >> 0\},$$

where $\mathbf{R}$ is the set of real numbers. For functions f and g as above:

$$\gamma(f+g) \leq \max\{\gamma(f), \gamma(g)\},$$

$$\gamma(fg) \leq \gamma(f) + \gamma(g),$$

$$\gamma(f \circ g) \leq \gamma(f)\gamma(g),$$

where $f \circ g$ is the composition of the functions and in the last inequality g is an integer valued function.

Let A be an affine algebra with generators $a_1, \dots, a_s$. Then A is equipped with a standard finite dimensional filtration F: $A = \cup_{i \geq 0} A_i$, where $A_0 = K$, $A_1 = K + \sum_{i=1}^s K a_i$, $A_i := (A_1)^i$, $i \geq 2$. Let M be a finitely generated A-module and M_0 be a finite dimensional generating subspace of M, the module M can be equipped with a filtration $\{M_i = A_i M_0\}$. The degree $\mathrm{GK}(A)$ and $\mathrm{GK}(M)$ of the function $i \mapsto \dim A_i$ and $i \mapsto \dim M_i$ respectively do not depend on the choice of the algebra and the module filtrations and is called **the Gelfand-Kirillov** dimension of the algebra A and of the A-module M respectively.

If the algebra A is **simple**, define the **return function** $\nu_F : \mathbf{N} \to \mathbf{N} \cup \{\infty\}$ of A associated with F in the following way

$$\nu_F(i) = \inf\{j \in \mathbf{N} \cup \{\infty\} : \, A_j a A_j \ni 1 \text{ for all } 0 \neq a \in A_i\},$$

where $A_j a A_j$ is the subspace of A generated by the products xay, for all $x, y \in A_j$.

Define the **left return function** $\lambda_F : \mathbf{N} \to \mathbf{N} \cup \{\infty\}$ of the algebra A as

$$\lambda_F(i) = \min\{j \in \mathbf{N} \cup \{\infty\} : \, A a A_j \ni 1 \;\text{ for all }\; 0 \neq a \in A_i\},$$

where AaA_j is the left ideal of A generated by aA_j.

Definition. ([Bav1, 3]). The **filter dimension** and **the left filter dimension** of the simple affine algebra A are defined as $\mathrm{fd}\, A = \gamma(\nu_F)$ and $\mathrm{lfd}\, A = \gamma(\lambda_F)$ respectively.

These dimensions do not depend on the choice of F.

In similar manner the right filter dimension rfd A of A is defined. Clearly, the following inequality holds:

$$\operatorname{lfd} A \leq \operatorname{fd} A.$$

Let $\mathrm{d}(A) = \operatorname{fd} A$ or $\mathrm{d}(A) = \operatorname{lfd} A$.

Theorem 1.2. ([Bav1, 3]). (An analog of the Bernstein's inequality for simple affine algebras).

Let A be a simple affine algebra; then

$$\mathrm{GK}(A) \leq \mathrm{GK}(M)(\mathrm{d}(A) + \max\{\mathrm{d}(A), 1\}) \tag{1.1}$$

for every nonzero finitely generated A-module M. Moreover, if A is infinite dimensional, then

$$\operatorname{fd} A \geq 1/2. \blacksquare$$

Proof. Let $\lambda = \lambda_F$ be as above and $0 \neq a \in A_i$. It follows from

$$AaM_{\lambda(i)} = AaA_{\lambda(i)}M_0 \supseteq 1M_0 = M_0$$

that the linear map

$$A_i \to \mathrm{Hom}(M_{\lambda(i)}, M_{\lambda(i)+i}), a \mapsto (m \mapsto am),$$

is injective, so dim $A_i \leq$ dim $M_{\lambda(i)}$ dim $M_{\lambda(i)+i}$. Using the above elementary properties of the degree (see also [MR], 8.1.7), we have

$$\begin{aligned}
\mathrm{GK}(A) &= \gamma(\dim A_i) \\
&\leq \gamma(\dim M_{\lambda(i)}) + \gamma(\dim M_{\lambda(i)+i}) \\
&\leq \gamma(\dim M_i)\gamma(\lambda) + \gamma(\dim M_i)\max\{\gamma(\lambda), 1\} \\
&= \mathrm{GK}(M)(\operatorname{lfd} A + \max\{\operatorname{lfd} A, 1\}). \blacksquare
\end{aligned}$$

We say that an algebra S is *(left) finitely partitive* ([Jos 1] or [MR], 8.3.17) if, given any finitely generated S-module M, there is an integer $n > 0$ such that for every chain

$$M = M_0 \supset M_1 \supset \ldots \supset M_m$$

with $\mathrm{GK}(M_i/M_{i+1}) = \mathrm{GK}(M)$, one has $m \leq n$.

Lemma 1.3. *Let A be a finitely partitive algebra with* $\mathrm{GK}(A) < \infty$. *Let $a \in \mathbf{N}$, $b \geq 0$ and suppose, for all finitely generated A-modules M with* $\mathrm{K.dim}(M) = a$, *that* $\mathrm{GK}(M) \geq a + b$ *and , for all finitely generated A-modules N with* $\mathrm{K.dim}(N) \geq a$, *that* $\mathrm{GK}(N) \in \mathbf{N}$. *Then* $\mathrm{GK}(A) \geq \mathrm{K.dim}(A) + b$.

Proof. We use induction on $n = \mathrm{K.dim}(M)$. The base of induction, $n = a$, is true. Let $n > a$. There exists a descending chain of submodules $M = M_1 \supseteq M_2 \supseteq \cdots$ with $\mathrm{K.dim}(M_i/M_{i+1}) = n-1$ for $i \geq 1$. By iduction , $\mathrm{GK}(M_i/M_{i+1}) \geq n - 1 + b$ for $i \geq 1$. The algebra A is finitely partitive, so there exists i such that $\mathrm{GK}(M) > \mathrm{GK}(M_i/M_{i+1})$, thus $\mathrm{GK}(M) - 1 \geq \mathrm{GK}(M_i/M_{i+1}) \geq n - 1 + b$, since $\mathrm{GK}(M) \in \mathbf{N}$, and hence $\mathrm{GK}(A) \geq \mathrm{K.dim}(A) + b$. $\blacksquare$

Theorem 1.4. ([Bav2, 3]). *Let A be a finitely partitive simple affine algebra with $GK(A) < \infty$. Suppose that the Gelfand-Kirillov dimension of every finitely generated A-module is a natural number. Then, for any nonzero finitely generated A-module M, the Krull dimension*

$$\mathrm{K.dim}(M) \leq \mathrm{GK}(M) - \mathrm{GK}(A)/(\mathrm{d}(A) + \max\{\mathrm{d}(A), 1\}). \tag{1.2}$$

In particular,

$$\mathrm{K.dim}(A) \leq \mathrm{GK}(A)(1 - 1/(\mathrm{d}(A) + \max\{\mathrm{d}(A), 1\})). \blacksquare \tag{1.3}$$

Proof. Apply the lemma above to the family of finitely generated A-modules of Krull dimension 0. By Theorem 1.2, we can put $a = 0$ and $b = \mathrm{GK}(A)/(\mathrm{d}(A) + \max\{\mathrm{d}(A), 1\})$, and the result follows. ∎

Let K be a field of characteristic zero and B be a commutative K-algebra. The ring of (K-linear) *differential operators* $\mathcal{D}(B)$ on B is defined as $\mathcal{D}(B) = \cup_{i=0}^{\infty} \mathcal{D}^i(B)$ where $\mathcal{D}^0(B) = \mathrm{End}_R(B) \simeq B$, $((x \mapsto bx) \leftrightarrow b)$,

$$\mathcal{D}^i(B) = \{u \in \mathrm{End}_K(B) : [u, r] \in \mathcal{D}^{i-1}(B) \text{ for each } r \in B\}.$$

Note that $\{\mathcal{D}^i(B)\}$ is the filtration for $\mathcal{D}(B)$. The subalgebra $\Delta(B)$ of $\mathrm{End}_K(B)$ generated by $B \equiv \mathrm{End}_R(B)$ and by the set $\mathrm{Der}_K(B)$ of all K-derivations of B is called the *derivation ring* of B. The derivation ring $\Delta(B)$ is a subring of $\mathcal{D}(B)$.

Let the algebra B be a *regular commutative Noetherian affine domain of Krull dimension $n < \infty$*. In geometric terms, B is the coordinate ring $\mathcal{O}(X)$ of a smooth irreducible affine variety X of dimension n. Then

- $\mathrm{Der}_K(B)$ *is a finitely generated projective B-module of rank n*;
- $\mathcal{D}(B) = \Delta(B)$;
- $\mathcal{D}(B)$ *is a simple (left and right) Noetherian domain with* $\mathrm{GK}\,\mathcal{D}(B) = 2n$ $(n = \mathrm{GK}(B) = \mathrm{K.dim}(B))$;
- $\mathcal{D}(B)$ *is an almost centralizing extension of B*;
- *the associated graded ring* $\mathrm{gr}\,\mathcal{D}(B) = \oplus \mathcal{D}^i(B)/\mathcal{D}^{i-1}(B)$ *is a commutative domain*;
- *the Gelfand-Kirillov dimension of every finitely generated $\mathcal{D}(B)$-module is a natural number.*

For the proofs the reader is referred to [MR], Chapter 15. $\mathcal{D}(B)$ is a simple affine infinite dimensional Noetherian algebra ([MR], ch. 15).

Example. Let $P_n = K[X_1, \dots, X_n]$ be a polynomial ring. Then

$$\mathrm{Der}_K(P_n) = \oplus_{i=1}^n P_n \partial/\partial X_i,$$

and

$$\Delta(B) = K[X_1, \dots, X_n, \partial/\partial X_1, \dots, \partial/\partial X_n],$$

is the ring of differential operators with polynomial coefficients, i.e. the n'th Weyl algebra A_n.

Let $\{B_i\}$ and $\{\mathcal{D}(B)_i\}$ be standard finite dimensional filtrations on B and $\mathcal{D}(B)$ respectively such that $B_i \subseteq \mathcal{D}(B)_i$ for all $i \geq 0$. Then the enveloping algebra

$\mathcal{D}^e(B) := \mathcal{D}(B) \otimes \mathcal{D}(B)^o$ can be equipped with a standard finite dimensional filtration $\{\mathcal{D}^e(B)_i\}$ which is the tensor product of filtrations $\{\mathcal{D}(B)_i\}$ and $\{\mathcal{D}(B)_i^o\}$, where $\mathcal{D}(B)^o$ is the opposite algebra to $\mathcal{D}(B)$. The algebra $\mathcal{D}(B)$ with the filtration $\{\mathcal{D}(B)_i\}$ is the left filtered $\mathcal{D}^e(B)$-module.

Lemma 1.5. *([Bav 2]). There exist natural numbers a and b such that for any $d \in \mathcal{D}(B)_m$ there exists $w \in \mathcal{D}^e(B)_{am+b}$ satisfying $wd = 1$.* ∎

Theorem 1.6. ([Bav2]).

$$\mathrm{fd}\,(\mathcal{D}(B)) = \mathrm{lfd}(\mathcal{D}(B)) = 1.$$

Proof. Let $\mathrm{d} = \mathrm{fd}\,,\mathrm{lfd}$. By Lemma 1.5, $\nu_F(i) \leq ai + b$, where $F = \{\mathcal{D}(B)_i\}$, hence $\mathrm{d}(\mathcal{D}(B)) \leq 1$. It follows from the direct decomposition of B-modules:

$${}_B\mathcal{D}(B) = B \oplus \mathcal{D}(B)\mathrm{Der}_K(B)$$

that the Gelfand-Kirillov dimension of the $\mathcal{D}(B)$-module

$$\mathcal{D}(B)/\mathcal{D}(B)\mathrm{Der}_K(B) \simeq B$$

is $\mathrm{GK}(B)$. Since $\mathrm{GK}(\mathcal{D}(B)) = 2\mathrm{GK}(B)$ and $\mathrm{GK}(\mathcal{D}(B)) \leq \mathrm{GK}(B)(\mathrm{d}(\mathcal{D}(B)) + \max\{\mathrm{d}(\mathcal{D}(B)), 1\})$, we have the opposite inequality $\mathrm{d}(\mathcal{D}(B)) \geq 1$. ∎

As an application we compute the Krull dimension of $\mathcal{D}(B)$.

Theorem 1.7. ([MR], ch. 15).

$$\mathrm{K.dim}\,\mathcal{D}(B) = \mathrm{GK}(\mathcal{D}(B))/2 = \mathrm{K.dim}(B).$$ ∎

Proof. The second equality is clear ($\mathrm{GK}(\mathcal{D}(B)) = 2\mathrm{GK}(B) = 2\mathrm{K.dim}(B)$). It follows from Theorems 1.4, 1.6 that $\mathrm{K.dim}\,\mathcal{D}(B) \leq \mathrm{GK}(\mathcal{D}(B))/2 = \mathrm{K.dim}(B)$. The map $I \to \mathcal{D}(B)I$ from the set of left ideals of B to the set of left ideals of $\mathcal{D}(B)$ is injective, thus $\mathrm{K.dim}\,(B) \leq \mathrm{K.dim}\,\mathcal{D}(B)$. ∎

Theorem 1.8. ([MR], 15.4.3). *Let M be a nonzero finitely generated $\mathcal{D}(B)$-module. Then*

$$\mathrm{GK}(M) \geq \mathrm{GK}(\mathcal{D}(B))/2 = \mathrm{K.dim}(B).$$

Proof. Follows from Theorems 1.2, 1.6. ∎

Theorem 1.9. *Let $\mathrm{char}\,K = 0$. For the n'th Weyl algebra A_n:*

$$\mathrm{d}(A_n) = 1.$$

Proof. Denote by $a_1, \dots, a_{2n}$ the canonical generators of the Weyl algebra A_n and denote by $F = \{A_{n,i}\}_{i\geq 0}$ the standard filtration associated with the canonical generators. The associated graded algebra $\mathrm{gr}\,A_n := \oplus_{i\geq 0} A_{n,i}/A_{n,i-1}$, $(A_{n,-1} = 0)$ is a polynomial ring in $2n$ variables, so

$$\mathrm{GK}(A_n) = \mathrm{GK}(\mathrm{gr}\,A_n) = 2n. \tag{1.4}$$

For every $i \geq 0$, we have

$$\mathrm{ad}\,a_j : A_{n,i} \to A_{n,i-1},\; x \mapsto \mathrm{ad}\,a_j(x) := a_j x - x a_j.$$

The algebra A_n is central ($Z(A_n) = K$), so

$$\operatorname{ad} a_j(x) = 0 \text{ for all } j = 1, \dots, 2n \Leftrightarrow x \in Z(A_n) = K.$$

Using these two facts, we obtain $\nu_F(i) \leq i$ for $i \geq 0$, so $\mathrm{d}(A_n) \leq 1$.

The A_n-module $P_n := K[X_1, \dots, X_n] \simeq A_n/(A_n\partial_1 + \cdots + A_n\partial_n)$ has Gelfand-Kirillov dimension n. By (1.1), applied to P_n, we have

$$2n = \mathrm{GK}(A_n) \leq n(\mathrm{d}(A) + \max\{\mathrm{d}(A), 1\}),$$

hence $\mathrm{d}(A_n) \geq 1$, thus $\mathrm{d}(A_n) = 1$. ∎

PROOF OF THE BERNSTEIN'S INEQUALITY (THEOREM 1.1).

Put $\mathrm{GK}(A_n) = 2n$ and $\mathrm{fd}\,(A_n) = 1$ in (1.1). ∎

Theorem 1.10. *([RG]). If* $\operatorname{char} K = 0$ *then*

$$\mathrm{K.dim}(A_n) = n.$$

Proof. Putting $\mathrm{GK}(A_n) = 2n$ and $\mathrm{fd}\,(A_n) = 1$ into the formula (1.3) we have $\mathrm{K.dim}(A_n) \leq n$. The polynomial ring $P_n = K[X_1, \dots, X_n]$ is a subalgebra of A_n such that A_n is a free right P_n-module. The map $I \to A_n I$ from the set of left ideals of P_n to the set of left ideals of A_n is injective, thus $n = \mathrm{K.dim}(P_n) \leq \mathrm{K.dim}(A_n)$, so $\mathrm{K.dim}(A_n) = n$. ∎

2. KRULL, GELFAND-KIRILLOV AND FAITHFUL DIMENSIONS OF AFFINE ALGEBRAS

The material of this section is contained in [BL1].

Let K be a field and let A be an affine K-algebra equipped with a standard finite dimensional filtration $F : A = \cup_{i\geq 0} A_i$ as in Section 1. Let M be a finitely generated A-module with a finite dimensional subspace M_0 such that $M = AM_0$. There is a standard finite dimensional filtration, $\{M_i\}$, of M given by $M_i := A_i M_0$.

Assume that M is a faithful A-module. For any nonzero element $a \in A$, there exists a least integer i such that $aM_i \neq 0$. Set $n_{M,F,M_0}(a)$ to be this least integer i. For convenience, set $n_{M,F,M_0}(0) := 0$.

For any subset V of A, set

$$n_{M,F,M_0}(V) := \sup\{n_{M,F,M_0}(v) \mid v \in V\}.$$

Lemma 2.1. *Let A be an affine algebra and let M be a finitely generated faithful A-module. Then, $n_{M,F,M_0}(V) < \infty$ for any finite dimensional subspace V of A.*

Proof. Suppose that the result is false, so that $n_{M,F,M_0}(V) = \infty$ for some finite dimensional subspace V of A. Choose a sequence $a_1, a_2, \dots$ of elements of A such that $n(a_1) < n(a_2) < \dots$. For each $i \geq 1$, set $V_i = \sum_{i\leq j} Ka_j$. Thus, there is a descending sequence of subspaces

$$V \supseteq V_1 \supseteq V_2 \supseteq \dots .$$

Since V is finite dimensional, this descending sequence must terminate, say at $V_m = V_{m+1} \ldots$. Since $a_m \in V_m = V_{m+1}$, there are scalars α_i, for $i > m$ and with only finitely many $\alpha_i \neq 0$, such that $a_m = \sum_{i>m} \alpha_i a_i$.

Hence,

$$0 \neq a_m M_{n(a_m)} \subseteq \sum_{i>m} \alpha_i a_i M_{n(a_m)} \subseteq \sum_{i>m} \alpha_i a_i M_{n(a_i)-1} = 0,$$

a contradiction. ▪

Since each of the subspaces A_i of the standard filtration of A is finite dimensional, the following definition makes sense.

Definition. ([BL1]).

$$n(i) := n_{M,F,M_0}(A_i),\ i \geq 0.$$

In other words, $n(i)$ is the least integer j such that the K-linear map

$$A_i \to \mathrm{Hom}_K(M_j, M_{j+i}), \quad a \mapsto (m \mapsto am), \tag{2.1}$$

is injective.

The next lemma shows that the degree for the function $n(i)$, just defined, does not depend on the particular choices of filtrations and generating subspaces for A and M.

Lemma 2.2. *Let F and F' be standard filtrations of an affine algebra A, and let M_0 and M_0' be finite dimensional generating subspaces of a faithful A-module M. Then*

$$\gamma(n_{M,F,M_0}) = \gamma(n_{M,F',M_0'}).$$

Proof. Set $n(i) = n_{M,F,M_0}(i)$ and $n'(i) = n_{M,F',M_0'}(i)$. Choose integers α, β such that $A_1 \subseteq A'_\alpha$ and $A'_1 \subseteq A_\alpha$, while $M_0 \subseteq M'_\beta$ and $M'_0 \subseteq M_\beta$. Then, $M_i \subseteq M'_{\beta+\alpha i}$ and $M'_i \subseteq M_{\beta+\alpha i}$, for $i \geq 0$. From this one has

$$n'(i) \leq \beta + \alpha n(i\alpha) \text{ and } n(i) \leq \beta + \alpha n'(i\alpha),$$

for each $i \geq 0$, and the result follows. ▪

As a consequence of the previous lemma the following definition becomes appropriate.

Definition. ([BL1]). The **left faithful dimension** , $\mathrm{lf}(M)$, is defined to be

$$\gamma(n_{M,F,M_0}),$$

for any standard filtration F of the factor algebra $A/\mathrm{ann}(M)$ (where $\mathrm{ann}(M)$ is the annihilator of the A-module M) and for any finite dimensional generating subspace M_0 of M.

The behaviour of this growth on passing to faithful submodules is interesting: it cannot decrease, as the following lemma shows.

Lemma 2.3. *Let A be an affine algebra and let M, N be finitely generated faithful modules with finite dimensional generating subspaces M_0 and N_0, respectively. If either (i) N is a submodule of M with $N_0 \subseteq M_0$, or (ii) N is an epimorphic image of M with N_0 being the image of M_0 under an epimorphism, then*

$$n_{M,F,M_0}(i) \leq n_{N,F,N_0}(i),$$

for $i \geq 0$, and, consequently,

$$\mathrm{lf}(M) \leq \mathrm{lf}(N).$$

Proof. This follows easily, for example in case (i) from the fact that if $a \in A$ and $aM_j = 0$ then $aN_j = 0$, also, since $N_j \subseteq M_j$. ∎

Corollary 2.4. *Let A be an affine algebra and let M, N be finitely generated faithful modules with finite dimensional generating subspaces M_0 and N_0, respectively. Suppose that N is a subfactor of M (that is, $N \cong X/Y$ for some submodules $Y \subseteq X$ of M), and suppose that $N_0 \subseteq M_0 + Y$. Then*

$$n_{M,F,M_0}(i) \leq n_{N,F,N_0}(i),$$

for $i \geq 0$ and, consequently,

$$\mathrm{lf}(M) \leq \mathrm{lf}(N). \blacksquare$$

We are now able to establish the following estimate on the relationship between the growth of the algebra A and the growth of the module M.

Theorem 2.5. *Let A be an affine algebra and let M be a finitely generated faithful A-module. Then*

$$\mathrm{GK}(A) \leq \mathrm{GK}(M) \times (\mathrm{lf}(M) + \max\{\mathrm{lf}(M), 1\}). \tag{2.2}$$

Proof. The linear map

$$A_i \to \mathrm{Hom}_K(M_{n(i)}, M_{n(i)+i}), \qquad a \mapsto (m \mapsto am)$$

is injective, by (2.1); so that $\dim(A_i) \leq \dim(M_{n(i)}) \times \dim(M_{n(i)+i})$. Using elementary properties of Gelfand-Kirillov dimension, this inequality gives

$$\begin{aligned}
\mathrm{GK}(A) &= \gamma(\dim(A_i)) \\
&\leq \gamma(\dim(M_{n(i)})) + \gamma(\dim(M_{n(i)+i})) \\
&\leq \gamma(\dim(M_i))\gamma(n(i)) + \gamma(\dim(M_i))\gamma(n(i)+i) \\
&= \mathrm{GK}(M) \times (\mathrm{lf}(M) + \max\{\mathrm{lf}(M), 1\}). \blacksquare
\end{aligned}$$

Let $\hat{A}$ be the set of isomorphism classes of simple A-modules, and let $\mathrm{Prim}(A)$ be the set of primitive ideals of A. For a given J in $\mathrm{Prim}(A)$, let $\widehat{(A, J)}$ denote the subset of $\hat{A}$ consisting of modules with annihilator equal to J.

Definition.

$$S_A := \inf\left\{\frac{\mathrm{GK}(A/J)}{\mathrm{lf}(M) + \max\{\mathrm{lf}(M), 1\}} \mid J \in \mathrm{Prim}(A), M \in \widehat{(A, J)}\right\}.$$

Corollary 2.6. *Let A be an affine algebra. Then*

$$\mathrm{GK}(M) \geq S_A,$$

for any nonzero simple A-module M. ∎

Theorem 2.7. *Let A be an affine left finitely partitive algebra such that $\mathrm{GK}(A) < \infty$ and that the Gelfand-Kirillov dimension of every finitely generated A-module is a natural number. Then,*

$$\mathrm{K.dim}(M) \leq \mathrm{GK}(M) - S_A,$$

for any finitely generated left A-module M.
In particular,

$$\mathrm{K.dim}(A) \leq \mathrm{GK}(A) - S_A.$$

Proof. Applying Lemma 1.3 to the family of finitely generated A-modules of Krull dimension 0 ($a = 0$), we can set $b = S_A$ and the result follows. ∎

Lemma 2.8. *Let A be a simple affine algebra and let M_0 be a finite dimensional generating subspace for an A-module M. Then $n_{M,F,M_0}(i) \leq \lambda_F(i)$, for all $i \geq 0$; hence,*

$$\mathrm{lf}(M) \leq \mathrm{lfd}(A) \quad \text{and} \quad \frac{\mathrm{GK}(A)}{\mathrm{lfd}(A) + \max\{\mathrm{lfd}(A), 1\}} \leq S_A.$$

Proof. In the proof of Theorem 1.2 we proved that the linear map

$$A_i \to \mathrm{Hom}_K(M_{\lambda(i)}, M_{\lambda(i)+i}), \qquad a \mapsto (m \mapsto am),$$

is injective. Thus, $n_{M,F,M_0}(i) \leq \lambda_F(i)$, for all $i \geq 0$, as required. ∎

Let B be a commutative regular integral domain of Krull dimension n, affine over a field K of characteristic zero. Let $\mathcal{D}(B)$ be its ring of differential operators.

Definition. A finitely generated $\mathcal{D}(B)$-module M is called a **holonomic** module provided $\mathrm{GK}(M) = \mathrm{GK}(\mathcal{D}(B))/2$.

Corollary 2.9. *Let M be a holonomic $\mathcal{D}(B)$-module. Then* $\mathrm{lf}(M) = 1$.

Proof. It follows from Theorem 2.5 and $\mathrm{GK}\,\mathcal{D}(B) \leq 2\mathrm{GK}(M)$ that $\mathrm{lf}(M) \geq 1$. That $\mathrm{lf}(M) \leq \mathrm{fd}\,(\mathcal{D}(B)) = 1$ follows from Lemma 2.8 and Theorem 1.6. ∎

Examples. Let K be an algebraically closed field and let $D := K[H_1^{\pm 1}, \dots, H_n^{\pm 1}]$ be the commutative Laurent polynomial ring in n indeterminates. Let $\lambda_1, \dots, \lambda_n \in K$ be such that the multiplicative subgroup of K generated by $\lambda_1, \dots, \lambda_n$ is free abelian of rank n. Set $A = A_n := D[X^{\pm 1}; \sigma]$, the skew-Laurent polynomial ring, where $\sigma(H_i) = \lambda_i^{-1} H_i$, so that $XH_i = \lambda_i^{-1} H_i X$. Alternatively, we can present A as $A = K[X^{\pm 1}][H_1^{\pm 1}, \dots, H_n^{\pm 1}; \sigma_1, \dots, \sigma_n]$ where $\sigma_i(X) = \lambda_i X$.

The algebra $A = \oplus_{i \in \mathbf{Z}} A^i$ is a central simple $\mathbf{Z}$-graded algebra, where $A^i := DX^i$, for $i \in \mathbf{Z}$.

The map

$$\mathrm{Maxspec}(D) \to (K^*)^n, \quad \mathcal{M}_\mu := \langle H_1 - \mu_1, \dots, H_n - \mu_n \rangle \mapsto \mu = (\mu_1, \dots, \mu_n),$$

is a bijection. For each μ, consider the simple D-module $V \equiv V_\mu := D/\mathcal{M}_\mu \cong K$. The induced module

$$A(V) := A \otimes_D V = \bigoplus_{i\in\mathbf{Z}} Ke_i, \quad e_i := X^i \otimes \bar{1}, \quad \bar{1} = 1 + \mathcal{M}_\mu \in V,$$

is $\mathbf{Z}$-graded:

$$A(V) = \bigoplus_{i\in\mathbf{Z}} A(V)^i, \quad A(V)^i := Ke_i.$$

The elements of A act on this induced module as follows: $Xe_i = e_{i+1}$ and $H^\alpha e_i = \mu^\alpha \lambda^{i\alpha} e_i$, for all $\alpha = (\alpha_1, \dots, \alpha_n) \in \mathbf{Z}^n$, where $H^\alpha = \prod H_j^{\alpha_j}$, $\mu^\alpha = \prod \mu_j^{\alpha_j}$, $\lambda^{i\alpha} = \prod \lambda_j^{i\alpha_j}$.

Lemma 2.10. *For every simple D-module V_μ, the A-module $A(V_\mu)$ is a simple A-module with $\mathrm{GK}(A(V_\mu)) = 1$ and $\mathrm{lf}(A(V_\mu)) = n$. Two such A-modules $A(V_\mu)$ and $A(V_\nu)$ are isomorphic if and only if $\mathcal{M}_\nu = \sigma^i(\mathcal{M}_\mu)$, for some $i \in \mathbf{Z}$; that is, $\nu_j = \lambda_j^i \mu_j$, for all j and for some $i \in \mathbf{Z}$.* ∎

The algebra A can also be described as the skew Laurent extension

$$A = K[X^{\pm 1}][H_1^{\pm 1}, \dots, H_n^{\pm 1}; \sigma_1, \dots, \sigma_n]$$

where $\sigma_i(X) = \lambda_i X$. Set $R := K[X^{\pm 1}]$. A is a $\mathbf{Z}^n$-graded algebra via

$$A = \bigoplus_{\alpha\in\mathbf{Z}^n} A_\alpha, \quad A_\alpha = RH^\alpha.$$

For $\mu \in K^*$, consider the A-module induced from the simple R-module $U_\mu := R/\langle X - \mu\rangle \cong K$. So,

$$A(U_\mu) := A \otimes_R U_\mu = \bigoplus_{\alpha\in\mathbf{Z}^n} Ke_\alpha, \quad e_\alpha = H^\alpha \otimes \bar{1}, \quad \bar{1} = 1 + \langle X - \mu\rangle.$$

The module $A(U_\mu)$ is $\mathbf{Z}^n$-graded by $A(U_\mu)_\alpha := Ke_\alpha$, for $\alpha \in \mathbf{Z}^n$. The action of elements of A on $A(U_\mu)$ is defined by $H^\beta e_\alpha = e_{\alpha+\beta}$ and $Xe_\alpha = \mu\lambda^{-\alpha} e_\alpha$, for $\alpha, \beta \in \mathbf{Z}^n$. As a D-module, $A(U_\mu)$ is free of rank 1. The standard filtration on $A(U_\mu) = \cup_{i=0}^\infty A(U_\mu)_i$, where $A(U_\mu)_i = A_i e_0 = D_i e_0$, "coincides" with the standard filtration of the algebra D, so that $\dim A(U_\mu)_i = \dim D_i = \frac{2^n}{n!} i^n + \dots$, and so $\mathrm{GK}(A(U_\mu)) = n$.

Lemma 2.11. *The A-module $A(U_\mu)$ is simple with $\mathrm{GK}\, A(U_\mu) = n$ and $\mathrm{lf}\, A(U_\mu) = n^{-1}$, for each $\mu \in K^*$. Two such modules $A(U_\mu)$ and $A(U_\nu)$ are isomorphic if and only if $\nu = \lambda^\alpha \mu$, for some $\alpha \in \mathbf{Z}^n$.* ∎

3. Schur Dimension

The material of this section is contained in [BL1] except Proposition 3.6. Let K be an algebraically closed field, and let A be an affine K-algebra. An A-module M is called **schurian** if $\mathrm{End}_A(M) = K$. An algebra A is said to be **schurian** if each simple A-module is schurian.

The class of schurian algebras is a wide class of algebras containing many interesting and important rings. For example, if the field K is uncountable and algebraically closed then all affine algebras are schurian. (In fact, they satisfy the Nullstellensatz, a stronger requirement, see [MR], Chapter 9.) In addition, for any algebraically closed field k, any **constructible** algebra is schurian, again, see [MR], Chapter 9.

Let A be an affine algebra with a standard finite dimensional filtration $F = \{A_i\}$ and let M be a faithful schurian simple module with the standard finite dimensional filtration $\{M_i = A_i M_0\}$, where M_0 is a finite dimensional generating subspace of M. The map

$$A \to \mathrm{Hom}_K(M, M), \qquad a \mapsto (m \mapsto am),$$

is injective, since the module M is faithful. We identify A with its image under this injection, and note that A acts as a dense ring of K-linear transformations, since M is a faithful schurian simple A-module. This terminology will be used throughout this section.

Consider the following function.

Definition. ([BL1]).

$$\mu(i) := \mu_{M,F,M_0}(i) = \min\{j \mid \mathrm{Hom}_K(M_i, M_i) \subseteq A_j\},$$

where the inclusion above means that for any K-linear map $\phi : M_i \to M_i$ there exists an element $a \in A_j$ such that $\phi(m) = am$, for all $m \in M_i$.

In this notation,

$$\mathrm{Hom}_K(M_i, M_i) \subseteq A_{\mu(i)}, \tag{3.1}$$

and so

$$(\dim(M_i))^2 \le \dim(A_{\mu(i)}). \tag{3.2}$$

As usual, we need to check that the rate of growth of this function is independent of the standard filtrations involved.

Lemma 3.1. *Let F and F' be standard filtrations of an affine algebra A, and let M_0 and M_0' be finite dimensional generating subspaces of a faithful simple schurian A-module M. Then*

$$\gamma(\mu_{M,F,M_0}) = \gamma(\mu_{M,F',M_0'}).$$

Proof. Set $\mu = \mu_{M,F,M_0}$ and $\mu' = \mu_{M,F',M_0'}$. Let α, β be as in the proof of Lemma 2.2. Then $M_i \subseteq M'_{\beta+\alpha i}$ and $M'_i \subseteq M_{\beta+\alpha i}$, and similarly, $A_i \subseteq A'_{\alpha i}$ and $A'_i \subseteq A_{\alpha i}$, for all $i \ge 0$.

Thus,

$$\mathrm{Hom}_K(M_i, M_i) \subseteq \mathrm{Hom}_K(M'_{\beta+\alpha i}, M'_{\beta+\alpha i}) \subseteq A'_{\mu'(\beta+\alpha i)} \subseteq A_{\alpha\mu'(\beta+\alpha i)}.$$

Hence,

$$\mu(i) \le \alpha\mu'(\beta + \alpha i),$$

for $i \ge 0$; and so $\gamma(\mu) \le \gamma(\mu')$. The opposite inequality follows by a symmetrical argument. ∎

Definition. ([BL1]). The **Schur dimension**, $\mathrm{sd}(M)$, of a module M is given by

$$\mathrm{sd}(M) := \gamma(\mu_{M,F,M_0}).$$

Theorem 3.2. *Let A be an affine algebra and let M be a faithful simple schurian A-module. Then*

$$\mathrm{GK}(M) \leq \frac{\mathrm{GK}(A)}{2}\mathrm{sd}(M).$$

Proof. Using (3.2), we have

$$\begin{aligned} 2\mathrm{GK}(M) &= \gamma((\dim(M_i))^2) \leq \gamma(\dim(A_{\mu(i)})) \\ &\leq \gamma(\dim(A_i))\gamma(\mu) = \mathrm{GK}(A)\mathrm{sd}(\mu). \blacksquare \end{aligned}$$

Corollary 3.3. *Let A be an affine algebra and let M be a faithful simple schurian A-module, and suppose that M is not a finite dimensional A-module, so that* $\mathrm{GK}(M) > 0$. *Then*

$$2 \leq \mathrm{sd}(M)\left\{\mathrm{lf}(M) + \max\{\mathrm{lf}(M), 1\}\right\}.$$

Proof. The claim follows from the previous theorem and Theorem 2.5 ($\mathcal{D}(B)$ is schurian by [MR], 15.1.22). ■

Let B be a commutative regular integral domain of Krull dimension n, affine over an algebraically closed field K of characteristic zero. Let $\mathcal{D}(B)$ be its ring of differential operators. Then $\mathcal{D}(B)$ is a schurian algebra.

Corollary 3.4. *If M is a simple $\mathcal{D}(B)$-module then* $\mathrm{sd}(M) \geq 1$. *If* $\mathrm{sd}(M) = 1$, *then M is holonomic.*

Proof. The statement follows from the inequalities

$$\frac{\mathrm{GK}(\mathcal{D}(B))}{2} \leq \mathrm{GK}(M) \leq \frac{\mathrm{GK}(\mathcal{D}(B))}{2}\mathrm{sd}(M). \blacksquare$$

Remark. In contrast, the famous simple *nonholonomic* modules for the Weyl algebra A_n, for $n \geq 2$, discovered by Stafford, [St], have $\mathrm{sd}(M) > 1$ (by Corollary 3.4).

CONJECTURE. *Let M be a simple $\mathcal{D}(B)$-module. Then M is holonomic if and only if* $\mathrm{sd}(M) = 1$.

Let K be a field of characterisric zero. Denote by $\mathrm{Aut}_F(A_n)$ the group of automorphisms of the Weyl algebra A_n which preserve the natural filtration $F = \{A_{n,i}\}$ of A_n. The group $\mathrm{Aut}_F(A_n)$ can be identified with the subgroup of $\mathrm{GL}(A_{n,1}) \simeq \mathrm{GL}_{2n+1}(K)$ ($\dim A_{n,1} = 2n+1$) consisting of elements which preserve the bilinear antisymmetric form

$$[\cdot,\cdot] : A_{n,1} \times A_{n,1} \to K, \quad (u,v) \mapsto [u,v] = uv - vu.$$

Let M be a finitely generated A_n-module and M_0 be a finite dimensional generating subspace of M. Then the module M is equipped with the filtration $\{M_i = A_{n,i}M_0\}$. Since the associated graded algebra $\mathrm{gr}\, A_n$ is a polynomial ring , it follows from the

Hilbert-Serre theorem that there exists a polynomial $H_M(t) = e(M)t^d/d! + \cdots$ with rational coefficients such that

$$\dim M_i = H_M(i) \text{ for } i >> 0.$$

The polynomial H_M is called the *Hilbert* polynomial of M, its degree $d = d(M)$ coincides with the Gelfand-Kirillov dimension of M and the *natural* number $e(M)$ is called the *multiplicity* of M. The multiplicity $e(M)$ does not depend on the choice of M_0.

Example. For the A_n-module

$$P_n = K[X_1, \dots, X_n] \simeq A_n/(A_n\partial_1 + \cdots + A_n\partial_n)$$

with the generating subspace $P_{n,0} = K\bar{1}$, $\bar{1} = 1 + A_n\partial_1 + \cdots + A_n\partial_n$, the filtration $\{P_{n,i} = A_{n,i}P_{n,0}\}$ coincides with the natural filtration of the polynomial ring P_n defined by degree of variables. So,

$$\dim P_{n,i} = (i+1)(i+2)\cdots(i+n)/n! = i^n/n! + \cdots .$$

thus $d(P_n) = n = \mathrm{GK}(P_n)$ and $e(P_N) = 1$.

The next theorem classifies the holonomic A_n-modules with multiplicity 1.

Theorem 3.5. *([Bav5]). Let M be an A_n-module. Then $\mathrm{GK}(M) = n$ and $e(M) = 1$ if and only if the module M is isomorphic to the twisted A_n-module ${}^{\tau}P_n$ for some $\tau \in \mathrm{Aut}_F(A_n)$.* ∎

Proposition 3.6. *Let M be an A_n-module with $\mathrm{GK}(M) = n$ and $e(M) = 1$. Then $\mathrm{sd}(M) = 1$, in particular, $\mathrm{sd}(P_n) = 1$.* ∎

Proof. Since $\mathrm{sd}(M) = \mathrm{sd}({}^{\tau}M)$ for all $\tau \in \mathrm{Aut}_F(A_n)$, by Theorem 3.5, we may assume that $M = P_n$. The polynomial ring $P_n = \oplus_{i\geq 0} P_n^{[i]}$ is a graded ring by the degree of variables and $P_{n,i} = \oplus_{0\leq j\leq i} P_n^{[j]}$, for $i \geq 0$. The set of monomials $X^\alpha = X_1^{\alpha_1}\cdots X_n^{\alpha_n}$ ($\alpha \in \mathbf{N}^n$) with $i = |\alpha| = \alpha_1 + \cdots + \alpha_n$ is a base of the vector space $P_n^{[i]}$, for $i \geq 0$. Set $e^\alpha = \partial^\alpha/\alpha!$ where $\partial^\alpha = \partial_1^{\alpha_1}\cdots\partial_n^{\alpha_n}$ and $\alpha! = \alpha_1!\cdots\alpha_n!$. Then

$$e^\alpha \cdot x^\beta = \delta_{\alpha,\beta} \text{ for all } \alpha, \beta \in \mathbf{N}^n \text{ such that } |\alpha| \geq |\beta|.$$

Hence $e^\alpha \cdot P_{n,|\alpha|-1} = 0$. For $i \geq 0$, set

$$Q_i = \sum \{KX^\beta e^\alpha : |\beta| \leq |\alpha| \leq i\},$$

then $Q_i \subseteq A_{n,2i}$. We claim that for every K-linear map $\phi : P_{n,i} \to P_{n,i}$ there exists an element $a \in Q_i$ such that $\phi(m) = am$ for all $m \in P_{n,i}$. Suppose that we are done then $\mathrm{Hom}_K(P_{n,i}, P_{n,i}) \subseteq A_{n,2i}$, so $\mu(i) \leq 2i$, hence $\mathrm{sd}(P_n) \leq 1$ and, by Theorem 3.4, $\mathrm{sd}(P_n) = 1$.

It suffices to find such an element $a = a^i_{\alpha,\beta} \in Q_i$ for every map $\phi^i_{\alpha,\beta}$ which maps the monomial X^α to X^β and any other monomial to zero. We use induction on i. The case when $i = 0$ is evident. Suppose that the result is true for all $i' < i$.

If $|\alpha| = i$, then take $a^i_{\alpha,\beta} = X^\beta e^\alpha$. If $|\alpha| < i$ and $|\beta| < i$, then the map $\phi^i_{\alpha,\beta}$ restricted to $P_{n,i-1}$ coincides with $\phi^{i-1}_{\alpha,\beta}$. By induction, there exists the corresponding element $b = a^{i-1}_{\alpha,\beta}$. Then take $a^i_{\alpha,\beta} = b - \sum_\gamma \{(b \cdot X^\gamma)e^\gamma \mid |\gamma| = i\}$. If $|\alpha| < i$ and $|\beta| = i$, then $X^\beta = X_j X^{\beta'}$ for some j and some $\beta' \in \mathbf{N}^n$ with $|\beta'| = i - 1$. Then take $a^i_{\alpha,\beta} = X_j a^i_{\alpha,\beta'}$ where $a^i_{\alpha,\beta'}$ is defined in the previous case. ▪

4. Krull dimension of skew Laurent extensions and generalized Weyl algebras

Let R be a ring with an automorphism $\sigma \in \text{Aut}(R)$ and a central element a. The *generalized Weyl algebra* $T = R(\sigma, a)$ of degree 1 (GWA, for short), is the ring generated by R and two indeterminates X an Y subject to the relations:

$$Xr = \sigma(r)X \text{ and } Yr = \sigma^{-1}(r)Y, \text{ for } r \in R, \quad YX = a \text{ and } XY = \sigma(a).$$

The *skew Laurent extension* $S = R[X, X^{-1}; \sigma]$ is the GWA $S = R(\sigma, a)$ with $a = 1$. We fix this notation in this section.

Theorem 4.1. *([Ho]). Let a commutative Noetherian ring R have finite Krull dimension; then*

$$\mathcal{K}(R[X, X^{-1}; \sigma]) = \sup\{\mathcal{K}(R), ht\,\mathbf{q} + 1 \mid \mathbf{q} \textit{ is a } \sigma - \textit{semistable prime of } R\}. ▪$$

Here $ht\,\mathbf{p}$ is the *height* of a prime ideal $\mathbf{p}$ and $\mathbf{p}$ is *σ-semistable*, if $\sigma^n(\mathbf{p}) = \mathbf{p}$ for some integer n. If there is no such n, the ideal is called *σ-unstable.*

Example. Let $S = K[H, H^{-1}][X, X^{-1}]$, where $\sigma(H) = \lambda H$ for some $\lambda \in K$. If λ is a root of 1, then $\sigma^n = \text{id}$ for some natural n, hence all ideals of the Laurent polynomial ring $K[H, H^{-1}]$ are σ-semistable and, by Theorem 4.1, $\mathcal{K}(S) = \mathcal{K}(K[H, H^{-1}]) + 1 = 2$. If λ is not a root of 1, then there is no a σ-semistable nonzero ideal in $K[H, H^{-1}]$, so $\mathcal{K}(S) = \mathcal{K}(K[H, H^{-1}]) = 1$.

Theorem 4.2. *([BVO]). Let R be a commutative Noetherian ring with $\mathcal{K}(R) < \infty$ and $T = R(\sigma, a)$ be a generalized Weyl algebra. Then*

$$\mathcal{K}(T) = \sup\{\mathcal{K}(R),\ ht\,\mathbf{p}{+}1,\ ht\,\mathbf{q}{+}1 \mid \mathbf{p} \textit{ is a } \sigma{-}\textit{unstable prime ideal of } R \textit{ for which}$$
$$\textit{there exist infinitely many integers } i \textit{ with } a \in \sigma^i(\mathbf{p});$$
$$\mathbf{q} \textit{ is a } \sigma - \textit{semistable prime ideal of } R\}. ▪$$

Example. The first Weyl algebra is isomorphic to the GWA

$$A_1 \simeq K[H](\sigma, H),\ \partial \leftrightarrow Y,\ X \leftrightarrow X,\ \partial X \leftrightarrow H,$$

where $\sigma(H) = H - 1$. If $\text{char}\,K = p > 0$, then $\sigma^p = \text{id}$, hence all ideals of the polynomial ring $K[H]$ are σ-semistable and, by Theorem 4.2, $\mathcal{K}(A_1) = \mathcal{K}(K[H]) + 1 = 2$. If $\text{char}\,K = 0$, then all nonzero ideals of $K[H]$ are σ-unstable, so, by Theorem 4.2, $\mathcal{K}(A_1) = \mathcal{K}(K[H]) = 1$ (see the paper of Tom Lenagan in the present volume for an alternative proof).

Let K be a field and $Usl(2)$ be the universal enveloping algebra (over K) of the Lie algebra

$$sl(2) =< X, Y, H \mid [H, X] = X,\ [H, Y] = -Y,\ [X, Y] = 2H >,$$

The Casimir element $C = YX + H(H + 1)$ is central in $Usl(2)$).

$$Usl(2) \simeq K[H, C](\sigma, a = C - H(H + 1)),\ X \to X,\ Y \to Y,\ H \to H,$$

where the automorphism σ of the polynomial ring $K[H, C]$ is given by $H \to H - 1$, $C \to C$. Applying Theorem 4.2 we obtain the following corollary.

Corollary 4.3. *([Sm] for $K = \mathbf{C}$; see also [MR], 8.6.15). $\mathcal{K}(Usl(2))$ is either two when char $K = 0$, or three when char $K = p > 0$.*

Proof. If char $K = 0$, then all maximal ideals of $K[H, C]$ are σ-unstable. If a subset $I \subseteq \mathbf{Z}$ is infinite, then the K-subspace of $K[H, C]$ generated by $\{\sigma^i(a) = C - (H - i)(H - i + 1), i \in I\}$ contains 1, thus Theorem 4.2 implies that $\mathcal{K}(Usl(2)) = \mathcal{K}(K[H, C]) = 2$. When char $K = p > 0$, then $\sigma^p = 1$, thus $\mathcal{K}(Usl(2)) = 3$. ∎

Till the end of this section R denotes a left Noetherian ring.

Definition. ([GL]). A critical R-module M is called S-**clean** if M is $S \otimes_R M$ is a critical S-module.

A proper factor module of a submodule of a module M is called a *minor* subfactor of M.

Definition. ([GL]). Let A be a simple R-module and let M be an arbitrary R-module. Then the **height** $h(A : M)$ of A in M is defined to be the supremum of those non-negative integers n for which there exists a sequence $A = A_0, A_1, \ldots, A_n$ of S-clean R-modules such that A_i is isomorphic to a minor subfactor of A_{i+1}, for $i = 0, \ldots, n - 1$, while A_n is isomorphic to a subfactor (not necessarily minor) of M.

Theorem 4.4. *([GL]). Let R be a left Noetherian ring with finite Krull dimension. Then $\mathcal{K}(S) = \mathcal{K}(R)$ unless there exists a simple R-module A such that*

$$\mathcal{K}(S \otimes_R A) = 1 \text{ and } h(A : R) = \mathcal{K}(R),$$

in which case $\mathcal{K}(S) = \mathcal{K}(R) + 1$. ∎

If R is a fully bounded Noetherian ring, then the values $h(A : R)$ for simple R-modules M may be replaced by the heights of maximal ideals of R.

Corollary 4.5. *([GL]). Let R be a fully bounded Noetherian ring with finite Krull dimension. Then $\mathcal{K}(S) = \mathcal{K}(R)$ unless there exists a maximal ideal $\mathbf{p}$ of R such that* ht $\mathbf{p} = \mathcal{K}(R)$ *and $\mathbf{p}$ is σ-semistable. In this latter case, $\mathcal{K}(S) = \mathcal{K}(R) + 1$.* ∎

In the case of GWA there are not enough clean modules in the sense above. Three types of T-clean T-modules were introduced in [BL2] in order to obtain similar to [GL] formulas for the Krull dimension of GWA with left Noetherian coefficients.

Definition. ([BL2]). Let A be a simple R-module and let M be an arbitrary R-module.

1. If A is a-torsionfree, then $h(A : M)$ is defined to be the supremum of those non-negative integers n for which there exists a sequence $A = A_0, A_1, \dots, A_n$ of T-clean R-modules such that A_i is isomorphic to a minor subfactor of A_{i+1}, for $i = 0, \dots, n-1$, while A_n is isomorphic to a subfactor (not necessarily minor) of M.

2. If A is a-torsion, then $h_+(A : M)$, $(h_-(A : M))$, is defined to be the supremum of those non-negative integers n for which there exists a sequence $A = A_0, A_1, \dots, A_n$ of R-modules such that each of the R-modules $A = A_0, A_1, \dots, A_i$, for some $i \geq 0$, is $(T, +)$-clean, $((T, -)$-clean), while each of the R-modules $A_{i+1}, \dots, A_n$ is T-clean; and A_j is isomorphic to a minor subfactor of A_{j+1}, for $j = 0, \dots n-1$, while A_n is isomorphic to a subfactor (not necessarily minor) of M.

Theorem 4.6. *([BL2]). Let R be a left noetherian ring with finite Krull dimension. Then $\mathcal{K}(T) = \mathcal{K}(R)$ unless there exists either*
(i) a simple a-torsionfree R-module A such that $h(A : R) = \mathcal{K}(R)$ and ${}^{\sigma^i}A \simeq A$ for some i; or
(ii) a simple $(T, +)$-clean R-module A such that $h_+(A : R) = \mathcal{K}(R)$ and the set $St_+(A)$ is infinite; or
(iii) a simple $(T, -)$-clean R-module A such that $h_-(A : R) = \mathcal{K}(R)$ and the set $St_-(A)$ is infinite;
in which case $\mathcal{K}(T) = \mathcal{K}(R) + 1$, where $St_\pm(A) = \{\pm i \geq 0 \,|\, \sigma^{\mp i}(a)A = 0\}$. ∎

Corollary 4.7. *([BL2]). Let R be a fully bounded noetherian ring with finite Krull dimension. Then $\mathcal{K}(T) = \mathcal{K}(R)$ unless there exists a maximal ideal $\mathbf{p}$ of R such that $\mathrm{ht}(\mathbf{p}) = \mathcal{K}(R)$ and either $\mathbf{p}$ is invariant under some nonzero power of σ or there are infinitely many $i \in \mathbf{Z}$ with $\sigma^i(a) \in \mathbf{p}$.* ∎

Examples. Let R be a ring. Suppose that we are given the following data: $\sigma \in \mathrm{Aut}(R)$ and $b, \rho \in Z(R)$, the centre of R, with ρ being a σ-stable unit; that is, $\sigma(\rho) = \rho$. We form an overing, $E = R\langle\sigma; b, \rho\rangle$, by adjoining symbols X and Y to R subject to the relations:

$$Xr = \sigma(r)X, \; Yr = \sigma^{-1}(r)Y, \text{ for all } r \in R; \quad XY - \rho YX = b.$$

As an example, observe that in the case $R = K[H]$, $\rho = 1$, $b = 2H$ and σ is defined by $\sigma(H) = H - 1$, E is the enveloping algebra $Usl(2)$.

We may view E as the iterated skew polynomial ring $E = R[Y; \sigma^{-1}][X; \sigma, \partial]$, where ∂ is a σ-derivation of $R[Y; \sigma^{-1}]$ and $\partial R = 0$, while $\partial Y = b$; moreover, σ is extended from R to $R[Y; \sigma^{-1}]$ by setting $\sigma(Y) = \rho Y$.

When R is left Noetherian, it is known, [MR], 6.5.4, that

$$\mathcal{K}(R\langle\sigma; b, \rho\rangle) = \mathcal{K}(R) + 1 \text{ or } \mathcal{K}(R) + 2.$$

It is not trivial to decide which of the two cases actually happens.

The rings $E = D\langle\sigma; b, \rho\rangle$ are generalized Weyl algebras.

Lemma 4.8. *Let R be a ring; then $R\langle\sigma; b, \rho\rangle \simeq R[H](\sigma, H)$ and σ is extended from R to $R[H]$ by $\sigma(H) = \rho H + b$.* ∎

Recall that a K-algebra A is said to have the **endomorphism property** over K if, for each simple A-module M, the endomorphism ring $\mathrm{End}_K(M)$ is algebraic over K, see, for example [MR], 9.1.4. Every countably generated algebra over an uncountable field has the endomorphism property, [MR], 9.1.7.

Corollary 4.9. *([BL2]). Let K be an algebraically closed field of characteristic zero, and let A be a left noetherian schurian K-algebra which is a domain and has Krull dimension equal to one. Let $E = A\langle\sigma; b, \rho\rangle$, with $\sigma \in \mathrm{Aut}_K(A)$ and $b, \rho \in K^* = K\backslash\{0\}$. Then $\mathcal{K}(E) = 3$ if and only if at least one of the following cases holds.*
1. $\rho \neq 1$ and there is a simple A_1-module M such that ${}^{\sigma^n}M \simeq M$, for some $n \geq 1$, or
2. $\rho \neq 1$, but $\rho^n = 1$, for some $n \geq 2$.
In all other cases, $\mathcal{K}(A\langle\sigma; b, \rho\rangle) = 2$. ∎

A particular class of algebras to which we could apply this corollary would be the McConnell-Pettit algebras, [MP] (and the first Weyl algebra). Let K be an uncountable algebraically closed field of characteristic zero, and let $\Lambda = (\lambda_{ij})$ be a multiplicatively antisymmetric matrix of nonzero elements of K. The McConnell-Pettit algebra, $P(\Lambda)$, associated with this data, is the K-algebra generated by $X_1^{\pm 1}, \dots, X_n^{\pm 1}$, subject to the commutation relations $X_iX_j = \lambda_{ij}X_jX_i$. In the case that the multiplicative subgroup of K^* generated by the λ_{ij} has maximal rank, $n(n-1)/2$, the algebra $P(\Lambda)$ is known to have Krull dimension one, [MP], Corollary 3.10; and so the Corollary can be applied.

References

[Bav1] V. V. Bavula, Filter dimension of algebras and modules, a simplisity criterion of generalized Weyl algebras, Comm. Algebra, **24** (1996), no. 6, 1971-1992.

[Bav2] V. V. Bavula, Krull, Gelfand-Kirillov and filter dimensions of simple affine algebras, J. Algebra, **206** (1998), no. 1, 33–39.

[Bav3] V. V. Bavula, Krull and Gelfand-Kirillov dimensions of simple almost commutative algebras. J. of Algebra, (to appear).

[Bav4] V. V. Bavula, Filter dimension and its application, *Trends in Ring Theory* (Miskolc, 1996), 1–12, CMS Conf. Proc., **22**, Amer. Math. Soc., Providence, RI, 1998.

[Bav5] V. V. Bavula, Generalized Weyl algebras, kernel and tensor-simple algebras, their simple modules, Proccedings of the 6th Int. Conf. on Represent. of Algebras (V.Dlab and H.Lenzing Eds), CMS Conf. Proc., v. 14, 1993, 83–107.

[BVO] V. V. Bavula and F. Van Oystaeyen, Krull dimension of generalized Weyl algebras and iterated polynomial rings. Commutative coefficients, J. of Algebra, **208** (1998), 1–35.

[BL1] V. V. Bavula and T. H. Lenagan, A Bernstein-Gaber-Joseph theorem for affine algebras, Proc. of the Edinburgh Math. Soc., **42** (1999), 311–332.

[BL2] V. V. Bavula and T. H. Lenagan, Krull dimension of generalized Weyl algebras with non-commutative coefficients, Preprint, Univ. of Edinburgh, (1998); Tr. AMS, (to appear).

[Be] J. Bernstein, Modules over the ring of differential operators. A study of fundamental solution to equations with constant coefficients, Funk. Anal. i Pril., **5** (1971), no. 2, 1-16.

[BK] W. Bohro and H. Kraft, Über die Gelfand-Kirillov Dimension, Math. Ann, **220**, (1976), 1–24.

[Jo] D. Jordan, Krull and global dimension of certain iterated skew polynomial rings, Cont. Math., **130** (1992), 201–213.

[Jos] A. Joseph, Dimension en algèbre non-commutative, Cours de troisième cycle, Univ. de Paris 6, mimeographed notes, 1980.

[GHL] K. R. Goodearl, T. J. Hodges and T. H. Lenagan, Krull and global dimensions of Weyl algebras over division rings, J. Algebra, **91** (1984), 334–359.

[GK] G.R. Krause and T.H. Lenagan, Growth of algebras and Gelfand-Kirillov dimension, Pitman, 1985.

[GK] I. M. Gelfand and A. A. Kirillov, Sur les corps liés aux algèbres enveloppantes des algèbres de Lie, Inst. Hautes Etudes Sci. Publ. Math., **31** (1966), 5–19.

[GL] K. R. Goodearl and T. H. Lenagan, Krull dimension of skew-Laurent extensions, Pacific J. of Math., **114** (1984), no. 1, 109–147.

[Go-Ro] R. Gordon and J. C. Robson, Krull dimension, Memoirs Amer. Math. Soc., no. 133 (1973).

[Ho] T. J. Hodges, The Krull dimension of skew Laurent extensions of commutative Noetherian rings, Commun. Algebra, **12** (1984), 1301–1310.

[M] J. C. McConnell, On the global and Krull dimensions of Weyl algebras over affine coefficients rings, J. London Math. Soc., **29** (1984), 249–253.

[MP] J. C. McConnell and J. J. Pettit, Crossed products and multiplicative analogues of Weyl algebras, J. London Math. Soc., **38** (1988), 47-55.

[MR] J. C. McConnell and J. C. Robson, Non-commutative Noetherian rings, Wiley, 1987.

[RG] R. Rentschler and P. Gabriel, Sur la dimension des anneaux est ensembles ordonnés, C. R. Acad. Sci. Paris, Sér. A, **265** (1967), 712–715.

[Sm] S. P. Smith, Krull dimension of the enveloping algebra of $sl(2)$, J. Algebra, **71** (1981), 89–94.

[St] J. T. Stafford, Non-holonomic modules over Weyl algebras and enveloping algebras, Invent. Math.,**79** (1985), 619-638.

DEPARTMENT OF MATHEMATICS AND COMPUTER SCIENCE, UNIVERSITY OF ANTWERP (U.I.A), UNIVERSITEITSPLEIN 1, B-2610 WILRIJK, BELGIUM

E-mail address: bavula@uia.ua.ac.be, bavula@maths.ed.ac.uk, sveta@kinr.kiev.ua

Trends in Mathematics, © 2000 Birkhäuser Verlag Basel/Switzerland

COHEN - MACAULAY MODULES AND APPROXIMATIONS

ALEX MARTSINKOVSKY

Dedicated to Helmut Lenzing on the occasion of his 60-th birthday

1. Cohen - Macaulay modules and approximations

1.1. **Introduction.** The purpose of this write-up is to provide a quick, "bare bones" introduction to maximal Cohen - Macaulay (mCM, for short) approximations. The reader I had in mind does not have much time (or, perhaps, desire) to learn in-depth yet another formalism, but she/he is curious about this technique and is willing to experiment with it if the necessary basic tools are provided on the fly. There are two such tools to be found in the first part of these notes: one, due to Auslander - Buchweitz, is called the *pitchfork construction*, and the other, appearing in joint work with Jürgen Herzog, is known as the *gluing construction*. To learn that much, the reader may need to set aside 2 - 3 hours for what is essentially a bed-time reading. At the end of the first lecture, I have included an explicit example of gluing for the residue field of a complete intersection, followed by an application to quasihomogeneous singularities. That segment is more technical and is not required for understanding later material.

In the second lecture, new homological invariants of rings and modules are introduced. They generalize the invariants related to mCM approximations over Gorenstein local rings that were originally introduced by Auslander. That generalization is justified by immediate applications given in the first half of the lecture. That part can be read by an undergraduate student who knows the definitions of projective resolution and of homotopy of chain maps. In particular, knowledge of mCM approximations is neither assumed nor needed. Finally, the same invariants are introduced again, this time using Pierre Vogel's construction of Tate cohomology.

These notes are not intended as a comprehensive survey. As a consequence, many topics (and, inevitably, names) have been omitted. In particular, outside the scope of this write-up are the results concerning (weak) liftings of representations as well as the results of Ding and others on the Auslander index of a Gorenstein ring. Besides being more technical, these topics can also be treated in a more general setup using the new invariants, as will be indicated (at least for the index) in the second lecture.

1.2. **Cohen - Macaulay rings and modules.** Throughout these lectures $(R, \mathfrak{m}, k)$ will denote a commutative noetherian local ring with maximal ideal $\mathfrak{m}$ and residue field k.

1.2.1. *CM rings.* First, recall the definitions of Krull dimension and depth.

Definition 1. *The Krull dimension of R, denoted* $\dim R$, *is the supremum of the lengths d of strictly increasing chains $\mathfrak{p}_0 \subsetneq \mathfrak{p}_1 \subsetneq \ldots \subsetneq \mathfrak{p}_d$ of prime ideals of R.*

The Krull dimension is a "geometric size" of the ring.

Definition 2. *The depth of R, denoted* $\operatorname{depth} R$, *is the supremum of the lengths t of regular sequences $x_1, \ldots, x_t$ contained in $\mathfrak{m}$, i.e., the sequences with the property that x_{i+1} is not a zerodivisor for $R/(x_1, \ldots, x_i), i = 1, \ldots, t-1$. If no such sequence exists, the depth of R is set equal to zero.*

The depth is an "arithmetic size" of the ring.

The proof of the following lemma can be found in ([14], Ch. IV, B.1)

Lemma 3. $\operatorname{depth} R \leq \dim R$ *for any local ring R.* ∎

Definition 4. *The ring R is called Cohen - Macaulay (CM) if* $\operatorname{depth} R = \dim R$.

Proposition 5. *If R is CM and x is a non-zerodivisor for R then $R/(x)$ is also CM.*

To prove this, one needs a cohomological description of depth (see below).

1.2.2. *Examples of CM rings.*

1. Regular local rings, i.e., local rings of finite global dimension. ([14], Ch. IV, D.1, Cor. 3, p.103)
2. Hypersurface rings, i.e., local rings of the form $S/(x)$, where S is a regular local ring and x is a non-zero element of the maximal ideal of S. Since S is necessarily a domain ([14], Ch. IV, D.1, Cor. 3, p.105), such x is a non-zerodivisor.
3. Complete intersections, i.e., local rings of the form $S/(x_1, \ldots, x_n)$, where S is a regular local ring and $x_1, \ldots, x_n$ is a regular system.
4. Gorenstein rings, i.e., local rings that are of finite injective dimension over itself. The rings in the examples above are all Gorenstein. (See [7], Prop. 3.1.20).
5. Let k be an algebraically closed field and G a linearly reductive group over k acting linearly on the polynomial ring $k[X_1, \ldots, X_n]$. Then the ring of invariants $k[X_1, \ldots, X_n]^G$ is CM. This is a theorem of Hochster - Roberts. Notice that the invariant ring is not local; in such a case being CM means, by definition, that the localization of that ring at each maximal ideal is CM. In general, the rings of invariants described above are not Gorenstein.
6. Let Δ be a simplicial complex on vertices $X_1, \ldots, X_n$. The Stanley - Reisner ring (or face ring) $k[\Delta]$ of Δ is defined as $k[X_1, \ldots, X_n]/I_\Delta$, where I_Δ is the ideal generated by all the monomials $X_{i_1} \cdot \ldots \cdot X_{i_s}$ with $i_1 < \ldots < i_s$ such that

$\{X_{i_1}, ..., X_{i_s}\}$ is not a face of Δ. Suppose the geometric realization of Δ is homeomorphic to a sphere; it can then be shown that $k[\Delta]$ is CM. This fact was used by Stanley in his proof of the Upper Bound Conjecture for spheres. Again, Stanley - Reisner rings can be CM without being Gorenstein.

1.2.3. *CM modules.* The concepts of dimension and depth can easily be extended to modules. Let M be a non-zero finitely generated R-module.

Definition 6. *The dimension of M, denoted* $\dim M$, *is defined as* $\dim(R/\operatorname{Ann} M)$.

Definition 7. *The depth of M, denoted* $\operatorname{depth} M$, *is the supremum of the lengths t of M-regular sequences $x_1, ..., x_t$ contained in $\mathfrak{m}$, i.e., the sequences with the property that x_{i+1} is not a zerodivisor for $M/(x_1, ..., x_i), i = 1, ..., t-1$. If no such sequence exists the depth of M is set equal to zero.*

The proof of the following lemma can be found in ([14], Ch. IV, B.1).

Lemma 8. $\operatorname{depth} M \leq \dim M \leq \dim R$ *for any R-module M.*

Definition 9. *The module M is called Cohen - Macaulay (CM) if* $\operatorname{depth} M = \dim M$. *If* $\operatorname{depth} M = \dim M = \dim R$ *then M is called a maximal Cohen - Macaulay (mCM) module.*

1.2.4. *Examples of CM modules.*

1. Suppose that M is an R-module of finite length. Then $\operatorname{Ann} M$ is $\mathfrak{m}$-primary (as the intersection of a finite number of $\mathfrak{m}$-primary ideals $\operatorname{Ann} x_i$, where the x_i's are generators of M). This shows that $\dim M = 0$. By the lemma above, $\operatorname{depth} M \leq \dim M$ and therefore M is CM. In particular, the residue field k is a CM module. Thus any ring admits a CM module.
2. It is not known whether any ring R admits a maximal CM module.
3. Suppose R is a CM ring. Then R is an mCM module over itself.
4. Let M be a CM R-module and x a non-zerodivisor for M. Then M/xM is also a CM R-module. To show this one needs a cohomological description of depth (see the next subsection).

1.2.5. *A cohomological description of depth.*

Proposition 10. *Let M be a non-zero module. Then*

$$\operatorname{depth} M = \inf\{i \mid \operatorname{Ext}^i(k, M) \neq 0\}$$

For the proof, see ([14], Ch. IV, A.4, Prop. 6). ∎

Combining this proposition with Lemma 8 we have

Corollary 11. *The category of mCM R-modules is closed under finite direct sums, direct summands, and kernels of epimorphisms.*

As another consequence, we have

Corollary 12. *Suppose* $0 \to K \to L \to M \to 0$ *is exact in* $\operatorname{mod} R$. *If* $\operatorname{depth} L > \operatorname{depth} M$ *then* $\operatorname{depth} K = \operatorname{depth} M + 1$. ∎

The last result can be used to construct mCM modules starting with any module over a CM ring. More precisely, we have

Corollary 13. *Suppose that R is CM of dimension d. Then, for any finitely generated R-module M, the d-th syzygy module $\Omega^d M$ is mCM.* ∎

Remark: An interesting problem is to classify CM rings which have only finitely many isomorphism classes of indecomposable mCM modules. As of now (October 1998), the problem is still open. The answer is known if R is a complete Gorenstein algebra over an algebraically closed field: such rings are exactly the simple hypersurface singularities A_n, D_n, E_6, E_7, and E_8.

1.3. **MCM modules over Gorenstein rings.** Until the end of this lecture the ring R will be assumed to be Gorenstein of dimension d (i.e., R is of injective dimension d over itself). Let $\mathrm{mCM}(R)$ denote the full subcategory of $\mathrm{mod}\, R$ consisting of mCM modules.

We shall now review the duality for CM modules (see [7], Prop. 3.3.10). Let M be a CM R-module of codepth n (i.e., $n = \operatorname{depth} R - \operatorname{depth} M$). Then:

1. $\mathrm{Ext}^i(M, R) = 0$ if $i \neq n$.
2. $M^\vee := \mathrm{Ext}^n(M, R)$ is also CM of codepth n.
3. M is isomorphic to $M^{\vee\vee}$.

This leads to the following characterizations of mCM modules over a Gorenstein ring.

Proposition 14. *Over a Gorenstein ring R, a finitely generated module M is mCM if and only if $\mathrm{Ext}^i(M, R) = 0$ for all $i \geq 1$.*

Proof. The "only if" part follows from 1 above. Conversely, suppose $\mathrm{Ext}^i(M, R) = 0$ for all $i \geq 1$, and let $\ldots \to P_1 \to P_0 \to M \to 0$ be a projective resolution of M. Applying the functor $(-)^* := \mathrm{Hom}(-, R)$ we obtain an exact complex

$$0 \to M^* \to P_0^* \to P_1^* \to \ldots$$

By Cor. 13, both M^* and $\mathrm{Tr}\, M := \mathrm{Coker}(P_0^* \to P_1^*)$ are mCM. In particular, $\mathrm{Ext}^i(\mathrm{Tr}\, M, R) = 0$ for all $i \geq 1$. The short exact sequence (see [3], Prop. 2.6)

$$0 \to \mathrm{Ext}^1(\mathrm{Tr}\, M, R) \to M \to M^{**} \to \mathrm{Ext}^2(\mathrm{Tr}\, M, R) \to 0$$

now shows that M is reflexive. Since M^* is mCM, $\mathrm{Ext}^i(M^*, R) = 0$ for all $i \geq 1$. Applying the beginning of the argument to M^* we have that $M^{**} \cong M$ is mCM. ∎

Corollary 15. *Over a Gorenstein ring R, a finitely generated module M is mCM if and only if $\mathrm{Ext}^i(M, L) = 0$ for all $i \geq 1$ and for all modules L of finite projective dimension.*

Proof. Induction on the projective dimension of L. ∎

1.3.1. *Complete resolutions.* The proof of the proposition above shows that a projective resolution $P_M \to M$ of the mCM module M and the dual $(P_{M^*} \to M^*)^*$ of a projective resolution of M^* can be spliced together, because $M \cong M^{**}$, in a doubly infinite exact complex of projective modules. We shall refer to this exact complex C_M as a *complete resolution* of M. To assure minimality, we have to require that M have no free summands. If this is not the case, the first syzygy module of M will have no free summands and we can construct a minimal complete resolution for that module, which will still be called a complete resolution of M. The following important property of complete resolutions is an immediate consequence of Prop. 14.

Lemma 16. *Let M be an mCM module over a Gorenstein ring R and C_M a complete resolution of M. Then the complex* $\mathrm{Hom}(C_M, R)$ *is exact.* ∎

1.4. **MCM approximations over Gorenstein rings.** The problems of lifting and extending (various classes of) morphisms are of fundamental importance in any category. The approximations we are going to discuss are a particular example of that kind of problems.

Definition 17. *([4]) An mCM approximation of $C \in \mathrm{mod}\, R$ is a short exact sequence*

$$0 \to Y_C \to X_C \xrightarrow{f} C \to 0$$

where X_C is mCM and Y_C is of finite projective dimension. It is called minimal if the map $f : X_C \to C$ is right minimal.

Recall that $f : X_C \to C$ is said to be *right minimal* if any endomorphism $h : X_C \to X_C$ that makes the diagram

$$\begin{array}{ccc} X_C & \xrightarrow{f} & C \\ \downarrow^{h} & \nearrow^{f} & \\ X_C & & \end{array}$$

commutative is an automorphism.

Remark: It is possible to define mCM approximations over CM rings which admit a dualizing module. The only difference is that the module Y_C above is required to be of finite injective dimension. Notice that over a Gorenstein ring a module is of finite injective dimension if and only if it is of finite projective dimension.

Suppose now that X is mCM. Then the functor $\mathrm{Ext}^1(X, -)$ vanishes on modules of finite projective dimension (Cor. 15). This implies that any morphism $g : X \to C$, where X is mCM, can be lifted over the approximation $f : X_C \to C$. It is now easy to show that minimal approximations are unique up to (non-canonical) isomorphism. In particular, any invariant of a minimal approximation is also an invariant of the module.

It can also be shown that if a right minimal morphism $f' : X'_C \to C$, where X'_C is mCM, has the lifting property for the morphisms $Y \to C$ with Y mCM, then the kernel of f' is of finite projective dimension. As a result, *minimal* mCM approximations can be defined either as short exact sequences or as morphisms with the above lifting property.

We now come to the main existence theorem.

Theorem 18. *(Auslander - Buchweitz, ca. 1986, see* [4], *) For any module C in* mod R *there exists an mCM approximation.*

Remark: It can be shown that any module in mod R has a minimal mCM approximation.

In fancier terms, the theorem above says that the subcategory mCM(R) is contravariantly finite in mod R. We shall give two proofs of this theorem when we discuss the pitchfork construction of Auslander - Buchweitz and the gluing construction of J. Herzog - Martsinkovsky. These constructions are different from the one appearing in the paper cited in the theorem.

At the end of this subsection we want to mention how one can recognize minimal morphisms. The recipe is contained in the following propositions.

Proposition 19. *A morphism in* mod R *is minimal if and only if the* $\mathfrak{m}$*-adic completion of that morphism is minimal in* mod $\widehat{R}$, *where* $\widehat{R}$ *is the* $\mathfrak{m}$*-adic completion of* R.

Proposition 20. *Suppose R is complete in the* $\mathfrak{m}$*-adic topology. A morphism in* mod R *is minimal if and only if no direct summand of the domain is sent by that morphism to zero.*

The proofs are left to the reader as an exercise. (Hint: in the proof of the second proposition one may pass to the quotients of R by the powers of $\mathfrak{m}$, use the Fitting Lemma, and pass to the limit.)

1.5. **Hulls of finite injective (projective) dimension.** Auslander and Buchweitz also showed [4] that any module in mod R can be approximated, in a slightly different way, by a module of finite projective dimension (we are still assuming that R is Gorenstein) with cokernel being mCM.

Definition 21. *A hull of finite projective dimension of $C \in$* mod R *is a short exact sequence*

$$0 \to C \xrightarrow{f} Y^C \to X^C \to 0$$

where X^C is mCM and Y^C is of finite projective dimension. It is called minimal if the map $f : C \to X^C$ is left minimal.

The definition of a *left minimal* morphism is obtained from its dextrous counterpart above by reversing the directions of the arrows. Like approximations, minimal hulls exist and are unique up to (non-canonical) isomorphisms.

Remark: Similar to mCM approximations, the definition of the hull can be extended to CM rings admitting a dualizing module: just require the module Y^C be of finite injective dimension. The resulting construct is called the hull of finite injective dimension.

As of now, there have been almost no applications of hulls. This fact is especially surprising since the approximations continue to show up in a variety of different settings. Part of the reason may be the fact that, for the hulls, the modules at the other end of the approximating arrow are of finite projective dimension rather than the better behaved mCM's. We shall see later, when we describe the gluing construction, that hulls and approximations are very closely related to each other, a fact already noticed (and cleverly used!) by Auslander - Buchweitz in their proof of the existence theorem. This relationship is best understood in the language of derived categories. In other words, thinking of those constructs only in terms of arrows at the expense of short exact sequences (also known as *triangles*) is a methodologically deficient approach.

1.6. **An example: the Auslander module of a two-dimensional normal singularity.** The results described in this section are taken from [2]. Let R be a two-dimensional complete local algebra over an algebraically closed field k of characteristic zero. As before, we assume that R is Gorenstein. In addition, we assume that R is an isolated singularity. Algebraically this means that the localization of R at any non-maximal ideal is regular. Under the above assumptions, an R-module is mCM if and only if it is reflexive. It is not difficult to show that the dual $\mathrm{Ext}^2(k, R)$ of the CM module k is isomorphic to k. Thus, interpreting the elements of $\mathrm{Ext}^2(k, R)$ as four-term exact sequences, we have that all such non-split extensions are equivalent (as they differ from each other by invertible factors). Choose one of them and write it in the form

$$0 \to R \to A \to R \to k \to 0$$

It follows from the just mentioned equivalence that the module A, nowadays called the Auslander module, is unique up to isomorphism. It can also be shown that A is reflexive and therefore mCM. Truncating, we have that the short exact sequence

$$0 \to R \to A \to \mathfrak{m} \to 0$$

is an mCM approximation of the maximal ideal. It is minimal because $\mathfrak{m}$ is not mCM and hence R cannot be a direct summand of A.

By a result of Auslander [1], the category of mCM's over an isolated CM singularity has almost split sequences. The four-term exact sequence above is called the *fundamental almost split sequence.* The name comes from the fact that any almost split sequence in the category of reflexive modules can be obtained from this sequence by means of the reflexive tensor product.

Specializing further, we now assume that R is a rational double point, i.e., the defining equation of R is one of Arnold's simple singularities A_n, D_n, E_6, E_7, and E_8. Then there are only finitely many isomorphism classes of indecomposable mCM's. The isoclasses of *non-projective* mCM's are in one-to-one correspondence with the vertices of the corresponding Dynkin diagram. To incorporate the unique indecomposable projective R one has to consider the *extended* Dynkin diagrams and identify the maximal root with R. In this language, the Auslander module corresponds to the vertex (vertices) adjacent to the maximal root. This indicates that, figuratively speaking, the Auslander module is close to being projective.

In fact, utilizing the concept of *preprojective partition*, invented by Auslander and Smalø, we can make the last observation very precise. Returning now to the more general setting described in the beginning of this section, we recall that, by another result of Auslander, the category of mCM's over an isolated singularity has a preprojective partition. Let us examine the first non-projective cell of that partition. Suppose M is an mCM without free summands. Then any map from M to R lands in the maximal ideal and therefore, by the lifting property of mCM approximations, it can be lifted over the map $A \to R$ in the fundamental sequence. It is not difficult to check that the uniqueness of the fundamental sequence implies that it must be self-dual. This means that dualizing the approximation of the maximal ideal into R we recover the fundamental sequence. Notice that, upon that dualization, the extension problem over the map $R \to A$ for any map $R \to M^*$ becomes the lifting problem that has just been solved. Moreover, because mCM's over a Gorenstein ring are reflexive, as M runs through all mCM's without free summands M^* runs through the same class. Now that the extension problem is solvable for that class of modules and since R covers all those modules, (a direct sum of) the Auslander module(s) covers any mCM without free summands. This argument shows that the Auslander module is the sole occupant of the first non-projective cell in the preprojective partition of the category of mCM modules over R. We have thus formalized the statement that the Auslander module is close to being projective.

1.7. **The pitchfork construction.** We are now ready to describe the first method for constructing mCM approximations. Let C be an R-module of codepth n. This means that $n = \operatorname{depth} R - \operatorname{depth} M$. By Cor. 12, $\Omega^n C$ is the first syzygy module of C which is mCM. The projective resolution $\mathbf{P}$ of C and the complete resolution $\mathbf{Q}$ of $\Omega^n C$ truncated in degree 0 can be put together into a commutative pitchfork diagram:

$$
\begin{array}{ccccccccccc}
 & & & & P_{n-1} & \to & \dots & \to & P_0 & \to & C & \to & 0 \\
 & & & \nearrow & & & & & & & & & \\
\dots & \to & P_n & & \uparrow & & & & \uparrow & & \uparrow & & \\
 & & & \searrow & & & & & & & & & \\
 & & & & Q_{n-1} & \to & \dots & \to & Q_0 & \to & \underline{X}_C & \to & 0
\end{array}
$$

The vertical maps can be constructed inductively, starting with the isomorphism $\Omega^n \underline{X}_C \xrightarrow{\simeq} \Omega^n C$, then breaking the complete resolution into short exact sequences, and using the fact that the functor $\operatorname{Ext}^1(-, R)$ vanishes on mCM's. As a result, we have a map $f : \underline{X}_C \to C$ and a chain map $f' : \mathbf{Q} \to \mathbf{P}$ that lifts f. Adding, if necessary, trivial complexes of projectives to $\mathbf{Q}$ we may assume, without loss of generality, that f' is an epimorphism (hence a split epimorphism) in each degree and, therefore, f is also an epimorphism. This gives rise to a short exact sequence of complexes:

$$0 \to \mathbf{L} \to \mathbf{Q} \xrightarrow{f'} \mathbf{P} \to 0$$

The long homology exact sequence shows that the homology of $\mathbf{L}$ is concentrated in degree 0. Since f' is split in each degree, $\mathbf{L}$ is a complex of projectives and since f' is the identity map starting in degree k, this complex is zero in degrees n and up. As a consequence, $\mathbf{L}$ is a finite projective resolution of the module Y_C defined by the short exact sequence

$$0 \to Y_C \to X_C \xrightarrow{f} C \to 0$$

(and pd $Y_C \leq n-1$). Since the Ext^1-functor with coefficients in a module of finite projective dimension vanishes on mCM's, the map f is an approximation. Notice that have simultaneously constructed approximations for all syzygy modules of C. ∎

If one uses minimal resolutions only (including the construction of the complete resolution) then $\underline{X}_C$ will have no free summands. Thus the approximation of C is of the form

$$0 \to Y_C \to \underline{X}_C \amalg F_C \to C \to 0$$

where $\underline{X}_C$ has no free summands and F is a free module. If $f|_{F_C}$ is a projective cover then it can be shown that this approximation is right minimal and is therefore unique up to non-canonical isomorphisms.

1.8. **Auslander's delta and index.** The pitchfork construction shows that a minimal complete resolution of a syzygy module of C is not enough to construct the approximation: it may be necessary to add trivial complexes of projectives. Our next goal is to provide a quantitative tool for describing that phenomenon.

Definition 22 (Auslander). *The delta invariant is defined as* $\delta(C) := \operatorname{rk} F_C$, *where* $\underline{X}_C \amalg F_C \to C$ *is a minimal mCM approximation. Similarly,* $\delta^i(C) := \delta(\Omega^i C)$.

The following properties of δ can easily be verified.

Lemma 23. *Let C and D be R-modules. Then:*

1. $0 \leq \delta^i(C) \leq \beta_i(C)$ *for all* i. *Here* $\beta_i(C)$ *is the* i*-th betti number of* C.
2. *If* C *is projective then* $\delta^i(C) = \beta_i(C)$ *for all* i.
3. $\delta^i(C \amalg D) = \delta^i(C) + \delta^i(D)$ *for all* i.
4. *If* $C \to D$ *is an epimorphism, then* $\delta(C) \geq \delta(D)$. ∎

It should be pointed out that the last property, unlike the first three, holds only for $\delta = \delta^0$.

Based on δ, Auslander also introduced another invariant of a Gorenstein local ring:

Definition 24. $\operatorname{index}(R) := \inf\{i \mid \delta(R/\mathfrak{m}^i) \neq 0\}$

Applications were found for these invariants almost immediately after their introduction. Here is a partial list:

- Suppose R is an isolated hypersurface singularity: $R = S/(f)$, where $S := k[|X_1, ..., X_n|]$ and k is algebraically closed of characteristic 0. Then R is quasihomogeneous if and only if $\delta(R/\overline{j(f)}) = 0$. Here $j(f)$ is the ideal of S generated by the partials of f (See [11].)
- A Gorenstein local ring R is non-regular if and only if $\delta^i(k) = 0$ for all i. (Auslander, unpublished, ca. 1990.)
- If R is Gorenstein, then

$$\operatorname{mult} R \geq (\operatorname{edim} R - \dim R) + \operatorname{index}(R) - 1$$

(See [8].)
- Let x be a non-zerodivisor for R and $\overline{R} := R/(x)$. If an $\overline{R}$-module C is weakly liftable to R, then $\sum(-1)^i \delta^i_{\overline{R}}(C) = 0$. (See [5].)

It became clear that the delta-invariant conveyed useful information. In particular, conditions for the vanishing of it became important. Here is a partial list of what is now known.

- If $R := S/(f)$ is a hypersurface ring and a (minimal) projective resolution (P_C, d) of the module C is known, then the problem is completely solved. Pass to the Gulliksen - Eisenbud operator $\tau := d^2/f$, where the ratio is computed as follows: lift d to S, compute d^2 over S, divide by f, and reduce the result modulo f. This degree -2 operator is defined uniquely up to homotopy. The vertical maps $f_i^!$ mentioned in the pitchfork construction can now be chosen as $\tau_{i+2} \circ \tau_{i+4} \circ \tau_{i+6} \circ \ldots$. Beyond the codepth of C the operator τ becomes the identity, hence the composition makes sense. Thus the δ's can be computed in terms of the resolution only. (See [13].)
- Suppose that $R := S/(f)$, where S is a Gorenstein ring and f is regular on S. If $f \in \mathfrak{m}_S \operatorname{Ann}_S C$, then $\delta^i_R(C) = 0$ for all i. If C is cyclic then the converse is also true. More precisely, $\delta_R(C) = 0 \Leftrightarrow f \in \mathfrak{m}_S \operatorname{Ann}_S C \Leftrightarrow \delta^i_R(C) = 0$ for all i. (See [8].)
- If, again, R is a hypersurface ring, C is a CM module and minimal projective resolutions of both C and $C^\vee$ are known, then the deltas can be expressed via the betti numbers of the two modules by a closed formula. (See [9].)
- Suppose $R = S/(f)$, where S is a Gorenstein ring and f is regular on S. Then the minimality of the Shamash resolution of C implies that all $\delta^i(C)$ vanish. That generalizes the result of Ding. If S is regular, then the converse is true. (See [10].) For details on the Shamash construction, see [15]

1.9. The gluing construction. We shall now describe a second method for constructing mCM approximations. It is perhaps not as direct, compared with the pitchfork construction, but it has several advantages of its own. First, it yields at the same time both the approximations and hulls of the module and all of its syzygy modules. Secondly, it establishes connections with derived categories and, consequently, explains (at least over Gorenstein rings) why approximations and hulls are closely related. Thirdly, it codifies the module-theoretic machinery of approximations and hulls in terms traditionally associated with homotopy theory, thus giving new perspectives on the original algebraic constructs; in particular, one is again led to the formalism of stable homotopy theory. Fourthly, the delta-invariants are exhibited as the ranks of the homogeneous components of the gluing map (see below) modulo the maximal ideal, offering yet another perspective on the deltas. Loosely speaking, the advantage of the gluing construction is similar to that of a generating function compared with the original discrete set of numbers.

We begin by recalling some basic homotopy theory. The diagram shown below starts with a continuous map $f : X \to Y$. The middle row relates X, the mapping cylinder $\mathrm{Cyl}(f)$, and mapping cone $\mathrm{Con}(f)$ of f, while the top row relates Y, $\mathrm{Con}(f)$, and the suspension $\sum X$ of X. The inclusion $i : Y \to \mathrm{Cyl}(f)$ and the retraction $r : \mathrm{Cyl}(f) \to Y$ are homotopy inverses of each other. Furthermore, in a naive sense, the middle and top rows are "exact".

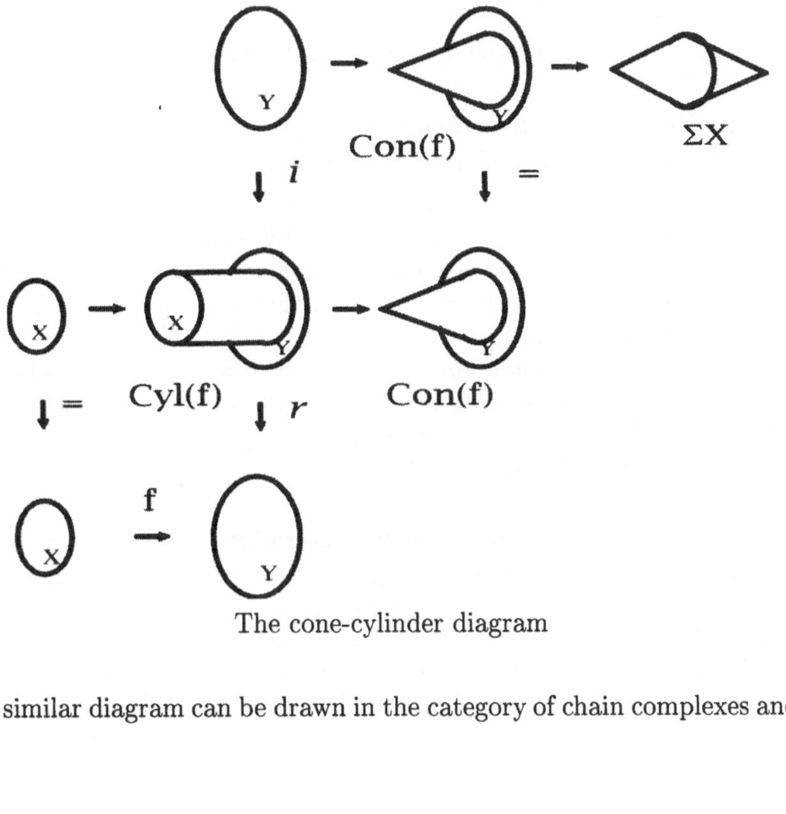

The cone-cylinder diagram

A similar diagram can be drawn in the category of chain complexes and chain maps:

$$\begin{array}{ccccccccc}
 & & 0 & \to & L & \to & \mathrm{Con}(f) & \to & \sum L \to 0 \\
 & & & & \downarrow^{i} & & \downarrow^{=} & & \\
0 & \to & K & \to & \mathrm{Cyl}(f) & \to & \mathrm{Con}(f) & \to & 0 \\
 & & \downarrow^{=} & & \downarrow^{r} & & & & \\
 & & K & \xrightarrow{f} & L & & & &
\end{array}$$

Here K and L are chain complexes, f is a chain map, the retraction r and injection i are homotopy inverses of each other, and $\sum L := L[-1]$. In addition, the middle row and the top row are exact.

We are now ready to describe the gluing construction (for more details see [9]). As before, R is assumed to be Gorenstein. For simplicity, we also assume that the module C is CM of codepth n. Let $\mathbf{P} \to C$ and $\mathbf{Q} \to C^{\vee}$ be projective resolutions of C and, respectively, $C^{\vee}$ (recall that $C^{\vee} := \mathrm{Ext}^n(C, R)$). Since C is CM, the complexes $\mathbf{P}$ and $\mathbf{Q}^*[-n]$, where $(-)^* := \mathrm{Hom}(-, R)$, have isomorphic homology which are both concentrated in degree 0. Since $\mathbf{P}$ is projective and $\mathbf{Q}^*[-n]$ is exact in positive degrees, this isomorphism gives rise to a quasi-isomorphism $f : \mathbf{P} \to \mathbf{Q}^*[-n]$, which is called a *gluing map* for C. Notice that the mapping cone of f is an exact complex of projective modules. Applying the cone - cylinder construction to f, we have that mCM approximations of C and its syzygy modules come from truncations of the top row, whereas hulls of finite injective (i.e., finite projective) dimension of C and its syzygy modules come from truncations of the middle row. If $\mathbf{P}$ and $\mathbf{Q}$ are minimal then the obtained mCM approximation is also minimal. Furthermore, $\delta^i(C) = \dim_k(f_i \otimes k)$.

Remark: If C is not CM then $\mathbf{Q}$ should be replaced with a projective resolution of the homologically bounded complex $\mathbf{P}^*$.

For the reader who is familiar with the basics of derived categories it should now become clear why mCM approximations and hulls of finite projective dimension are closely related: they are *triangles* obtained from each other by rotations. Notice that one extra rotation places the approximated object in the middle of a triangle. Even though such sequences do not seem to have a counterpart in topology (or do they?), they are reminiscent of an algebraic construct called torsion theory. It is therefore of interest to extend the notion of torsion to derived categories. We should also remark that, from a different point of view, such sequences have recently been investigated by K. Kato.

1.10. **An example: the residue field of a complete intersection.** What follows is an explicit construction of a gluing map for the residue field of a complete intersection. Let $R := S/(h_1, ..., h_m)$, where the h's form a regular sequence and S is a regular local ring whose maximal ideal is minimally generated by $x_1, ..., x_n$. Tate ([16]) constructed a projective resolution, with an additional algebra structure, for any cyclic module over a local ring. Later Gulliksen showed that the Tate resolution of the residue field is always minimal. Tate's construction consists of consecutive adjoining of new variables (and their products), with the Koszul complex on the x's as the first step. If R is a complete intersection, then this process already terminates in degree 2; the degree 2 variables correspond to the relations (over S)

$$h_j = \sum_{i=1}^{n} h_{ij} x_i, j = 1, ..., m$$

Now let F be a free R-module with basis $f_1, ..., f_m$ and G a free R-module with basis $g_1, ..., g_n$. (The symmetric algebra S_*F is just a polynomial algebra in m indeterminates and we may view $G \otimes S_*F$ as a free module over S_*F.) Since exterior powers commute with ring extension, there is a natural isomorphism

$$\dot{\wedge} G \otimes S.F \cong \dot{\wedge}(G \otimes S.F)$$

The Tate resolution of k is the total space of the double complex $\mathbf{T}$ with

$$\mathbf{T}_{ij} := (\wedge^i G \otimes S_j F)^*$$

To describe the differentials, we introduce the elements $x := \sum_{i=1}^{n} \overline{x}_i g_i \otimes 1 \in \wedge^1 G \otimes S_0 F$ and $y := \sum_{i=1}^{n} (g_i^* \otimes \sum_{j=1}^{m} \overline{h}_{ij} f_j) \in \wedge^1 G^* \otimes S_1 F$, where overbar indicates the image under the canonical surjection $S \to R$. The matrix $(\overline{h}_{ij})$ gives rise to the R-linear map

$$\varphi : F^* \to G^* : f_j^* \longmapsto \sum_{i=1}^{n} \overline{h}_{ij} g_i^*, j = 1, ..., m$$

Let $\mu_x : \wedge^i G \otimes S_j F \to \wedge^{i+1} G \otimes S_j F, i = 0, 1, ...$ denote the exterior multiplication by x in $\dot{\wedge}(G \otimes S.F)$. Furthermore, let $\partial_y = \mu_y^* : \wedge^i G \otimes S_j F \to \wedge^{i-1} G \otimes S_{j+1} F$ or, explicitly, $\partial_y(g_{l_1} \wedge ... \wedge g_{l_i}) := \sum_{j=1}^{i} (-1)^{j+1} \varphi^*(g_{l_j}) g_{l_1} \wedge ... \wedge \widehat{g_{l_j}} \wedge ... \wedge g_{l_i}$. The maps μ_x and ∂_y are the differentials of the double complex $\mathbf{T}$.

We are now ready to describe the gluing map $\nu : \mathbf{T} \to \mathbf{T}^*[-r]$, where $r := n - m$. To this end, we choose orientations (i.e., isomorphisms) $\gamma : \wedge^m F^* \to R$ and $\delta : \wedge^n G^* \to R$ and define $\nu_i : \wedge^i G^* \to (\wedge^{r-i} G^*)^*$ by

$$(\nu_i(u))(v) := \varepsilon \delta(u \wedge v \wedge ((\wedge^m \varphi)(z)))$$

for all $u \in \wedge^i G^*, v \in \wedge^{r-i} G^*$, where $z := \gamma^{-1}(1)$, and $\varepsilon = 1$ when $i \equiv 0, 3 \bmod 4$ and $\varepsilon = -1$ when $i \equiv 1, 2 \bmod 4$. Notice that for each $i = 1, ..., r$ the map v_i gives rise to the homomorphism

$$(\wedge^i G \otimes S_0 F)^* \simeq \wedge^i G^* \xrightarrow{\nu_i} (\wedge^{r-i} G^*)^* \simeq \wedge^{r-i} G \otimes S_0 F$$

Extending, for each $i = 1, ..., r$ this map by 0 to the whole component $\mathbf{T}_i$ we obtain the family of maps $\nu : \mathbf{T} \to \mathbf{T}^*[-r]$ which is a gluing map for k.

The following diagram illustrates the situation for $r = 2$ (the symbol $\wedge^i \otimes S_j$ stands for $\wedge^i G \otimes S_j F$):

$$
\begin{array}{ccccccccccc}
 & & & & 0 & & \cdots & & \cdots & & \\
 & & & & \uparrow & & \uparrow{\scriptstyle \mu_x} & & \uparrow{\scriptstyle \mu_x} & & \\
 & & 0 & \to & (\wedge^0 \otimes S_0)^* & \overset{\nu_0}{\to} & \wedge^2 \otimes S_0 & \overset{\partial_y}{\to} & \wedge^1 \otimes S_1 & \to & \cdots \\
 & & \uparrow & & \uparrow{\scriptstyle \mu_x^*} & & \uparrow{\scriptstyle \mu_x} & & \uparrow{\scriptstyle \mu_x} & & \\
0 & \to & (\wedge^0 \otimes S_1)^* & \overset{\partial_y^*}{\to} & (\wedge^1 \otimes S_0)^* & \overset{\nu_1}{\to} & \wedge^1 \otimes S_0 & \overset{\partial_y}{\to} & \wedge^0 \otimes S_1 & \to & 0 \\
 & & \uparrow{\scriptstyle \mu_x^*} & & \uparrow{\scriptstyle \mu_x^*} & & \uparrow{\scriptstyle \mu_x} & & \uparrow & & \\
\cdots & \to & (\wedge^1 \otimes S_1)^* & \overset{\partial_y^*}{\to} & (\wedge^2 \otimes S_0)^* & \overset{\nu_2}{\to} & \wedge^0 \otimes S_0 & \to & 0 & & \\
 & & \uparrow{\scriptstyle \mu_x^*} & & \uparrow{\scriptstyle \mu_x^*} & & \uparrow{\scriptstyle \mu_x} & & & & \\
 & & \cdots & & \cdots & & 0 & & & &
\end{array}
$$

1.11. **Quasihomogeneity of two-dimensional isolated CM singularities.** Returning to the fundamental almost split sequence of a complete two-dimensional isolated CM singularity, we notice that Auslander module A is the only module in that sequence that has not been identified. It can be shown that if R is a rational double point then A is isomorphic to the module of Zariski differentials $D_k(R)^{**}$.

Conjecture 25 (Martsinkovsky). *Let R be a two-dimensional integrally closed analytic k-algebra over an algebraically closed field k of characteristic* 0. *Then $A \simeq D_k(R)^{**}$ if and only if R is quasihomogeneous.*

Quasihomogeneity means that after a suitable change of coordinates the defining equations of R can be made weighted homogeneous (polynomials).

The "if" part of the conjecture is true. The other direction seems to be very hard. It has been proved for the following classes of singularities:

- Quotient singularities and hypersurfaces with defining equation $f(x, y, z) = z^n + g(x, y)$. (See [12].)
- Minimally elliptic singularities and rational singularities with reduced fundamental cycle. (See [6].)
- Hypersurfaces with defining equations $f = g + H(g)$, where $g = X^p + Y^q + Z^r$ and $H(g)$ is the Hessian of g. (See [17].)

1.12. **Quasihomogeneity in higher dimensions.** Can the above conjecture be generalized to arbitrary dimensions? To answer this question, we shall assume for a moment that $R := S/(f)$ is a hypersurface. Then it can be shown that $D_k(R)^{**} \simeq D_k(R)^*$ and, by the pitchfork construction, that A is isomorphic to $\Omega^3 k$ (here we used Eisenbud's result that minimal projective resolutions of mCM modules over hypersurface rings are periodic of period at most two). But

$$D_k(R)^* \simeq \Omega^2 \operatorname{Tr} D_k(R) \simeq \Omega^2 R/\overline{j(f)}$$

where $j(f)$ is the jacobian ideal of f, i.e., $j(f)$ is the ideal generated by the partial derivatives of f. Since R is assumed to be an isolated singularity, the *moduli algebra* $R/\overline{j(f)}$ is a module of finite length. Thus if R is quasihomogeneous then the finite length modules k and $R/\overline{j(f)}$ have the same, up to a shift, high enough syzygy modules.

This should not be too surprising if we recall the Euler identity. Assuming that f is quasihomogeneous of degree d with x_i being of degree d_i we have

$$\sum_{i=1}^{n} d_i x_i \frac{\partial f}{\partial x_i} = d \cdot f$$

Viewed modulo f (i.e., over R) this identity has two interpretations: either as a relation between the generators of $\overline{j(f)}$ with coefficients generating the maximal ideal $\mathfrak{m}$ of R or as a relation between the generators of $\mathfrak{m}$ with coefficients generating the ideal $\overline{j(f)}$ of R. Whence the appearance of the residue field and the moduli algebra.

We now propose another

Conjecture 26. *Let R be a Gorenstein analytic algebra which is an isolated singularity of positive dimension. Then R is quasihomogeneous if and only if high enough syzygy modules of k and of* $\operatorname{Tr} D_k(R)$ *are isomorphic.*

The gluing construction for the residue field of a complete intersection allows us to prove the "only if" part for that class of algebras. The idea is that up to homotheties the matrix h_{ij} defined above is, in the quasihomogeneous case, the jacobian matrix. The two different truncations of the complete resolution of k, providing projective resolutions of k and, respectively, of $\operatorname{Tr} D_k(R)$, have the same "tail", as is shown in the figure below. For more details, see [9].

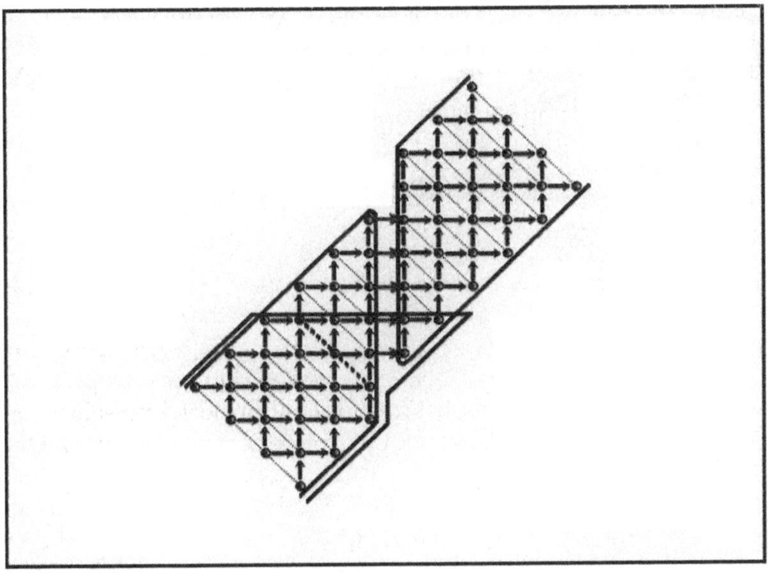

Projective resolutions of k and $\operatorname{Tr} D_k(R)$

References

[1] M. Auslander, *Isolated singularities and existence of almost split sequences*, in Representation Theory, II (Ottawa, Ont., 1984), 194–242, Lecture Notes in Math., 1178, Springer, Berlin-New York, 1986.

[2] M. Auslander, *Rational singularities and almost split sequences*, Trans. Amer. Math. Soc. 293 (1986), 511-531.

[3] M. Auslander and M. Bridger, *Stable Module Theory*, Mem. Amer. Math. Soc. vol. 94, 1969.

[4] M. Auslander and R.-O. Buchweitz, *The homological theory of maximal Cohen - Macaulay approximations*, Mém. Soc. Math. France, No. 38 (1989), 5-37.

[5] M. Auslander, S. Ding, and Ø. Solberg, *Liftings and weak liftings of modules*, J. Alg. 156 (1993), 273-317.

[6] K. Behnke, *On Auslander modules of normal surface singularities*, manuscripta math. 66 (1989), 205–223.

[7] W. Bruns and J. Herzog, *Cohen - Macaulay rings*, Cambridge University Press, Cambridge, 1993.

[8] S. Ding, *Cohen - Macaulay approximations and multiplicity*, J. Alg. 153 (1992), 271-288.

[9] J. Herzog and A. Martsinkovsky, *Gluing Cohen - Macaulay modules with applications to quasihomogeneous complete intersections with isolated singularity*, Comment. Math. Helvetici 68 (1993), 365-384.

[10] K. Kato, *Vanishing of free summands in Cohen - Macaulay approximations*, Comm. Alg. 23 (1995), 2697-2717.

[11] A. Martsinkovsky, *Almost split sequences and Zariski differentials*, Ph. D. Thesis, Brandeis University, 1987.

[12] A. Martsinkovsky, *Almost split sequences and Zariski differentials*, Trans. Amer. Math. Soc. 319 (1990), 285-307.

[13] A. Martsinkovsky, *Free summands in maximal Cohen - Macaulay approximations and Eisenbud operators over hypersurface rings*, Proc. Amer. Math. Soc. 115 (1992), 915-921.

[14] J.-P. Serre, *Algèbre locale. Multiplicités*, 3-rd edition, Lecture Notes in Mathematics 11, Springer-Verlag, Berlin, 1975.

[15] J. Shamash, *The Poincaré series of a local ring*, J. Alg. 12 (1969), 453-470.

[16] J. Tate, *Homology of Noetherian rings and local rings*, Illinois J. Math. 1 (1957), 14-27.

[17] Y. Yoshino and K. Kato, *Auslander modules and quasi-homogeneity of local rings*, Publ. RIMS 30 (1994), 1009-1038.

2. New homological invariants of rings and modules

2.1. **Introduction.** In the first lecture we discussed mCM approximations over Gorenstein rings and two invariants associated with them: Auslander's delta and index. It should be pointed out that most applications of the approximations are based on these two invariants. Around 1990 Maurice Auslander asked whether the delta-invariant was of cohomological nature. In the spring of 1991 Ragnar-Olaf Buchweitz ([5]) made the crucial observation that the delta is the dimension of the kernel of the map from the usual Ext to Tate cohomology (with coefficients in k):

$$\delta^i(M) = \dim_k \operatorname{Ker}(\operatorname{Ext}^i_R(M,k) \to \underline{\operatorname{Ext}}^i_R(M,k))$$

With this observation as the starting point, we shall define ξ, a new homological invariant over an arbitrary local ring, which in the case of a Gorenstein ring reduces to δ. Using ξ we shall also extend the definition of index. Our main technical tool is a remarkable construction of Pierre Vogel (unpublished) that greatly simplified and generalized the definition of Tate (co)homology[1].

The new invariants are not abstract concoctions, they are a useful tool. In the first and completely elementary part of this lecture we shall see some applications of those invariants. In the second part I will explain Vogel's construction and we shall see the new invariants in a cohomological context. In the third part I will define the index for an arbitrary local ring and show that it is always finite.

For more information and further results, see [12], [13], and [14].

2.2. **The (co)syzygy modules of the residue field.** Throughout this lecture, $(R, \mathfrak{m}, k)$ will be a commutative noetherian local ring.

Conjecture 27 (Kaplansky). *If some power $\mathfrak{m}^n$ of the maximal ideal $\mathfrak{m}$ is different form zero and is of finite projective dimension then R is regular (i.e., of finite global dimension).*

For $n = 1$ this was a result of Auslander - Buchsbaum and Serre.

This conjecture was proved by Kaplansky's student G. Levin in his thesis. His result was actually stronger (see [11]):

Theorem 28. *If M is a finitely generated R-module such that $\mathfrak{m}M \neq 0$ and pd $\mathfrak{m}M < \infty$ then R is regular.*

More recently, Dutta, working on homological conjectures for local rings ([8]) stated the following result:

Theorem 29. *Suppose that, for some n, the n-th syzygy module of k has a free summand. Then R is regular.*

We shall show that these results are particular cases of a much more general result. Let M and N be finitely generated R-modules. The module $\operatorname{Hom}_R(M,N)$ will be denoted simply (M,N). The symbols $\beta(M) := \beta_0(M)$ will denote $\dim_k(M,k)$.

---

[1] At about the same time, in the mid-1980's, Buchweitz came up with the same construction (see his unpublished manuscript [4]). It was later rediscovered by other people.

By Nakayama's Lemma this is the minimal number of generators of M. Similarly, the i-th betti number $\beta_i(M)$ is defined as $\beta_0(\Omega^i M)$, where $\Omega^i M$ denotes the i-th syzygy module of M in a minimal projective resolution of M. It is convenient at this point to choose and fix a minimal projective resolution for each module in $\operatorname{mod} R$.

Definition 30. *Let $V(M)$ denote the subset of the finite-dimensional k-vector space (M,k) consisting of all maps $f : M \to k$ that admit a bounded lifting to the projective resolutions of M and k.*

It is immediate that $V(M)$ is a vector subspace of (M,k).

Definition 31. *Let $\xi(M) := \xi^0(M) := \dim_k V(M)$ and $\xi^i(M) := \dim_k V(\Omega^i(M))$.*

Since all minimal resolutions of a given module are isomorphic, ξ is an invariant of the module and not of the chosen resolutions. The working formalism for the ξ's is captured in the following lemma.

Lemma 32.
1. *$\xi^i(M) = \xi(\Omega^i M)$ for all $i \geq 0$.*
2. *$0 \leq \xi^i(M) \leq \beta_i(M)$ for all i.*
3. *If $\operatorname{pd} M < \infty$ then $\xi^i(M) = \beta_i(M)$ for all i.*
4. *$\xi^i(M \amalg N) = \xi^i(M) + \xi^i(N)$ for all i.*
5. *If $M \xrightarrow{f} N$ is an epimorphism, then $\xi(M) \geq \xi(N)$.*

Proof. The first three properties are immediate from the definition. To prove the fourth property, we choose the direct sum of the minimal resolutions of M and N as the minimal resolution of $M \amalg N$ and remark that a map from a direct sum is eventually zero if and only if it is eventually zero on each direct summand. To prove the last property, notice first that the diagram

$$\begin{array}{ccc} V(N) & \to & (N,k) \\ \downarrow^{(f,k)|_{V(N)}} & & \downarrow^{(f,k)} \\ V(M) & \to & (M,k) \end{array}$$

where the horizontal maps are the canonical inclusions, is commutative. The subtle point here is that the image of the left vertical map is in $V(M)$; this follows from the fact that the composition of a bounded map with any map is again bounded. (We have just proved that $V(-)$ is a subfunctor of $(-,k)$.) The desired result now follows from the left-exactness of the Hom-functor. ∎

We can now prove the main result of this subsection. For Gorenstein rings this result, stated in terms of Auslander's δ-invariant, was originally proved by Auslander ([2]).

Theorem 33. *If the ring R is non-regular then $\xi^i(k) = 0$ for all i.*

Proof. Suppose $\xi^i(k) \neq 0$ for some i. Then there are a map $\theta : \Omega^i k \to k$ and a number N such that in a certain lifting $\theta.$ of θ we have that $\theta_n = 0$ for all $n \geq N$. We shall now view θ as (a representative of) a degree i element of the Yoneda algebra $\operatorname{Ext}^*(k,k)$ of R. Then for any $\phi \in \operatorname{Ext}^*(k,k)$ with $\deg \phi \geq N$ we have that the Yoneda product $\phi\theta$ is zero. Thus θ is annihilated by all elements of sufficiently high degree.

On the other hand, the Yoneda algebra $\operatorname{Ext}^*(k,k)$ of R is a universal enveloping algebra UL of a uniquely defined graded Lie algebra L ([15], [1], [16]). It was shown in Lemma 3.1 (i) of [9] that if UL is infinite-dimensional then no element of UL can be annihilated by the ideal of elements of positive degrees. The same argument shows that no element of UL can be annihilated by the ideal $UL_{\geq N}$ consisting of elements of degree at least N. It goes as follows. If $(UL_{\geq N})\theta = 0$ then, by the graded Poincaré-Birkhoff-Witt theorem, the algebra L must be concentrated in odd degrees, thus making UL into an exterior algebra. If UL were infinite-dimensional, then, by the Poincaré-Birkhoff-Witt theorem again, no element of UL could be annihilated by $UL_{\geq N}$. Thus we conclude that UL must be finite-dimensional. Since R was assumed to be non-regular, we have that UL is infinite-dimensional. The obtained contradiction finishes the proof of the theorem. ∎

As an immediate application, we have the following

Proposition 34. *Let $f : \amalg_{i\in I}(\Omega^i k)^{j_i} \to L$ be a surjective homomorphism of finitely generated R-modules, where $L \neq 0$ and pd $L < \infty$. Then R is regular.*

Proof. Suppose R is non-regular. On one hand, by property 3 above, $\xi(L) = \beta_0(L) > 0$. On the other hand, by properties 5, 4, 1, and by the preceding theorem we have that $\xi(L) \leq 0$. A contradiction. ∎

Now we can deduce the aforementioned results of Levin and Dutta. The latter is the particular case of the preceding proposition with $j_i = 1$ when $i = n$ and $j_i = 0$ otherwise, and f being the canonical projection from the n-th syzygy module onto its free direct summand.

The former result corresponds to the case $j_i = \beta_0(M)$ when $i = 1$ and $j_i = 0$ otherwise, and $f = \mathfrak{m}p$, where p is the projective cover $R^{\beta_0(M)} \to M$. To see that just multiply the last map by $\mathfrak{m}$.

Surprisingly, with very little effort, the last proposition can be substantially strengthened.

Proposition 35. *Let $f : \amalg_{i\in I}(\Omega^i k)^{j_i} \to L$ be a homomorphism between finitely generated R-modules, where $L \neq 0$ and pd $L < \infty$. If $f \otimes k \neq 0$ then R is regular.*

This result is an immediate consequence of the following

Lemma 36. *Let $(R, \mathfrak{m}, k)$ be a commutative noetherian local ring and $f : M \to N$ a homomorphism of finitely generated R–modules M and N such that $\xi(M) = 0$ and pd $N < \infty$. Then $f \otimes k = 0$.*

Proof. Suppose $f \otimes k \neq 0$. Then there exists a surjective composition $M \xrightarrow{f} N \to N/\mathfrak{m}N \to k$, where the map in the middle is the canonical surjection. But then, contrary to the assumption, we would have that $\xi(M) \neq 0$ since pd $N < \infty$. ∎

2.2.1. *Dual results.* Replacing now projective resolutions with injective ones, reversing the directions of arrows, and replacing the syzygy modules in projective resolutions by cosyzygy modules in injective resolutions, we arrive at another series of invariants $\chi^i(M)$. They enjoy properties completely analogous to those of $\xi^i(M)$. However the betti numbers $\dim_k \operatorname{Ext}^i(M,k)$ should be replaced by the Bass numbers $\dim_k \operatorname{Ext}^i(k,M)$, finite projective dimension should be replaced by finite injective dimension, and surjective homomorphisms in property 5 of Lemma 32 should be replaced by injections (the new property just says that $\chi = \chi^0$ does not decrease under injections). In addition, to accommodate infinitely generated cosyzygy modules, we want to allow χ to take on the value ∞. We can now state a result dual to Theorem 33.

Theorem 37. *If the ring R is non-regular then $\chi^i(k) = 0$ for all i.*

The proof is completely analogous to that of the prototype. ∎

As an application we have the following

Proposition 38. *Let $g : L \to \amalg_{i\in I}(\Omega^{-i}k)^{j_i}$ be an injective homomorphism of R-modules, where $\operatorname{Soc} L \neq 0$ and id $L < \infty$. Then R is regular. (Here Ω^{-i} stands for the i-th cosyzygy module.)* ∎

A strengthening of the last result is given by the following

Proposition 39. *Let $g : L \to \amalg_{i\in I}(\Omega^{-i}k)^{j_i}$ be a homomorphism of R-modules, where $\operatorname{Soc} L \neq 0$ and id $L < \infty$. If $f(\operatorname{Soc} L) \neq 0$ then R is regular.*

This result is an immediate consequence of

Lemma 40. *Let $(R, \mathfrak{m}, k)$ be a commutative noetherian local ring and $f : M \to N$ a homomorphism of R-modules such that $\chi(N) = 0$ and id $M < \infty$. Then $f(\operatorname{Soc} M) = 0$.*

Proof. Suppose $f(\operatorname{Soc} M) \neq 0$. Then there exists a composite monomorphism $k \to \operatorname{Soc} M \to M \xrightarrow{f} N$, where the middle map is the canonical injection. But then we would have, contrary to the assumption, that $\chi(N) \neq 0$ since id $M < \infty$. ∎

2.3. **Defining ξ via Vogel cohomology.** What we just saw was an elementary definition of a new invariant whose properties are identical to those of Auslander's delta. One of the goals of this subsection is to show that the two invariants coincide when the ring R is Gorenstein. This will be done by giving yet another definition of ξ using a cohomological technique. This direction of research was discovered by the author after Buchweitz had made the crucial observation ([5]) that, for a module M over a Gorenstein ring, $\delta^i(M)$ equals the dimension of the kernel of the natural map

$$\operatorname{Ext}^i(M,k) \to \underline{\operatorname{Ext}}^i(M,k)$$

from Ext to Tate cohomology. The latter can be defined similar to the usual Ext, except that one uses a complete resolution of M rather than its projective resolution (see [4] for more details). The claim follows immediately from the pitchfork construction (the natural map above is derived from the vertical maps in that construction). Buchweitz's observation came as an answer to a question of Auslander (ca. 1990) whether the delta-invariant was of homological nature.

Our goal is to prove a similar statement for the ksi-invariant over *arbitrary* local rings, for which one obviously needs a generalization of Tate cohomology (along with a natural transformation from the Ext-functor). This will be done by employing Pierre Vogel's cohomology that extends (and greatly simplifies) Tate cohomology and that works for modules over arbitrary associative rings.

2.3.1. *Vogel's construction.* Let Λ be an associative ring and M and N (left) Λ-modules. Choose and fix projective resolutions (P_M, d_M) and (P_N, d_N) of M and, respectively, N. Erasing the differential in the chosen resolutions we obtain diagrams P_M and P_N in the category of Λ-modules. We reserve the notation (P_M, P_N) for the diagram of abelian groups whose degree n term is given by

$$(P_M, P_N)_n := \Pi_{i\in \mathbf{Z}} \operatorname{Hom}_{\Lambda}(A_i, B_{i+n})$$

A differential on (P_M, P_N) can be defined by the standard formula

$$D(f) = \partial \circ f - (-1)^{\deg f} f \circ \partial$$

where f is a degree n element of (P_M, P_N), thus making it into a complex. The reader will immediately verify that in this complex:

- the chains are exactly the *homogeneous maps* from P_M to P_N;
- the 0-cycles are exactly the *chain maps* from (P_M, d_M) to (P_N, d_N);
- the 0-boundaries are exactly the *null-homotopic chain maps* from (P_M, d_M) to (P_N, d_N).

An important property of this complex is that its homology groups are $\operatorname{Ext}^*(M, N)$. This is a consequence of the following lemma (see [3], §5, no. 2, Prop. 4 for a complete statement):

Lemma 41. *Let $u : C' \to C$ be a quasi-isomorphism of complexes and P a complex of projectives. If P is bounded on the right or if both C and C' are bounded on the right then*

$$(1, u) : (P, C') \to (P, C)$$

is also a quasi-isomorphism. ∎

Let $(P_M, P_N)_b$ denote the subdiagram of (P_M, P_N) consisting of bounded maps, i.e., the subdiagram whose elements are homogenous maps from P_M to P_N with only finitely many non-zero components. Symbolically,

$$((P_M, P_N)_b)_n := \amalg_{i\in \mathbf{Z}} \operatorname{Hom}_{\Lambda}(A_i, B_{i+n})$$

We now have the short exact sequence of diagrams of abelian groups

$$0 \to (P_M, P_N)_b \to (P_M, P_N) \xrightarrow{p} (\widehat{P_M, P_N}) \to 0 \tag{1}$$

Suppose that f is abounded map. Then $D(f) = \partial \circ f - (-1)^{\deg f} f \circ \partial$ is again a bounded map, which means that the differential D restricts to $(P_M, P_N)_b$ thus making it a subcomplex of the complex (P_M, P_N). As a result, (1) becomes a short exact sequence of *complexes*. We can now define Vogel cohomology groups $\widehat{\operatorname{Ext}}^i(M, N)$.

Definition 42. $\widehat{\operatorname{Ext}}^i(M, N) := H_{-i}(\widehat{P_M, P_N})$.

It remains to check that the just defined cohomology groups do not depend on the chosen projective resolutions. To this end, observe that $(-,-)_b$ can be viewed (after the reader has given appropriate definitions) as an additive sub-bifunctor of the bifunctor $(-,-)$. As a consequence, $\widehat{(-,-)}$ is an additive bifunctor. Hence, homotopy of chain complexes, being an additive notion, is preserved by $\widehat{(-,-)}$. Finally, recall that any two projective resolutions of a module are homotopy equivalent, which shows that a different choice of projective resolutions would result in a homotopy equivalent complex, and therefore in the same homology.

2.3.2. *The new* ξ. As was pointed out above, the homology of the middle complex in (1) is just $\operatorname{Ext}^*(M, N)$, and therefore the corresponding long cohomology exact sequence gives rise to a natural map from Ext to Vogel cohomology. We can now give another definition of ξ.

Definition 43. *Let $(R, \mathfrak{m}, k)$ be an arbitrary commutative noetherian local ring and M a finitely generated R-module. Then*

$$\xi^i(M) := \dim_k \operatorname{Ker}(\operatorname{Ext}^i(M, k) \to \widehat{\operatorname{Ext}}^i(M, k))$$

To justify the double usage of the letter ξ we shall now give an interpretation of the elements in the kernel of the natural map above. Let f_* be such an element of degree zero. Then f_* is a chain map from the projective resolution of M to that of k. In particular, it lifts a homomorphism $f : M \to k$. Since f_* represents the zero class in Vogel cohomology, there is a null-homotopic chain map g_* such that

$$f_* = g_* + \text{ a bounded map}$$

and therefore $f_n = g_n$ for all n larger than some natural number N. Since g_* is null-homotopic, $f_{N+1} = g_{N+1} = sd + ds$, where s is the contracting homotopy. Now let $f'_N := f_N - ds$ and $f'_{N+1} := f_{N+1} - sd - ds = 0$. Now we have a new chain map which still lifts f and which is zero in degree $N+1$. In other words, f is homotopic to a bounded chain map. This shows that the new definition of ξ is equivalent to the one given earlier. The details are left to the reader.

2.3.3. ξ *vs.* δ. As part of his construction, Vogel proved that the cohomology $\widehat{\operatorname{Ext}}^*(M, N)$ of the factor-complex $(\widehat{P_M, P_N})$ coincides with Tate cohomology $\underline{\operatorname{Ext}}^*(M, N)$ when Λ is a group ring of a finite group. In fact, his proof, as we shall see in a moment, works for arbitrary Gorenstein rings. Moreover we shall also see that, over Gorenstein rings, the ksi-invariant coincides with the delta-invariant.

Theorem 44. *([17]) Suppose the local ring R is Gorenstein. Then $\widehat{\mathrm{Ext}}^*_R(M,N)$ is naturally isomorphic to the Tate cohomology module $\underline{\mathrm{Ext}}^*_R(M,N)$ for any R-modules M and N.*

Proof. Let C_M be a complete resolution of M. Forfeiting, if necessary, the minimality of C_M we may assume that the map $q : C_M \to P_M$ is surjective. Thus we have a short exact sequence of complexes of projective modules

$$0 \to L \to C_M \xrightarrow{q} P_M \to 0$$

which we want to consider as a split-exact sequence of *diagrams*. Applying the functor $(-, P_N)$ we have the commutative diagram in the category of diagrams of abelian groups

$$\begin{array}{ccccccccc}
 & & 0 & & 0 & & 0 & & \\
 & & \downarrow & & \downarrow & & \downarrow & & \\
0 & \to & (P_M, P_N)_b & \to & (P_M, P_N) & \to & \widehat{(P_M, P_N)} & \to & 0 \\
 & & \downarrow & & \downarrow & & \downarrow & & \\
0 & \to & (C_M, P_N)_b & \to & (C_M, P_N) & \xrightarrow{r} & \widehat{(C_M, P_N)} & \to & 0 \\
 & & \downarrow & & \downarrow & & \downarrow & & \\
0 & \to & (L, P_N)_b & \to & (L, P_N) & \to & \widehat{(L, P_N)} & \to & 0 \\
 & & \downarrow & & \downarrow & & \downarrow & & \\
 & & 0 & & 0 & & 0 & &
\end{array}$$

with exact rows and columns (the left column is exact since the middle column is a split-exact sequence of diagrams). Since L is bounded on the left and P_N is bounded on the right, any map between these two complexes is bounded. Thus $\widehat{(L, P_N)} = 0$ and it remains to show that the map r is a quasi-isomorphism. This is equivalent to saying that $(C_M, P_N)_b$ is exact. This complex is isomorphic to $(C_M, R) \otimes_R P_N$. Since R is Gorenstein, the first factor is exact by Lemma 16. The second factor is a complex of projective modules bounded on the right. The desired assertion now follows. ∎

As an immediate corollary of the *proof* above we have the following

Proposition 45. *If the ring R is Gorenstein, then $\xi^i(M) = \delta^i(M)$ for all finitely generated R-modules M and for all i.* ∎

> **Remark:** It is convenient to think of $\xi^i(M)$ as the dimension of the vector subspace of $\mathrm{Ext}^i(M,k)$ consisting of (the classes of) all maps $\Omega^i M \to k$ that, upon lifting to the projective resolutions, are homotopic to a bounded map.

2.3.4. *The tau-invariant.* If, instead of Vogel cohomology, one chooses to work with Vogel homology then a functor $\widehat{\mathrm{Tor}}_i(M,N)$ appears (see [10] for details) along with natural maps $\widehat{\mathrm{Tor}}_i(M,N) \to \mathrm{Tor}_i(M,N)$. If the ring R is Gorenstein then the new functor coincides with Tate homology. Setting now $N = k$ we can define another invariant:

Definition 46. *Let M be a finitely generated module over a noetherian local ring R. We set*

$$\tau_i(M) := \dim_k \operatorname{Coker}(\widehat{\operatorname{Tor}}_i(M,k) \to Tor_i(M,k))$$

Again if R is Gorenstein then the new invariant coincides with Auslander's $\delta^i(M)$. In the general case, the new invariant τ has properties similar to those of ξ and listed in Lemma 32. Because of the nice interpretation of ξ given in the preceding remark, this invariant is usually more convenient to work with than τ.

2.4. **The index of a local ring.** In this section we want to generalize the notion of the index of a Gorenstein local ring $(R, \mathfrak{m}, k)$, due to Auslander, to an arbitrary local noetherian ring. According to Auslander's definition,

$$\operatorname{index}(R) := \inf\{i \mid \delta(R/\mathfrak{m}^i) \neq 0\}$$

Assuming now that R is an arbitrary noetherian local ring and replacing δ with ξ we introduce the following

Definition 47. $\operatorname{index}(R) := \inf\{i \mid \xi(R/\mathfrak{m}^i) \neq 0\}$.

Proposition 48. *The ring R is regular if and only if* $\operatorname{index}(R) = 1$.

Proof. The "only if" part is obvious. Suppose now that $\xi(k) \neq 0$. Then there exists a non-zero map $\theta : k \to k$ with a bounded lifting. Since θ must be an isomorphism, we may also find a lifting which is an isomorphism in each degree. The difference of the two liftings is null-homotopic and is an isomorphism in sufficiently high degrees. Since we could have started with minimal resolutions and the homotopy is an R-module homomorphism, we conclude, by Nakayama's lemma, that the projective resolution of k is eventually zero. ∎

Proposition 49. *If R is a noetherian local ring then* $\operatorname{index}(R) < \infty$.

Proof. Let d be the Krull dimension of R. We shall make use of local cohomology whose definition we briefly recall. Let M be an R-module. The functor $\operatorname{Ext}^i_R(-, M)$ applied to the inverse system

$$\ldots \to R/\mathfrak{m}^n \to R/\mathfrak{m}^{n-1} \to \ldots \to R/\mathfrak{m}^2 \to R/\mathfrak{m} = k$$

gives rise to the direct system $\operatorname{Ext}^i_R(R/\mathfrak{m}^n, M), n = 1, 2, \ldots$. The i-th local cohomology module $H^i_{\mathfrak{m}}(M)$ is defined as $\varinjlim \operatorname{Ext}^i_R(R/\mathfrak{m}^n, M)$. An important property of local cohomology is that it vanishes in degrees above the Krull dimension, in our case d. (This is due to the fact that local cohomology can also be defined via the Čech complex.) In particular,

$$\varinjlim \operatorname{Ext}^{d+1}(R/\mathfrak{m}^n, \Omega^{d+1}k) = H^{d+1}_{\mathfrak{m}}(\Omega^{d+1}k) = 0$$

Let $\alpha \in \operatorname{Ext}^{d+1}(k, \Omega^{d+1}k)$ be the class of the identity map

$$\Omega^{d+1}k \xrightarrow{id} \Omega^{d+1}k$$

Since the image of α in $H^{d+1}_{\mathfrak{m}}(\Omega^{d+1}k)$ is zero, we have that for a high enough value of i, the image of α in $\mathrm{Ext}^{d+1}(R/\mathfrak{m}^i, \Omega^{d+1}k)$ under the canonical surjection $R/\mathfrak{m}^i \longrightarrow k$ is also zero, i.e., lifting a representative $\Omega^{d+1}(R/\mathfrak{m}^i) \longrightarrow \Omega^{d+1}k$ of that image to the projective resolutions we obtain a null-homotopic map. Thus we can assume that the lifting is zero. On the other hand, α can be thought of as a lifting of the identity map on k. Combining this with the preceding statement, we have a lifting of the canonical surjection $R/\mathfrak{m}^i \longrightarrow k$ that vanishes beginning with degree $d+1$. Thus $\xi(R/\mathfrak{m}^i) \neq 0$ and the proposition is proved. ∎

This new definition of index yields the following result, which is an immediate generalization of a result of Ding proved for Gorenstein ring (see Prop. 1.6 of [6])

Proposition 50. *Let $(R, \mathfrak{m}, k)$ be a Cohen - Macaulay ring of multiplicity* $\mathrm{mult}(R)$. *Suppose that k is infinite. Then*

$$\mathrm{mult}(R) \geq \mathrm{index}(R) + (e\dim R - \dim R) - 1$$

If R is regular then $\mathrm{mult}(R) = \mathrm{index}(R) = 1$.

Proof. The proof is a verbatim repetition of Ding's argument which we reproduce here. Let a be an ideal of R generated by a system of parameters and such that $\mathrm{mult}(R) = l(R/a)$. Let n be the smallest number i such that $\mathfrak{m}^i \subset a$. Consider the filtration $R \supset \mathfrak{m} = \mathfrak{m} + a \supset \mathfrak{m}^2 + a \supset \ldots \supset \mathfrak{m}^n + a = a$. Then, by Nakayama's lemma, we have that

$$l(R/a) \geq 1 + (e\dim R - \dim R) + n - 2$$

On the other hand, $R/\mathfrak{m}^n$ surjects onto R/a, the latter being a module of finite projective dimension. By properties 3 and 5 of Lemma 32, we have that $\xi(R/\mathfrak{m}^n) > 0$ and therefore $\mathrm{index}(R) \leq n$, which proves the first claim. The last assertion of the lemma is straightforward. ∎

Remark: If R is Cohen-Macaulay, one can still define the index using δ (the latter understood as the rank of the largest free summand in the minimal approximation of a module). However, as was shown by Ding ([7]), with that definition, the index of a ring need no longer be finite, thus making a generalization similar to ours impossible.

As a consequence of the preceding proposition, we have a generalization of another result of Ding (see Th. 3.3 of [6]).

Corollary 51. *Let R be a complete non-regular Cohen-Macaulay local ring with infinite residue field. Then R is a hypersurface ring if and only if* $\mathrm{mult}(R) = \mathrm{index}(R)$.

The argument is again identical to the one given by Ding in the Gorenstein case and the reader is referred to [ibid.] for details.

References

[1] M. André, *Hopf algebras with divided powers*, J. Algebra 18 (1971), 19-50.
[2] M. Auslander, *Private communication,* 1991.
[3] N. Bourbaki, *Algèbre*, Ch. X, Masson, Paris, 1980
[4] R.-O. Buchweitz, *Maximal Cohen - Macaulay modules and Tate-cohomology over Gorenstein rings*, preprint, ca. 1986.
[5] R.-O. Buchweitz, *Private communication*, 1991.
[6] S. Ding, *Cohen - Macaulay approximations and multiplicity*, J. Algebra 153 (1992), 271-288.
[7] S. Ding, *A note on the index of Cohen - Macaulay local rings*, Comm. Alg. 21 (1993), 53-71.
[8] S. P. Dutta, *Syzygies and homological conjectures,* in Commutative Algebra (M. Hochster, C. Huneke, J. D. Sally, eds.) MSRI Publications, pp. 139-156 (1989).
[9] Felix, Y., Halperin, S., Jacobsson, C., Löfwall, C., Thomas, J.-C., *The radical of the homotopy Lie algebra*, Am. J. Math. 110(1988), 301-322.
[10] F. Goichot, *Homologie de Tate - Vogel équivariante*, J. Pure Appl. Alg. 82 (1992), 39-64.
[11] G. Levin and W.V. Vasconcelos, *Homological dimensions and Macaulay rings*, Pac. J. Math 25(1968), 315-323.
[12] A. Martsinkovsky, *New homological invariants for modules over local rings*, I, J. Pure Appl. Alg. 110 (1996), 1-8.
[13] A. Martsinkovsky, *A remarkable property of the (co)syzygy modules of the residue field of a nonregular local ring*, J. Pure Appl. Alg. 110 (1996), 9-13.
[14] A. Martsinkovsky, *New homological invariants for modules over local rings*, II, to appear in J. Pure Appl. Alg.
[15] J.W. Milnor and J.C. Moore, *On the structure of Hopf algebras*, Ann. Math. 81 (1965), 211-264.
[16] G. Sjödin, *Hopf algebras and derivations*, J. Algebra 64 (1980), 218-229.
[17] P. Vogel, *Private communication*, 1991

Mathematics Department, Northeastern University, Boston, MA 02115, USA
E-mail address: alexmart@neu.edu

Trends in Mathematics, © 2000 Birkhäuser Verlag Basel/Switzerland

THE GENERIC REPRESENTATION THEORY OF FINITE FIELDS: A SURVEY OF BASIC STRUCTURE

NICHOLAS J. KUHN

ABSTRACT. If $\mathbf{F}_q$ is the finite field of characteristic p and order $q = p^s$, let $\mathcal{F}(q)$ be the category whose objects are functors from finite dimensional $\mathbf{F}_q$–vector spaces to $\mathbf{F}_q$–vector spaces, and with morphisms the natural transformations between such functors. We survey the basic structure of this category and its close connections to the finite general linear groups, the symmetric groups, classical Schur algebras, algebraic K-theory, and the Steenrod algebra.

CONTENTS

1. Introduction 194
2. First observations and basic examples 194
3. Basic structure 196
3.1. Tensor products 196
3.2. Duality 196
3.3. Composition 196
3.4. Scalar decomposition 197
3.5. Frobenius twist 197
4. Finite and polynomial functors, and a finiteness conjecture 197
5. Connections with $GL_k(\mathbf{F}_q)$ and $M_k(\mathbf{F}_q)$ 198
6. The structure of the fundamental injective 201
7. Connections with Σ_r and Schur algebras 204
8. A Kunneth theorem for Ext groups 206
9. Two deep properties of the symmetric powers 206
10. Connections with algebraic K–theory and group cohomology 207
11. Connections with rational cohomology 208
12. Connections with the Steenrod algebra 209
References 211

1991 *Mathematics Subject Classification.* Primary 20G05, 19D55, 55S10, Secondary 16P60.
Partially supported by the N.S.F..

1. Introduction

If $\mathbf{F}_q$ is the finite field of characteristic p and order $q = p^s$, let $\mathcal{F}(q)$ be the category with objects the functors

$$F : \text{finite dimensional } \mathbf{F}_q\text{–vector spaces} \longrightarrow \mathbf{F}_q\text{–vector spaces},$$

and with morphisms the natural transformations between such functors. Thus $\mathcal{F}(q)$ is the world in which 'natural' $\mathbf{F}_q$–linear algebra is done.

A first observation about this category is that one can view an object $F \in \mathcal{F}(q)$ as a 'generic representation' of the general linear groups over $\mathbf{F}_q$: $F(V)$ becomes an $\mathbf{F}_q[GL(V)]$–module for all $\mathbf{F}_q$–vector spaces V. As we will see, the tight relationship between $\mathcal{F}(q)$ and the categories of $\mathbf{F}_q[GL_n(\mathbf{F}_q)]$–modules, for all n, makes the study of $\mathcal{F}(q)$ of great representation theoretic interest.

As one investigates this relationship more carefully, one discovers close connections between $\mathcal{F}(q)$ and many other fundamental and diverse mathematical objects:

- the symmetric groups
- the classic Schur algebras
- Topological Hochshild Homology (which arises in algebraic K–theory)
- the Steenrod algebra (which arises in algebraic topology)

We will survey these connections, illustrated by examples. Included among these are various gadgets of 'infinite length': important objects with infinite composition series, and even infinite filtrations of the category itself.

In the case q is prime, the category $\mathcal{F}(q)$ first came to the author's attention in 1987 work of Jean Lannes, Lionel Schwartz, and Hans-Werner Henn [HLS] who used $\mathcal{F}(q)$ to help organize our understanding of the many beautiful results about the Steenrod algebra that had been discovered in the 1980's by algebraic topologists. Much of the material in §§2–5,8,9,11 appeared in our series of papers [K:I, K:II, K:III], and represents work done in the early 1990's. More recent are the theorems of a K–theoretic flavor, surveyed in §§10,11, with key papers including [FLS], [FS], [K3], and [FFSS]. The work in §6 appeared in [K1]. Finally, though versions of the main result in §7 have been known to the author for about 6 years, the formulation given here is new in 1998.

This paper is an expanded version of a three talk series I gave at the Conference on Infinite Length Modules held at the University of Bielefeld, September 7 – 11, 1998, and I have tried to present results with this audience in mind. It is a pleasure to thank the conference organizers for their hospitality.

2. First observations and basic examples

It is illuminating to think of a functor category like $\mathcal{F}(q)$ as analogous to the category of modules over a ring. In fact, this becomes more than just an analogy when one remembers that a ring can be regarded as an additive category with a single object, so that an additive category is a "ring with many objects". Given such a multiobject ring R, an R–module is just an additive functor from R to abelian groups. If $\mathbf{F}$ is some ground field, there are obvious $\mathbf{F}$–linear flavors of these concepts.

In our case, let L_q be the $\mathbf{F}_q$–linear extension of the category of finite dimensional $\mathbf{F}_q$–vector spaces . Explicitly, L_q is the multiobject $\mathbf{F}_q$–algebra with objects finite dimensional $\mathbf{F}_q$–vector spaces, and with $L_q(W,V) = \mathbf{F}_q[\mathrm{Hom}(W,V)]$.[1] Then $\mathcal{F}(q)$ is the category of L_q–modules.

Explicitly, the structure defining $F \in \mathcal{F}(q)$ is the following:

- an $\mathbf{F}_q$–vector space $F(V)$ for each V,
- linear maps $\mathbf{F}_q[\mathrm{Hom}(W,V)] \otimes F(W) \to F(V)$ for each V and W,

and these linear maps need to be compatible in the obvious ways.

Viewing $\mathcal{F}(q)$ as a category of modules reminds us that this category is abelian in an obvious way:

$$0 \to F \to G \to H \to 0$$

is exact if

$$0 \to F(V) \to G(V) \to H(V) \to 0$$

is exact for all V. Thus the standard concerns of homological algebra and module theory apply: one can study the simple objects, extension groups, etc.

Let's introduce some basic examples of objects in $\mathcal{F}(q)$.

Definitions 2.1.
(1) T^n is defined by $T^n(V) = V^{\otimes n}$ (the n fold tensor product).
(2) Λ^n is defined by $\Lambda^n(V) =$ the nth exterior product of V.
(3) S^n is defined by $S^n(V) = (V^{\otimes n})_{\Sigma_n}$ (the nth symmetric product on V).
(4) S_n is defined by $S_n(V) = (V^{\otimes n})^{\Sigma_n}$ (the nth symmetric invariants on V).

Note that $T^0 = \Lambda^0 = S^0 = S_0 = \mathbf{F}_q$, the constant functor that sends each V to $\mathbf{F}_q$, and every map to the identity. Also $T^1 = \Lambda^1 = S^1 = S_1 =$ the identity functor.

Example 2.2. With $q = p = 2$, there are natural short exact sequences

$$0 \to \Lambda^1(V) \to S^2(V) \to \Lambda^2(V) \to 0.$$

These are not split, and represent a nonzero element in $\mathrm{Ext}^1_{\mathcal{F}(2)}(\Lambda^2, \Lambda^1)$. Furthermore, Λ^2, Λ^1 are both simple functors.

The next two families of examples are quite different than those above.

Definitions 2.3.
(1) For each W, $P_W \in \mathcal{F}(q)$ is defined by

$$P_W(V) = L_q(W,V) = \mathbf{F}_q[\mathrm{Hom}(W,V)].$$

[1]If S is a set, we use the notation $\mathbf{F}_q[S]$ to denote the $\mathbf{F}_q$–vector space with basis S. Also, unlabelled Homsets and tensor products will be over $\mathbf{F}_q$.

By Yoneda's lemma, $\operatorname{Hom}_{\mathcal{F}(q)}(P_W, F) = F(W)$, and thus P_W is projective.
(2) For each W, $I_W \in \mathcal{F}(q)$ is defined by

$$I_W(V) = \mathbf{F}_q^{\operatorname{Hom}(W,V^*)}.$$

By Yoneda's lemma, $\operatorname{Hom}_{\mathcal{F}(q)}(F, I_W) = F(W^*)^*$, and thus I_W is injective.

3. Basic structure

3.1. **Tensor products.** Given $F, G \in \mathcal{F}(q)$, $F \otimes G \in \mathcal{F}(q)$ is defined by

$$(F \otimes G)(V) = F(V) \otimes G(V).$$

Example 3.1. There are natural isomorphisms

$$P_{W_1 \oplus W_2} \simeq P_{W_1} \otimes P_{W_2}.$$

One concludes that the functors $P_{\mathbf{F}_q}^{\otimes k}, k \geq 0$, form a set of projective generators for $\mathcal{F}(q)$. Similar things hold for the I_W's.

Example 3.2. For all n, and all q, the Koszul complex yields exact sequences

$$0 \to \Lambda^n \to \Lambda^{n-1} \otimes S^1 \to \Lambda^{n-2} \otimes S^2 \to \ldots \to \Lambda^1 \otimes S^{n-1} \to S^n \to 0.$$

3.2. **Duality.** We have an exact contravariant duality functor on $\mathcal{F}(q)$: given $F \in \mathcal{F}(q)$, $DF \in \mathcal{F}(q)$ is defined by

$$DF(V) = F(V^*)^*.$$

For all $F, G \in \mathcal{F}(q)$, one has

$$\operatorname{Hom}_{\mathcal{F}_q}(F, DG) = \operatorname{Hom}_{\mathcal{F}_q}(G, DF).$$

Example 3.3. $DS^n = S_n$, $DP_W = I_W$, and T^n, Λ^n are self dual.

An important property of duality goes as follows.

Proposition 3.4. [K:II, Theorem 7.1] *If S is simple, then $DS \simeq S$.*

3.3. **Composition.** If $G \in \mathcal{F}(q)$ takes finite dimensional values, then the composition $F \circ G$ is defined for all F. Precomposition with very innocent looking functors can do strange things.

Example 3.5. Let $sh \in F(q)$ be the 'shift' functor: $sh(V) = V \oplus \mathbf{F}_q$. Then $S^n \circ sh = \bigoplus_{k=0}^n S^k$, and $\Lambda^n \circ sh = \Lambda^n \oplus \Lambda^{n-1}$.

Example 3.6. $P_W = P_{\mathbf{F}_q} \circ \operatorname{Hom}(W, \ )$.

3.4. **Scalar decomposition.** $\mathbf{F}_q$, viewed as a multiplicative semigroup, acts centrally on the category $\mathcal{F}(q)$. Given $r = 0, \dots, q-1$, and $F \in \mathcal{F}(q)$, let

$$F_r(V) = \{x \in F(V) \mid F(\lambda)(x) = \lambda^r x \ \forall \lambda \in \mathbf{F}_q\}.$$

Then F decomposes as a direct sum

$$F = F_0 \oplus \cdots \oplus F_{q-1},$$

and we get a categorical decomposition

$$\mathcal{F}(q) = \mathcal{F}_0(q) \oplus \cdots \oplus \mathcal{F}_{q-1}(q).$$

If $F \in \mathcal{F}_r(q)$, we say the F has *scalar degree* r.[2]

Example 3.7. [K:I, Lemma 3.6] Under the scalar decomposition, $P_{\mathbf{F}_q}$ and $I_{\mathbf{F}_q}$ each decompose into a direct sum of q indecomposable functors, each having one dimensional endomorphism rings.

3.5. **Frobenius twist.** If V is an $\mathbf{F}_q$–vector space, there is a natural decomposition of $\mathbf{F}_q$–vector spaces

$$\mathbf{F}_q \otimes_{\mathbf{F}_p} V \simeq V \oplus V_\xi \oplus \cdots \oplus V_{\xi^{s-1}},$$

where V_{ξ^k} is a copy of V with scalars twisted by the p^kth power map: $\lambda \cdot v_{\xi^k} = (\lambda^{p^k} v)_{\xi^k}$. Given $F \in \mathcal{F}(q)$, we define $\mathcal{F}_\xi$ by $F_\xi(V) = F(V_\xi)$. One can iterated this construction, and $F_{\xi^s} = F$.

Example 3.8. The pth power map in polynomial algebras can be viewed as a map

$$\Phi : (S^n)_\xi \to S^{pn}.$$

4. Finite and polynomial functors, and a finiteness conjecture

Since $\mathcal{F}(q)$ is an abelian category, one can define the following useful concepts.

Definitions 4.1.
(1) $F \in \mathcal{F}(q)$ is *finite* if it has only a finite number of simple composition factors.
(2) $F \in \mathcal{F}(q)$ is *locally finite* if it is the union of its finite subfunctors.

Example 4.2. For all n, Λ^n is simple while S^n, S_n, and T^n are all finite. The injectives I_W are locally finite (but not finite, unless $W = 0$), while the projectives P_W have no finite subfunctors (except the direct summand $P_0 = \mathbf{F}_q$) [K:II, Appendix B].

From a practical point of view, it turns out to be very easy to decide when a functor is finite, by using the notion of a 'polynomial' functor: a concept perhaps first formally defined by algebraic topologists S. Eilenberg and S. MacLane in their 1954 study of cohomology operations and 'stable' group cohomology [EM].

---

[2]This is what is called "strict polynomial degree" in [FS].

Note that $F(V)$ is canonically a direct summand in $F(V \oplus \mathbf{F}_q)$. We let $(\Delta F)(V)$ be the complementary summand. We thus obtain an exact functor

$$\Delta : \mathcal{F}(q) \longrightarrow \mathcal{F}(q),$$

and this can be iterated.

Definition 4.3. F has *polynomial degree* $\leq n$ if $\Delta^{n+1}F = 0$.

Proposition 4.4. [K:I] *Given $F \in \mathcal{F}(q)$, the following are equivalent:*
(1) F is finite.
(2) F takes finite dimensional values and is polynomial.
(3) The function $k \longmapsto dim_{\mathbf{F}_q} F(\mathbf{F}_q^k)$ is a polynomial function of k.

We remark that, in the situation of this proposition, the degree of the growth function in *(3)* will equal the polynomial degree of F.

Following the authors of [HLS], we call a functor *analytic* if it is the union of its polynomial subfunctors.

Corollary 4.5. *F is locally finite if and only if it is analytic.*

We let $\mathcal{F}^n(q) \subset \mathcal{F}(q)$ denote the subcategory of functors of degree $\leq n$, and $\mathcal{F}^\omega(q) \subset \mathcal{F}(q)$ denote the subcategory of locally finite functors.

We end this section with a discussion of an open conjecture due to Lionel Schwartz, which dates from the late 1980's.

Conjecture 4.6. (**Artinian Conjecture**) The functors I_W are Artinian objects. Equivalently, the functors P_W are Noetherian.

In [K:II, Proposition 3.13], we give many equivalent formulations of the conjecture. One goes as follows: call a projective *small* if it is a summand of a finite direct sum of P_W's. (This agrees with the usual categorical meaning of 'small'.) The conjecture is equivalent to the statement that every F with finite head $(F/\operatorname{Rad}(F))$ has a projective resolution consisting of small projectives. (Call such a resolution *small.*)

Lionel Schwartz has a clever argument establishing the following partial result.

Theorem 4.7. *(See* [S, Theorem 5.3.8] *or* [FLS, Proposition 10.8]*) Given $F \in \mathcal{F}(q)$, if there exists an n such that $\Delta^{n+1}F$ is a small projective, then F has a small projective resolution. In particular, every finite F has a small projective resolution.*

We will return to this conjecture in §6.

5. Connections with $GL_k(\mathbf{F}_q)$ and $M_k(\mathbf{F}_q)$

Recall the observation made in the introduction: given $F \in \mathcal{F}(q)$, one gets a family of modules over the general linear groups via evaluation. More precisely, $F(\mathbf{F}_q^k)$ will be a module for $M_k(\mathbf{F}_q)$, the multiplicative semigroup of $k \times k$ matrices over $\mathbf{F}_q$, and, by further restriction, a $GL_k(\mathbf{F}_q)$-module.

Let $\mathcal{M}_k(q), \mathcal{GL}_k(q)$ respectively denote the categories of modules over the semigroup ring $\mathbf{F}_q[M_k(\mathbf{F}_q)]$ and the group ring $\mathbf{F}_q[GL_k(\mathbf{F}_q)]$. In this section, we describe very precise relationships between $\mathcal{F}(q)$ and these categories.

We need some categorical preliminaries. (See [K:II] for more details.) A *recollement* diagram is a diagram of abelian categories

$$\mathcal{A} \underset{\xleftarrow{p}}{\overset{\xleftarrow{q}}{\xrightarrow{i}}} \mathcal{B} \underset{\xleftarrow{r}}{\overset{\xleftarrow{l}}{\xrightarrow{e}}} \mathcal{C}$$

satisfying the following conditions:

- l is left adjoint to e, which is left adjoint to r. Thus e is exact.
- q is left adjoint to i, which is left adjoint to p. Thus i is exact.
- i is a full embedding, and $e(B) \simeq 0$ if and only if $B \simeq i(A)$ for some A in $\mathcal{A}$.
- The adjunctions $\epsilon_C : er(C) \to C$ and $\eta_C : C \to el(C)$ are isomorphisms for all C in $\mathcal{C}$.

The existance of e, r, i with the cited properties is equivalent to saying that $\mathcal{A}$ can be viewed (via i) as a localizing subcategory of $\mathcal{B}$, and $\mathcal{B}/\mathcal{A} \simeq \mathcal{C}$. Here one says that a subabelian category is *localizing* if it is *thick* (or *epaisse*) - i.e. closed under extensions, and taking subobjects and quotients - and also is closed under directed colimits.

Lots of nice things happen when one has a recollement diagram. For example, the simple objects of $\mathcal{B}$ are determined by those of $\mathcal{A}$ and $\mathcal{C}$:

Example 5.1. [K:II, Proposition 4.7] In this setting, there is a bijection between sets of isomorphism classes of simple objects:

$$\{\text{simples in } \mathcal{A}\} \amalg \{\text{simples in } \mathcal{C}\} \longleftrightarrow \{\text{simples in } \mathcal{B}\}$$

given by sending a simple $A \in \mathcal{A}$ to $i(A)$ and a simple $C \in \mathcal{C}$ to $c(C)$. Here $c : \mathcal{C} \to \mathcal{B}$ is the 'intermediate extension' defined as follows. First, let $\gamma_C : l(C) \to r(C)$ be adjoint to $(\epsilon_C)^{-1} : C \to erC$. Then let $c(C) = Im\{\gamma_C : l(C) \to r(C)\}$. This satisfies $ec(C) \simeq C$, naturally, for all $C \in \mathcal{C}$.

Here is a typical situation leading to recollement. Given an idempotent e in a ring R, one gets a recollement diagram

$$R/ReR\text{–modules} \underset{\xleftarrow{p}}{\overset{\xleftarrow{q}}{\xrightarrow{i}}} R\text{–modules} \underset{\xleftarrow{r}}{\overset{\xleftarrow{l}}{\xrightarrow{e}}} eRe\text{–modules},$$

with $e(M) = eM, l(M) = Re \otimes_{eRe} M,$ and $r(M) = \mathrm{Hom}_{eRe}(eR, M)$.

(I suspect that if $\mathcal{A}, \mathcal{B},$ and $\mathcal{C}$ are Morita equivalent to categories of modules over finite dimensional algebras over a field, then a recollement diagram for $\mathcal{A}, \mathcal{B},$ and $\mathcal{C}$ would be equivalent to one of this special form.)

Specialize to the case $R = \mathbf{F}_q[M_k(\mathbf{F}_q)]$ and e is $e_{k-1} = \begin{pmatrix} I_{k-1} & 0 \\ 0 & 0 \end{pmatrix}$. Then $eRe = \mathbf{F}_q[M_{k-1}(\mathbf{F}_q)]$ and ReR is the span of the noninvertible matrices, so that $R/ReR = \mathbf{F}_q[GL_k(\mathbf{F}_q)]$. One concludes

Theorem 5.2. [K:II, Theorem 2.3] *There is a recollement diagram*

$$\mathcal{GL}_k(q) \underset{\overset{\longleftarrow}{p}}{\overset{\overset{q}{\longleftarrow}}{\xrightarrow{i}}} \mathcal{M}_k(q) \underset{\overset{\longleftarrow}{r}}{\overset{\overset{l}{\longleftarrow}}{\xrightarrow{e}}} \mathcal{M}_{k-1}(q).$$

It is compelling (at least for an algebraic topologist!) to assemble these diagrams into a tower

$$\begin{array}{ccccccccc} & & \mathcal{GL}_3(q) & & \mathcal{GL}_2(q) & & \mathcal{GL}_1(q) & & \mathcal{GL}_0(q) \\ & & \downarrow & & \downarrow & & \downarrow & & \downarrow \\ \cdots & \xrightarrow{e_3} & \mathcal{M}_3(q) & \xrightarrow{e_2} & \mathcal{M}_2(q) & \xrightarrow{e_1} & \mathcal{M}_1(q) & \xrightarrow{e_0} & \mathcal{M}_0(q). \end{array}$$

Proposition 5.3. [K:II] *$\mathcal{F}(q)$ is the 'inverse limit' of the tower of $\mathcal{M}_k(q)$'s.*

As a consequence, one learns

Corollary 5.4. *The simple functors in $\mathcal{F}(q)$ are in bijective correspondance with the simple $GL_k(\mathbf{F}_q)$–modules, for all $k \geq 0$.*

Related to these results, but *not* just a formal consequence of recollement is

Proposition 5.5. [K:II] *For $F \in \mathcal{F}(q)$, the following are equivalent:*
(1) *F is simple.*
(2) *$F(V)$ is a simple End(V)–module, whenever it is not 0.*
(3) *$F(V)$ is a simple $GL(V)$–module, whenever it is not 0.*

Schwartz's finiteness theorem, Theorem 4.7, has the following consequence for Ext groups.

Theorem 5.6. [K:II, Theorem 3.10] *Given $F \in \mathcal{F}(q)$ finite, and $s \geq 0$, there exists K such that, for all $k > K$ and for all $G \in \mathcal{F}(q)$, the evaluation map*

$$\operatorname{Ext}^s_{\mathcal{F}(q)}(F, G) \to \operatorname{Ext}^s_{\mathcal{M}_k(q)}(F(\mathbf{F}_q^k), G(\mathbf{F}_q^k))$$

is an isomorphism.

We will later see that $\mathcal{M}_k(q)$ can be replaced by $\mathcal{GL}_k(q)$ in this theorem (although K will also depend on G).

Remark 5.7. If we let $M_\infty(\mathbf{F}_q)$ be the union of the multiplicative semigroups $M_k(\mathbf{F}_q)$ (with one included in the next by adding a last row and column of 0's), then $\mathcal{F}(q)$ is equivalent to the full subcategory of $\mathbf{F}_q[M_\infty(\mathbf{F}_q)]$–modules M such that $\bigcup_k e_k M = M$. (Note that $M_\infty(\mathbf{F}_q)$ is a semigroup without unit.)

6. The structure of the fundamental injective

Recall that $P_{\mathbf{F}_q}(V) = \mathbf{F}_q[V]$ and $I_{\mathbf{F}_q}(V) = \mathbf{F}_q^{V^*}$. In this section, we describe the complete lattice of subobjects of these fundamental objects. These functors (and thus their lattices of subobjects) are dual to each other. We will focus on the injective $I_{\mathbf{F}_q}$, and we let $\mathcal{L}(I_{\mathbf{F}_q})$ denote its lattice of subobjects.

We begin by rewriting $\mathbf{F}_q^{V^*}$. As $\mathbf{F}_q^{V^*}$ is a commutative $\mathbf{F}_q$–algebra, the inclusion of the linear functions $V \longrightarrow \mathbf{F}_q^{V^*}$ extends to a natural $\mathbf{F}_q$–algebra map $S^*(V) \longrightarrow \mathbf{F}_q^{V^*}$. The kernel of this map contains all elements of the form $x^q - x$, as the relation $x^q - x = 0$ holds in $\mathbf{F}_q$.

Lemma 6.1. $S^*(V)/(x^q - x) \simeq I_{\mathbf{F}_q}(V)$.

Recall that there is a decomposition into indecomposable summands

$$I_{\mathbf{F}_q} \simeq I_{\mathbf{F}_q,0} \oplus I_{\mathbf{F}_q,1} \oplus \cdots \oplus I_{\mathbf{F}_q,q-1}.$$

From the lemma, it is evident that $I_{\mathbf{F}_q,0} = S^0 = \mathbf{F}_q$, and for $1 \le r \le q-1$, $I_{\mathbf{F}_q,r}$ is the image in $I_{\mathbf{F}_q}$ of $\bigoplus_{n=0}^{\infty} S^{r+n(q-1)}$. (Note also that the lemma shows that $I_{\mathbf{F}_q}$ is locally finite, as asserted earlier.)

Recall that $q = p^s$. Let $\mathbf{N}^s$ be the additive monoid of s–tuples $I = (i_0, \dots, i_{s-1})$ of nonnegative integers. Given $I = (i_0, \dots, i_{s-1}) \in \mathbf{N}^s$, let $r(I) = \sum_{k=0}^{s-1} i_k p^k$, and define $S^I, \tilde{S}^I, \bar{S}^I \in \mathcal{F}(q)$ as follows. First defining $\bar{S}^n$ to be the degree n component of $\bar{S}^*(V) = S^*(V)/(x^p)$, we let

$$S^I = S^{i_0} \otimes S_\xi^{i_1} \otimes \cdots \otimes S_{\xi^{s-1}}^{i_{s-1}},$$

and

$$\bar{S}^I = \bar{S}^{i_0} \otimes \bar{S}_\xi^{i_1} \otimes \cdots \otimes \bar{S}_{\xi^{s-1}}^{i_{s-1}}.$$

Now let $\Phi^I : S^I \longrightarrow S^{r(I)}$ be the composite

$$S^{i_0} \otimes S_\xi^{i_1} \otimes \cdots \otimes S_{\xi^{s-1}}^{i_{s-1}} \xrightarrow{p^{th} \text{ powers}} S^{i_0} \otimes S^{p i_1} \otimes \cdots \otimes S^{p^{s-1} i_{s-1}} \xrightarrow{\text{multiply}} S^{r(I)}.$$

Then $\tilde{S}^I \subset I_{\mathbf{F}_q}$ is defined to be the image of the composite $S^I \xrightarrow{\Phi^I} S^{r(I)} \xrightarrow{\pi} I_{\mathbf{F}_q}$, where π is the inclusion $S^{r(I)} \hookrightarrow S^*$ followed by the projection $S^* \longrightarrow I_{\mathbf{F}_q}$.

Let $R_0, \dots, R_{s-1} \in \mathbf{Z}^s$ be the following vectors: if $s = 1$, $R_0 = p - 1$, and, if $s > 1$, $R_0 = (-1, 0, \dots, 0, p)$, $R_1 = (p, -1, 0, \dots, 0)$, $R_2 = (0, p, -1, 0, \dots, 0), \dots,$ and $R_{s-1} = (0, \dots, 0, p, -1)$. Let $\mathcal{J}(p, s)$ be the poset $(\mathbf{N}^s, \le)$, where "$\le$" is the partial ordering generated by the inequalities $J < I$, if $I = J + R_r$, for some r.

As illustrated in the next example, it is not hard to see that if $J < I$ then $\tilde{S}^J \subset \tilde{S}^I$.

Example 6.2. Let $q = p^3$, let $J = (2, 1, 1)$ and $I = J + R_0 = (1, 1, p+1)$. Then $\tilde{S}^J(V)$ is the span in $I_{\mathbf{F}_q}(V)$ of all the monomials of the form $wxy^p z^{p^2}$, with $w, x, y, z \in V$, while $\tilde{S}^I(V)$ is the span of all the monomials of the form $xy^p z^{p^2} w_1^{p^2} \dots w_p^{p^2}$, with $x, y, z, w_1, \dots, w_p \in V$. A monomial of the former form will be equivalent (in $I_{\mathbf{F}_q}(V)$) to $w^{p^3}xy^pz^{p^2} = xy^pz^{p^2}(w^{p^2})^p$, a monomial of the latter form. Thus $\tilde{S}^J(V) \subset \tilde{S}^I(V)$.

Among other things, the next proposition shows that the composition factors of $I_{\mathbf{F}_q}$ are exactly the $\bar{S}^I$, and each occurs with multiplicity 1.

Proposition 6.3. [K1]
(1) $I_{\mathbf{F}_q} = \sum_I \tilde{S}^I$.
(2) $\tilde{S}^I/\operatorname{Rad}(\tilde{S}^I) \simeq \bar{S}^I$, *and is simple.*
(3) $\operatorname{Rad}(\tilde{S}^I) = \sum_{J<I} \tilde{S}^J$.
(4) $\bar{S}^I \simeq \bar{S}^J$ *if and only if* $I = J$.

To now elegantly describe $\mathcal{L}(I_{\mathbf{F}_q})$, we need a standard construction from lattice theory (as in, e.g. [Gra, p.72]). If $\mathcal{J}$ is a poset, call $\mathcal{K} \subseteq \mathcal{J}$ an *order ideal* if $K \in \mathcal{K}$ and $J \leq K$ implies $J \in \mathcal{K}$. Let $\mathcal{L}(\mathcal{J}) = \{\mathcal{K} \subseteq \mathcal{J} \mid \mathcal{K}$ is an order ideal $\}$. This becomes a distributive lattice using union and intersection for the join and meet lattice operations.

Proposition 6.3, together with general lattice theory, implies

Theorem 6.4. [K1] *The assignment* $I \longmapsto \tilde{S}^I$ *induces an isomorphism of lattices*

$$\mathcal{L}(\mathcal{J}(p,s)) \simeq \mathcal{L}(I_{\mathbf{F}_q}).$$

Thus the finite subobjects of $I_{\mathbf{F}_q}$ having a simple head are precisely the $\tilde{S}^I$, $I \in \mathcal{J}(p,s)$, and every subobject $G \subseteq I_{\mathbf{F}_q}$ has a representation

$$G = \sum_{I \in \mathcal{K}} \tilde{S}^I,$$

for a unique order ideal $\mathcal{K} \subseteq \mathcal{J}(p,s)$.

The indecomposable summand version of Theorem 6.4 is easily stated. Let $\mathcal{L}(I_{\mathbf{F}_q,r})$ denote the lattice of subobjects of $I_{\mathbf{F}_q,r}$. There is a decomposition

$$\mathcal{J}(p,s) = \coprod_{r=0}^{q-1} \mathcal{J}(p,s)_r$$

into indecomposable posets, where $\mathcal{J}(p,s)_0 = \{\mathbf{0}\}$, and, for $1 \leq r \leq q-1$, $\mathcal{J}(p,s)_r = \{I \in \mathcal{J}(p,s) - \{\mathbf{0}\} \mid r(I) \equiv r \mod q-1\}$.

Theorem 6.5. [K1] *For* $0 \leq r \leq q-1$, *there is an isomorphism of lattices*

$$\mathcal{L}(\mathcal{J}(p,s)_r) \simeq \mathcal{L}(I_{\mathbf{F}_q,r}).$$

When $s = 1$ (i.e. $q = p$), one learns from this theorem that, for each $r > 0$, $I_{\mathbf{F}_p,r}$ is infinite uniserial. For example, $I_{\mathbf{F}_2,1}$ has composition factors $\Lambda^n, n \geq 1$, which are assembled into a uniserial tower. Figure 1 below shows the lower portion of the infinite poset $\mathcal{J}(2,2)_1$, i.e. the poset that describes the subobject structure of $I_{\mathbf{F}_4,1}$, the injective envelope of S^1 in $\mathcal{F}(4)$. In general, $\mathcal{J}(p,s)_r$ would have a diagram that would look roughly like an s dimensional cone.

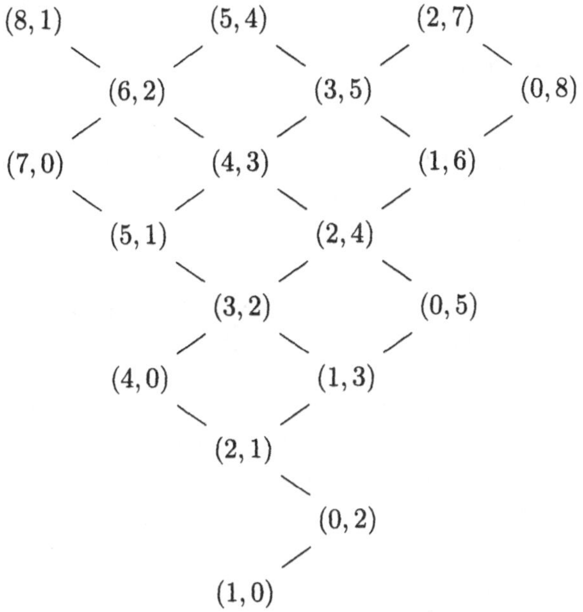

Figure 1. The poset $\mathcal{J}(2,2)_1$ in degrees less than 10.

One can have some fun analyzing the posets $\mathcal{J}(p,s)$. For example, one discovers that the socle series filtration of $I_{\mathbf{F}_q}$ coincides with the polynomial degree filtration. (In the example above, nodes in horizontal rows represent composition factors of the same degree: if $I = (i_0, \dots, i_{s-1})$, the polynomial degree of $\bar{S}^I$ is $i_0 + \dots + i_{s-1}$.)

Slightly deeper, one discovers that for a fixed $I \in \mathcal{J}(p,s)_r$, all but a finite number of $J \in \mathcal{J}(p,s)_r$ satisfy $J > I$ [K1, Proposition 3.6]. We thus conclude

Corollary 6.6. *Every proper subobject of $I_{\mathbf{F}_p,r}$ is finite. Thus $I_{\mathbf{F}_q}$ is an Artinian object in $\mathcal{F}(q)$.*

This corollary suggests a refined form of the Artinian Conjecture. We need a recursively defined notion of 'dimension' for locally finite objects[3].

Definition 6.7. In an abelian category $\mathcal{A}$, say that dim $\mathbf{0} = -\mathbf{1}$. If $A \in \mathcal{A}$ is locally finite and α is an ordinal, say that dim $A = \alpha$ if dim $A \not< \alpha$, but in any ascending chain of subobjects of A, all but finitely many factors have dimension less than α.

With this definition, finite F have dim $F = 0$, while the corollary shows that dim $I_{\mathbf{F}_q} = 1$. If a locally finite object F has a well defined dimension, then F is Artinian: this is proved by straightforward transfinite induction.

Conjecture 6.8. (Strong Artinian Conjecture) dim $I_W = \dim_{\mathbf{F}_q} W$.

[3]This is the Artinian version of Krull dimension [L].

A proof along the lines of the corollary is presumably impossible: it seems unreasonable to try to describe all the subobjects of I_W when $\dim_{\mathbf{F}_q} W > 1$. However, when $q = p = 2$, and $\dim_{\mathbf{F}_2} W = 2$, Geoff Powell has verified the conjecture [P1]. See [P2] for a discussion of this.

7. Connections with Σ_r and Schur algebras

In this section, I describe lots of subcategories of $\mathcal{F}(q)$, in fact, a lattice of subcategories corresponding to the lattice $\mathcal{L}(I_{\mathbf{F}_q})$. (Details of this work will appear in [K3].)

Given $G \subset I_{\mathbf{F}_q}$, let $\mathcal{J}(G) \subset \mathcal{J}(p,s)$ be the corresponding order ideal. Let $\mathcal{J}_{\max}(G)$ be the set of maximal elements in $\mathcal{J}(G)$, so that $\mathcal{J}_{\max}(G) = \mathcal{J}(G) - \mathcal{J}(\mathrm{Rad}(G))$.

Note that, since G is a subobject of $I_{\mathbf{F}_q}$, DG will be a quotient object of $P_{\mathbf{F}_q}$. Now define $\mathcal{F}^G(q)$ to be the following full subcategory of $\mathcal{F}(q)$: a functor F will be in $\mathcal{F}^G(q)$ if, for all W, V, the structure map

$$\mathbf{F}_q[\mathrm{Hom}_{\mathbf{F}_q}(W,V)] \otimes F(W) \to F(V)$$

factors through the quotient

$$\mathbf{F}_q[\mathrm{Hom}_{\mathbf{F}_q}(W,V)] \otimes F(W) \to DG(\mathrm{Hom}_{\mathbf{F}_q}(W,V)) \otimes F(W).$$

Example 7.1. Let $soc^n I_{\mathbf{F}_q} \subset I_{\mathbf{F}_q}$ be the nth stage of the socle filtration of $I_{\mathbf{F}_q}$. As discussed in the last section, $soc^n I_{\mathbf{F}_q}$ is also the maximal subfunctor of degree $\leq n$. It follows quite easily that $\mathcal{F}^{soc^n I_{\mathbf{F}_q}}(q)$ can be identified with $\mathcal{F}^n(q)$, the category of functors of degree $\leq n$.

Example 7.2. $\mathcal{F}^{I_{\mathbf{F}_q,r}}(q) = \mathcal{F}_r(q)$.

Example 7.3. If $q \geq r$, S^r is a natural subobject of $I_{\mathbf{F}_q}$. The resulting category $\mathcal{F}^{S^r}(q)$ is the category $\mathcal{P}^r$ introduced by E. Friedlander and A. Suslin in [FS]. Moreover, as they observe, $\mathcal{P}^r$ is easily seen to be Morita equivalent to the category of modules over the Schur algebra $S(k,r)$, for any $k \geq r$, where we recall that $S(k,r) = S_r(M_k(\mathbf{F}_q))$ with multiplication induced by that on $M_k(\mathbf{F}_q)$.

Given $I = (i_0, \ldots, i_{s-1}) \in \mathcal{J}(p,s)$, let $n(I) = i_0 + \cdots + i_{s-1}$ and let $\Sigma_I = \Sigma_{i_0} \times \cdots \times \Sigma_{i_{s-1}}$.

Theorem 7.4.
(1) *Each $\mathcal{F}^G(q)$ is a localizing subcategory of $\mathcal{F}(q)$.*
(2) *For each finite $G \subset I_{\mathbf{F}_q}$, there is a recollement diagram*

$$\mathcal{F}^{\mathrm{Rad}\,G}(q) \underset{\longleftarrow}{\overset{\longleftarrow}{\xrightarrow{i}}} \mathcal{F}^G(q) \underset{\longleftarrow}{\overset{\longleftarrow}{\xrightarrow{e}}} \prod_{I \in \mathcal{J}_{max}(G)} \mathbf{F}_q[\Sigma_I]\text{-modules.}$$

Example 7.5. Corresponding to Example 7.1, we learn that there is a recollement diagram

$$\mathcal{F}^{n-1}(q) \underset{\longleftarrow}{\overset{\longleftarrow}{\xrightarrow{\;i\;}}} \mathcal{F}^{n}(q) \underset{\longleftarrow}{\overset{\longleftarrow}{\xrightarrow{\;e\;}}} \prod_{n(I)=n} \mathbf{F}_q[\Sigma_I]\text{-modules.}$$

This special case has been known to the author since 1992, and can be viewed as a strengthening for finite fields of a general result of T.Pirashvili [Pi] about polynomial functors on R-modules. (See also [S, §5.5] for a discussion of a version of the case when q is prime due to Lannes and Schwartz, and dating from 1986.)

Example 7.6. Corresponding to Example 7.3, we learn that there is a recollement diagram

$$\mathcal{F}^{\mathrm{Rad}\, S^r}(q) \underset{\longleftarrow}{\overset{\longleftarrow}{\xrightarrow{\;i\;}}} \mathcal{P}^r \underset{\longleftarrow}{\overset{\longleftarrow}{\xrightarrow{\;e\;}}} \mathbf{F}_q[\Sigma_r]\text{-modules.}$$

In this case, the exact functor e is the classic Schur functor [Gre, Chapter 6].

We indicate the ingredients of the proof of Theorem 7.4. If $I = (i_0, \dots, i_{s-1})$, let

$$T^I = T^{i_0} \otimes T^{i_1}_{\xi} \otimes \cdots \otimes T^{i_{s-1}}_{\xi^{s-1}}.$$

It is easy to show that $\mathrm{End}_{\mathcal{F}(q)}(T^I) = \mathbf{F}_q[\Sigma_I]$. One then proves that, if $I \in \mathcal{J}_{\max}(G)$, then T^I is both projective and injective in $\mathcal{F}^G(q)$. In the theorem, the Ith component of e,

$$e^I : \mathcal{F}^G(q) \to \mathbf{F}_q[\Sigma_I]\text{-modules,}$$

is defined by $e^I(F) = \mathrm{Hom}_{\mathcal{F}(q)}(T^I, F)$.

Example 7.7. Let $G \subset I_{\mathbf{F}_4,1}$ correspond to the order ideal:

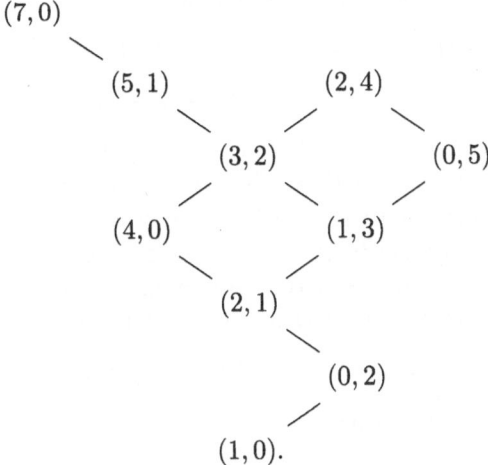

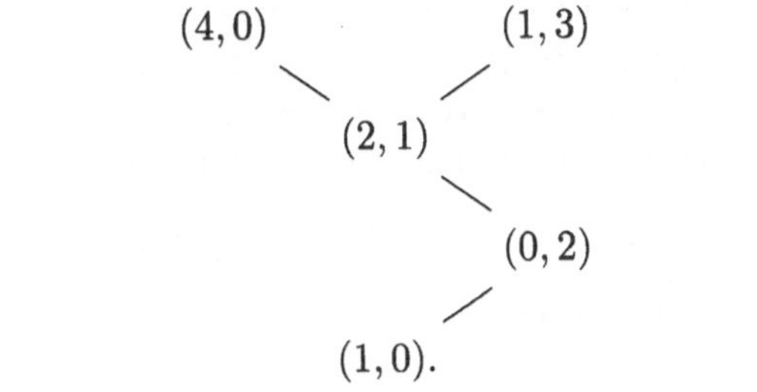

Then the category $\mathcal{F}^G(4)$ admits two exact 'Schur functors': one to the category of Σ_7–modules and one to the category of $\Sigma_2 \times \Sigma_4$–modules.

8. A KUNNETH THEOREM FOR EXT GROUPS

One of the beautiful things about the category $\mathcal{F}(q)$ is that, just using standard homological tools (but with lots of ingenuity!), many complete calculations have been done of extension groups between classic functors. The most macho paper along these lines is [FFSS], where the authors, following the lead of [FLS], give a complete calculation of the trigraded objects $\operatorname{Ext}^*_{\mathcal{F}(q)}(\Lambda^*, \Lambda^*)$, $\operatorname{Ext}^*_{\mathcal{F}(q)}(S^*, S^*)$, and $\operatorname{Ext}^*_{\mathcal{F}(q)}(S_*, S^*)$.

There are two elementary but powerful results used over and over again in doing such calculations. The first goes as follows.

Theorem 8.1. [K:III] *Suppose a graded functor F^* is 'exponential': there are natural isomorphisms $F^*(V \oplus W) \simeq F^*(V) \otimes F^*(W)$. Then, for all functors G and H, there is a natural isomorphism*

$$\operatorname{Ext}^*_{\mathcal{F}(q)}(F^*, G \otimes H) \simeq \operatorname{Ext}^*_{\mathcal{F}(q)}(F^*, G) \otimes \operatorname{Ext}^*_{\mathcal{F}(q)}(F^*, H).$$

The proof is so simple it begs to be sketched: If $G = I_V$ and $H = I_W$, the isomorphism of the conclusion becomes that of the hypothesis. One then notes that the tensor product of injective resolutions for G and H will yield one for $G \otimes H$.

The second result is closely related to the first. Say that G is *constant free* if $G(\mathbf{0}) = \mathbf{0}$.

Theorem 8.2. *If F is polynomial of degree less than n, and $G_1, \dots, G_n$ are all constant free, then*

$$\operatorname{Ext}^*_{\mathcal{F}(q)}(F, G_1 \otimes \cdots \otimes G_n) = 0.$$

Results like this originated in the work of T. Pirashvili (see e.g. [JP]). The theorem as stated here appears in both [K2] and [FFSS].

Example 8.3. Recall that, for each n, one has the Koszul complex exact sequence

$$0 \to \Lambda^n \to \Lambda^{n-1} \otimes S^1 \to \Lambda^{n-2} \otimes S^2 \to \dots \to \Lambda^1 \otimes S^{n-1} \to S^n \to 0.$$

One can view this sequence as $n-1$ short exact sequences spliced together. Applying $\operatorname{Ext}^*_{\mathcal{F}(q)}(S^1, \ )$ to any of these short exact sequences, yields a long exact sequence of Ext-groups. The theorem in the case $F = S^1$ (a degree 1 functor) implies that every third term in these long exact sequences is zero, and one gets isomorphisms of degree 1 between the remaining terms. Stringing these isomorphisms together, one deduces that there are isomorphisms

$$\operatorname{Ext}^s_{\mathcal{F}(q)}(S^1, S^n) \simeq \operatorname{Ext}^{s+n-1}_{\mathcal{F}(q)}(S^1, \Lambda^n).$$

This shifting trick is at the heart of [FLS] (and then [FS, FFSS, K2]).

9. TWO DEEP PROPERTIES OF THE SYMMETRIC POWERS

There are two important and related theorem about the S^n's and S_n's.

Theorem 9.1. *(Embedding Theorem)* [K:I] *Every finite F embeds in a finite sum of S^n's.*

Equivalently (via duality), the theorem says that the S_n's, $n \geq 0$ generate the locally finite category $F^{\omega}(q)$.

Let $\Phi^{-1}S^n$ denote the locally finite functor

$$\Phi^{-1}S^n = \operatorname{colim}\,\{S^n \xrightarrow{\Phi} S^{qn} \xrightarrow{\Phi} S^{q^2n} \to \dots\}.$$

Theorem 9.2. *(Vanishing Theorem)* [K:III] *For all n, $\Phi^{-1}S^n$ is injective in $\mathcal{F}_{\omega}(q)$.*

These two theorems can be given simultaneous proofs, with the key point being a very explicit construction of a natural embedding

$$S^*(V)/(x^q - x) \hookrightarrow \Phi^{-1}S^*(V).$$

Using Theorem 4.7, one deduces from the Vanishing Theorem that if F is finite and $s > 0$, then $\operatorname{colim}_t \operatorname{Ext}^s_{\mathcal{F}(q)}(F, S^{q^t n}) = 0$. V. Franjou and collaborators [FLS, F] then show that this colimit is achieved, thus we have

Corollary 9.3. *For fixed $s > 0$ and finite F, $\operatorname{Ext}^s_{\mathcal{F}(q)}(F, S^{q^t n}) = 0$, for all $t >> 0$.*

10. Connections with algebraic K–theory and group cohomology

In [MacL] (dating from 1956), MacLane defined a very explicit chain complex of R modules, given a ring R, and studied the resulting homology. We will denote this $H^{ML}_*(R)$. Much later, in the 1980's, M. Bökstedt defined $THH_*(R)$, a variant on classical Hochschild Homology called Topological Hochschild Homology. It seems to have been an observation of Pirashvili and his collaborators [JP, PW] that both of these are equal to each other, and are isomorphic to $\operatorname{Tor}^{\mathcal{F}(R)}_*(\mathrm{Id}, \mathrm{Id})$, where $\mathcal{F}(R)$ is the category of functors from finitely generated free R–modules to R–modules, and Id is the inclusion (or identity) functor. Thus we have

$$\operatorname{Tor}^{\mathcal{F}(q)}_s(\mathrm{Id}, \mathrm{Id}) = H^{ML}_*(\mathbf{F}_q) = THH_*(\mathbf{F}_q).$$

Over a field, Ext and Tor groups are dual, and thus, given $F, G \in \mathcal{F}(q)$, we might interpret $\operatorname{Ext}^s_{\mathcal{F}(q)}(F, G)$ as Topological Hochschild Cohomology with twisted coefficients. This makes the techniques for calculation of Ext groups outlined above of even greater interest.

Recall that one can view $\mathcal{F}(q)$ as the category of $M_{\infty}(\mathbf{F}_q)$–modules. Thus one can write

$$\operatorname{Ext}^*_{\mathcal{F}(q)}(F, G) = H^*(M_{\infty}(\mathbf{F}_q); \operatorname{Hom}(F, G)),$$

where $\operatorname{Hom}(F, G)$ denotes the obvious bifunctor on $\mathbf{F}_q$–vector spaces. There is a restriction map from this to

$$H^*(GL_{\infty}(\mathbf{F}_q); \operatorname{Hom}(F, G))$$

and it is tempting to ask if this is an isomorphism.

One can relate this question to the cohomology of the finite general linear groups. Theorem 5.6 said that, for finite F and G,

$$H^*(M_\infty(\mathbf{F}_q); \operatorname{Hom}(F, G)) \simeq H^*(M_k(\mathbf{F}_q); \operatorname{Hom}(F(\mathbf{F}_q^k), G(\mathbf{F}_q^k))),$$

for $k >> 0$. Analogously, W. Dwyer [D] has proved that

$$H^*(GL_\infty(\mathbf{F}_q); \operatorname{Hom}(F, G)) \simeq H^*(GL_k(\mathbf{F}_q); \operatorname{Hom}(F(\mathbf{F}_q^k), G(\mathbf{F}_q^k))),$$

for $k >> 0$. Thus the question asks if

$$\operatorname{Ext}^*_{\mathcal{F}(q)}(F, G) \simeq H^*(GL_k(\mathbf{F}_q); \operatorname{Hom}(F(\mathbf{F}_q^k), G(\mathbf{F}_q^k))), \tag{10.1}$$

for $k >> 0$.

Example 10.1. If $F = G = \mathbf{F}_q$ (the constant functor), the left side of (10.1) is 0 for $* > 0$. So is the right side: this is D. Quillen's theorem [Q] that $H^*(GL_\infty(\mathbf{F}_q); \mathbf{F}_q) = 0$ in positive degrees.

Example 10.2. If $F = G = \mathrm{Id}$, (10.1) is a special case of a theorem of B. Dundas and R. McCarthy [DM] that says that "stable K–theory equals topological Hochschild homology".

Very recently, Suslin [FFSS, Appendix] and Betley [B] have independently proved

Theorem 10.3. *(10.1) holds for all finite F and G in $\mathbf{F}(q)$.*

11. Connections with rational cohomology

In Example 7.3, we briefly discussed Friedlander and Suslin's categories $\mathcal{P}^r$. One has a Frobenius twist map $\Phi : \mathcal{P}^r \to \mathcal{P}^{pr}$, and we will denote $\Phi(F)$ by F_ξ. For $F, G \in \mathcal{P}^r$ there are obvious 'restriction' (or 'inflation') maps $\operatorname{Ext}^*_{\mathcal{P}^r}(F, G) \to Ext^*_{\mathcal{F}(q)}(F, G)$, and these are compatible as q varies, and also with the Frobenius twist on both categories. (Note that on $\mathcal{F}(q)$ twisting is an equivalence.)

The following theorem is proved by the author in [K2], and independently by Franjou, Friedlander, Suslin, and Skorichenko in [FFSS].

Theorem 11.1. *Given s, and $F, G \in \mathcal{P}^r$,*

$$\operatorname{Ext}^s_{\mathcal{P}^{p^t r}}(F_{\xi^t}, G_{\xi^t}) \to Ext^s_{\mathcal{F}(q)}(F, G)$$

is an isomorphism for $t >> 0$ and $q >> 0$.

Both proofs of this theorem use all of the ideas presented above in §8 and §9.

The theorem is a 'stable' analogue of a theorem in the theory of algebraic groups [CPSvK]: If M, N are finite dimensional modules over an algebraic group G defined over $\mathbf{F}_p$, then, given s,

$$\operatorname{Ext}^s_G(M_{\xi^t}, N_{\xi^t}) \to Ext^s_{G(\mathbf{F}_q)}(M, N)$$

is an isomorphism for $t >> 0$ and $q >> 0$.

12. Connections with the Steenrod algebra

There is a remarkable connection, noticed by Henn, Lannes, and Schwartz in [HLS], between the category $\mathcal{F}(q)$, and a category of interest to topologists: the category of "unstable modules over the Steenrod algebra". In this section, we give a very quick introduction to this topic. For those interested in learning more about these connections to topology, Lionel Schwartz's book [S] and my papers [K:I, K:III] offer more gentle and complete entries into the literature.

Let $\mathcal{S}^*(q) \subset \mathcal{F}(q)$ be the full subcategory with objects $S^n, n \geq 0$. Similarly define $\mathcal{S}_*(q)$ to be the full subcategory with objects $S_n, n \geq 0$, and note that $\mathcal{S}^*(q) \simeq (\mathcal{S}_*(q))^{op}$. We let $\mathrm{Rep}(\mathcal{S}^*(q))$ denote the category of $\mathcal{S}^*(q)$–modules, i.e. $\mathbf{F}_q$–linear functors

$$M : \mathcal{S}^*(q) \longrightarrow \mathbf{F}_q\text{–vector spaces.}$$

Unraveling this definition, we see that $M \in \mathrm{Rep}(\mathcal{S}^*(q))$ is a graded vector space M^* equipped with appropriately compatible linear maps $a : M^n \longrightarrow M^{n+i}$ for every $a : S^n \longrightarrow S^{n+i}$.

Note that, for all V, $S^*(V) \in \mathrm{Rep}(\mathcal{S}^*(q))$. Another source of representations of $\mathcal{S}^*(q)$ is the following: given $F \in \mathcal{F}(q)$, let $r(F) \in \mathrm{Rep}(\mathcal{S}^*(q))$ have $r(F)^n = \mathrm{Hom}_{\mathcal{F}(q)}(S_n, F)$. A map $a : S^n \longrightarrow S^{n+i}$ corresponds to a map $a : S_{n+i} \longrightarrow S_n$, and thus induces $a : r(F)^n \longrightarrow r(F)^{n+i}$.

This construction r is natural, and, indeed, r is the right adjoint of a pair of functors

$$\mathrm{Rep}(\mathcal{S}^*(q)) \underset{r}{\overset{l}{\rightleftarrows}} \mathcal{F}^{\omega}(q),$$

where the left adjoint, l, has a couple of useful descriptions:

$$l(M)(V) = M \otimes_{\mathcal{S}_*(q)} S_*(V) = \mathrm{Hom}_{\mathrm{Rep}(\mathcal{S}^*(q))}(M, S^*(V'))'.$$ [4]

Note that we have restricted ourselves to the locally finite functors: this is due to the fact that the S_n's are finite.

One formulation of the Embedding Theorem (Theorem 9.1) is that the functors $S_n, n \geq 0$, generate $\mathcal{F}^{\omega}(q)$. Using a variant of the Gabriel–Popescu Theorem [GP] (itself a variant of the classical Morita theorems), the Embedding Theorem is *equivalent* to each of the following two theorems [K:I].

Theorem 12.1. *l is exact, and, for all $F \in \mathcal{F}^{\omega}(q)$, the natural map $l(r(F)) \longrightarrow F$ is an isomorphism.*

Theorem 12.2. *For all V, $S^*(V)$ is an injective object in Rep$(\mathcal{S}^*(q))$. Furthermore, for all V, W, the natural map*

$$\mathbf{F}_q[Hom(W,V)] \longrightarrow \mathrm{Hom}_{Rep(\mathcal{S}^*(q))}(S^*(W), S^*(V))$$

is an isomorphism.

---

[4] In this formula, V' denotes the dual of V, while $\mathrm{Hom}_{\mathrm{Rep}(\mathcal{S}^*(q))}(M, S^*(V'))'$ denotes the continuous dual of $\mathrm{Hom}_{\mathrm{Rep}(\mathcal{S}^*(q))}(M, S^*(V'))$, which has a natural profinite topology [K:I, Remark 6.7].

I now explain why these theorems are of interest to topologists. For simplicity, I will restrict my attention to the case when $q = 2$.

If X is a topological space, its mod 2 cohomology, $H^*(X; \mathbf{Z}/\mathbf{2})$ is (tautologically) a module over the algebra A of natural cohomology operations. By the late 1940's, Steenrod had constructed certain explicit operations, the so-called Steenrod (reduced) squares,

$$Sq^i : H^n(X; \mathbf{Z}/\mathbf{2}) \to H^{n+i}(X; \mathbf{Z}/\mathbf{2}),$$

and by the mid 1950's, it was known that A is generated by the $Sq^i, i \geq 1$, and these are subject to certain explicit relations. The A–module $H^*(X; \mathbf{Z}/\mathbf{2})$ automatically satisfies an extra "unstable" condition:

$$\text{If } x \in H^n(X; \mathbf{Z}/\mathbf{2}) \text{ then } Sq^i x = 0 \text{ for all } i > n.$$

We let $\mathcal{U}$ be the full subcategory of the category of A–modules satisfying this extra condition.

Given a group G, there is an associated classifying space BG. In particular, $S^*(V) = H^*(BV'; \mathbf{Z}/\mathbf{2})$ is an A–module. The classifying space construction is functorial in G, and thus we conclude that $a \in A^i$ will induce $a : S^n \to S^{n+i}$ for all n. A calculation of $\mathrm{Hom}_{\mathcal{F}(2)}(S^*, S^*)$ reveals that all maps between symmetric products arise in this way, and that A acts freely on the S^n's, except for being subject to the unstable condition. One concludes

Proposition 12.3. *Rep*$(\mathcal{S}^*(2)) = \mathcal{U}$.

From the above, we can thus conclude that, for all V,

$$H^*(BV; \mathbf{Z}/\mathbf{2}) \text{ is a } \mathcal{U}\text{–injective},$$

and, for all V and W,

$$\mathbf{F}_2[Hom(W, V)] \simeq \mathrm{Hom}_{\mathcal{A}}(H^*(BV; \mathbf{Z}/\mathbf{2}), H^*(BW; \mathbf{Z}/\mathbf{2})).$$

Both of these results were proved first using other methods, and were heavily used in some of the most striking work of the 1980's in homotopy theory. The injectivity of $H^*(BV; \mathbf{Z}/\mathbf{2})$, first (essentially) proved by G. Carlsson, was crucial to work of H. Miller and Lannes on the Sullivan Conjecture. The calculation of the A–module maps between the $H^*(BV; \mathbf{Z}/\mathbf{2})$'s was part of a more general theorem proved by J. F. Adams, J. Gunawardena, and Miller in their work on the Segal Conjecture. In both cases, the algebraic theorems were used as input into various sorts of Adams Spectral Sequence to deduce topological consequences.

The Vanishing Theorem (Theorem 9.2) also is equivalent to a theorem about unstable $\mathcal{A}$–modules.

Let $\mathcal{U}(q) = \mathrm{Rep}(\mathcal{S}^*(q))$, and $A(q)$ the q–analogue of A [K:I]. Then Theorem 12.1 and Proposition 12.3 combine to show that there is a partial recollement diagram:

$$\mathcal{N}(q) \underset{\longleftarrow}{\overset{\longrightarrow}{}} \mathcal{U}(q) \underset{\overset{r}{\longleftarrow}}{\overset{l}{\longrightarrow}} \mathcal{F}^{\omega}(q),$$

with $N \in \mathcal{N}(q)$ if $\mathrm{Hom}_{A(q)}(N, S^*(V)) = 0$ for all V [HLS, K:I].

There is the *suspension* operation $\Sigma : \mathcal{U}(q) \longrightarrow \mathcal{U}(q)$, satisfying $(\Sigma M)^n = M^{n-1}$. Let $\mathcal{NIL}(q) \subset \mathcal{U}(q)$ be the smallest localizing category containing all suspensions. ($N \in \mathcal{NIL}(q)$ is said to be *nilpotent.*[5])

The Vanishing Theorem turns out to be equivalent to the following theorem of Lannes and Schwartz [LS].

Theorem 12.4. $\mathcal{N}(q) = \mathcal{NIL}(q)$.

The point of this theorem is that it gives an 'internal' criterion for checking when a module N satisfies $\mathrm{Hom}_{A(q)}(N, S^*(V)) = 0$ for all V.

References

[B] S. Betley, *Stable K-theory of finite fields*, K-Theory **17**(1999), 103–111.

[CPSvK] E. Cline, B. Parshall, L. Scott, and W. van der Kallen, *Rational and generic cohomology*, Inv. Math. **39** (1977), 143–163.

[DM] B. Dundas and R. McCarthy, *Stable K-theory and Hochschild homology*, Ann. Math. **140** (1994), 685–701; erratum, Ann. Math. **142** (1995), 425–426.

[D] W. G. Dwyer, *Twisted homological stability for general linear groups*, Ann. Math. **111** (1980), 239–251.

[EM] S. Eilenberg and S. MacLane, *On the groups $H(\pi, n)$, II*, Ann. Math. **60** (1954), 49–139.

[F] V. Franjou, *Extensions entre puissances extérieures et entre puissances symétriques*, J. Algebra **179**(1996), 501–522.

[FFSS] V. Franjou, E. M. Friedlander, A. Skorichenko, and A. Suslin, *General linear and functor cohomology over finite fields*, Ann. Math. **150**(1999), 663–728.

[FLS] V. Franjou, J. Lannes, and L. Schwartz, *Autour de la cohomologie de MacLane des corps finis*, Invent. Math. **115**(1994), 513–538.

[FS] E. M. Friedlander and A. Suslin *Cohomology of finite group schemes over a field*, Inv. Math. **127**(1997), 209–270.

[GP] P. Gabriel and N. Popescu, *Charactérisattion des catégories abeéliennes avec générateurs et limites inductives exactes*, C. R. Acad. Sci. Paris **258** (1964), 4188–4190.

[Gra] G. Grätzer, *Lattice Theory: First Concepts and Distributive Lattices*, W. H. Freeman and Co., San Francisco, 1971.

[Gre] J. A. Green, *Polynomial Representations of GL_n*, Springer L. N. Math. **830**, New York, 1980.

[HLS] H.-W. Henn, J. Lannes, and L. Schwartz *The categories of unstable modules and unstable algebras modulo nilpotent objects*, Amer. J. Math. **115**(1993), 1053–1106.

[JP] Mamuka Jibladze, Teimuraz Pirashvili, *Cohomology of algebraic theories*, J. Algebra **137**(1991), 253–296.

[K:I] N. J. Kuhn, *Generic representation theory of the finite general linear groups and the Steenrod algebra: I*, Amer. J. Math. **116**(1994), 327–360.

[K:II] N. J. Kuhn, *Generic representation theory of the finite general linear groups and the Steenrod algebra: II*, K-Theory J. **8**(1994), 395–428.

[K:III] N. J. Kuhn, *Generic representation theory of the finite general linear groups and the Steenrod algebra: III*, K-Theory J. **9**(1995), 273–303.

[5]When $q = 2$ this terminology can be explained as follows. For $M \in \mathcal{U}$, let $Sq_0 : M \longrightarrow M$ be the 'top' operation $Sq_0 x = Sq^{|x|}x$. It is not hard to see that N is in $\mathcal{NIL}$ iff every element in N is nilpotent under iterating Sq_0. In $H^*(X; \mathbf{Z}/2)$, a ring using the cup product, $Sq_0 x = x^2$. Thus $N \subset H^*(X; \mathbf{Z}/2)$ is in $\mathcal{NIL}$ iff every element in N is nilpotent in the usual algebraic sense.

[K1] N. J. Kuhn, *Invariant subspaces of the ring of functions on a vector space over a finite field*, J. Algebra **191**(1997), 212–227.

[K2] N. J. Kuhn, *Rational cohomology and cohomological stability in generic representation theory*, Amer. J. Math. **120**(1998), 1317–1341.

[K3] N. J. Kuhn, *Filtrations generic representations, and generalized Schur functors*, in preparation.

[LS] J.Lannes and L.Schwartz, *Sur la structure des A-modules instables injectifs,* Topology **28**(1989), 153–169.

[L] T. H. Lenagan, *Dimension theory of Noetherian rings*, to appear in the proceedings of the conference on infinite length modules, Bielefeld, Germany, 1998.

[MacL] S. MacLane, *Homologie des anneaux et des modules*, Coll. topologie algébrique, Louvain, (1956), 55–80.

[Pi] T. Pirashvili, *Higher additivisations* [Russian, English summary], Trudy Tbiliss. Mat. Inst. Razmodze Akad. Nauk Gruzin. SSR **91**(1988), 44–54.

[PW] T. Pirashvili and F. Waldhausen, *MacLane homology and Topological Hochshild homology*, J. Pure App. Alg. **82** (1992), 81–98.

[P1] G. M. L. Powell, *The Artinian conjecture for $I \otimes I$*, J. Pure App. Alg. **128** (1998), 291–310.

[P2] G. M. L. Powell, *On Artinian objects in the category of functors between* $\mathbf{F}_2$*–vector spaces*, to appear in the proceedings of the conference on infinite length modules, Bielefeld, Germany, 1998.

[Q] D. Quillen, *On the cohomology and K–theory of the general linear groups over a finite field*, Ann. Math. **96** (1972), 552–586.

[S] L. Schwartz, *Unstable modules over the Steenrod algebra and Sullivan's fixed point conjecture,* Chicago Lectures in Math., University of Chicago Press, 1994.

Department of Mathematics, University of Virginia, Charlottesville, VA 22903

Trends in Mathematics, © 2000 Birkhäuser Verlag Basel/Switzerland

ON ARTINIAN OBJECTS IN THE CATEGORY OF FUNCTORS BETWEEN $\mathbb{F}_2$-VECTOR SPACES

GEOFFREY M.L. POWELL

ABSTRACT. This note surveys the study of the artinian conjecture in the category $\mathcal{F}(k)$ of functors between vector spaces over a finite field k and the study of the structure of injective cogenerators of this category. This is illustrated by examples, for which the field is taken to be $\mathbb{F}_2$.

1. INTRODUCTION

The category $\mathcal{F}(k)$ of functors from the category $\mathcal{V}_k^f$ of finite-dimensional vector spaces over a field k to the category $\mathcal{V}_k$ of all k-vector spaces arises in many contexts. More generally, one can consider the category $\mathcal{F}(R)$ of functors from the category of finitely-generated free left R-modules to the category of all left R-modules, when R is a ring [JP]. The restriction of the domain to the category of finite-dimensional spaces is standard and imposes little limitation: the category $\hat{\mathcal{F}}(k)$ of functors $\mathcal{V}_k \to \mathcal{V}_k$ is equipped with a restriction functor $\hat{\mathcal{F}}(k) \to \mathcal{F}(k)$ and there is a left Kan extension $\mathcal{F}(k) \to \hat{\mathcal{F}}(k)$ so that $\mathcal{F}(k)$ may be viewed as a subcategory of $\hat{\mathcal{F}}(k)$. In particular, the left Kan extension allows the composition of functors which do not take finite-dimensional values. The condition that functors should be allowed to take infinite-dimensional values is essential so that $\mathcal{F}(k)$ contains all colimits, which is a necessary condition if one is motivated by the study of unstable modules over the Steenrod algebra [S0].

When k is an infinite field, the category $\mathcal{F}(k)$ has been little studied. Instead, it has been usual to study polynomial representations of the general linear groups, as in [G], and categories of polynomial functors [MacD] (in characteristic zero). This is linked to the study of representations of Schur algebras [M], which, in characteristic zero, goes back to Schur's thesis. The recent work of Friedlander and Suslin studies categories of strict polynomial functors, $\mathcal{P}^n(k)$, which are essentially categories of modules over Schur algebras.

When $k = \mathbb{F}_q$, a finite field with q elements, the categories $\mathcal{F}(k)$ have received much attention; various motivations for this are described in Kuhn's survey article [K] and, equally, in Schwartz's survey [S0] in this volume. In this case, there are readily-described projective generators, which are given by Yoneda's lemma, and which take finite-dimensional values; it should be stressed at this point that the functors considered are in general non-additive. A consequence of this observation is that the simple functors take finite-dimensional values. Kuhn's survey [K] describes the connections between the functor categories $\mathcal{F}(\mathbb{F}_q)$ and representations of the general linear groups and also the symmetric groups, by using the formal *recollement* framework. These both yields parametrizations of the simple functors in $\mathcal{F}(\mathbb{F}_q)$. Good examples of simple functors are provided by the exterior power functors, $\Lambda^n : V \mapsto V^{\wedge n}$, for $n \geq 0$, where Λ^0 is understood to be the constant functor $V \mapsto \mathbb{F}_q$.

A functor in $\mathcal{F}(\mathbb{F}_q)$ is said to be *finite* if it has a finite composition series; by the above remarks, a finite functor takes finite-dimensional values. Conversely, a functor taking finite-dimensional values is finite if and only if it is *polynomial* in the sense of Eilenberg and MacLane (see [K]). Finite functors are induced by finite-dimensional modules over finite-dimensional algebras of the form $\mathbb{F}[\mathrm{End}(\mathbb{F}^t)]$, for some t. Less familiar are the functors which are *not* finite and, more particularly, not polynomial; these form the object of this note.

The category $\mathcal{F}(\mathbb{F}_q)$ has enough injectives; in particular, there are injective cogenerators I_V, indexed over $V \in \mathcal{V}_k$, which are functors which take finite-dimensional values (see [K]). These objects are *not* finite when $\dim V > 1$; however, they are *locally finite*, namely they are the colimit of their finite sub-objects. The full sub-category of locally finite functors is written as $\mathcal{F}_\omega(\mathbb{F}_q)$ and most of what is considered here takes place within this category.

Remark 1.0.1 It should be noted that a choice is made here to consider the structure of the injective objects and, hence, locally finite objects; one could equally study the duals of the locally finite objects (with respect to the duality functor $D : \mathcal{F} \to \mathcal{F}^{\mathrm{op}}$ recalled in Section 2). The category $\mathcal{F}_\omega$ is more natural for topologists, since it arises in connection with the Steenrod algebra [HLS, S0].

There are straightforward examples of functors in $\mathcal{F}(\mathbb{F}_q)$ which are neither locally finite nor dual to locally finite functors. For example, the tensor product of a non-finite indecomposable projective functor and a non-finite indecomposable injective functor has no finite sub-objects or finite quotients[1].

The reader is warned that the restriction to the category of locally finite functors can pose problems; for example, Suslin's appendix to [FFSS] addresses properties of *Ext* groups in $\mathcal{F}$ between finite functors; these are calculable in the category $\mathcal{F}_\omega$. However, his proof uses a functor $\tilde{a} : \mathcal{F} \to \mathcal{F}$ which, even on the most innocent of all non-trivial functors $\mathbb{F}$, the constant functor, has image which is neither locally finite nor dual to a locally finite functor. □

The purpose of this note is to explain the artinian conjecture, due to Lionel Schwartz, which is introduced below.

[1] An example of this type was suggested to the author by Lionel Schwartz.

AC0 *THE ARTINIAN CONJECTURE.* [K1, Appendix] The functors I_V are artinian objects in the category $\mathcal{F}$ (satisfy the descending chain condition for sub-objects).

In addition, specializing to the case $\mathbb{F}_q = \mathbb{F}_2$, the note illustrates results on the structure of the injectives I_V when $\dim V > 1$. At the time of writing, for the case $\mathbb{F}_q = \mathbb{F}_2$, this conjecture is only known for the special cases $\dim V \leq 2$. By contrast, the special case $\dim V = 1$, over an arbitrary field, admits an explicit description, which is given in [K].

The limitation of knowledge is exemplified by the following two questions:

1. If G is artinian, is the functor ΔG artinian, where $\Delta : \mathcal{F} \to \mathcal{F}$ denotes the difference functor $\Delta F(V) = F(V \oplus \mathbb{F})/F(V)$?
2. If G is artinian, is the functor $G \otimes \Lambda^1$ artinian, where Λ^1 denotes the first exterior power functor?

Neither of these questions is known to have an affirmative answer, whilst both are straightforward consequences of the artinian conjecture.

1.1. **Motivation for the artinian conjecture.** The artinian conjecture may be given a number of equivalent formulations, some of which may be found in [K2]. Say that a functor F is *i-small* if it embeds in an injective functor of the form I_V. This is equivalent to stating that the functor F is locally finite and has a finite socle. In particular, an injective is i-small if and only if it is a finite direct sum of injectives.

The statements labelled '*AC*.' below are all equivalent to the artinian conjecture.

AC1 The category $\mathcal{F}$ is locally Noetherian.

AC2 Every i-small functor F has a presentation $0 \to F \to I^0 \to I^1$ by i-small injectives I^0, I^1.

The artinian conjecture was first posed as a question in the form above, motivated by the work in [HLS], as is explained in [S0].

AC3 Every i-small functor F has an injective resolution $0 \to F \to I^\bullet$ by i-small injectives $I^\bullet$.

These formulations already serve to illustrate the importance of the artinian conjecture, given the connection between the calculation of *Ext* groups in the functor category and the calculation of the MacLane cohomology and stable K-theory of a field, with coefficients in suitable bifunctors (see [K, Section 10] and the references given there). As another example there is a functor $\Phi^{-1}S^n$ which is defined as the colimit of the diagram $S^n \to S^{nq} \to S^{nq^2} \to \ldots$, in which each map is the Frobenius q^{th} power map. This functor does not take finite-dimensional values for $n > 0$, although its socle does. The vanishing theorem of Kuhn reviewed in [K] asserts that this functor is injective in the category of locally finite functors. It may be expressed as an infinite direct sum of injective objects; the artinian conjecture would imply that $\Phi^{-1}S^n$ is injective in $\mathcal{F}(\mathbb{F}_q)$.

Another example of a problem in which the artinian conjecture arises naturally is due to Lionel Schwartz (see [S0]); for the convenience of the reader, this is recalled here. There is an unstable module over the Steenrod algebra which is associated to any locally finite functor F; this may be considered as a non-negatively graded vector space, which is given in dimension n by the vector space $\mathrm{Hom}_{\mathcal{F}}(S_n, F)$, where S_n is the nth symmetric invariant functor $V \mapsto (V^{\otimes n})_{\Sigma_n}$. One may consider the Poincaré series of this graded vector space; this can only be expected to yield reasonable information on the functor when some finiteness hypothesis is imposed, namely that F is i-small. Schwartz [S0] asks (in a precise form) whether the Poincaré series of F is sufficient to determine F as an element of the 'Grothendieck group' of i-small functors. Unfortunately, this Grothendieck group is only defined if the artinian conjecture is known.

The artinian conjecture may also be related to the problem of the consideration of functors taking finite-dimensional values. The full sub-category $\mathcal{F}_{\mathrm{fd}}(\mathbb{F}_q)$ of functors taking finite-dimensional values is a Serre sub-category which contains the indecomposable injectives. However, the example $\Phi^{-1}S^n$ suffices to show that $\mathcal{F}_{\mathrm{fd}}(\mathbb{F}_q)$ does not have enough injectives. More generally, one has:

Lemma 1.1.1. *Suppose that $V \in \mathcal{V}^f_{\mathbb{F}_q}$ and $\{S_{\mu_i}\}$ is an infinite set of simple functors (indexed over the non-negative integers) which are composition factors of I_V. Then the injective envelope of $\bigoplus_i S_{\mu_i}$ in $\mathcal{F}(\mathbb{F}_q)$ does not take finite-dimensional values.*

This gives a version of the artinian conjecture in terms of functors taking finite-dimensional values:

AC4 Every i-small functor F has an injective resolution $0 \to F \to J^\bullet$ by injectives which take finite-dimensional values.

The conjecture may clearly be related to the description of $\mathcal{F}$ given in terms of infinite-dimensional matrices [K, K2]. Following Kuhn, write $M_\infty(\mathbb{F}_q)$ for

$$\operatorname{colim} M_n(\mathbb{F}_q),$$

where the embedding $M_n(\mathbb{F}_q) \hookrightarrow M_{n+1}(\mathbb{F}_q)$ is given by $\boxed{A} \mapsto \begin{pmatrix} \boxed{A} & 0 \\ 0 & 0 \end{pmatrix}$ and $M_n(\mathbb{F}_q)$ denotes the semi-group of $n \times n$ matrices with entries in $\mathbb{F}_q$. Then $\mathcal{M}_\infty(\mathbb{F}_q)$ is defined as the category of left $\mathbb{F}_q[M_\infty(\mathbb{F}_q)]$-modules N such that $N = \operatorname{colim} e_n N$, where e_n is the idempotent in $\mathbb{F}_q[M_\infty(\mathbb{F}_q)]$ associated to $1_n \in M_n(\mathbb{F}_q)$. Kuhn shows that $\mathcal{M}_\infty(\mathbb{F}_q)$ is isomorphic to the inverse limit $\lim_{\leftarrow} \mathcal{M}_n(\mathbb{F}_q)$, where $\mathcal{M}_n(\mathbb{F}_q)$ is the category of left $\mathbb{F}_q[M_n(\mathbb{F}_q)]$-modules. This inverse limit is tautologically isomorphic to $\mathcal{F}(\mathbb{F}_q)$, hence the category of functors may be identified with $\mathcal{M}_\infty(\mathbb{F}_q)$. In particular, the artinian conjecture may be formulated in terms of a *locally* Noetherian[2] condition on the algebra $\mathbb{F}_q[M_\infty(\mathbb{F}_q)]$.

[2]This is a *Noetherian* condition since the algebra $\mathbb{F}_q[M_\infty(\mathbb{F}_q)]$ corresponds to a projective functor.

AC5 Suppose that $I \lhd \mathbb{F}_q[M_\infty(\mathbb{F}_q)]$ is a left ideal such that $Ie_n = I$ for some n, then I is a finitely-generated left ideal.

Remark 1.1.2 To connect this with known terminology, observe that a left $\mathbb{F}_q[M_\infty(\mathbb{F}_q)]$-module is finitely-generated if and only if the associated functor is generated in finite dimension, in the terminology of [FLS, Définition 1.2]. □

It is clear that a functor defines a representation of the finitary symmetric group; it would be possible to formulate a strong conjecture for the group algebra of the finitary symmetric group which would *imply* the Artinian conjecture. However, there is an open problem [Z, page 612]:

Problem: Is it true that the *two-sided* ideals of the group algebra of the finitary symmetric group satisfy the ascending chain condition?

This suggests that any conjecture in this context which is strong enough to imply the artinian conjecture would be hard to address[3].

Remark 1.1.3 *(The artinian conjecture for other functor categories).* It is natural to ask whether the artinian conjecture may be formulated meaningfully in any other related functor categories. Essentially this questions whether the artinian conjecture makes sense over an infinite field. At present, the categories $\mathcal{F}(k)$, where k is an infinite field, are not sufficiently understood to formulate the conjecture. Conversely, the categories $\mathcal{P}^n(k)$ of *strict polynomial functors* are defined as module categories over finite-dimensional algebras and, in this setting, the artinian conjecture is trivial.

1.2. **Krull codimension.** The size of an artinian object in $\mathcal{F}$ may be measured by the Krull codimension [Le]: so as to be consistent with the article [Ps], the following terminology is used:

Definition 1.2.1 A functor F is said to be *artinian of type* -1 if it is zero. The functor F is *simple artinian of type* n, for $n \geq 0$, if every proper sub-object is artinian of type $n-1$ and F is not artinian of type $n-1$. The functor F is *artinian of type* n if it has a finite filtration of which the sub-quotients are simple artinian of type $\leq n$. The abbreviation *SAT1* will be used for a functor which is simple artinian of type one. □

Remark 1.2.2 A functor which is artinian of type n is certainly artinian, whereas the strong artinian conjecture (Section 2) would imply that an artinian functor is of type n, for some finite n. A functor F is simple if and only it is simple artinian

[3]With thanks for the anonymous suggestion that this reference may be of relevance.

of type zero; however, for $n > 0$ a functor which is simple artinian of type n is *not* simple in $\mathcal{F}(\mathbb{F}_q)$; it is simple in the category $\mathcal{F}(\mathbb{F}_q)$ localized away from the full sub-category of artinian functors of type $\leq n-1$. The simple artinian functors correspond to the critical objects for the Krull codimension. □

The artinian conjecture may be refined by introducing co-Weyl objects, motivated by the study of highest weight categories introduced by Cline, Parshall and Scott. The co-Weyl objects are conjecturally simple artinian objects. The Krull codimension may also be compared with a notion of dimension which is given by nilpotence under the action of certain quotients of the difference functor, Δ. This material is treated in following sections.

1.3. **Artinian functors of type one.** A recent advance in the study of artinian objects in the category $\mathcal{F}(\mathbb{F}_2)$ is provided by the following Theorem. If S is a simple functor, then S occurs as a composition factor of some injective I_V. Let $\overline{I}$ denote the injective envelope of the functor Λ^1, then the functor $\overline{I} \otimes S$ occurs as a sub-quotient of the functor $I_{V \oplus \mathbb{F}}$. The artinian conjecture would imply that the functor $\overline{I} \otimes S$ is artinian, so that the study of such functors is a natural test case for the artinian conjecture.

Theorem 1.3.1. [P3] *Suppose that $\mathbb{F}_q = \mathbb{F}_2$, S is a simple functor and let $\overline{I}$ denote the injective envelope of Λ^1, then*

1. *$\overline{I} \otimes S$ is artinian of type one.*
2. *There exists an explicit sub-object $X_S \hookrightarrow \overline{I} \otimes S$ such that every non-finite sub-functor of $\overline{I} \otimes S$ contains X_S; in particular, X_S is SAT1.*

A crucial related problem is:

Problem: What are the SAT1 functors? (One can only hope to classify these modulo finite sub-objects). □

Underlying this question is the problem of studying the category $\mathcal{F}_\omega$/finites of locally finite functors localized away from the finite functors, which form a full sub-category $\{\text{finites}\} \subset \mathcal{F}_\omega$. The category $\mathcal{F}_\omega$/finites has the same objects as $\mathcal{F}_\omega$ whereas the morphisms between two objects X, Y are given by

$$\lim_{\substack{\rightarrow \\ X' \subset X, Y' \subset Y}} \operatorname{Hom}(X', Y/Y'),$$

where the colimit is taken over $X' \subset X, Y' \subset Y$, where X/X' and Y' are finite functors. It is straightforward to observe that every non-zero simple object in $\mathcal{F}_\omega$/finites is isomorphic in this category to a SAT1 functor.

Theorem 4.1.2 gives some very slim evidence to suggest that the following property may hold.

Question 1.3.2 Is every SAT1 functor in $\mathcal{F}(\mathbb{F}_2)$ a quotient of some functor X_S, defined as a sub-object of $\overline{I} \otimes S$ by Theorem 1.3.1?

1.4. **The 'type n' approach to the artinian conjecture.** The usual strategy for attacking the artinian conjecture is based on an induction upon the dimension of V for the injectives I_V; this strategy motivates the approach described in following sections. The progress in understanding SAT1 functors outlined above suggests an alternative approach.

Notation 1.4.1 Suppose that F is an i-small functor. Write $\mathcal{I}(F)$ for the set of isomorphism classes of simple functors S such that $\mathrm{Ext}^1(S, F) \neq 0$.

AC6 The set $\mathcal{I}(F)$ is finite for i-small functors F.

The key role played by the simple artinian functors of type n is shown by the following result, which is readily established.

Proposition 1.4.2. *The artinian conjecture is true if and only if $\mathcal{I}(F)$ is finite for every simple artinian functor of type n, for all $n \geq 0$.*

The case $n = 0$ is implied by the following, more general, result of Schwartz:

Theorem 1.4.3. [S, S0] *Suppose that F is a finite functor, then F has an injective resolution in which each injective is a finite direct sum of indecomposable injectives.*

The case $n = 1$ seems to be within reach; it will clearly require the classification of the SAT1 functors (up to finite functors). A full understanding of these functors will be necessary before passing to the consideration of simple artinian functors of type 2.

2. Co-Weyl objects in the category $\mathcal{F}$

This section defines certain *co-Weyl* objects in the category $\mathcal{F}(\mathbb{F}_2)$. In order to do this, recall some basic structure which is used throughout this note; (henceforth the field is always taken to be $\mathbb{F}_2$, which will be written as $\mathbb{F}$). The injective object I_V is defined[4] by $I_V(W) := \mathbb{F}^{\hom(W,V^*)}$, the $\mathbb{F}$-vector space of set maps to $\mathbb{F}$ from the set of linear maps $\hom(W, V^*)$. This satisfies the relation $\mathrm{Hom}_{\mathcal{F}}(F, I_V) \cong DF(V)$, for a functor $F \in \mathcal{F}$, where D denotes the duality functor $D : \mathcal{F} \to \mathcal{F}^{\mathrm{op}}$, which is defined by $DF(V) = F(V^*)^*$, where $*$ indicates vector space duality.

The injective $I_{\mathbb{F}}$ decomposes as $I_{\mathbb{F}} \cong \mathbb{F} \oplus \overline{I}$, where $\overline{I}$ is the injective envelope of the functor Λ^1. The functor $\overline{I}$ is uniserial (has a unique composition series); there are subfunctors $p_n\overline{I} \hookrightarrow \overline{I}$ for $n \geq 0$ with $p_0\overline{I} = 0$ and (non-split) short exact sequences

$$p_{n-1}\overline{I} \to p_n\overline{I} \to \Lambda^n,$$

[4]Warning: the variance in V varies in the literature!

for $n \geq 1$. The notation is chosen since this filtration coincides with the polynomial filtration of $\overline{I}$; namely $p_n\overline{I}$ is the largest sub-functor of $\overline{I}$ which has polynomial degree $\leq n$. (For the basic notions on polynomial degree, see [K]).

The functor $\overline{I}$ gives the first example of a co-Weyl object; the term co-Weyl object is taken from the study of highest weight categories (see [M], for example) and should not be taken to indicate what are variously called dual Weyl functors or Schur functors, encountered in the representation theory of the finite general linear groups and symmetric groups.

To define the co-Weyl objects in general[5], recall [K2] that there is an evaluation functor $E_n : \mathcal{F} \to \mathcal{M}_n$, $F \mapsto F(\mathbb{F}^n)$, where $\mathcal{M}_n$ indicates the category of $\mathrm{End}(\mathbb{F}^n)$-modules. The functor E_n admits a right adjoint $R_n : \mathcal{M}_n \to \mathcal{F}$. If N is a $\mathrm{GL}_n(\mathbb{F})$-module, N may be regarded as an object of $\mathcal{M}_n$ via the surjective algebra map $\mathbb{F}[\mathrm{End}(\mathbb{F}^n)] \to \mathbb{F}[\mathrm{GL}_n(\mathbb{F})]$ sending singular morphisms to zero.

Definition 2.0.1 [P2] If S is a simple $\mathrm{GL}_n(\mathbb{F})$-module, let $J(S)$ denote the co-Weyl object associated to the pair (S, n), given by $J(S) := R_n S$. □

The duality functor D permits a (typographically) simpler description of $J(S)$: $DJ(S)(V) = \mathbb{F}[\hom(\mathbb{F}^n, V)] \otimes_{\mathrm{End}(\mathbb{F}^n)} S$. (This relies on the fact that the simple functors are self dual and that the projective functors take finite-dimensional values).

The functor $J(S)$ has simple socle in $\mathcal{F}$, which is also denoted by S, so that there are embeddings:

$$J(S) \hookrightarrow \mathrm{Inj.env.}_{\mathcal{F}}(S) \hookrightarrow I_{\mathbb{F}^n}.$$

The defining property of $J(S)$ states informally that $J(S)$ is the largest sub-functor of $I_{\mathbb{F}^n}$ such that any injection $S \hookrightarrow I_{\mathbb{F}^n}$ extends *uniquely* to a map $J(S) \to I_{\mathbb{F}^n}$.

The most fundamental example of a co-Weyl object for $\mathrm{GL}_n(\mathbb{F})$ is given by the trivial module $\mathbb{F}$. In this case, the co-Weyl object is written as $\overline{D}(n)$; the notation is derived from the fact that $\overline{D}(n)$ embeds inside the Dickson invariants of $I_{\mathbb{F}^n}$, namely the invariants under the natural $GL_n(\mathbb{F})$-action, obtained by regarding I_V as a covariant functor in V to $\mathcal{F}$. Write $\mathrm{Gr}_n(V)$ for the set of n-planes in the vector space V; the set $\mathrm{Gr}_n(V) \amalg *$ admits a natural action by $\mathrm{End}(V)$, regarding $\mathrm{Gr}_n(V) \amalg *$ as a quotient of the set $\hom(\mathbb{F}^n, V)$, with canonical $\mathrm{End}(V)$-action, by sending morphisms of rank less that n to the point $*$ and identifying maps conjugate under the right $GL_n(\mathbb{F})$-action. Then $\mathbb{F}[\mathrm{Gr}_n(V) \amalg *]$ is naturally an $\mathrm{End}(V)$-module, equipped with an inclusion of the trivial module $\mathbb{F} \cong \mathbb{F}[*]$. (Here, for a set X, $\mathbb{F}[X] := \bigoplus_{x \in X} \mathbb{F}$). Thus, define:

$$\overline{D}(n)(V) = D(\mathbb{F}[\mathrm{Gr}_n(V) \amalg *]/\mathbb{F}[*]),$$

[5]The co-Weyl objects arise very naturally in 'functor categories'; in particular they are used to construct simple objects in $\mathcal{F}$. A similar construction of simple objects is used in various categories of Mackey functors, eg [TW].

with the induced action as an End(V)-module. The functor $\overline{D}(1)$ identifies with the injective $\overline{I}$, whereas $\overline{D}(n)$ is not injective, for $n \geq 2$.

The importance of the co-Weyl objects is shown by:

Theorem 2.0.2. [P2] *The indecomposable injectives in $\mathcal{F}$ are J-good; namely admit a finite filtration with filtration quotients which are co-Weyl objects.*

For example, the injective envelope of Λ^2 in $\mathcal{F}$ may be represented as:

$$\begin{array}{ccc} \overline{D}(2) & & \overline{I} \\ \backslash & & / \\ & \overline{D}(2). & \end{array}$$

In particular, this shows that the category $\mathcal{F}_\omega$ of locally finite objects is not a highest weight category. There is, however, a homological characterization of J-good functors, inspired by results on the existence of good filtrations in the representation theory of algebraic groups [Jant].

Theorem 2.0.3. [P2, Theorem 5.0.1] *An i-small functor F is J-good if and only if it admits an injective resolution $0 \to F \to I^\bullet$ such that there exists an integer d so that each I^k embeds in a finite direct sum $\bigoplus I_{\mathbb{F}^d}$.*

This allows a simple proof of the following two assertions:

1. If F is a J-good functor, then ΔF is J-good.
2. If F, G are J-good functors, then $F \otimes G$ is J-good.

The introduction of co-Weyl objects permits the formulation of a strengthened version of the artinian conjecture:

The Strong Artinian Conjecture: A co-Weyl object induced from a simple $\mathrm{GL}_n(\mathbb{F})$-module is simple artinian of type n.

2.1. **A filtration of $\overline{D}(2)$.** The functor $\overline{D}(2)$ is an illustration of a 'large' non-finite functor. To describe its structure, use the following special case of Theorem 1.3.1, first proved by Piriou [Pi].

Theorem 2.1.1. [P1] *For $n \geq 0$ there exist functors $\overline{K}_n$ with $\overline{K}_0 = \overline{I}$ and, for $n > 0$, short exact sequences $\overline{K}_n \to \overline{I} \otimes \Lambda^n \to \overline{K}_{n-1}$. Moreover, every non-finite subfunctor of $\overline{I} \otimes \Lambda^n$ contains $\overline{K}_n$; in particular, the functor $\overline{K}_n$ is SAT1.*

Remark 2.1.2 This statement indicates the difficulty of considering the structure of a functor $G \otimes \Lambda^1$, even when G is SAT1; for example, consider the functor $\overline{K}_1 \otimes \Lambda^1 \subset \overline{I} \otimes \Lambda^1 \otimes \Lambda^1$. The functors $\Lambda^1 \otimes \Lambda^1$ has socle series $\Lambda^2, \Lambda^1, \Lambda^2$, hence one concludes readily that $\overline{K}_1 \otimes \Lambda^1$ has a filtration with sub-quotients $\overline{K}_2, \overline{K}_1, \overline{K}_2, \overline{K}_1$. □

There is a natural action of $\mathrm{GL}_2(\mathbb{F})$ on $I_{\mathbb{F}^2}$. The functor $I_{\mathbb{F}^2}$ decomposes as $(\overline{I} \otimes \overline{I}) \oplus \overline{I}^{\oplus 2} \oplus \mathbb{F}$ and the sum of all the $\mathrm{GL}_2(\mathbb{F})$-maps induces a 'trace' map

$\overline{I} \otimes \overline{I} \to \overline{D}(2)$, which is surjective. Hence, the filtration $p_n\overline{I} \otimes \overline{I} \subset \overline{I} \otimes \overline{I}$ induces a filtration of $\overline{D}(2)$ by:

$$f_n\overline{D}(2) := \text{Image}\{p_n\overline{I} \otimes \overline{I} \to \overline{D}(2)\}.$$

One may establish:

Proposition 2.1.3. [Ps] *For $n \geq 1$ there are isomorphisms $f_n\overline{D}(2)/f_{n-1}\overline{D}(2) \cong \overline{K}_{n-1}/\text{finite}$, for some (undetermined) finite functor.*

The structure of the functor $\overline{D}(2)$ may be represented as follows

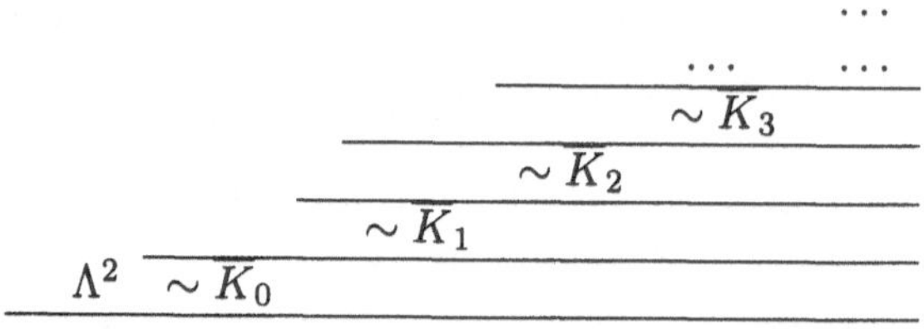

The layers represent the sub-quotients $\overline{K}_n$, with the $\sim$ implying the condition 'modulo a finite sub-functor', and the Λ^2 indicates the simple socle of $\overline{D}(2)$. The shift to the right of the successive sub-quotients indicates the increasing connectivity of the sub-quotients. (The connectivity of a functor F is defined to be the least integer c for which $F(\mathbb{F}^c)$ is non-zero).

This is not sufficient to establish that the functor $\overline{D}(2)$ is artinian but does show that the Krull codimension of $\overline{D}(2)$ is greater than one.

3. The functors $\widetilde{\nabla}_n$ and the simplicity theorem

The difference functor $\Delta : \mathcal{F} \to \mathcal{F}$ is a useful tool; in particular it is exact and $\Delta F = 0$ if and only if F is a constant functor. It is simple to show:

Lemma 3.0.1. *Suppose that there exists an integer $t \geq 0$ such that $F(\mathbb{F}^t)$ is finite-dimensional and $\Delta^{t+1}F$ is artinian, then F is artinian.*

Unfortunately, this is seldom useful, since Δ does not reduce the size of non-finite functors: for example, $\Delta(\overline{I} \otimes \Lambda^n)$ contains a direct summand of $\overline{I} \otimes \Lambda^n$. A solution is to use a suitable quotient of the functor Δ. The functor Δ is right adjoint to the functor $- \otimes D\overline{I}$, hence the following definition is natural:

Definition 3.0.2 [P]

1. Set $[p_0\Delta] := 0$ and, for $n > 0$, take $[p_n\Delta]$ to be the right adjoint to the functor $- \otimes Dp_n\overline{I}$.
2. For $n > 0$ define $\widetilde{\nabla}_n := \Delta/[p_{n-1}\Delta]$, where the natural transformation $[p_{n-1}\Delta] \to \Delta$ is induced by the inclusion $p_{n-1}\overline{I} \hookrightarrow \overline{I}$.

□

Thus, $\widetilde{\nabla}_n$ is a functor $\mathcal{F} \to \mathcal{F}$; to illustrate a motivating property, one may calculate that $\widetilde{\nabla}_2(\overline{I} \otimes \Lambda^k) = I_{\mathbf{F}} \otimes \Lambda^{k-1}$, for $k \geq 1$. The functors $\widetilde{\nabla}_n$ have good properties when applied to the category $\mathcal{F}_\omega$ and the action of $\widetilde{\nabla}_n$ upon the simple objects in $\mathcal{F}$ is illuminating. In order to state this, recall one parametrization of the simple objects below.

Kuhn [K] has shown that there is a recollement diagram which permits the filtration of the category $\mathcal{F}(\mathbb{F})$ so that the sub-quotients are module categories for the finite symmetric groups. Over the field $\mathbb{F}_2$, this filtration coincides with the polynomial filtration of the category $\mathcal{F}(\mathbb{F}_2)$[6]. Hence, the construction of the simple modules for the finite symmetric groups [J, JK] may be used to show that the simple functors of $\mathcal{F}$ are indexed by the 2-regular partitions, which are partially ordered by the dominance order. (Here, a 2-regular partition may be taken to be a strictly decreasing sequence of positive integers). In particular, the simple functor indexed by $\lambda_1 > \ldots > \lambda_s > 0$ is the highest weight (with respect to the dominance order) composition factor of $\Lambda^{\lambda_1} \otimes \ldots \otimes \Lambda^{\lambda_s}$.

1. $\widetilde{\nabla}_n$ preserves surjections and injections (it is not exact).
2. If $S_{(\lambda_1,\ldots,\lambda_s)}$ is the simple functor indexed by $\lambda_1 > \ldots > \lambda_s > 0$, then $\widetilde{\nabla}_n S_{(\lambda_1,\ldots,\lambda_s)}$ is zero if $n > s$ and isomorphic to $S_{(\lambda_1-1,\ldots,\lambda_s-1)}$ if $n = s$.

Notation 3.0.3 Say that the functor F is $\widetilde{\nabla}_n$-nilpotent if there exists t such that $(\widetilde{\nabla}_n)^t F = 0$. Write $\widetilde{\nabla}\mathrm{nil}(F)$ for the least n such that F is $\widetilde{\nabla}_n$-nilpotent. □

For example, if F takes finite-dimensional values, then $\widetilde{\nabla}\mathrm{nil}(F) = 1$ if and only if F is a finite functor. The invariant $\widetilde{\nabla}\mathrm{nil}(F)$ has the important property:

Theorem 3.0.4. [P, P4]

1. *If $F \to G \to H$ is a short exact sequence of locally finite functors, then* $\widetilde{\nabla}\mathrm{nil}(G) = \max\{\widetilde{\nabla}\mathrm{nil}(F), \widetilde{\nabla}\mathrm{nil}(H)\}$.
2. *If F, G are locally finite functors, then* $\widetilde{\nabla}\mathrm{nil}(F \otimes G) = \widetilde{\nabla}\mathrm{nil}(F) + \widetilde{\nabla}\mathrm{nil}(G) - 1$.

One may thus define a dimension theory on the category $\mathcal{F}_\omega$, associated to the operators $\widetilde{\nabla}_\bullet$ by setting:

$$\dim_{\widetilde{\nabla}} F = \widetilde{\nabla}\mathrm{nil} F - 1.$$

The strongest form of the artinian conjecture which is proposed is:

The Very Strong Artinian Conjecture: If F is a locally finite functor with finite socle, then Artinian type$(F) = \dim_{\widetilde{\nabla}} F$. □

Some justification for this conjecture is shown by the *simplicity theorem*:

[6] Results of this form first arose in the work of Pirashvili on *higher additivisations*. The original work is not readily available; see [B, BP].

Theorem 3.0.5. [P2] *Suppose that J is a co-Weyl object induced from a simple $\mathrm{GL}_n(\mathbb{F})$-module, then:*

1. $J \hookrightarrow \widetilde{\nabla}_n J$*; in particular, J is not $\widetilde{\nabla}_n$-nilpotent.*
2. *If $F \hookrightarrow J$ is a proper sub-functor, then there exists t such that $(\widetilde{\nabla}_n)^t F = 0$.*

This result is extremely powerful; an example of a weak consequence is:

Corollary 3.0.6. *Suppose that $F \subset J$ is a sub-functor, then J/F is indecomposable.*

Proof: Suppose that J is induced from a $\mathrm{GL}_n(\mathbb{F})$-module and that J/F admits a non-trivial direct sum decomposition $J/F \cong G \oplus H$ and write $\hat{G}, \hat{H}$ for the pull-backs of G, H respectively along $J \to J/F$. Thus $\hat{G}, \hat{H}$ are proper sub-functors of J and are hence $\widetilde{\nabla}_n$-nilpotent. In particular, the functor H is $\widetilde{\nabla}_n$-nilpotent.

There is a short exact sequence $\hat{G} \to J \to H$, which shows that J is $\widetilde{\nabla}_n$-nilpotent, since the category of $\widetilde{\nabla}_n$-nilpotent functors is thick. This is a contradiction. □

Question 3.0.7 If F is a locally finite functor, then the associated unstable module over the Steenrod algebra [HLS, K1] may be regarded as a non-negatively graded vector space, by forgetting the action of the Steenrod algebra. Does the 'averaged growth' of this graded vector space determine $\dim_{\widetilde{\nabla}} F$, when F has a finite socle?

4. The 'dimension' approach to the artinian conjecture

The standard approach to the artinian conjecture, as mentioned above, is an induction upon the dimension of V for I_V. The simplicity theorem, Theorem 3.0.5, reduces one to the study of certain explicit 'small' sub-functors of the co-Weyl objects. If J is induced from a simple $\mathrm{GL}_n(\mathbb{F})$-module, then define:

$$g_k J := \ker\{J \to (\widetilde{\nabla}_n)^k J \otimes \overline{I}^{\otimes k}\},$$

where the map is the adjoint to the natural surjection $\Delta^k J \to (\widetilde{\nabla}_n)^k J$. For formal reasons, $g_k J$ is the largest sub-functor of J which is zero under $(\widetilde{\nabla}_n)^k$. It suffices to establish that the functors $g_k J$ are artinian, for all $k \geq 1$.

A possible strategy is the following, based upon an induction upon n:

1. Prove that $g_1 J$ is artinian; intuitively, the structure of $g_1 J$ should be closely linked to the structure of $I_{\mathbb{F}^{n-1}}$.
2. Devise an induction on k, based on the functor $\widetilde{\nabla}_n$, to show that $g_k J$ is artinian.

Unfortunately, the functor $\widetilde{\nabla}_n$ loses so much information that it seems implausible that this naive approach will work. To illustrate the non-triviality of the first step, consider the following example:

Example 4.0.1 For all $n \geq 1$, there is a map $\overline{D}(n) \to \overline{D}(n+1)$; this necessarily factors through $g_1\overline{D}(n+1)$. In the case $n = 1$, the image of $\overline{D}(1)$ in $\overline{D}(2)$ identifies with $g_1\overline{D}(2)$. Moreover, the functor $\Delta g_1\overline{D}(2)$ embeds in $I_{\mathbf{F}}$.

In the case $n = 2$, it is not known whether $g_1\overline{D}(3)$ identifies with the image of $\overline{D}(2)$; hence it is not clear whether $g_1\overline{D}(3)$ is artinian. One naturally asks whether $\Delta g_1\overline{D}(3)$ embeds in $I_{\mathbf{F}^2}$, so as to generalize the second property given above. This does *not* happen, since such an embedding would have to factor through $\overline{D}(2)$, for elementary reasons. Since $\overline{D}(2)$ 'almost' embeds in $g_1\overline{D}(3)$ and $\Delta\overline{D}(2)$ is considerably larger than $\overline{D}(2)$, this is easily shown to be impossible.

4.1. **Remarks on the proof that $\overline{D}(2)$ is artinian.** The difficulties of studying the artinian conjecture are shown by the study of $\overline{D}(2)$. Two natural filtrations have been defined on $\overline{D}(2)$, namely $f_*\overline{D}(2)$ induced from the filtration $(p_*\overline{I}) \otimes \overline{I}$ of $\overline{I} \otimes \overline{I}$ and the $\widetilde{\nabla}_2$-nilpotency filtration $g_*\overline{D}(2)$. For each n, there are inclusions $f_n\overline{D}(2) \hookrightarrow g_n\overline{D}(2)$ and the functor $f_n\overline{D}(2)$ is artinian. The simplicity theorem implies that it is sufficient to establish the strong property:

- For every n, there exists $N(n)$ and an inclusion $g_n\overline{D}(2) \hookrightarrow f_{N(n)}\overline{D}(2)$.

The proof presented in [Ps] is an induction upon n, using the functors $\widetilde{\nabla}_2$; this is possible since the functors G for which $\widetilde{\nabla}_2 G = 0$ are controlled by:

Proposition 4.1.1. [Ps] *Suppose that G is a SAT1 functor such that $\widetilde{\nabla}_2 G = 0$, then G is isomorphic to a quotient of $\overline{I}$.*

There is a non-trivial map $\overline{I} \to \overline{D}(2)$, with image isomorphic to $\overline{I}/\Lambda^1$; the key ingredient to the proof is the assertion that this is essentially the only quotient of $\overline{I}$ which occurs as a sub-quotient of $\overline{D}(2)$. This requires the use of the following result due to V. Franjou:

Theorem 4.1.2. [Ps, Theorem 6.2] *Suppose that F is a finite functor, then* $\mathrm{Ext}^*_{\mathcal{F}}(\overline{I}, F) = 0$.

This result is of interest in its own right; it would be interesting to obtain a proof that $\overline{D}(2)$ is artinian which does not appeal to this result.

Theorem 4.1.2 has the following generalization:

Proposition 4.1.3. *Suppose that G is a finite functor and $s \geq 0$, then* $\mathrm{Ext}^*_{\mathcal{F}}(\overline{K}_s, G) = 0$.

Proof: An induction upon s, using the short exact sequences $\overline{K}_s \to \overline{I} \otimes \Lambda^s \to \overline{K}_{s-1}$ reduces one to showing that $\mathrm{Ext}^*_{\mathcal{F}}(\overline{I} \otimes \Lambda^s, G) = 0$. A standard argument using the embedding theorem of Kuhn [K1] and dimension shifting reduces to the case $G = S^n$, the nth symmetric power functor. These extension groups may be expressed as direct sums of $\mathrm{Ext}^*_{\mathcal{F}}(\overline{I}, S^a) \otimes \mathrm{Ext}^*_{\mathcal{F}}(\Lambda^s, S^b) = 0$, with $a + b = n$, using the exponential property of S^n and the Künneth formula. These groups are zero, by Theorem 4.1.2. □

The results of [Ps] together with Proposition 4.1.3 establish the following result:

Corollary 4.1.4. *Suppose that F is a proper sub-functor of $\overline{D}(2)$, then F admits a finite filtration $0 = F_k \subset F_{k-1} \subset \ldots \subset F_0 \subset F$ such that*

1. *F/F_0 is a finite functor.*
2. *For $0 \leq n \leq k-1$, F_n/F_{n+1} is SAT1 and isomorphic to $\overline{K}_{s(n)}/\text{finite}$, for some $s(n) \geq 0$ and some finite functor.*
3. *The numbers $s(n)$ are pairwise distinct.*

In particular, if $F \subset \overline{D}(2)$ is SAT1, then F is isomorphic modulo to $\overline{K}_s/\text{finite}$, for some s. There is a non-trivial map $\overline{I} \otimes \Lambda^2 \to \overline{D}(2)$ induced by the unique non-trivial map from $\overline{I} \otimes \Lambda^2$ to the injective envelope of Λ^2. This factors through a non-trivial map $\overline{K}_1 \to \overline{D}(2)$.

Proposition 4.1.5. *Suppose that $F \subset \overline{D}(2)$ is SAT1, then either $F \cong \overline{I}/\Lambda^1$ or $F \cong \text{Image}\{\overline{K}_1 \to \overline{D}(2)\}$ for the unique such non-trivial map.*

Proof: It suffices to consider $\text{Hom}_{\mathcal{F}}(\overline{K}_s, \overline{D}(2))$. This group is zero unless $s \in \{0,1\}$, since $\overline{K}_s(\mathbb{F}^2) = 0$ for $s > 1$. For $s = 0, 1$, there are unique non-trivial maps in these groups. □

An inductive argument utilising the functor $\widetilde{\nabla}_2$ gives an indication of the structure of the functor $\overline{D}(2)$. One should note that this result proves:

Corollary 4.1.6. *The functor $\overline{D}(2)$ is not uniserial in the category $\mathcal{F}_\omega/\text{finite}$.*

5. The functors $\overline{I} \otimes S$

This final section sketches the proof of Theorem 1.3.1, as an illustration of the techniques used, since a study of the functors of the form $\overline{I} \otimes F$, with F finite, underlines some of the limitations of arguments derived from the functors $\widetilde{\nabla}_\bullet$. It seems plausible that one should be able to prove that such a functor is artinian by induction on the polynomial degree of F, using the functors $\widetilde{\nabla}_2$. However, this is not sufficient; the proof presented by the author requires a more explicit argument.

Suppose that S is a simple functor; the functor $\overline{I} \otimes S$ has a filtration $p_{n-1}\overline{I} \otimes S \subset p_n\overline{I} \otimes S \subset \ldots$, which has filtration quotients $\Lambda^n \otimes S$, for $n \geq 1$. Suppose that $n \gg 0$; the functor $\Lambda^n \otimes S$ has a unique highest weight composition factor; moreover, there is a single composition factor of this simple functor in $\overline{I} \otimes S$. Hence, if $P_{(n,S)}$ denotes the projective cover of this simple functor, there is a unique non-trivial map $P_{(n,S)} \to \overline{I} \otimes S$; write $X_{(n,S)}$ for the image of this map. The subfunctor X_S is then defined as $X_S = \bigcup_n X_{(n,S)}$.

The key to showing that X_S is SAT1 is a detection argument with respect to a suitable functor $\widetilde{\nabla}_t$. If F is a functor, then the surjection $\Delta F \to \widetilde{\nabla}_n F$ has adjoint $F \to \widetilde{\nabla}_n F \otimes \overline{I}$. The detection result is the following:

Theorem 5.0.1. [P3] *Suppose that F is a finite functor and t is an integer such that the adjunction map $F \to \widetilde{\nabla}_{t-1}F \otimes \overline{I}$ is an inclusion. If $G \hookrightarrow \overline{I} \otimes F$ is a sub-functor such that $\widetilde{\nabla}_t G$ is finite, then G is finite.*

The proof that X_S is SAT1 then proceeds by an induction on the polynomial degree of S; this works because of the explicit description of the functor X_S and establishes the stronger result that every non-finite sub-object of $\overline{I} \otimes S$ contains X_S. This is not sufficient to prove that $\overline{I} \otimes S$ is artinian; for this one uses the following general result:

Proposition 5.0.2. [P3] *Suppose that Y is a functor taking finite-dimensional values such that there is a surjection $\Delta Y \to Y$. This surjection induces a surjection $\Delta^t Y \to Y$ for all $t \geq 0$. Suppose that there exists a sub-functor $X \hookrightarrow Y$ such that*

1. *X is artinian,*
2. *the kernel of $\Delta^t Y \to Y$ is artinian, for all t,*
3. *there exists an integer t such that the composite $\Delta^t X \to \Delta^t Y \to Y$ is surjective,*

then Y is artinian.

This is applied to the functor $Y = \overline{I} \otimes S$, taking $X = X_S$. An induction on polynomial degree and the argument sketched above show that the first two hypotheses hold. The third hypothesis is established by direct calculation, again relying on the explicit definition of X_S.

Acknowledgement: *This paper is a revised version of the author's talk at the Bielefeld Euroconference on Infinite Length Modules. The author's participation was supported by the European network 'Invariants and Representations of Algebras'. The author would like to thank the organizers for their invitation to present these results and participate at this conference.*

References

[B] S. BETLEY, Homology of $GL(R)$ with coefficients in a functor of finite degree, *J. Alg.* **150** (1992), 73-86.

[BP] S. BETLEY and T. PIRASHVILI, Twisted (co)homological stability for monoids of endomorphisms, *Math. Ann.* **295** (1995), 709-720.

[FFSS] V. FRANJOU, E.M. FRIEDLANDER, A. SCORICHENKO and A. SUSLIN, General linear and functor cohomology over finite fields, *Preprint* (1997).

[FLS] V. FRANJOU, J. LANNES and L. SCHWARTZ, Autour de la cohomologie de MacLane des corps finis, *Invent. Math.* **115** (1994), 513-538.

[FS] E.M. FRIEDLANDER and A. SUSLIN, Cohomology of finite group schemes over a field, *Invent. Math.* **127** (1997), 209-270.

[G] J.A. GREEN, **Polyonomial Representations of GL_n**, Lecture Notes in Math. **830**, Springer-Verlag (1980).

[HLS] H.-W. HENN, J. LANNES and L. SCHWARTZ, The categories of unstable modules and unstable algebras over the Steenrod algebra modulo nilpotent objects, *Amer. J. Math.* **115** (1993), 1053-1106.

[J] G.D. JAMES, **The Representation Theory of the Symmetric Groups**, *Lect. Notes in Math.*, **682**, Springer-Verlag, (1978).

[JK] G.D. JAMES and A. KERBER, **The Representation Theory of the Symmetric Groups**, *Ency. Math. Appl., Addison-Wesley*, Vol. **16**, 1981.

[Jant] J.C. JANTZEN, **Representations of Algebraic Groups**, *Academic Press*, Boston, 1987.

[JP] M. JIBLADZE and T. PIRASHVILI, Cohomology of algebraic theories, *J. Alg.* **137** (1991), 253-296.

[K] N.J. KUHN, The generic representation theory of finite fields: a survey of basic structure, *this volume.*

[K1] N.J. KUHN, Generic representations of the finite general linear groups and the Steenrod algebra: I, *Amer. J. Math.* **116** (1993), 327-360.

[K2] N.J. KUHN, Generic representations of the finite general linear groups and the Steenrod algebra: II, *K-Theory* **8** (1994), 395-426.

[K3] N.J. KUHN, Generic representations of the finite general linear groups and the Steenrod algebra: III, *K-Theory* **9** (1995), 273-303.

[Le] T.H. LENAGAN, Dimension theory of Noetherian rings, *this volume.*

[MacD] I.G. MacDONALD, **Symmetric Functions and Hall Polynomials**, *Oxford Math. Monographs*, Oxford University Press.

[M] S. MARTIN, **Schur Algebras and Representation Theory**, *Cambridge Tracts in Math.*, Cambridge University Press, (1993).

[Pi] L. PIRIOU, Sous-objets de $\overline{I} \otimes \Lambda^n$ dans la catgorie des foncteurs entre $\mathbb{F}_2$-espaces vectoriels, *J. Algebra* **194** (1997), 53–78.

[Ps] G.M.L. POWELL *with an Appendix by* L. SCHWARTZ, The Artinian conjecture for $I \otimes I$, *J. Pure Appl. Alg.*, **128** (1998), 291-310.

[P] G.M.L. POWELL, Polynomial filtrations and Lannes' T-functor, *K-Theory* **13**, (1998), 279-304.

[P1] G.M.L. POWELL, The structure of $\overline{I} \otimes \Lambda^n$ in generic representation theory, *J. Alg.* **194**, (1997), 455-466.

[P2] G.M.L. POWELL, The structure of the indecomposable injectives in generic representation theory, *Trans. AMS.* **350** (1998), 4167–4193.

[P3] G.M.L. POWELL, On the structure of $\overline{I} \otimes F$ in the category of functors between $\mathbb{F}$-vector spaces, *Preprint* (1998).

[P4] G.M.L. POWELL, The tensor product theorem for $\widetilde{\nabla}$-nilpotence and the dimension of unstable modules, *Preprint* (1998).

[S0] L. SCHWARTZ, Unstable modules over the Steenrod algebra, functors and the cohomology of spaces, *this volume.*

[S] L. SCHWARTZ, **Unstable Modules over the Steenrod Algebra and Sullivan's Fixed Point Set Conjecture**, *Chicago Lecture Notes in Mathematics, Univ. Chicago Press, Chicago and London* (1994).

[TW] J. THEVENAZ and P. WEBB, The structure of Mackey functors. *Trans. Am. Math. Soc.* **347** (1995), 1865-1961.

[Z] A.E. ZALESSKI, Modular groups rings of the finitary symmetric group, *Israel J. Math.* **96** (1996), 609-621.

LAGA, UMR 7539, INSTITUT GALILÉE, UNIVERSITÉ PARIS 13, VILLETANEUSE, FRANCE
E-mail address: powell@math.univ-paris13.fr

Trends in Mathematics, © 2000 Birkhäuser Verlag Basel/Switzerland

UNSTABLE MODULES OVER THE STEENROD ALGEBRA, FUNCTORS, AND THE COHOMOLOGY OF SPACES.

LIONEL SCHWARTZ

ABSTRACT. This report presents the connections discovered in the 80's and the 90's between homotopy theory, specifically unstable modules over the Steenrod algebra and homotopy classes of maps from classifying spaces, and certain categories of functors. These categories of functors are deeply linked to modular representation theory of symmetric or general linear groups.

In section 1 to section 4 we will present the topological background concerning modules over the Steenrod algebra. In section 5 to section 9 we will define and study the categories of functors in question and explain relations and applications. Here is the summary :

- **1. Unstable modules and algebras**
- **2. Free objects in the category $\mathcal{U}$**
- **3. Injective objects and representability**
- **4. Injectivity of the mod 2 cohomology of elementary abelian group and Lannes' functor $\mathbf{T}_V$**
- **5. $\mathcal{U}/\mathcal{N}il$ and analytic functors**
- **6. The functor $p_n : \mathcal{U}/\mathcal{N}il \rightarrow \mathcal{M}od_{\mathbf{F}_2[S_n]}$, the filtration on $\mathcal{U}/\mathcal{N}il$, and simple objects**
- **7. The decomposition of the Carlsson modules $K(i)$**
- **8. The Krull filtration on $\mathcal{U}$**
- **9. The Grothendieck ring of $\mathcal{U}$**

1. UNSTABLE MODULES AND ALGEBRAS

The main problem of homotopy theory is to classify maps from a space X to a space Y up to homotopy; let $[X, Y]$ be the set of homotopy classes. This cannot be solved in general, except for example when Y is an Eilenberg-Mac Lane space. In this last case the set of maps is a cohomology group of X. This case is very special and the result is intimately linked to the structure of Eilenberg-Mac Lane spaces, which are constructed to satisfy this property.

A very important case has been studied in the 80's. This is the case where $X = BV$, V being a finite dimensional vector space over the field with two elements. The result, conjectured by H. Miller and proved by J. Lannes, is a landmark in homotopy theory and has had numerous applications. We are concerned with, and will describe, the algebraic structures on the category of unstable modules over the Steenrod algebra that were discovered and used in this work.

Let us denote by H^*X the singular mod-2 cohomology of the space X. If X is the classifying space of a group G, or, equivalently, an Eilenberg-MacLane space

$K(G,1)$, this cohomology is the cohomology of the group G with coefficients in the finite field $\mathbf{F}_2$ considered as a trivial module. In fact, after Kan and Thurston, one knows that, for any space X, there exists a group $KT(X)$ having the same cohomology as X. Thus all statements we are going to consider could be stated either in terms of cohomology of groups or in terms of cohomology of spaces.

The mod-2 cohomology of a space is a commutative $\mathbf{N}$-graded $\mathbf{F}_2$-algebra. The fundamental result of Steenrod and Adem is that it is a natural graded module over a certain algebra $\mathcal{A}$ called the Steenrod algebra.

The Steenrod algebra $\mathcal{A}$ is the quotient of the free associative unital $\mathbf{N}$-graded $\mathbf{F}_2$-algebra generated by elements, of degree i, denoted Sq^i, $i > 0$, by the ideal generated by certain relations, known as the Adem relations. They are as follows [St] :

$$\mathrm{Sq}^i\mathrm{Sq}^j = \sum_{0}^{[i/2]} C_{j-k-1}^{i-2k}\mathrm{Sq}^{i+j-k}\mathrm{Sq}^k$$

for all $i, j > 0$ such that $i < 2j$.

Here C_b^a stands for the binomial coefficient $\frac{b!}{(b-a)!a!}$.

In these formulas Sq^0 is understood to be the unit.

The operation Sq^1 is the classical Bockstein homomorphism associated to the short exact sequence :

$$0 \to \mathbf{Z}/2 \to \mathbf{Z}/4 \to \mathbf{Z}/2 \to 0$$

The $\mathcal{A}$-module structure of the cohomology is natural in the sense that for any continuous map $f : X \to Y$ the induced map $f^* : H^*Y \to H^*X$ is $\mathcal{A}$-linear.

N. E. Steenrod constructed the operations on the cohomology of spaces, then J. Adem showed that these operations satisfy the aboved mentioned relations. The next theorem, which is a consequence of the work by H. Cartan and J.-P. Serre on the cohomology of the Eilenberg-Mac Lane spaces, shows that the Steenrod algebra is an intrisic object.

Theorem 1.1. *The Steenrod algebra is the algebra of all natural stable transformations of mod 2 cohomology.*

Recall that a cohomology operation τ of degree $d \geq 0$ is a family of natural transformations τ_i, $i \geq 0$ from the functor H^i (i-th cohomology group) to the functor H^{i+d} ($(i+d)$-th cohomology group). A cohomology operation is said to be stable if it commutes with the suspension isomorphism $\sigma_i : \bar{H}^iX \to \bar{H}^{i+1}\Sigma X$, i.e. if $\sigma_i\tau_i = \tau_{i+1}\sigma_i$. Here ΣX denotes the reduced suspension of the pointed space X. Stable cohomology operations are necessarily additive.

The Steenrod algebra has been a major tool in homotopy theory since its discovery. The aim of this report is to explain why it has lead to connexions with the work on and around the Sullivan conjecture to the consideration of a certain category of functors. It is worth adding that these categories of functors occur in different contexts in algebraic topology and, in particular, in algebraic K-theory (see the paper of N. Kuhn in the same volume).

The mod 2 cohomology of a space X has, as a module over the Steenrod algebra, a particular property called instability : if $x \in H^*X$ and $i > |x|$, then

$$\mathrm{Sq}^i x = 0 \quad .$$

Here $|x|$ denotes the degree of x.

An $\mathcal{A}$-module M is unstable if it satisfies the preceding property. The category of $\mathbf{N}$-graded modules over the Steenrod algebra with morphisms $\mathcal{A}$-linear maps of degree zero will be denoted by $\mathcal{M}$. The full subcategory whose objects are unstable modules will be denoted by $\mathcal{U}$. This is an abelian category, the quotient of an unstable module is unstable, a sub-module of an unstable module is unstable. However it is not true that if one has a short exact sequence :

$$0 \to M' \to M \to M'' \to 0$$

with M' and M'' both unstable that M is unstable.

The mod-2 cohomology of a space has some more structure. It is an unstable $\mathcal{A}$-algebra : the cup product determines the structure of a $\mathbf{N}$-graded $\mathbf{F}_2$-algebra in a natural way on the mod-2 cohomology of a space X. This algebra structure is related to the unstable module structure by two properties.

The first one is known as the Cartan formula :

$$\mathrm{Sq}^i(xy) = \sum_{k+\ell=i} \mathrm{Sq}^k x \, \mathrm{Sq}^\ell y \quad \text{for all } x \quad \text{and any } y \in H^*X \quad .$$

The second one relates the cup square to the action :

$$\mathrm{Sq}^{|x|} x = x^2 \quad \text{for} \quad \text{any} \quad x \in H^*X \quad .$$

An unstable $\mathcal{A}$-algebra K is an unstable module provided with maps $\mu : K \otimes K \to K$ and $\eta : \mathbf{F}_2 \to K$ which determine a commutative, unital, $\mathbf{F}_2$-algebra structure on K and such that the two properties stated above hold.

We will denote by $\mathcal{K}$ the category of unstable $\mathcal{A}$-algebras.

The most fundamental example is the mod 2 cohomology of the space $B\mathbf{Z}/2$, $H^*B\mathbf{Z}/2$ is the polynomial algebra $\mathbf{F}_2[u]$ on one generator u of degree 1. The action of $\mathcal{A}$ on $\mathbf{F}_2[u]$ is completely determined by the fact that it is an unstable algebra; one proves easily by induction on n that

$$\mathrm{Sq}^i u^n = C_n^i u^{n+i} \quad .$$

The cohomology of this space is not finitely generated as an $\mathcal{A}$-module. In fact the classes 1, and u^{2^h-1} form a minimal set of generators.

It is worth noting the following easy consequence : $\mathrm{Sq}^i u^{2^h} \neq 0$ if and only if $i = 0$ or $i = 2^h$. This is nothing other than the description of the unstable module $F(1)$ which follows.

Consider a somewhat more general case. Let V be a finite dimensional $\mathbf{F}_2$-vector space. The mod 2 cohomology of BV will be denoted by H^*V, the reduced mod 2 cohomology of BV will be denoted by $\bar{H}^*V$. As a functor of V the cohomology H^*V identifies with the graded symmetric algebra on the dual vector space V^*, $S^*(V^*)$, the elements of V^* having degree 1. If we denote by d the dimension of the vector space V and by $\{x_1, \dots, x_d\}$ a basis of the dual V^* it identifies with the polynomial algebra $\mathbf{F}_2[x_1, \dots, x_d]$. Although these algebras are very simple from the point of view of products, their structure is very rich as modules over the Steenrod algebra, and even more taking into account the unstable algebra structure.

A map from BV to X induces an unstable algebra map f^* from H^*X to H^*BV, this map respects the products and the module structure. One has :

$$[BV, X] \to \mathrm{Hom}_{\mathcal{K}}(H^*X, H^*BV), \ f \mapsto f^* \quad .$$

The conjecture of Miller proved by Lannes is that this map is a bijection under nice assumptions on X.

Let us now observe that H^*V is a polynomial algebra. Thus the radical $\sqrt{\{0\}}$ of H^*X is sent to zero. It is possible to prove that the radical is stable under the action of the Steenrod algebra. Thus one gets :

$$\mathrm{Hom}_{\mathcal{K}}(H^*X, H^*BV) \cong \mathrm{Hom}_{\mathcal{K}}(H^*X/\sqrt{\{0\}}, H^*BV) \quad .$$

In other words the set of homotopy classes of maps from BV to X depends only on $H^*X/\sqrt{\{0\}}$. Let us define a *nilpotent* module M to be one such that the operation $\mathrm{Sq}_0 \ : \ x \mapsto \mathrm{Sq}^{|x|}x$ is locally nilpotent, *i.e.* for any $x \in M$ there exists c_x such that $\mathrm{Sq}_0^{c_x} x = 0$. If M is a submodule of an unstable algebra because of the cup-square formula it is equivalent to the usual notion : all elements in M are nilpotent. Recall also the suspension ΣM of an unstable module M; it is defined by $(\Sigma M)^{n+1} \cong M^n$, and $\mathrm{Sq}^i(\Sigma x) = \Sigma(\mathrm{Sq}^i x)$. The instability condition forces the map Sq_0 to be trivial on a suspension.

The full subcategory of $\mathcal{U}$ whose objects are nilpotent modules is denoted by $\mathcal{N}il$. It is the smallest Serre class stable under colimits and containing suspensions.

A module is *reduced* if the largest nilpotent submodule contained in M is trivial, equivalently if the largest suspension contained in M is trivial. The cohomology H^*BV is reduced.

2. Free objects in the category $\mathcal{U}$

The category $\mathcal{U}$ has enough projective objects. This is implied by :

Proposition 2.1 (St). *There is, up to isomorphism, a unique unstable module $F(n)$ with a class ι_n of degree n such that the natural transformation :*

$$\mathrm{Hom}_{\mathcal{U}}(F(n), M) \to M^n, \quad f \mapsto f(\iota_n)$$

is an equivalence of functors.

As the functor $M \mapsto M^n$ from $\mathcal{U}$ to the category F of $\mathbf{F}_2$-vector spaces is right exact, the unstable module $F(n)$ is projective. The unstable module $F(n)$ is called the free unstable module on one generator of degree n.

As said above, the unstable module $F(1)$ can be identified with the sub-module of $H^*\mathbf{Z}/2$ generated by the class u. The set $\{u, u^2, u^4, \dots\}$ is a basis over $\mathbf{F}_2$ of $F(1)$.

If M and N are unstable modules, using the Cartan formula, one defines on an unstable module structure on $M \otimes N$, by :

$$\mathrm{Sq}^k(x \otimes y) = \sum_{i+j=k} \mathrm{Sq}^i x \otimes \mathrm{Sq}^j y$$

The unstable module $F(1)^{\otimes n}$ is of dimension 1 in degree n. Therefore, there is just one non trivial map $\omega_n : F(n) \to F(1)^{\otimes n}$. The symmetric group $\mathcal{S}_n$ acts on $F(1)^{\otimes n}$ by permutation of the coordinates. The non trivial class of degree n of $F(1)^{\otimes n}$ is S_n-invariant. Therefore, the map ω_n takes values in the invariants under the action of S_n, which are denoted $(F(1)^{\otimes n})^{\mathcal{S}_n}$.

Here is a description of $F(n)$:

Theorem 2.2 (LZ). *The map ω_n is an isomorphism from $F(n)$ onto $(F(1)^{\otimes n})^{\mathcal{S}_n}$.*

3. Injective objects and representability

We describe certain injective objects in the category $\mathcal{U}$. Recall that $\{C_\alpha\}$, $\alpha \in A$, is a set of cogenerators for a category $\mathcal{C}$ if any object of the category embeds in a product of C_α's. Any product of injective objects is again injective. As the category $\mathcal{U}$ is locally noetherian and has exact filtered direct limits one has :

Proposition 3.1. *Any filtered direct limit, in particular any direct sum, of injective objects in $\mathcal{U}$ is injective.*

See [Ga] for example.

An $\mathcal{A}$-module which is finite dimensional as an $\mathbf{F}_2$-vector space in every degree will be said to be of *finite type.*

In order to construct injective objects in $\mathcal{U}$ we construct contravariant functors from $\mathcal{U}$ to the category $\mathcal{E}$ of $\mathbf{F}_2$-vector spaces. The functors that we will consider will be representable and left exact. Thus the representing modules will be injective. The following classical lemma tells us when a contravariant functor $R : \mathcal{U} \to \mathcal{E}$ is representable.

Lemma 3.2. *The functor R is representable if and only if it is right exact and transforms direct sums into products.*

That the conditions are necessary is obvious. We discuss briefly the other implication. Suppose that the problem is solved and that the functor is represented by an unstable module $B(R)$. Thus we have a natural equivalence

$$\gamma : R \to \mathrm{Hom}_{\mathcal{U}}(-, B(R))$$

In particular we have

$$R(F(n)) \cong \mathrm{Hom}_{\mathcal{U}}(F(n)), B(R)) = B(R)^n \quad ,$$

which defines $B(R)$ as a graded vector space. An unstable module structure is defined on this graded vector space as follows. The operation $\theta \in \mathcal{A}$ acts as $R(u_\theta)$, where $u_\theta : F(n+|\theta|) \to F(n)$ is the $\mathcal{A}$-linear map associated to $\theta\iota_n \in F(n)^{n+|\theta|}$.

One defines the natural transformation as follows. Let M be in $\mathcal{U}$ and x be in $R(M)$. An element $y \in M$ can be identified with a map $\hat{y} : F(|y|) \to M$. We set $\gamma_M(x)$ to be the ($\mathcal{A}$-linear) map which sends $y \in M$ to the element $R(\hat{y})(x) \in R(F(|y|) = B(R)^{|y|}$. By the very construction $\gamma_{F(n)}$ is an isomorphism. To show that γ_M is an isomorphism, M being given in $\mathcal{U}$, one considers the beginning of a projective resolution of M :

$$\to P_1 \to P_0 \to M \to 0 \quad ,$$

where P_0 and P_1 are direct sums of $F(n)$'s and one considers an appropriate diagram using the properties of R.

The functor

$$J_i : M \mapsto (M^i)^* \quad ,$$

is an example of such a functor. However the representing module, called a Brown-Gitler module is not of interest for us. It is not reduced, in fact it is finite and trivial in degree 0 (if $i > 0$). The example that follows yields, by construction, reduced injective modules.

Consider the direct limit of the vector spaces $M^{2^q i}$ along the maps $\mathrm{Sq}^{2^q i}$:

$$\ldots \longrightarrow M^{2^q i} \xrightarrow{\mathrm{Sq}^{2^q i}} M^{2^{q+1} i} \longrightarrow \ldots$$

then take the dual vector space, and define a functor K_i by :

$$K_i : M \mapsto (\mathrm{dirlim}_q \, \{M^{2^q i}, \mathrm{Sq}^{2^q i}\})^* \quad .$$

It is obviously right exact and transforms direct sums into products. Therefore it is representable. As it is also left exact, the representing module is injective. The Carlsson module $K(i)$ is the representing module for the functor K_i.

The unstable module $K(i)$ can be interpreted as the inverse limit of Brown-Gitler modules. One can show that the unstable modules $K(i)$ are of finite type.

Proposition 3.3. *The unstable modules $K(i)$ are reduced.*

Proof : It is enough to show that there are no non-trivial nilpotent modules in $K(i)$. Let M be a nilpotent module contained in $K(i)$. By definition, the above direct limit is trivial for a nilpotent module. Hence $\mathrm{Hom}_{\mathcal{U}}(M, K(i))$ is trivial. Therefore M is trivial and $K(i)$ is reduced.

These modules are injective and are cogenerators for reduced modules. It means that any reduced module embeds in a product of $K(i)$'s. Indeed, for any reduced module M the canonical map :

$$M \mapsto \prod_i \prod_{\alpha \in K_i(M)} K(i)$$

is injective.

Because of the isomorphism $K(i) \cong K(2i)$ one can define the module $K(i)$ for all rationals in $\mathbf{N}[1/2]$ by the following formulas

$$K(i) = K(2^q i) \quad \text{for} \quad 2^q i \in \mathbf{N} \quad .$$

Define now a $\mathbf{N} \times \mathbf{N}[1/2]$-bigraded object K_*^* by $K_i^j = K(i)^j$. This bigraded object K_*^* has an unstable module structure which preserves the lower degree. It has also a bigraded algebra structure, this structure is induced by maps from tensor products of Brown-Gitler modules to Brown-Gitler modules directly deduced from their definition.

The following theorem describes the structure of K_*^* :

Theorem 3.4 (Ca). , *[LZ] (Carlsson, Lannes, Zarati) The Carlsson algebra K_*^* is isomorphic to the polynomial algebra*

$$\mathbf{F}_2[\hat{x}_i, i \in \mathbf{Z}]$$

where

- - $\|\hat{x}_i\| = (1, 2^i)$,
- - *the relation* $\mathrm{Sq}^1 \hat{x}_i = \hat{x}_{i-1}^2$ *and the Cartan formula determine the unstable module structure.*

For a proof we refer to [Sc1] chapter 2.

4. Injectivity of the mod 2 cohomology of elementary abelian group and Lannes' functor $\mathbf{T_V}$

The following fundamental theorem is due to G. Carlsson and H. Miller.

Theorem 4.1 (Ca). , *[Mi] Let V be an elementary abelian 2-group. Then the unstable module H^*V is injective.*

Proof : We sketch a slightly modified version of the proof of Carlsson. One shows that $\bar{H}^*\mathbf{Z}/2$ is a direct summand in K_*^*. One defines a map by the formula (of H. Campbell and P. Selick [CS]) :

$$\gamma(u^n) = (\sum_{i \in \mathbf{Z}} \hat{x}_i)^{[n]}$$

where the term on the right is the (finite) sum of all monomials in the formal (infinite) sum $(\sum_{i \in \mathbf{Z}} x_i)^n$ whose second degree is 1. Next we let $g : K_*^* \to H^*\mathbf{Z}/2$ to be the unique algebra map such that $g(\hat{x}_j) = u$. This map is $\mathcal{A}$-linear.

Then, one shows that the map $g\gamma : \bar{H}^*\mathbf{Z}/2 \to \bar{H}^*\mathbf{Z}/2$ is the identity. For this, it is enough to show that $g\gamma$ is an isomorphism in degree 1, and thus on $F(1)$. The reason being that in some sense (in $\mathcal{U}/\mathcal{N}il$) the socle of $\bar{H}^*\mathbf{Z}/2$ is $F(1)$.

One can proceed in an analogous way to prove the injectivity of H^*V. This is implicit in [Ca]. One can show that $\bar{H}^*V$ splits off $K(i)$ for an appropriate value of i.

We now move to the definition of the functor T_V. Let L be an unstable module of finite type. It follows from Freyd's adjoint functor theorem that the functor $M \mapsto L \otimes M$ from $\mathcal{U}$ to itself has a left adjoint. When $L = H^*V$ Lannes denoted it by T_V and showed :

Theorem 4.2 (L). *(Lannes) The functor T_V from $\mathcal{U}$ into itself is exact.*

For a proof see [Sc1] chapter 3.

In fact we will be only interested in the component of degree 0 of $T_V(M)$, which is just the continuous dual of $\mathrm{Hom}_{\mathcal{U}}(M, H^*V)$, and in this case the exactness is equivalent to the injectivity of H^*V. As H^*V is isomorphic to $\mathbf{F}_2 \oplus \bar{H}^*V$, the functor T_V is naturally equivalent to the direct sum of the identity functor and of the left adjoint of the tensor product by $\bar{H}^*V$. This last functor is denoted by $\bar{T}_V$ (resp. by $\bar{T}$ if $V \cong \mathbf{F}_2$).

After exactness the second main property of the functor T_V is its behaviour with respect to tensor product.

Theorem 4.3 (L). *(Lannes) For any unstable modules M_1 and M_2, there is a natural isomorphism*

$$T_V(M_1 \otimes M_2) \to T_V(M_1) \otimes T_V(M_2) \quad .$$

This is defined using the unit of the adjunction between T_V and the tensor product by H^*V, and the product on H^*V.

For a proof see [Sc1] chapter 3.

Again we will be mostly interested in the degree zero part of this statement. It is worth to add here that this functor computes, under nice assumptions on X the cohomology of the space of maps $\mathrm{map}(BV, X)$ [La].

We collect here a few examples of computations with T_V. The most basic example is the case of the mod 2-cohomology of an elementary abelian 2-group W :

$$T_V H^*W \cong (H^*W)^{\mathrm{Hom}(V,W)} \quad .$$

The above notation stands for the unstable module which in degree n is the space of *set theoretic* maps from $\mathrm{Hom}(V, W)$ to H^nW.

The degree zero part of this result is known as the Adams-Gunawardena-Miller, Lannes-Zarati theorem. After dualization, it may written as :

$$\mathrm{Hom}_{\mathcal{U}}(H^*W, H^*V) \cong \mathbf{F}_2[\mathrm{Hom}(V, W)] \quad .$$

In other words it means the following. Any unstable module map from H^*W to H^*V is, in a unique way, a linear combination of algebra maps from H^*W to H^*V. An algebra map from H^*W to H^*V is completely determined by its restriction to degree 1, which is nothing other than a map from W^* to V^*.

Let now G be a finite group, recall a theorem of Quillen. Let $\mathcal{C}(G)$ be the category whose objects are the abelian 2-subgroups E of G; the morphisms being subconjugacy. There is a natural map

$$q_G : H^*BG \to lim_{\mathcal{C}(G)} H^*BE \quad .$$

Quillen's theorem says that q_G is an F-isomorphism. It means that any element in the kernel of q_G is nilpotent, and that for any element y in the target, there exist n such that y^{2^n} is in the image. In our terminology this means that $\mathrm{Ker}\, q_G$ and $\mathrm{Coker}\, q_G$ are nilpotent unstable modules.

The following consequence of Quillen's theorem has been proved by various people, among them J. F. Adams, J .Lannes, H. Miller and C. Wilkerson.

Theorem 4.4. *The algebra $T_V^0 H^*BG$ is isomorphic to the Boolean algebra of functions from* $\mathrm{Rep}(V,G)$ *to* $\mathbf{F}_2$.

Se [Sc1] 3.10, for example.

Here $\mathrm{Rep}(V,G)$ means the set of homomorphisms of V to G, up to conjugacy.

5. $\mathcal{U}/\mathcal{N}il$ AND ANALYTIC FUNCTORS

Let $\mathcal{C}$ be an abelian category, $\mathcal{D}$ a full subcategory which has the property that, if we are given a short exact sequence in $\mathcal{C}$

$$0 \longrightarrow C' \longrightarrow C \longrightarrow C'' \longrightarrow 0 \quad ,$$

and if any two objects are in $\mathcal{D}$, so is the third. We shall say that $\mathcal{D}$ is a Serre class (in $\mathcal{C}$). With this assumption, it is possible to define the quotient category $\mathcal{C}/\mathcal{D}$ as follows. The objects are the same as in $\mathcal{C}$. If we are given $S, T \in \mathcal{C}$, $\mathrm{Hom}_{\mathcal{C}/\mathcal{D}}(S,T)$ is defined to be

$$\mathrm{dirlim}_{(S',T')}\mathrm{Hom}_{\mathcal{C}}(S', T/T') \quad ,$$

where (S',T') runs through the set of subobjects S' of S, T' of T such that $S/S' \in \mathcal{D}$ and $T' \in \mathcal{D}$. Roughly speaking, through this process a morphism (in $\mathcal{C}$) $\varphi : S \to T$ such that $\mathrm{Ker}(\varphi) \in \mathcal{D}$ (resp. $\mathrm{Coker}(\varphi) \in \mathcal{D}$) becomes a monomorphism (resp. an epimorphism) in $\mathcal{C}/\mathcal{D}$. The forgetful functor $\mathcal{C} \to \mathcal{C}/\mathcal{D}$ is exact.

Let $\mathcal{N}il$ be the full subcategory of $\mathcal{U}$ consisting of nilpotent objects. The category $\mathcal{N}il$ is a Serre class in $\mathcal{U}$. We can consider the quotient category $\mathcal{U}/\mathcal{N}il$. The forgetful functor $r \; : \; \mathcal{U} \to \mathcal{U}/\mathcal{N}il$ has a right adjoint s. It follows from the adjunction isomorphism that the unstable module $sr(M)$ is reduced and even $\mathcal{N}il$-closed using the terminology of [Ga].

Let $\mathcal{F}$ be the category of covariant functors from the category $\mathcal{E}_f$ of finite dimensional $\mathbf{F}_2$-vector spaces to the category $\mathcal{E}$. It is an abelian category. A short exact sequence of functors is a sequence $0 \to F' \to F \to F'' \to 0$, such that for any V,

$$0 \to F'(V) \to F(V) \to F''(V) \to 0 \quad .$$

is exact.

Let us introduce a functor $f : \mathcal{U} \longrightarrow \mathcal{F}$. It associates to an object M of $\mathcal{U}$ the functor

$$\mathcal{E}_f \to \mathcal{E}, \quad f(M) : V \mapsto T_V(M)^0 \quad .$$

Observe that $f(M)$ is just another instance of $T_V(M)^0$.

Theorem 5.1. *The functor f is exact.*

This is theorem 3.1. of [HLS].

The exactness of the functor T_V (in degree 0) is equivalent to the fact that the functor f is exact. The category $\mathcal{F}$ has a tensor product. The tensor product of the functors F and G is the functor $F \otimes G$ that associates to V the $\mathbf{F}_p$-vector space $F(V) \otimes G(V)$. A direct consequence of the natural isomorphism $T_V(M) \otimes T_V(N) \cong T_V(M \otimes N)$ (in degree zero) is :

Theorem 5.2. *For any unstable $\mathcal{A}$-modules M and N, there is a natural equivalence between the functors $f(M) \otimes f(N)$ and $f(M \otimes N)$.*

This is proposition 3.2. of [HLS].

Note also that the functor f commutes with colimits.

If M is the unstable $\mathcal{A}$-module $F(n)$, by the very definition,

$$f(F(n))(V) \cong H_n(V) \cong \Gamma^n(V) \quad ,$$

the n-th divided powers functor on V. One also gets

$$f(F(1)^{\otimes n})(V) \cong V^{\otimes n} \quad .$$

The following is a reformulation of the Adams-Gunawardena-Miller theorem. Let M be H^*E then $f(H^*E)(V)$ is equivalent to the functor I_E

$$V \mapsto \mathbf{F}_2^{\mathrm{Hom}(V,E)} \quad ,$$

the space of set maps from $\mathrm{Hom}(V, E)$ to $\mathbf{F}_2$.

We now describe the kernel of f and its image in $\mathcal{F}$. We begin with the kernel. If M is a nilpotent unstable $\mathcal{A}$-module $T_V(M)^0$ is trivial for any V. This factorization defines the functor f. Next :

Theorem 5.3 (LS). *(Lannes-Schwartz) The functor $f : \mathcal{U}/\mathcal{N}il \to \mathcal{F}$ is faithful.*

In fact, it is a consequence of [LS], and stated in [HLS] as theorem 4.2.

We are left with the task of describing the image of f. Let F be an object of $\mathcal{F}$, define ΔF to be the functor that associates to V the kernel of the linear map $F(\pi) : F(V \oplus \mathbf{F}_2) \to F(V)$, $\pi : V \oplus \mathbf{F}_2 \to V$ being the standard projection : the identity on V and zero on the summand $\mathbf{F}_2$.

Definition 5.4. *A functor F is of Eilenberg-MacLane degree less than (resp. exactly) k if and only if $\Delta^{k+1}F$ is trivial (resp. but $\Delta^k F$ is non trivial).*

It is easy to show that Δ is linked to $\bar{T}$ *via* the following formula, for any $M \in \mathcal{U}$ on has :

$$f(\bar{T}(M)) \cong \Delta f(M) \quad .$$

It is also easy to show that Δ is left adjoint to the functor from $\mathcal{F}$ into itself that sends F to $\bar{I} \otimes F$, where $\bar{I}$ is the functor that sends V to the space of set theoretic maps from V^* to $\mathbf{F}_2$ sending 0 to 0.

Definition 5.5. *An object F of $\mathcal{F}$ is polynomial if it is of degree less or equal to k for some k. An object F is analytic if it is the direct limit of its polynomial subfunctors.*

We denote by $\mathcal{F}_n$ the full sub-category of $\mathcal{F}$ of functors of degree less or equal to n, and by $\mathcal{F}_\omega$ the full sub-category of analytic functors. They are abelian subcategories of $\mathcal{F}$. They are stable under colimits.

The functors $V \mapsto V^{\otimes n}$ and $V \mapsto \Gamma^n(V)$ are polynomial of degree n.

Proposition 5.6. *The functor $I_E : V \mapsto \mathbf{F}_p^{\mathrm{Hom(V,E)}}$ is analytic for any E.*

Theorem 5.7. *For any unstable $\mathcal{A}$-module M the functor $f(M)$ is analytic.*

It is enough to check it on a set of generators of the category $\mathcal{U}$. But this is true for the unstable $\mathcal{A}$-module $F(n)$, for any n. Thus :

Theorem 5.8 (HLS). *The functor induced by f from $\mathcal{U}/\mathcal{N}il$ to $\mathcal{F}_\omega$ is an equivalence of categories.*

Here is a straightforward corollary :

Corollary 5.9. *The n-th divided powers functors Γ_n, $n \geq 0$, are generators for $\mathcal{F}_\omega$.*

This is classical result in the context of representation theory, see Kuhn's notes in this volume. Observe also also that they are not projective objects. The preceding result extends to the category $\mathcal{K}$, allowing generalizations of the Adams-Wilkerson theorems [AW]. In fact the artinian conjecture discussed by G. Powell in this volume finds its origin in the following observation, of the late 80's, of [HLS]. An unstable sub-algebra of H^*V is defined, as an object of $\mathcal{K}/\mathcal{N}il$, by a finite set of linear equations codified by what is called in [HLS] (theorem 2.4.) the Galois set : namely a sub-algebra of H^*V is defined, as an object of $\mathcal{K}/\mathcal{N}il$, as the kernel of a finite number of maps from H^*V into some H^*E for some $E \in \mathcal{E}_f$. It was natural to ask whether this also holds for sub-modules. In fact both questions were raised at the same time by Lannes. The one concerning sub-algebras was solved in 1988.

It is worth adding some more comments here. This problem is intimately linked to the classical Weyl problem of classifying n-uples of linear maps from a vector space to another vector space. Indeed recall that a map from H^*V into H^*E is given by a linear combination of the form $\sum_{1\leq i\leq n} \mu_i$ where the μ_i's are linear maps from V^* to E^*. When $n=1$ one gets an algebra map. When $n=2$ the classification of pairs is known, due to Kronecker and Weierstrass. This case corresponds to unstable algebras. Indeed the kernel of a map of the form $\mu+\mu'$ is an algebra. The general case (even if $n=3$) is known to be wild (see Ringel [Ri], sectio 3.2., and Gelfand, [Ge]).

The following conjecture claims that spaces yield in a natural way infinite socle length objects in $\mathcal{U}/\mathcal{N}il$. Recall that the socle series of an object A in an abelian category $\mathcal{C}$ is defined as follows. The sub-object A_0 is the largest semi-simple object (direct sum of simple objects) contained in A. Define A_n inductively as follows : the sub-object A_n of A is the inverse image, *via* the canonical epimorphism $A \to A/A_{n-1}$, of the largest semi-simple object of the quotient A/A_{n-1}. The object A is of infinite socle length if the series $A_0 \subset A_1 \subset \ldots \subset A_n \subset \ldots$ does not stabilize.

Conjecture : Let X be a space. Then, the mod-2 cohomology of X, as an object in $\mathcal{U}/Nil$ is either trivial or of infinite length, i.e. its socle series is infinite.

If X is finite H^*X is trivial in $\mathcal{U}/Nil$. If $X = B\mathbf{Z}/2$, the cohomology $H^*B\mathbf{Z}/2$ has an infinite socle series, having subquotients the modules $\Lambda^n(F(1))$. It is aconsequence of the work done in [HLS] that the conjecture is true for algebras of finite transcendence degree. In fact, the description of the algebras of finite transcendance degree allows one to reduce to H^*V.

We now come back on injective objects in $\mathcal{F}$

Proposition 5.10. *Let E be in $\mathcal{E}_f$. The functor $I_E : V \mapsto \mathbf{F}_2^{\mathrm{Hom}(V,E)}$ is injective in $\mathcal{F}$. It is a representing object for the functor $F \mapsto F(E)^*$, $\mathcal{F} \to \mathcal{E}$.*

This proposition shows that I_E is a natural injective object in $\mathcal{F}$. On the other hand, it is hard to show that the unstable module H^*E, which corresponds to I_E *via* f is injective.

We had natural injective objects in $\mathcal{U}$, the modules $K(i)$. One has :

Proposition 5.11 (K3). *The functor $f(K(i))$ is equivalent to the direct limit, denoted S_i, of the symmetric powers $S^{2^q i}$, under the squaring map. It follows that S_i is injective in $\mathcal{F}_\omega$.*

See theorem 1.3. and theorem 1.5. in the reference.

Note that the injectivity of the S_i's in the category $\mathcal{F}_\omega$ is not obvious. It is not known whether S_i is injective in the category $\mathcal{F}$; this is equivalent to the artinian question.

If $E = \mathbf{F}_2$ the functor $I_{\mathbf{F}_2}$ decomposes canonically as the direct sum of the constant functor (included in $I_{\mathbf{F}_2}$ as the set of constant functions from V^* to $\mathbf{F}_2$) and of a functor $\bar{I}$. By the very definition $\mathrm{Hom}(F, \bar{I})$ is canonically isomorphic to $\Delta F(0)$. More generally $\mathrm{Hom}_{\mathcal{F}}(F, \bar{I}^{\otimes k})$ is canonically isomorphic to $\Delta^k F(0)$.

Theorem 5.12. *The category $\mathcal{F}_\omega$ is locally noetherian. Consequently any direct sum $\bigoplus_\alpha E_{W_\alpha}$ is an injective object.*

A functor F in $\mathcal{F}_k$, such that $F(\mathbf{F}_2^{\oplus d})$ is finite for all d, has a finite filtration $\{0\} \subset F_0 \subset F_1 \subset \ldots \subset F_l = F$ whose quotient functors are simple. Indeed, one considers the polynomial defined by $P_F(d) = \dim_{\mathbf{F}_2} F(\mathbf{F}_2^{\oplus d})$. Then, observe that a polynomial that takes non-negative integer values on non-negative integers cannot be the sum of an arbitrary large number of such non-trivial polynomials.

All simple functors are polynomial. This follows from the fact that a simple functor is a subfunctor of some I_E.

Any analytic functor F has a socle which is semi-simple (isomorphic to a direct sum of simple objects), it is the largest semi-simple object contained in F. This socle $\mathrm{Soc}(F)$ has the property that any non-trivial subfunctor $G \subset F$ has a non-trivial intersection with it. This is proved as follows. Let G be a subfunctor of F, as G is analytic it contains a non-trivial polynomial subfunctor F such that $\dim_{\mathbf{F}_p} F(\mathbf{F}_p^{\oplus d})$ is finite for all integers d. This functor contains a non-trivial simple subfunctor. Hence G and $\mathrm{Soc}(F)$ have a non-trivial intersection.

Theorem 5.13. *An analytic functor F embeds in a direct sum $\bigoplus_\alpha I_{E_\alpha}$. If the socle of F is a finite direct sum of simple functors, F embeds in a finite direct sum.*

It is natural to ask whether or not this property can be extended to an injective resolution. This is another formulation of the artinian question. The following theorem can be seen as a weak form of the conjecture.

Theorem 5.14. *Let F be in $\mathcal{F}_k$, assume that $P_F(d)$ is finite for all d. There exists an injective resolution of F :*

$$0 \to F \to I_0 \to I_1 \to \ldots \to I_k \to \ldots$$

such that all the functors I_k are finite direct sums of I_E's.

In this theorem we could replace the condition on the I_k's by the somewhat more natural one that they are finite direct sums of indecomposable injective objects. But these are equivalent.

We give a proof a bit different than [Sc1], (5.3.) and [FLS] (section 10). One needs the following :

Lemma 5.15. *Let F be an analytic functor. Assume that* Soc(F) *is a finite direct sum of simple functors and that for some k $\Delta^k F$ is injective. Let $\gamma : F \to \bigoplus_\alpha E_{W_\alpha}$ be any embedding of F in a finite direct sum. Then the socle of the cokernel of γ is a finite direct sum of simple functors.*

The above condition on F holds for any polynomial subfunctor F that takes finite dimensional values.

Proof : As Δ is exact we get an exact sequence

$$0 \to \Delta^k F \to \oplus_\alpha \Delta^k E_{W_\alpha} \to \Delta^k \mathrm{Coker}(\gamma) \to 0 \quad .$$

In this exact sequence the two first terms are injective by hypothesis. Thus the functor $\Delta^k\mathrm{Coker}(\gamma)$ is also injective and is a direct summand in $\oplus_\alpha \Delta^k I_{E_\alpha}$. Its socle is contained in the socle of $\oplus_\alpha \Delta^k I_{E_\alpha}$, which is a finite direct sum. This is just because its ring of endomorphism of its injective hull is finite. Indeed, the ring of endomorphisms of I_E is $\mathbf{F}_2[\mathrm{End}(E)]$, which is finite. Moreover $\Delta^k I_E$ is isomorphic to $\mathbf{F}_2[E^{*\oplus k}] \otimes I_E$, a finite direct sum of I_E's. It follows that $\Delta^k\mathrm{Coker}(\gamma)$ is a a direct summand of a finite direct sum of I_E's. Hence the socle of $\Delta^k\mathrm{Coker}(\gamma)$ is a finite direct sum.

It remains to deduce the result for $\text{Coker}(\gamma)$ itself. This follows from :

Lemma 5.16. *Let F be an analytic functor. Then $\text{soc}(F)$ is finite if and only if $F(0)$ is a finite dimensional vector space and if $\text{soc}(\Delta F)$ is finite.*

This follows from the exact sequence, valid for any analytic functor :

$$F(0) \to F \to \bar{I} \otimes \Delta F \quad ,$$

the first map is induced by the canonical inclusion of the zero dimensional vector space and the second one is the adjoint of the identity of ΔF.

6. The functor $p_n : \mathcal{U}/\mathcal{N}il \to \mathcal{M}od_{\mathbf{F}_2[\mathcal{S}_n]}$, the filtration on $\mathcal{U}/\mathcal{N}il$, and simple objects

We push a step further the comparison of the category $\mathcal{U}/\mathcal{N}il$ and $\mathcal{F}_\omega$. We define a filtration on $\mathcal{U}/\mathcal{N}il$, and we identify the terms of the filtration with the filtration by the degree on $\mathcal{F}_\omega$. Then, we identify the quotient categories with the categories of modular representations of the symmetric groups. We finally describe simple objects in $\mathcal{U}/\mathcal{N}il$ and in $\mathcal{F}_\omega$. Denote by $\mathcal{M}od_{\mathbf{F}_2[\mathcal{S}_n]}$ the category of right $\mathbf{F}_2[\mathcal{S}_n]$-modules.

Let M be an unstable module, the vector space $\text{Hom}_{\mathcal{U}}(M, K(1)^{\otimes n})$ has, in a natural way an action of the symmetric group $\mathcal{S}_n$. The group acts by permutation of the factors on $K(1)^{\otimes n}$. It induces a left $\mathcal{S}_n$-action on $\text{Hom}_{\mathcal{U}}(M, K(1)^{\otimes n})$. Thus the dual $\text{Hom}_{\mathcal{U}}(M, K(1)^{\otimes n})^*$ is a right $\mathbf{F}_2[\mathcal{S}_n]$-module. Hence the assignment

$$M \mapsto \text{Hom}_{\mathcal{U}}(M, K(1)^{\otimes n})^*$$

determines a functor $p_n : \mathcal{U} \to \mathcal{M}od_{\mathbf{F}_2[\mathcal{S}_n]}$. Here one has to take a continuous dual, the topology one considers is a profinite one, obtained by considering M has the direct limit of its finitely generated submodules. The functor p_n commutes with colimits. It is right exact. The unstable module $K(1)^{\otimes n}$ is injective, therefore p_n is left exact. Hence it is exact.

It is trivial on the subcategory $\mathcal{N}il$ and factors uniquely as :

$$\mathcal{U} \to \mathcal{U}/\mathcal{N}il \to \mathcal{M}od_{\mathbf{F}_2[\mathcal{S}_n]} \quad .$$

This defines a functor

$$p_n : \mathcal{U}/\mathcal{N}il \to \mathcal{M}od_{\mathbf{F}_2[\mathcal{S}_n]} \quad .$$

Next, define the category $\mathcal{V}_n$ to be the full subcategory of $\mathcal{U}/\mathcal{N}il$ consisting of objects on which the functor induced by p_k is trivial for all $k > n$. By definition the functor p_n is trivial on $\mathcal{V}_{n-1}$

Theorem 6.1. *The functor p_n induces an equivalence of categories*

$$\mathcal{V}_n/\mathcal{V}_{n-1} \to \mathcal{M}od_{\mathbf{F}_2}[\mathcal{S}_n]$$

.

The following propositions are easy, and are main steps in the proof :

Proposition 6.2. *The representation $p_n(F(1)^{\otimes n})$ is isomorphic to $\mathbf{F}_2[\mathcal{S}_n]$.*

Proposition 6.3. *The ring* $\mathrm{End}_{\mathcal{U}}(F(1)^{\otimes n})$ *is isomorphic to* $\mathbf{F}_2[\mathcal{S}_n]$.

We are going to give now a more concrete description of the category $\mathcal{V}_n$, [FS].

Theorem 6.4. *The category* $\mathcal{V}_n$ *is the full subcategory of* $\mathcal{U}/\mathcal{N}il$ *consisting of objects* M *such that :*

$$M \in \mathcal{V}_n \Leftrightarrow s(M) \text{ is trivial in degrees } d \text{ such that } \alpha(d) > n \quad .$$

Recall that $\alpha(d)$ is the number of 1's in the 2-adic expansion of d and that $s : \mathcal{U}/\mathcal{N}il \to \mathcal{U}$ is the right adjoint of the forgetful functor $r = r_1 : \mathcal{U} \to \mathcal{U}/\mathcal{N}il$. It would be equivalent to require that any reduced unstable $\mathcal{A}$-module L which is isomorphic to M in $\mathcal{U}/\mathcal{N}il$ has the property stated for sM. One also shows that an unstable module such that $p_n(M)$ is trivial for all $n \geq 0$ is nilpotent.

In [FS] it is shown that if $M \in \mathcal{V}_n$, then

$$p_n(M) = \mathrm{Hom}_{\mathcal{F}}(F(1)^{\otimes n}, M) \quad .$$

In fact, in this case $p_k(M)$ is isomorphic to :

$$M^{2^{a_1}+\ldots+2^{a_n}}$$

(M finitely generated) is trivial as soon as $i \neq j$ implies $a_i \neq a_j$. Moreover this vector space is naturally a representation of the symmetric group $\mathcal{S}_n$. The action of the group can be made explicit in terms of Steenrod operations. The functor introduced in [FS] does not agree with the one introduced here on the whole of $\mathcal{U}/\mathcal{N}il$.

We compare the categories $\mathcal{V}_n$ and $\mathcal{F}_n$.

Theorem 6.5. *The equivalence* $f : \mathcal{U}/\mathcal{N}il \to \mathcal{F}_\omega$ *restricts to an equivalence* $\mathcal{V}_n \to \mathcal{F}_n$.

We claim that f induces a functor from $\mathcal{V}_n$ to $\mathcal{F}_n$. This is proved by induction on n by using that for all M in $\mathcal{U}/\mathcal{N}il$ one has

$$f(T(M)) \cong \Delta f(M) \quad .$$

We now give a more precise description of simple objects of $\mathcal{U}/\mathcal{N}il$ in terms of representation theory of the symmetric groups.

Let S be a simple object of $\mathcal{U}/\mathcal{N}il$ that is in $\mathcal{V}_n$ but not in $\mathcal{V}_{n-1}$.

Proposition 6.6. *The representation* $p_n(S)$ *is simple.*

Let R be a representation of the symmetric group $\mathcal{S}_n$. Recall the norm map

$$N \; : \; R \otimes F(1)^{\otimes n} \longrightarrow R \otimes F(1)^{\otimes n}$$

which is given by

$$N(r \otimes u^{2^{a_1}} \otimes \ldots \otimes u^{2^{a_n}}) = \sum_{\sigma \in \mathcal{S}_n} \sigma(r) \otimes u^{2^{a_{\sigma(1)}}} \otimes \ldots \otimes u^{2^{a_{\sigma(n)}}} \quad .$$

Theorem 6.7. *Let R be a simple $\mathbf{F}_2[\mathcal{S}_n]$-module. Then the image of the norm map N*

$$N \;:\; R\otimes F(1)^{\otimes n} \longrightarrow R\otimes F(1)^{\otimes n}$$

is a simple object of $\mathcal{V}_n - \mathcal{V}_{n-1}$. Any simple object of $\mathcal{V}_n - \mathcal{V}_{n-1}$ is isomorphic to one of this form. The representation $p_n(ImN)$ is equivalent to the representation R.

Moreover let $e \in \mathbf{F}_2[\mathcal{S}_n]$ be such that $e\mathbf{F}_2[\mathcal{S}_n]$ is isomorphic to R. Then the image of the norm map is isomorphic, as an unstable $\mathcal{A}$-module, to $eF(1)^{\otimes n}$.

As an example, if the representation of $\mathcal{S}_n$ is the trivial one, one gets the n-th exterior power $\Lambda^n(F(1))$.

In the proof one needs the following :

Proposition 6.8. *The largest subobject of $F(1)^{\otimes n}$ which is in $\mathcal{V}_{n-1}$ is trivial.*

7. The decomposition of the Carlsson modules $K(i)$

Let us be a bit more precise on the classification of simple objects. Recall a few facts about the modular representation theory of the symmetric groups $\mathcal{S}_n$ [JK]. The simple $\mathbf{F}_2$-representations of the symmetric group $\mathcal{S}_n$ are (up to isomorphism) indexed by partitions $\lambda = (\lambda_1, \ldots, \lambda_d)$ of n such that $\lambda_i > \lambda_{i+1}$, $1 \leq i \leq d-1$. Such partitions are called 2-regular. The conjugate partition of λ, λ' is such that $1 \geq \lambda'_i - \lambda'_{i+1} \geq 0$, *i.e.*is column 2-regular. Let R_λ be the simple $\mathbf{F}_2$-representation of $\mathcal{S}_n$ indexed by λ. It is isomorphic to $\epsilon_\lambda\mathbf{F}_2[\mathcal{S}_n]$ for some $\epsilon_\lambda \in \mathbf{F}_2[\mathcal{S}_n]$. The associated simple object S_λ in $\mathcal{U}/\mathcal{N}il$ is of the form $\epsilon_\lambda F(1)^{\otimes n}$. The corresponding simple functor, denoted also by S_λ is of the form :

$$V \mapsto \epsilon_\lambda V^{\otimes n} \quad .$$

For example the one dimensional trivial representation of $\mathcal{S}_n$ corresponds to the partition (n) and to the n-th exterior power functor.

Let us denote by I_λ the injective hull of S_λ in both categories. They form a set of representatives of indecomposable injective objects.

As we are in a locally noetherian category, the Azumaya-Krull-Scmidt theorem tells us that any injective object decomposes in a unique way as a direct sum of indecomposable injective objects. Recall that the categories in question are locally noetherian.

Consider the case of I_E (equivalently of H^*E). There exist unique integers (depending on the dimension of E) k_λ such that :

$$I_E \cong \oplus_\lambda I_\lambda^{\oplus k_\lambda(d)} \quad .$$

The sum is taken over all 2-regular partitions. Let d be the dimension of E.

It is shown in [CK1] that the integers $k_\lambda(d)$ are determined by the modular representation theory of the general linear groups.

To a 2-regular partition λ is associated, (*via* the Weyl correspondance, see [Scl]) a simple representation L_λ of the linear group $\mathrm{GL}_t(\mathbf{F}_2)$, where $t = \lambda_1$. If $t = d$ the integer $k_\lambda(d)$ is the dimension of this representation. If λ is the partition

$(\lambda_1, \dots, \lambda_h)$ with $\lambda_1 = d$ let us denote by $c(\lambda)$ the partition $(\lambda_2, \dots, \lambda_h)$. One has $k_\lambda(d) = k_{c(\lambda)}(d)$. The integer k_λ is zero in all other cases.

This is mainly a consequence of the Adams-Gunawardena-Miller, Lannes-Zarati theorem.

The analogous decomposition for the functors S_i, equivalently for the Carlsson modules $K(i)$, is much more complicated.

They are isomorphic to a direct sum $\oplus_\lambda I_\lambda^{\oplus a_{i,\lambda}}$. These integers can be computed in terms of modular representation theory invariants. Here is the formula established by Lannes and the author :

$$a_{i,\lambda} = \sum_\mu s(i,\mu)\dim_{\mathbf{F}_2}\mathbf{S}_\mu R_\lambda.$$

In this formula,

- $\mu = (\mu_1, \dots, \mu_\ell, \dots, \mu_t)$ runs through the set of all partitions of n (λ is a partition of n),
- $a(i,\mu)$ is the number of ways to write i as $\sum_\ell \mu_\ell 2^{q_\ell}$, $q_\ell \in \mathbf{Z}$ with $\ell < h \Rightarrow q_\ell < q_h$,
- $\mathcal{S}_\mu \subset \mathcal{S}_k$ is a Young subgroup isomorphic to $\mathcal{S}_{\mu_1} \times \dots \times \mathcal{S}_{\mu_t}$, $\mathbf{S}_\mu = \sum_{\sigma \in \mathcal{S}_\mu} \sigma \in \mathbf{F}_2[\mathbf{S}_n]$ and $\mathcal{S}_\mu R_\lambda$ is the image in R_λ of the norm with respect to $\mathcal{S}_\mu$ (multiplication by the element $\mathbf{S}_\mu$ of the group algebra of $\mathcal{S}_n$).

Here is an example of such a computation. Let (n) be the trivial partition of n, $I_{(n)}$ is not a direct summand in $K(1)$ if $n > 1$. The corresponding representation of $\mathcal{S}_n$ is the trivial one. Therefore, $\dim_{\mathbf{F}_2}\mathbf{S}_\mu\mathbf{F}_2$ is zero unless $\mu = (1, \dots, 1)$. In this case $s(1,\mu)$ is zero.

8. The Krull filtration on $\mathcal{U}$

An unstable module is locally finite if it is the direct limit of finite unstable modules. In other words it means that only a finite number of operations act non-trivially on any given element. The full subcategory $\mathcal{B}$ of locally finite modules is the first step of the Gabriel-Krull filtration of the category $\mathcal{U}$. Recall the definition of this filtration. The full subcategory $\mathcal{U}_0$ is the smallest Serre class in $\mathcal{U}$ which is stable under colimits and contains all simple objects in $\mathcal{U}$. As any simple objects in $\mathcal{U}$ is of the form $\Sigma^n\mathbf{F}_p$, one gets $\mathcal{U}_0 = B$. Suppose $\mathcal{U}_n$ has been defined. Define $\mathcal{U}_{n+1}$ as follows. Consider in the quotient category $\mathcal{U}/\mathcal{U}_n$ the smallest Serre class which is stable under colimits and contains all simple objects (of $\mathcal{U}/\mathcal{U}_n$). Then an object M in $\mathcal{U}$ belongs to $\mathcal{U}_{n+1}$, if and only if, as an object of $\mathcal{U}/\mathcal{U}_n$, it belongs to the subcategory we have just described.

Theorem 8.1. *The smallest abelian subcategory of $\mathcal{U}$ which contains all $\mathcal{U}_n$, $n \in \mathbf{N}$, and is stable under colimits, is $\mathcal{U}$ itself.*

See [Sc1], theorem 6.2.3.

Here is a characterisation of the Krull filtration in terms of Lannes' T functor.

Theorem 8.2. *Let M be an unstable $\mathcal{A}$-module. Then M is in $\mathcal{U}_n$ if and only if $T^{n+1}M$ is trivial.*

Here is an explicit description of objects in $\mathcal{U}_1$ (essentially stated in [Sc2]). Let $M \in \mathcal{U}_1$, there exists $K, L, J \in \mathcal{U}_0$ and an exact sequence :

$$\{0\} \to K \to M \to L \otimes F(1) \to J \quad .$$

In the next proposition we compare the Krull filtration on $\mathcal{U}$ to the polynomial filtration on $\mathcal{F}$.

Proposition 8.3 (Sc1). *The image by the functor f of the category $\mathcal{U}_n$ is equivalent to the category $\mathcal{F}_n$.*

N. Kuhn has conjectured in [K4] that the cohomology of a space X is either in $\mathcal{U}_0$, or does not belong to $\mathcal{U}_n$ for all n. If X is a finite space one is in the first case. If $X = B\mathbf{Z}_2$ one is in the second case. In fact Kuhn showed that if there is a space X so that the cohomology is in $\mathcal{U}_n$ but not in $\mathcal{U}_{n-1}$, $n > 1$, the same holds with $n = 1$. The author has shown in [Sc2] that this is impossible under the additionnal assumption that the cohomology of X is finitely generated as an $\mathcal{A}$-module. More precisely :

Theorem 8.4 (Sc2). *If the cohomology of a space is finitely generated as an $\mathcal{A}$-module it is finite.*

For example there is no space with reduced cohomology $F(1)$.

A proof in the general case will be, hopefully, given somewhere else. It depends on another filtration on the category $\mathcal{U}$, called the nilpotent filtration [Sc1]. There is a decreasing sequence of full sub-category of $\mathcal{U}$ denotes $\mathcal{N}il_s$. The category $\mathcal{N}il_0$ is just $\mathcal{U}$, $\mathcal{N}il_1$ is $\mathcal{N}il$. In general $\mathcal{N}il_s$ is the smallest Serre class stable under colimits and containing s-fold suspensions. The quotient categories $\mathcal{N}il_s/\mathcal{N}il_{s+1}$ are all equivalent to $\mathcal{F}_\omega$. This filtration is converging. We refer to [Sc1] for details.

Using the nilpotent filtration Kuhn made another conjecture [K2]. Let M be an unstable module there is a converging decreasing sequence of submodules in M :

$$\ldots M_s \subset M_{s-1} \subset \ldots \subset M \quad ,$$

where M_s is the largest subobject of M which is in $\mathcal{N}il_s$. The quotient M_s/M_{s+1} is easily shown to be the s-suspension of a reduced unstable module $R_s(M)$. The conjecture is as follows :

Conjecture 8.1. *Either all the unstable modules $R_s(M)$ are finite, and in fact concentrated in degree zero, or at least one, considered as an object in $\mathcal{U}/\mathcal{N}il$ has infinite socle series.*

9. The Grothendieck ring of $\mathcal{U}$

Let $G_0(\mathcal{U})$ be the quotient of the free abelian ring generated by isomorphism classes of noetherian unstable $\mathcal{A}$-modules by the ideal generated by elements $[E]-$

$[E'] - [E'']$, for any short exact sequence $0 \to E' \to E \to E'' \to 0$. For any $E \in G_0(\mathcal{U})$ denote by $P_E(q)$ the Poincaré series of E. One has :

Theorem 9.1. *The morphism of rings $G_0(\mathcal{U}) \to \mathbf{Z}[[q]]$, sending $[E]$ to $P_E(q)$, is injective.*

In fact behind this theorem is the following fact. There exist a character theory for finitely generated objects in $\mathcal{U}/\mathcal{N}il$ [Sc3]. The function φ_M replacing the character is the following one :

$$\mathbf{N}^* \to \mathbf{N}, \quad \varphi_M(i) = \dim_{\mathbf{F}_2}(K_i(M)) \quad .$$

We refer to the third section for the definition of the functor K_i. As this function satisfies $\varphi_M(2i) = \varphi_M(i)$ it is natural and useful to extend it to $\mathbf{N}[\frac{1}{2}]$, it shares all the expected properties of a character. A construction analogous to M. Atiyah's construction of the Adams operations is useful in this context.

It would be nice to extend the preceding result by removing the finitness hypothesis. Indeed, because of Kuhn's conjecture the preceding result does not say anything about the cohomology of spaces. To get informations on the nature of the cohomology of (infinite) spaces one should work in a Grothendieck ring of infinitely generated modules of finite type. It is hopeless to get something interesting without any additional hypothesis. The first sensible one to do, is to restrict attention, to unstable modules so that the injective hull is a finite direct sum of indecomposable injective objects. This holds for example for the cohomology of a finite group.

However, it is not clear that one can form the Grothendieck ring of such modules. Namely, this is equivalent to the artinian conjecture.

The second difficulty in this direction is that, it is necessary to take into account the nilpotent filtration on the module. With the above restriction this filtration is finite. Thus the most natural Poincare series type invariant to associate would be the finite sequence of the Poincare series of the quotients of the nilpotent filtration on the module.

Assuming the artinian conjectue is true we raise the following question :

Question 9.1. *Does the Grothendieck ring, of functors having as injective hull a finite direct sum of injective objects, embeds naturally in a ring of double formal power series?*

The only, very slim, evidence in this direction is the following (the author, unpublished). The composition factors in $\mathcal{U}/\mathcal{N}il$ of a submodule of $H^*(\mathbf{Z}/2^{\oplus 2})$ are determined by its Poincare series. This seems to be also true for $H^*(\mathbf{Z}/2^{\oplus 3})$.

In [CK2] Carlisle and Kuhn described the Grothendieck ring of finitely generated $\mathbf{F}_p[\mathrm{End}V]$-modules. Considering an adequate colimit process, and taking into account certain torsion conditions, it yields a computation of the Grothendieck ring of the category of noetherian objects in $\mathcal{F}_\omega$. Thus, using the identification between $\mathcal{U}/\mathcal{N}il$ and $\mathcal{F}_\omega$, it yields a computation of the Grothendieck ring of the category of noetherian objects in $\mathcal{U}/\mathcal{N}il$. One gets :

Theorem 9.2. *The Grothendieck ring of the category of noetherian objects in $\mathcal{U}/\mathcal{N}il$ is isomorphic to the quotient of the universal λ-ring on one generator quotiented by the ideal generated by the relations $\psi^2(x) = x$ for all x.*

Kuhn has also considered the Grothendieck ring of reduced injective objects of the category $\mathcal{U}$ [K1]. It is dual to the preceding object. Certain formulas concerning the products can be found in [CK2].

Aknowledgement : *This is a revised version of the author's talk at the Euroconference in Bielefeld on Infinite Length modules. The author was supported by the TMR network 'Invariants and Representations of Algebras'.*

REFERENCES

(AGM) J.F. ADAMS, J.H. GUNAWARDENA, H.R. MILLER, *The Segal conjecture for elementary abelian p-groups*, Topology 24 (1985), 435-460.

(AW) J.F. ADAMS, C.W. WILKERSON, *Finite H-spaces and algebras over the Steenrod algebra*, Ann. of Math. 111 (1980), 95-143.

(CS) H.E.A. CAMPBELL, S.P. SELICK, *Polynomial algebras over the Steenrod algebra*, Comment. Math. Helv. 65 (1990), 171-180.

(CK1) D. CARLISLE, N. KUHN, *Subalgebras of the Steenrod algebra and the action of matrices on truncated polynomial algebras*, J. Algebra 121 (1989), 370-387.

(CK2) D. CARLISLE, N. KUHN, *Smah products of summands of* $B(\mathbf{Z}/p)^n$, A.M.S Cont. Math. 96, 87-102.

(Ca) G. CARLSSON, *G.B. Segal's Burnside ring conjecture for* $(\mathbf{Z}/2)^k$, Topology 22 (1983), 83-103.

(CS) H.E.A. CAMPBELL, S.P. SELICK, *Polynomial algebras over the Steenrod algebra*, Comment. Math. Helv. 65 (1990), 171-180.

(FLS) V. FRANJOU, J. LANNES, L. SCHWARTZ, *Autour de la cohomologie de MacLane des corps finis*, Invent. Math. 115 (1994), 513-538.

(FS) V. FRANJOU, L. SCHWARTZ, 65 *Reduced unstable $\mathcal{A}$-modules and the modular representation theory of the symmetric groups*, Ann. Scient. Ec. Norm. Sup. 23 (1990), 593-624.

(Ga) P. GABRIEL, *Des catégories abéliennes*, Bull. Soc. Math. France, 90, (1962), 323-448.

(Ge) I. M. GELFAND, V. A. PONOMAREV *Problems in linerar algebra...* Colloquia Math. Soc. Janos Bolyai, Tihany Hungary, (1970) North Holland, (1972).

(HLS1) H.-W. HENN, J. LANNES, L. SCHWARTZ, *Analytic functors, unstable algebras and cohomology of classifying spaces*, Algebraic Topology Proc. 1988, Northwestern University, Cont. Math. 96 (1989), 197-220.

(HLS2) H.-W. HENN, J. LANNES, L. SCHWARTZ, *The categories of unstable modules and unstable algebras modulo nilpotent objects*, Am. J. of Math. (1993), Vol 115, Number 5, 1053-1106.

(JK) G. JAMES, A. KERBER *The representation theory of the symmetric group*, Encycl. Math. Appl. 16 (1981).

(K1,2,3) N. KUHN, *Generic representations of the finite general linear groups and the Steenrod algebra*, Am. Journal of Math. 116 (1993), 327-360; K-Theory 8 (1994) 395-426; K-Theory 9 (1995) 273-303.

(K4) N. KUHN, *On topologically realizing modules over the Steenrod algebra*, Ann. of Math.141 1995 321-347.

(La) J. LANNES, *Sur les espaces fonctionnels dont la source est le classifiant d'un p-groupe abélien élémentaire*, Publ. I.H.E.S. 75 (1992), 135-244.

(LS) J. LANNES, L. SCHWARTZ, *Sur la structure des $\mathcal{A}$-modules instables injectifs*, Topology 28 (1989), 153-169.

(LZ) J. LANNES, S. ZARATI, *Sur les $\mathcal{U}$-injectifs*, Ann. Scient. Ec. Norm. Sup. 19 (1986), 1-31.

(Mi) H.R. MILLER, *The Sullivan conjecture on maps from classifying spaces*, Annals of Math. 120 (1984), 39-87; and corrigendum, Annals of Math. 121 (1985), 605-609.

(Ri) C. RINGEL , Tame algebras and Integral Quadratic Forms, SLNM 1099, (1984).

(Sc1) L. SCHWARTZ, *Unstable modules over the Steenrod algebra and Sullivan's fixed pointset conjecture*, Chicago Lectures in Mathematics Series 1994.

(Sc2) L. SCHWARTZ, *A propos de la conjecture de non-réalisation due à N. Kuhn*, Invent. Math. 134 (1998) 211-227.

(Sc3) L. SCHWARTZ, *L'anneau de Grothendieck de la categorie $\mathcal{U}$* , Notes 1991.

(Se) J.-P. SERRE, *Cohomologie modulo 2 des complexes d'Eilenberg-Mac-Lane*, Comm. Math. Helv. 27 (1953), 198-232.

(St) N. E. STEENROD, D. B. EPSTEIN, *Cohomology Operations*, Annals of Math. Studies, 50, Princeton Univ. Press (1962).

LAGA, UMR 7539 DU CNRS, INSTITUT GALILÉE UNIVERSITÉ PARIS NORD, 93430 VILLETANEUSE, FRANCE

E-mail address: schwartz@math.univ-paris13.fr

INFINITE DIMENSIONAL MODULES FOR FINITE GROUPS

D. J. BENSON

ABSTRACT. In the first lecture, I give an introduction to some of the questions currently under investigation in the theory of finite dimensional modular representations of finite groups. In the second lecture, I give some idea of how the theory of infinite dimensional modules has recently had an impact on our understanding of the category of finite dimensional modules. In the third lecture, I present recent progress on understanding phantom maps. These may be viewed as the aspects of infinite dimensional module theory which are invisible from the point of view of the finite dimensional modules.

These notes form a reasonably faithful account of the three lectures I gave at the Euroconference on Infinite Length Modules in Bielefeld in September 1998. I hope they form a useful introduction to the role played by infinite dimensional modules in the representation theory of finite groups. I would like to thank Claus Ringel for his enthusiastic efforts in bringing this conference together and making it so worthwhile.

1. Blocks and Cohomology

This first lecture is intended as a gentle introduction to modular representation theory, with emphasis on describing some of the problems which have motivated recent developments in the subject. In recent years, progress has been made on a number of these problems, and the second lecture will describe some aspects of recent progress which have depended on developments in the theory of infinite dimensional modules. Rickard's idempotent modules and functors have played a central role in these developments.

Let k be a field and G be a finite group. If the characteristic of k is zero or a prime not dividing the group order, then the representation theory is fairly well understood:

• Every kG-module can be written as a direct sum of simple (or irreducible) modules, so that simple and indecomposable are equivalent.

• Every kG-module is projective.

• $H^n(G, M) = 0$ for $n > 0$ for every kG-module M.

• The category $\mathsf{mod}(kG)$ of finitely generated kG-modules is easy to describe. Everything is determined by *character theory:* If k has characteristic zero, the

The author is partly supported by a grant from the NSF.

character of a finitely generated kG-module M is the function on G sending the element g to the trace $\mathrm{Tr}(g, M)$. This is constant on conjugacy classes, and modules are determined by their characters. If k has characteristic p, it is necessary to lift eigenvalues to characteristic zero before adding, to get a "Brauer trace". If p does not divide the group order, then modules are again determined by their Brauer characters.

If p does divide the group order, the Brauer character exactly determines the set of composition factors of a finitely generated kG-module, but it does not indicate how these composition factors extend each other. The phrase "modular representation" refers to a representation in characteristic p with p dividing the group order.

As an example, let G be the group $\mathbb{Z}/2 \times \mathbb{Z}/2$, of order four, generated by two commuting elements g_1 and g_2 of order two. Let k be a field of characteristic two. Then the representations

$$g_1 \mapsto \begin{pmatrix} 1 & 1 \\ 0 & 1 \end{pmatrix} \qquad g_2 \mapsto \begin{pmatrix} 1 & \lambda \\ 0 & 1 \end{pmatrix}$$

are indecomposable but not simple or projective. Different values of λ give pairwise nonisomorphic modules, so that if k is an infinite field then we get an infinite collection of isomorphism classes of two dimensional indecomposable modules this way.

There are also infinite dimensional indecomposable modules, and almost every conceivable pathology occurs in the infinite dimensional representation theory. For example, Krull–Schmidt fails for finite direct sums of indecomposables, and even cancellation fails. There are modules with no indecomposable direct summands, and there exists a module M which is isomorphic to $M \oplus M \oplus M$ but not to $M \oplus M$. Examples like this occur whenever there are infinitely many isomorphism classes of indecomposable finitely generated kG-modules, and that happens precisely when the Sylow p-subgroups of G are noncyclic, where p is the characteristic. This is related to the subject of Paul Eklof's article in these proceedings.

Blocks

Decompose the group algebra kG as a Cartesian product of indecomposable two-sided ideals:

$$kG = B_0 \times B_1 \times \cdots \times B_s.$$

This corresponds to a decomposition of the identity

$$1 = e_0 + e_1 + \cdots + e_s$$

into a sum of primitive orthogonal central idempotents. If M is any kG-module, we have a direct sum decomposition

$$M = e_0M \oplus e_1M \oplus \cdots \oplus e_sM.$$

If M is indecomposable, then there exists a unique index j such that $M = e_jM$ and $e_iM = 0$ for $i \neq j$. We then say that M is "in the block B_j." We label so that the trivial module k (i.e., the field k with each group element acting as the identity) is in B_0, and we call B_0 the "principal block."

This way, we really regard a block as a receptacle for primitive central idempotents, two sided ideal direct summands of the group algebra, indecomposable modules, and so on. Another way to view it is that the category $\mathsf{mod}(kG)$, whose objects are the finitely generated kG-modules and whose morphisms are the module homomorphisms, decomposes as a sum of categories, one for each block.

Every block has associated to it a *defect group* D, which is a subgroup of G, minimal such that every module in B is a direct summand of a module induced from D. The defect groups of a block form a single conjugacy class of p-subgroups of G. For example, for B_0, the defect groups are the Sylow p-subgroups. The defect groups of a block B are trivial if and only if B is isomorphic to a matrix algebra over a division ring with k in its center. This is the case which most closely corresponds to the characteristic zero situation.

The first main theorem of Brauer says that there is a natural one-one correspondence between blocks of G with defect group D and blocks of $N_G(D)$ with defect group D. This is called the *Brauer correspondence.*

Cohomology

We view $\mathrm{Ext}^n_{kG}(M,N)$ in terms of n-fold extensions, which are exact sequences of the form

$$0 \to N \to M_1 \to \ldots \to M_n \to M \to 0.$$

We regard two such exact sequences as equivalent if there is a diagram of the form

$$\begin{array}{ccccccccccccc} 0 & \longrightarrow & N & \longrightarrow & M_1 & \longrightarrow & \cdots & \longrightarrow & M_n & \longrightarrow & M & \longrightarrow & 0 \\ & & \| & & \downarrow & & & & \downarrow & & \| & & \\ 0 & \longrightarrow & N & \longrightarrow & M_1' & \longrightarrow & \cdots & \longrightarrow & M_n' & \longrightarrow & M & \longrightarrow & 0 \end{array}$$

This is not quite an equivalence relation, but we complete it to one, and the equivalence classes are the elements of $\mathrm{Ext}^n_{kG}(M,N)$. The zero element is represented by such an exact sequence which is a splice of split short exact sequences. We denote $\mathrm{Ext}^n_{kG}(k,M)$ by $H^n(G,M)$.

There is a natural map from $H^*(G,k)$ to $\mathrm{Ext}^*_{kG}(M,M)$ given by tensoring long exact sequences with M. Usually, when we talk of tensor products of modules, we are referring to the tensor product over the field, with the diagonal action of the group. This natural map is a ring homomorphism, and lands in the graded center of $\mathrm{Ext}^*_{kG}(M,M)$. Via this map, we regard $\mathrm{Ext}^*_{kG}(M,M)$ as a module over $H^*(G,k)$. Then its annihilator as a module is the same as the annihilator of the identity element, namely the kernel of this ring homomorphism.

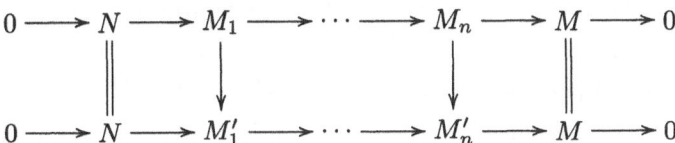

Another way of viewing cohomology is as follows. We denote by $\Omega(M)$ the kernel of a surjection from a projective module to M, and $\Omega^{-1}(M)$ denotes the cokernel of an injective map from M to an injective module. These operations are not quite well defined in general, but Schanuel's lemma shows that Ω is well defined up to adding and removing projective summands, and Ω^{-1} is well defined up to adding and removing injective summands. In the case we're interested in, namely kG-modules, projective modules are the same as injective modules, and the operations Ω and Ω^{-1} are inverse, to the extent that they are well defined. With these definitions, we have

$$\mathrm{Ext}^n_{kG}(M,N) \cong \underline{\mathrm{Hom}}_{kG}(\Omega^n M, N) \cong \underline{\mathrm{Hom}}_{kG}(M, \Omega^{-n} N)$$

where $\underline{\mathrm{Hom}}$ denotes homomorphisms modulo projective homomorphisms. A projective homomorphism is one which factors through some projective module.

If M and N are in different blocks, then $\mathrm{Ext}^n_{kG}(M,N) = 0$, because we can decompose any extension of M by N, using block idempotents, as a direct sum of an extension of M by the zero module and an extension of the zero module by N. This can also be seen using the $\underline{\mathrm{Hom}}$ description given above. In particular, if M is not in the principal block B_0 then $H^*(G,M) = 0$.

Since kG is a symmetric algebra, simple modules M and N are in the same block if and only if for some $n \geq 0$ there exists a chain of simple modules $M_1, \dots,$ M_n such that $\mathrm{Ext}^1_{kG}(M,M_1) \neq 0$, $\mathrm{Ext}^1_{kG}(M_1,M_2) \neq 0, \dots, \mathrm{Ext}^1_{kG}(M_n,N) \neq 0$.

Some Problems

The following list of problems includes some which provide major impetus for current research, as well as some which are intended to motivate the discussion in the next lecture. In particular, I have included some which have recently been solved using the theory of infinite dimensional modules.

1. If S is a simple module in B_0, is $H^n(G,S)$ necessarily nonzero for some $n \geq 0$? **Open**

References: Linnell [36], Linnell and Stammbach [37, 38].

2. Under what circumstances is it true that for all finitely generated nonprojective modules M in B_0 we have $H^n(G,M) \neq 0$ for some $n > 0$? **Solved**

References: Benson, Carlson and Robinson [11], Benson [5].

3. Does there exist an integer $n > 0$ such that if G is a finite group, k is a field and M is a *faithful* absolutely irreducible kG-module then $\dim_k H^1(G,M) \leq n$? Can we take $n = 2$? **Open**

References: Guralnick [29, 30].

4. Does there exist a value of $n > 0$ such that if $H^i(G,\mathbb{Z}) = 0$ for $i \leq n$ then G is the trivial group? **Open**

References: Loday originally conjectured that n can be taken to be 4. However, Milgram (1993) (see [1]) showed that $H^i(M_{23},\mathbb{Z}) = 0$ for $i \leq 5$.

5. How do you tell whether a kG-module is projective?

The reason why this is important is that a lot of effort goes into constructing projective resolutions for modules. For example, the construction of multiply periodic resolutions in [7] depends crucially on tests for projectivity.

References: Much is known. Chouinard's theorem [21], Dade's lemma in the finitely generated [23] and infinitely generated [9] cases, varieties for modules [4, 19], etc.

6. If M is a finitely generated indecomposable *periodic* kG-module (i.e., there is a periodic projective resolution of M) does there exist a projective module P such that $M \oplus P$ is isomorphic to a module induced from a suitable p-local subgroup?

Solved

[A p-local subgroup is defined to be a normalizer of a non-trivial p-subgroup of the group]

What about nonperiodic modules?

References: For the periodic case, see [5]. For the nonperiodic case, Carlson [20], Section 13, is probably about all that can be said.

7. Describe the *thick* subcategories of the stable category of finitely generated kG-modules. In other words, find all collections of modules closed under the following operations:

(i) finite direct sums and summands,

(ii) Ω and Ω^{-1}, and

(iii) extensions. **Solved** for p-groups but not in general

References: Benson, Carlson and Rickard [10].

8. (Broué's Conjecture): If the defect groups D of a block B are abelian, then the derived category of bounded complexes of finitely generated modules in B is equivalent to the derived category for the block b of $N_G(D)$ which corresponds to B under the Brauer correspondence:

$$D^b(\mathsf{mod}(B)) \simeq D^b(\mathsf{mod}(b)).$$

Solved for D cyclic (Rickard), and a few other cases. **Open** in general

References: Broué [15, 16, 17], Külshammer [34], Linckelmann [35], Rickard [42, 43]. [Jeremy Rickard's second talk, in these proceedings, contains an account of some new progress on this problem.]

Two blocks B_1 and B_2 are said to be *Morita equivalent* if their categories of finitely generated modules are equivalent:

$$\mathsf{mod}(B_1) \simeq \mathsf{mod}(B_2).$$

9. (Donovan's conjecture) Given a field k of characteristic p and a p-group D, do there only exist a finite number of Morita equivalence classes of blocks of group algebras with defect group D?

Known for D cyclic; known for G symmetric groups (Scopes); **Open** in general

References: Alperin [2], Scopes [48], Kessar [32].

10. (Alperin's conjecture) Is the number of simple kG-modules equal to the sum over the conjugacy classes of p-subgroups Q of G, of the number of projective

simple $k(N_G(Q)/Q)$-modules? Much studied, but still **Open**
(There is also a block by block version of Alperin's conjecture)

References: Alperin [3], Dade [24, 25], Knörr and Robinson [33], Külshammer [34], Robinson [45, 46], Robinson and Staszewski [47], Thévenaz [49, 50].

Rickard's Idempotent Modules

In the next lecture, we shall describe how some recent progress on some of these problems has depended on Rickard's infinite dimensional idempotent modules [44]. In preparation for this, we end this first lecture with a description of the easiest case of these idempotent modules,

Let

$$\begin{aligned}\zeta \in H^n(G,k) &\cong \mathrm{Ext}^n_{kG}(k,k)\\ &\cong \underline{\mathrm{Hom}}_{kG}(\Omega^n k, k)\\ &\cong \underline{\mathrm{Hom}}_{kG}(k, \Omega^{-n} k)\\ &\cong \underline{\mathrm{Hom}}_{kG}(\Omega^{-n} k, \Omega^{-2n} k)\\ &\cong \dots\end{aligned}$$

Then we can associate to ζ a filtered system of modules

$$k \xrightarrow{\hat{\zeta}} \Omega^{-n}k \xrightarrow{\hat{\zeta}} \Omega^{-2n}k \to \dots$$

by choosing representative maps, and we define $F(\zeta)$ to be the colimit of this sequence. There is a choice involved, but different choices give colimits which are isomorphic up to adding and removing projective summands.

The module $F(\zeta)$ constructed in this manner comes with an obvious map $k \to F(\zeta)$. We can make this surjective by adding a suitable projective module to k, and then the kernel is denoted $E(\zeta)$. These modules $E(\zeta)$ and $F(\zeta)$ are the Rickard idempotent modules associated to the element ζ in cohomology. Some of their properties are as follows.

$$\begin{aligned}&E(\zeta) \otimes E(\zeta) \cong E(\zeta) \oplus (\text{Projective})\\ &F(\zeta) \otimes F(\zeta) \cong F(\zeta) \oplus (\text{Projective})\\ &E(\zeta) \otimes F(\zeta) \text{ is projective}\end{aligned}$$

For a finitely generated kG-module M, the following are equivalent:

(i) Some power of ζ annihilates $\mathrm{Ext}^*_{kG}(M,M)$.

(ii) $M \otimes F(\zeta)$ is projective

(iii) $M \otimes E(\zeta) \cong M \oplus (\text{Projective})$

To give an explicit example, let $G = \mathbb{Z}/2 \times \mathbb{Z}/2$ with commuting generators g_1 and g_2 of order two, and k be a field of characteristic two. Then

$$H^*(G,k) \cong k[x,y],$$

a polynomial ring in two variables x and y in degree one. We represent $\Omega^{-1}k$ with the following diagram:

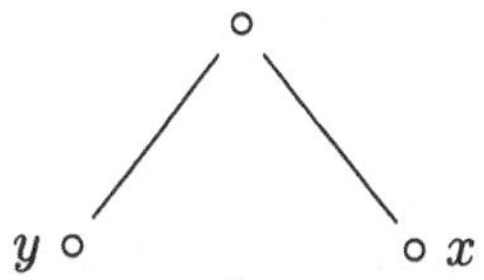

The vertices in this diagram represent basis elements as a k-vector space. Multiplication by $g_1 - 1$ acts on these basis elements by going down and to the left along a diagonal line, or by zero if there is no such diagonal line. Multiplication by $g_2 - 1$ acts similarly, but down and to the right.

The vertex marked x represents the image of the element

$$x \in H^1(G,k) \cong \mathrm{Hom}_{kG}(k, \Omega^{-1}k),$$

and similarly for y.

With this notation, $\Omega^{-2}(k)$ is represented by the diagram

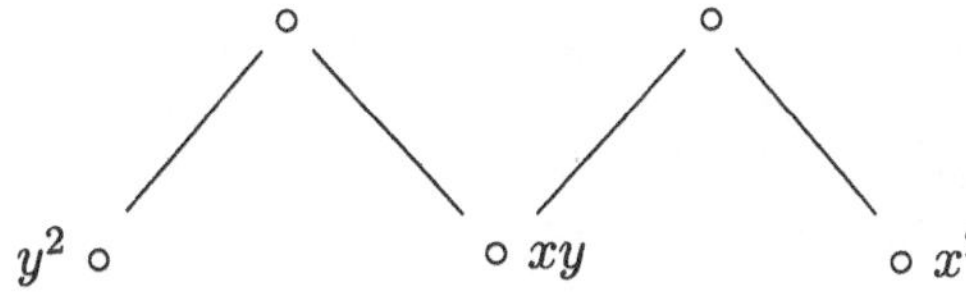

The map represented by x from $\Omega^{-1}k$ to $\Omega^{-2}k$ embeds the first of the above diagrams as the right hand part of the second. Continuing this way and taking the colimit, we see that $F(x)$ is the infinite dimensional module represented by the following diagram:

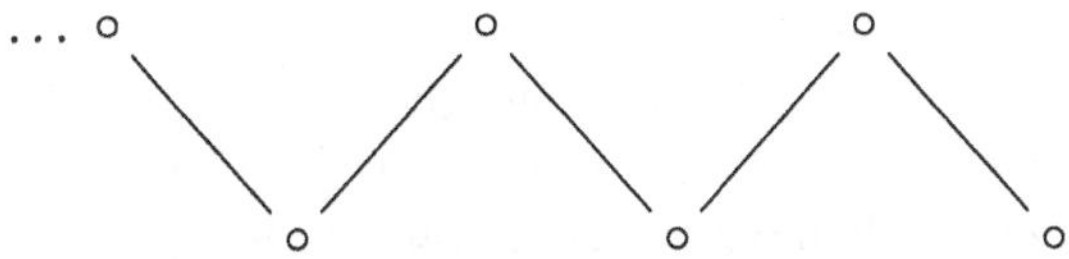

The map from k to $F(x)$ has the rightmost vertex as image. The module $E(x)$ is similar:

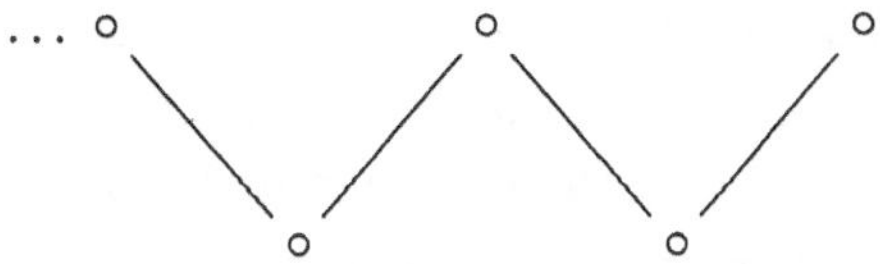

and the map from $E(x)$ to k sends all except the rightmost vertex to zero.

The endomorphism rings of these modules are quite interesting:

$$\underline{\mathrm{End}}_{kG}(F(x)) \cong k[t] \qquad \underline{\mathrm{End}}_{kG}(E(x)) \cong k[[t]].$$

2. VARIETIES AND THE NUCLEUS

In this section, we introduce the machinery of varieties for modules, and indicate the relevance of Rickard's idempotent modules to various questions about finite dimensional modules.

Varieties for Finitely Generated Modules

We denote by V_G the maximal ideal spectrum of the cohomology ring $H^*(G,k)$. This is not quite a commutative ring, but rather satisfies the graded commutativity relations $xy = (-1)^{\deg(x)\deg(y)}yx$. If $p = 2$ then it is strictly commutative. If p is odd then elements of odd degree square to zero, and are in the the nil radical. Modulo the ideal generated by the odd degree elements, the ring is commutative.

By a theorem of Evens [26], $H^*(G,k)$ is finitely generated as a k-algebra, and so by the above comments, its maximal ideal spectrum may be viewed as an affine variety over k. A choice of s generators for $H^*(G,k)$ (modulo its nil radical) gives an explicit embedding as a homogeneous subvariety of s-dimensional affine space over k, $\mathbb{A}^s(k)$.

As an example, if $G \cong (\mathbb{Z}/p)^r$ is an elementary abelian p-group of rank r and k has characteristic p, then the cohomology ring is given as follows. If $p = 2$, it is a polynomial ring in r degree one generators. If p is odd, it is a tensor product of a polynomial ring on r degree two generators and an exterior algebra on r degree one independent generators. In the latter case, the degree one generators square to zero, and generate the nil radical. Modulo these, the ring is polynomial on the r degree two generators. So in either case, the variety V_G is isomorphic to $\mathbb{A}^r(k)$.

An explicit description of V_G for a general finite group in terms of its elementary abelian p-subgroups was given by Quillen [40, 41]. For each conjugacy class of elementary abelian p-subgroup E of the group, the corresponding affine space V_E is quotiented by the action of the normalizer of E, and then these quotients are glued together according to the maps given by the inclusions between elementary abelian subgroups.

If M is a finitely generated kG-module, then $\operatorname{Ext}^*_{kG}(M,M)$ is finitely generated as a module over $H^*(G,k)$. Its annihilator $I_M \subseteq H^*(G,M)$ is a homogeneous ideal, so it defines a closed homogeneous subvariety of V_G, denoted $V_G(M)$. This is what is referred to as the *variety* of M. These varieties for modules have the following properties:

(i) $V_G(M \oplus N) = V_G(M) \cup V_G(N)$.

(ii) $V_G(M \otimes N) = V_G(M) \cap V_G(N)$.

(iii) $\dim V_G(M)$ is equal to an invariant of M called the *complexity*, which measures the polynomial rate of growth of the minimal projective resolution of M.

(iv) $V_G(M) = \{0\}$ if and only if M is projective.

(v) If $\hat{\zeta} : \Omega^n k \to k$ represents $\zeta \in H^n(G,k) \cong \underline{\mathrm{Hom}}_{kG}(\Omega^n k, k)$ then the module L_ζ defined by the short exact sequence

$$0 \to L_\zeta \to \Omega^n k \xrightarrow{\hat{\zeta}} k \to 0$$

satisfies

$$V_G(L_\zeta) = V_G\langle\zeta\rangle,$$

the hypersurface determined by ζ as an element of the coordinate ring of V_G.

(vi) If

$$0 \to M_1 \to M_2 \to M_3 \to 0$$

is a short exact sequence of kG-modules then for any permutation of the indices $\{i, j, k\} = \{1, 2, 3\}$ we have

$$V_G(M_i) \subseteq V_G(M_j) \cup V_G(M_k).$$

Properties (ii) and (v) show that every closed homogeneous subvariety is the variety of some module, namely an appropriate tensor product of modules of the form L_ζ.

It should be mentioned that it is possible to calculate the varieties $V_G(M)$ without knowing anything about cohomology. The idea is that it is enough to calculate on the elementary abelian p-subgroups, and for an elementary abelian p-group the variety of the module is essentially the same as the *rank variety* introduced by Carlson [19]. This is a subvariety of affine space defined by looking at the restriction to certain cyclic subgroups of the group algebra, called *cyclic shifted subgroups.* The points of affine space index these cyclic shifted subgroups, and a point is in the variety of the module if and only if its restriction to the corresponding cyclic shifted subgroup is not projective. The rank variety is not just a calculational tool; it is used, for example, in the proof of the tensor product property (ii) above.

An important ingredient in setting up the theory of varieties for finitely generated modules is *Dade's Lemma* [23], which says that a finitely generated kG-module is projective if and only if its restriction to every cyclic shifted subgroup is projective.

The Nucleus

A finite group G is said to be *p-nilpotent* if it has a normal subgroup $N \trianglelefteq G$ of order prime to p and index a power of p.

For any finite group G, the *nucleus* $Y_G \subseteq V_G$ is defined to be the union of the images of $V_H \to V_G$ as H runs over the subgroups whose centralizers are *not* p-nilpotent. If there is no such subgroup, Y_G is defined to be $\{0\}$. In the definition of the nucleus, it suffices to let H run over the elementary abelian p-subgroups whose centralizers are not p-nilpotent. This definition comes from [5, 10, 11].

Theorem 2.1. *The subvariety Y_G is equal to the union of the $V_G(M)$ as M runs over the finitely generated kG-modules in the principal block B_0 satisfying $H^n(G, M) = 0$ for all $n > 0$.*

This is the main theorem of [5], and in the remainder of this lecture, we shall try to give some idea of how Rickard's idempotent modules featured in the proof.

Corollary 2.2. *The following conditions on a group G are equivalent:*

(i) *For every finitely generated nonprojective kG-module in B_0, $H^n(G, M) \neq 0$ for all $n > 0$.*

(ii) $Y_G = \{0\}$.

(iii) *The centralizer of each element of order p in G is p-nilpotent.*

This corollary answers Problem **2** of the first lecture.

Thick Subcategories

The *stable category* of finitely generated kG-modules $\mathsf{stmod}(kG)$ has the same objects as the module category $\mathsf{mod}(kG)$, but for the arrows, instead of $\mathrm{Hom}_{kG}(M, N)$ we use the quotient by the projective homomorphisms, $\underline{\mathrm{Hom}}_{kG}(M, N)$. This means that $\mathsf{stmod}(kG)$ is not an abelian category, because we can no longer tell whether a morphism is injective, or whether it is surjective. Instead, it carries a structure of *triangulated category*, with the distinguished triangles corresponding to the short exact sequences in $\mathsf{mod}(kG)$. The octahedral axiom for triangulated categories corresponds to the third isomorphism theorem in the module category.

Similarly, if we do not restrict ourselves to finitely generated modules, we write $\mathsf{Mod}(kG)$ for the category of all kG-modules and module homomorphisms, and we write $\mathsf{StMod}(kG)$ for the corresponding stable category.

If V is a closed homogeneous subvariety of V_G, we define $\mathcal{C}(V)$ to be the subcategory of the stable module category $\mathsf{stmod}(kG)$ consisting of the finitely generated kG-modules M satisfying $V_G(M) \subseteq V$.

The category $\mathcal{C}(V)$ is a *thick subcategory* of $\mathsf{stmod}(kG)$, meaning that it is closed under Ω and Ω^{-1}, finite direct sums and direct summands, and extensions (i.e., completing an arrow to a triangle).

More generally, if $\mathcal{V}$ is a collection of closed homogeneous irreducible subvarieties of V_G which is closed under *specialization* (meaning that if $W \subseteq V$ and $V \in \mathcal{V}$ then $W \in \mathcal{V}$), then we have a thick subcategory $\mathcal{C}(\mathcal{V})$ of $\mathsf{stmod}(kG)$ consisting of the finitely generated kG-modules M such that $V_G(M)$ is a finite union of elements of $\mathcal{V}$.

Here are some more examples of thick subcategories of $\mathsf{stmod}(kG)$. If B is a block of kG then $\mathsf{stmod}(B)$ is a thick subcategory. The full subcategory of $\mathsf{stmod}(B_0)$ consisting of modules M satisfying $H^n(G, M) = 0$ for all $n > 0$ is a thick subcategory (the reason why it is closed under Ω and Ω^{-1} is that $H^n(G, \Omega M) \cong H^{n-1}(G, M)$, and if $H^n(G, M) \neq 0$ for some $n > 0$ then the same is true for infinitely many values of $n > 0$ [11]). A collection of finitely generated modules *generates* a thick subcategory, which is the smallest one which contains them. Further thick subcategories can be formed by taking intersections.

Idempotent Functors (Rickard [44])

If $\mathcal{C}$ is a thick subcategory of $\mathsf{stmod}(kG)$, then there are functors $\mathcal{E}_{\mathcal{C}}$ and $\mathcal{F}_{\mathcal{C}}$ on the stable category of *all* kG-modules $\mathsf{StMod}(kG)$ which fit into a triangle

$$\mathcal{E}_{\mathcal{C}}(M) \to M \to \mathcal{F}_{\mathcal{C}}(M).$$

If you are unfamiliar with triangulated categories, think of this as meaning that there is a short exact sequence

$$0 \to \mathcal{E}_{\mathcal{C}}(M) \to M \oplus (\text{projective}) \to \mathcal{F}_{\mathcal{C}}(M) \to 0.$$

The properties of these functors are:

(i) $\mathcal{E}_{\mathcal{C}}(M)$ is a filtered colimit of modules in $\mathcal{C}$.

(ii) $\mathcal{F}_{\mathcal{C}}(M)$ is $\mathcal{C}$-*local* in the sense that $\underline{\mathrm{Hom}}_{kG}(N, \mathcal{F}_{\mathcal{C}}(M)) = 0$ for any module N in $\mathcal{C}$.

The functors $\mathcal{E}_{\mathcal{C}}$ and $\mathcal{F}_{\mathcal{C}}$ are uniquely determined up to natural isomorphism by the above properties. This makes it easy in practice to check that candidates for these functors are what they are claimed to be.

Now consider the case where $\mathcal{C}$ is a *tensor ideal* thick subcategory, in the sense that the tensor product of a module in $\mathcal{C}$ with any module in $\mathsf{stmod}(kG)$ lands in $\mathcal{C}$. This is the case, for example, with the above defined categories $\mathcal{C}(\mathcal{V})$. In this case, the values of $\mathcal{E}_{\mathcal{C}}$ and $\mathcal{F}_{\mathcal{C}}$ are determined by the value on the trivial module. Namely, setting $E_{\mathcal{C}} = \mathcal{E}_{\mathcal{C}}(k)$ and $F_{\mathcal{C}} = \mathcal{F}_{\mathcal{C}}(k)$, in the stable category we have

$$\mathcal{E}_{\mathcal{C}}(M) \cong E_{\mathcal{C}} \otimes M \qquad \mathcal{F}_{\mathcal{C}}(M) \cong F_{\mathcal{C}} \otimes M.$$

As an example, if $\zeta \in H^n(G,k)$ and $\mathcal{V}$ is the set of closed homogeneous irreducible subvarieties of the hypersurface $V\langle\zeta\rangle$ then we get the modules $E(\zeta)$ and $F(\zeta)$ described at the end of the first lecture. More generally, if $\mathcal{C} = \mathcal{C}(\mathcal{V})$ we write $E(\mathcal{V})$ and $F(\mathcal{V})$ for $E_{\mathcal{C}}$ and $F_{\mathcal{C}}$. If $\mathcal{V}$ consists of all the closed homogeneous irreducible subvarieties of a given closed homogeneous subvariety V then we write $E(V)$ and $F(V)$.

We are now in a position to be able to define the variety of an infinite dimensional module. Rather than a single variety, this is a collection of varieties, and is defined as follows. If M is a (not necessarily finitely generated) kG-module, then $\mathcal{V}_G(M)$ is defined to be the collection of nonzero closed homogeneous irreducible subvarieties V of V_G such that

$$M \otimes E(V) \otimes F(\{\text{all proper subvarieties of } V\})$$

is not projective.[1]

Some of the properties of these varieties for not necessarily finitely generated modules are as follows:

(i) If M is finitely generated then $\mathcal{V}_G(M)$ is the set of closed homogeneous irreducible subvarieties of the previously defined cohomology variety $V_G(M)$.

(ii) M is projective if and only if $\mathcal{V}_G(M) = \emptyset$.

(iii) $\mathcal{V}_G(M \oplus N) = \mathcal{V}_G(M) \cup \mathcal{V}_G(N)$.

[1]It should be noted that the original definition of cohomology variety given in [8] was superseded by the definition given in [9], which is equivalent to the definition given here. The original definition did not satisfy the crucial tensor product property (iv). Note that when we write "all proper subvarieties of V", we really mean "all proper closed homogeneous irreducible subvarieties of V".

(iv) $\mathcal{V}_G(M \otimes N) = \mathcal{V}_G(M) \cap \mathcal{V}_G(N)$.

(v) If V is a closed homogeneous irreducible subvariety of V_G and

$$M = E(V) \otimes F(\{\text{all proper subvarieties of } V\})$$

then $\mathcal{V}_G(M) = \{V\}$.

Combining properties (iii) and (v) shows that each collection of closed homogeneous irreducible subvarieties of V_G occurs as the variety of some kG-module.

Just as in the finitely generated case, an important ingredient in the development of the varieties for infinitely generated modules is a version of Dade's lemma in this context. As it stands, Dade's lemma is false for infinitely generated modules. An easy example is given by the module $F(\mathcal{U})$ where $\mathcal{U}$ is the set of all proper subvarieties of V_G, with $G = \mathbb{Z}/2 \times \mathbb{Z}/2$ and where k has characteristic two. This is the module whose underlying k-vector space is $k(t) \oplus k(t)$ (t an indeterminate) with g_1 acting as $\left(\begin{smallmatrix}1&1\\0&1\end{smallmatrix}\right)$ and g_2 acting as $\left(\begin{smallmatrix}1&t\\0&1\end{smallmatrix}\right)$.

What is wrong in the above example is that despite the fact that k is algebraically closed, it is still not big enough. After forming a transcendental extension of k, we find that there are then cyclic shifted subgroups where the restriction is not projective. The infinite dimensional version of Dade's lemma says that a kG-module is projective if and only if its restriction to every cyclic shifted subgroup defined over every extension field is projective. In the above example, we need to pass to a transcendental extension field of k before we can find a cyclic shifted subgroup such that the restriction is not projective. In some sense, the variety of the above module is "V_G with all the proper subvarieties defined over k removed". The extension fields needed are the function fields of subvarieties, and then the cyclic shifted subgroups correspond to "generic points" for subvarieties. The details can be found in [9], and an informal account is given in [6].

We are now ready to sketch the proof of the thick subcategory theorem, which is a good illustration of how the infinitely generated idempotent modules can be used to get information about the category of finitely generated modules.

Theorem 2.3. *If M is a finitely generated kG-module then the thick subcategory of* $\mathsf{stmod}(kG)$ *generated by all modules of the form $M \otimes X$, as X ranges over finitely generated kG-modules, is equal to $\mathcal{C}(V_G(M))$.*

Proof. (Rickard) [10] Let $\langle\langle M\rangle\rangle$ denote the thick subcategory generated by all such $M \otimes X$. It is easy to see that $\langle\langle M\rangle\rangle$ is contained in $\mathcal{C}(V)$ where $V = V_G(M)$. The point is to prove the other inclusion.

For any finitely generated kG-modules N and X we have

$$\begin{aligned} 0 &= \underline{\mathrm{Hom}}_{kG}(M \otimes X, \mathcal{F}_{\langle\langle M\rangle\rangle}\mathcal{E}_V(N)) \\ &\cong \underline{\mathrm{Hom}}_{kG}(X, M^* \otimes \mathcal{F}_{\langle\langle M\rangle\rangle}\mathcal{E}_V(N)) \end{aligned}$$

and so $M^* \otimes \mathcal{F}_{\langle\langle M\rangle\rangle}\mathcal{E}_V(N)$ is projective. So we have

$$\mathcal{V}_G(M) \cap \mathcal{V}_G(\mathcal{F}_{\langle\langle M\rangle\rangle}\mathcal{E}_V(N)) = \emptyset.$$

But the triangle

$$\mathcal{E}_{\langle\langle M\rangle\rangle}\mathcal{E}_{\mathcal{V}}(N) \to \mathcal{E}_{\mathcal{V}}(N) \to \mathcal{F}_{\langle\langle M\rangle\rangle}\mathcal{E}_{\mathcal{V}}(N)$$

shows that

$$\mathcal{V}_G(\mathcal{F}_{\langle\langle M\rangle\rangle}\mathcal{E}_{\mathcal{V}}(N)) \subseteq \mathcal{V}_G(M)$$

and so we have

$$\mathcal{V}_G(\mathcal{F}_{\langle\langle M\rangle\rangle}\mathcal{E}_{\mathcal{V}}(N)) = \emptyset.$$

It follows that $\mathcal{F}_{\langle\langle M\rangle\rangle}\mathcal{E}_{\mathcal{V}}(N)$ is projective. So the map

$$\mathcal{E}_{\langle\langle M\rangle\rangle}(N) = \mathcal{E}_{\langle\langle M\rangle\rangle}\mathcal{E}_{\mathcal{V}}(N) \to \mathcal{E}_{\mathcal{V}}(N)$$

(where the first equality comes from the inclusion $\langle\langle M\rangle\rangle \subseteq \mathcal{C}(\mathcal{V})$) is a stable isomorphism, and hence $\langle\langle M\rangle\rangle$ is equal to $\mathcal{C}(\mathcal{V})$ as required. □

The tensor ideal version of the thick subcategory theorem follows immediately.

Theorem 2.4. *The tensor ideal thick subcategories of* $\mathsf{stmod}(kG)$ *are precisely the categories of the form* $\mathcal{C}(\mathcal{V})$*, as* $\mathcal{V}$*, as* $\mathcal{V}$ *runs over collections of closed homogeneous irreducible subvarieties of* V_G*, closed under specialization.* □

For a p-group, the fact that every finitely generated module has a filtration with quotients isomorphic to the trivial module implies that every thick subcategory of $\mathsf{stmod}(kG)$ is a tensor ideal, and so we get a classification of thick subcategories in this case.

For a more general finite group, if the nucleus Y_G is equal to $\{0\}$ then the classification of the thick subcategories of B_0 goes the same way as in the above theorem.

If $Y_G \neq \{0\}$ then the same techniques can be used to give a classification of:

(i) the thick subcategories of $\mathsf{stmod}(B_0)$ avoiding Y_G, and

(ii) the thick subcategories of the quotient category

$$\mathsf{stmod}(B_0)/\mathsf{stmod}(B_0) \cap \mathcal{C}(Y_G).$$

It appears from examples that in the case where $Y_0 \neq \{0\}$, each line through the origin in Y_G supports an infinite number of thick subcategories of $\mathsf{stmod}(B_0)$.

This gives a partial answer to Problem **7** of the first lecture.

Induction and Restriction

It is about time to explain the origin of the role of the nucleus Y_G in the theory. This comes from a detailed examination of induction and restriction of idempotent modules. In this context, restriction behaves better than induction. Let $i : H \to G$ denote the inclusion of a subgroup H of G, and let $i_* : V_H \to V_G$ be the corresponding map of cohomology varieties. If $\mathcal{V}$ is a collection of closed homogenous irreducible subvarieties of V_G, closed under specialization, then the triangle

$$E(\mathcal{V}) \to k \to F(\mathcal{V})$$

in $\mathsf{stmod}(kG)$ restricts to

$$E(i_*^{-1}(\mathcal{V})) \to k \to F(i_*^{-1}(\mathcal{V}))$$

in $\mathsf{stmod}(kH)$.

Induction is trickier, but we can get good information in a special case of interest. Let $A \leq G$ be an elementary abelian p-subgroup of G, and set $C = C_G(A)$ and $N = N_G(A)$. Let ℓ_0 be a line through the origin in V_A, which is not in the image of $V_{A'} \to V_A$ for any proper subgroup A' of A. Let D be the "diagonalizer" of ℓ_0; namely the subgroup of N consisting of the elements which fix ℓ_0 setwise. The quotient D/C acts faithfully on ℓ_0, so it is a cyclic p'-group.

Denote by ℓ the image of ℓ_0 in V_D, and by L the image of ℓ_0 in V_G.

Theorem 2.5. $E(\ell)\uparrow^G \cong E(L) \oplus$ (projective).

Proof. This theorem is proved in [5]. □

Corollary 2.6. *If* $\mathcal{V}_G(M) = \{L\}$ *then* $M \oplus$ (projective) *is induced from* D.

Proof. Modulo projectives, we have

$$M \cong M \otimes E(L) \cong M \otimes E(\ell)\uparrow^G \cong (M\downarrow_D \otimes E(\ell))\uparrow^G . \qquad \square$$

Since every periodic finitely generated module has a variety of this form for a suitable choice of A and ℓ_0, this corollary gives an answer to Problem **6** of the first lecture.

Actually, since [5] came out, Carlson [18] (see also section 13 of [20]) found a proof of this corollary which does not depend on the idempotent module technology. But at the same time, he wrote down a generalization for certain varieties of larger dimension which has not been proved without this technology.

Further analysis at the level of D gives the following, which explains where the nucleus fits into the picture. Under the circumstances of the above corollary, if C is p-nilpotent and M is in B_0, then $H^*(G, M) \neq 0$. If C is *not* p-nilpotent, then it is possible to construct a finitely generated periodic module M with $V_G(M) = L$ (so $\mathcal{V}_G(M) = \{L\}$) such that $H^*(G, M) = 0$.

We end this lecture with an open question. Consider the thick subcategory of $\mathsf{stmod}(B_0)$ consisting of the finitely generated modules M such that $\underline{\mathrm{Hom}}_{kG}(M, N) = 0$ for all modules N satisfying $H^n(G, N) = 0$ for all $n > 0$. Is this the same as the thick subcategory generated by the trivial module?

Jon Carlson has made some recent progress on this question, and he will report on this elsewhere in these proceedings.

3. Phantom Maps

In the previous two lectures, I have tried to give some indication of how certain infinite dimensional modules play a crucial role in understanding certain aspects of the category of finitely generated modules over a group algebra of a finite group. In this lecture, I want to switch attention to those properties of maps between infinitely generated modules that are invisible to the finite dimensional world.

The appropriate definition is borrowed from algebraic topology, so I shall begin by briefly reviewing the topological situation.

Phantoms in Topology

A map $f : X \to Y$ between CW complexes is said to be *phantom* if its restriction to each skeleton X^n is null homotopic. In other words, there exists a map $\phi : X^n \times [0,1] \to Y$ such that $\phi(x,0) = f(x)$ for all $x \in X^n$, and $\phi(x,1)$ is independent of $x \in X^n$.

This is not the only possible definition of phantom in topology. For example, one might want to use finite subcomplexes instead of the skeleta, or one might want to work in the world of spectra (Christensen and Strickland [22]). A good survey of phantom maps in topology is given in McGibbon [39], where the following theorems are discussed.

Theorem 3.1. *The set of homotopy classes of phantom maps from X to Y is in natural one-one correspondence with the elements of* $\varprojlim^1 [X^n; \Omega Y]$.

Note that on the category of filtered systems of abelian groups, $\varprojlim$ is left exact but not right exact, and $\varprojlim^i$ is used to denote its ith right derived functor. In the context of the above theorem, however, the groups in question need not be abelian. In the context of filtered systems of groups, $\varprojlim$ still has one well defined right derived functor $\varprojlim^1$. A discussion of the construction of this derived functor can be found in Chapter XI, §2 of Bousfield and Kan [14].

Theorem 3.2. *The composite of any two phantom maps is null homotopic.*

Theorem 3.3. *There is a (weakly) universal phantom map*

$$X \to \bigvee_{n=1}^{\infty} \Sigma X^n.$$

In other words, every phantom map out of X factors through this one, but not uniquely.[2]

In fact, the previous theorem follows from this one. Namely, if $X \to Y \to Z$ is a composite of two phantom maps, then it factors as $X \to \bigvee_{n=1}^{\infty} \Sigma X^n \to Y \to Z$, and any phantom map out of a wedge of finite dimensional complexes is null homotopic.

Example (Brayton Gray [28]). There are $2^{\aleph_0}$ homotopy classes of phantom maps from infinite dimensional complex projective space $\mathbb{C}P^\infty$ to the three dimensional sphere S^3.

---

[2] Mike Hopkins tells me that I should call such a map "versal": like universal but without the "uni".

Theorem 3.4. *If X and Y are nilpotent and of finite type, then a map $X \to Y$ is phantom if and only if its composite with the Sullivan profinite completion $Y \to \hat{Y}$ is null homotopic.*

Phantoms in Representation Theory

In representation theory, we make the following definition. Let G be a finite group and k a field of characteristic p. A map $M \to N$ of kG-modules is said to be *phantom* if its restriction to every finitely generated submodule of M factors through a projective module.

A couple of years ago, I asked my graduate student Gilles Gnacadja to look at this definition and see whether there were any interesting examples, and to what extent the topological theorems carried over. The results of his inquiries were more interesting than I had hoped. The first example he came up with was the following.

Example (Gilles Gnacadja [27]). Let G be the group $\mathbb{Z}/2\times\mathbb{Z}/2$ and let k be a field of characteristic two. We saw at the end of the first lecture that the cohomology ring is $H^*(G,k) = k[x,y]$, and we described the modules $E(x)$ and $F(x)$. It turns out that

$$\dim_k \underline{\mathrm{Hom}}_{kG}(F(x), F(y)) = 2^{\aleph_0}$$

and that all maps are phantom.

More generally, for G a finite group and k a field of characteristic p, if ζ, η is a regular sequence of length two consisting of homogeneous elements in $H^*(G,k)$ (actually, this places some restrictions on G), then

$$\dim_k \underline{\mathrm{Hom}}_{kG}(F(\zeta), \Omega^{-1}F(\eta)) = 2^{\aleph_0}$$

and all maps are phantom.

In the above example for $\mathbb{Z}/2\times\mathbb{Z}/2$, $\Omega^{-1}F(y)$ is (stably) isomorphic to $F(y)$.

Write $\mathrm{Ph}(M,N)$ for the space of phantom maps from M to N, and $\underline{\mathrm{Ph}}(M,N)$ for the quotient of $\mathrm{Ph}(M,N)$ by the projective maps. Then in contrast to Gray's example described above, we have the following.

Theorem 3.5 (Gnacadja [27]). *If N is finite dimensional then $\underline{\mathrm{Ph}}(M,N) = 0$.*

Theorem 3.6 (Gnacadja [27]). *Writing M as a filtered colimit of finitely generated kG-modules $M = \varinjlim_{\alpha} M_\alpha$, we have*

$$\underline{\mathrm{Ph}}(M,N) \cong \varprojlim_{\alpha}{}^1 \underline{\mathrm{Hom}}_{kG}(M_\alpha, \Omega N)$$

Universal Phantoms

If $M = \varinjlim_{\alpha} M_\alpha$ as above, then writing I_α for the injective hull of M_α, we define M' by the short exact sequence

$$0 \to \bigoplus_\alpha M_\alpha \to M \oplus \bigoplus_\alpha I_\alpha \to M' \to 0.$$

This is the same as completing $\bigoplus_\alpha M_\alpha \to M$ to a triangle in $\mathsf{StMod}(kG)$:

$$\bigoplus_\alpha M_\alpha \to M \to M'.$$

Then it is shown in [27] that the resulting map $M \to M'$ is a (weakly) universal phantom map out of M.

If either k is countable, or M is countably generated, or G has cyclic Sylow p-subgroups, then M' is a direct sum of finitely generated modules, and the argument used in the topological situation shows that the composite of any two phantom maps out of M is projective. Otherwise, M' is usually not a direct sum of finitely generated modules, and it seems likely that there are examples where the composite of two phantom maps is not projective, although this has not been proved.

As an explicit example, let $G = \mathbb{Z}/p \times \mathbb{Z}/p$ and k be an uncountable field of characteristic p. Let $\mathcal{U}$ be the collection of all proper subvarieties of V_G, and let $M = F(\mathcal{U})$. Then the module M' constructed above is definitely not a direct sum of finite dimensional modules, and there is probably a nonprojective composite of two phantom maps out of M.

Profinite Completion

Next, we look for the analog of Theorem 3.4 in the context of representation theory. To understand what should replace profinite completion, consider for a moment the situation for vector spaces. The collection of finite dimensional quotients of a vector space forms an inverse system, and we can form the inverse limit. This comes with a natural map from the original vector space. It turns out that this construction gives exactly the embedding of the vector space into its double dual. So the corresponding theorem is as follows.

Theorem 3.7. [12] *A map of kG-modules $M \to N$ is phantom if and only if the composite $M \to N \to N^{**}$ is projective.*

Purity

The key to understanding phantom maps turns out to be the concept of purity. The lectures of Birge Huisgen–Zimmermann in these proceedings give an excellent introduction to purity, which I shall not repeat. The relevant definitions are also recalled in [12], and a more extensive reference is the book of Jensen and Lenzing [31].

We just mention that the functor Pext^n is constructed in the same way as Ext^n, except that instead of taking an exact sequence of projective modules resolving M, we take a pure exact sequence of pure projective modules resolving

M. Then we take the cohomology of the complex formed by taking homomorphisms from this sequence into N. Just as in the case of Ext^n, this gives the same answer as if we had taken a pure exact sequence of pure injective modules resolving N, and then taken homomorphisms from M. The functor Pext^1 parametrizes pure extensions.

Theorem 3.8. [12] $\underline{\mathrm{Ph}}(M,N) \cong \mathrm{Pext}^1_{kG}(M,\Omega N)$.

In fact, a map of kG-modules $M \to N$ is phantom if and only if the corresponding extension

$$0 \to \Omega N \to X \to M \to 0$$

is pure exact, where this extension corresponds to the original homomorphism via

$$\underline{\mathrm{Hom}}_{kG}(M,N) \cong \mathrm{Ext}^1_{kG}(M,\Omega N).$$

Following this correspondence through, one finds that the following hold.

Theorem 3.9. [12] *For a kG-module M, the following are equivalent.*

(i) *M is pure projective.*

(ii) *M is a direct sum of finitely generated modules.*

(iii) *There are no phantom maps out of M.*

Theorem 3.10. [12] *For a kG-module M, the following are equivalent.*

(i) *M is pure injective.*

(ii) *The natural inclusion $M \to M^{**}$ splits.*

(iii) *M is a direct summand of a direct product of finitely generated modules.*

(iv) *M is a direct summand of the dual of some module.*

(v) *There are no phantom maps into M.*

Composites of Phantoms

We define $\underline{\mathrm{Ph}}^n(M,N)$ to be the space of maps from M to N which factorize as a composite of n phantoms, where the intermediate modules are not specified.

Theorem 3.11. [12] *$\underline{\mathrm{Ph}}^n(M,N)$ is isomorphic to the image of the natural map*

$$\mathrm{Pext}^n_{kG}(M,\Omega^n N) \to \mathrm{Ext}^n_{kG}(M,\Omega^n N)$$

via the natural isomorphism

$$\mathrm{Ext}^n_{kG}(M,\Omega^n N) \cong \underline{\mathrm{Hom}}_{kG}(M,N).$$

It is still not known whether the image of this map can ever be nonzero in degrees two and higher, although it seems likely that it can.

Cardinality

Theorem 3.12 (Gruson and Jensen). *If R is a ring of cardinality $\aleph_t$ then the pure global dimension of R is at most $t+1$.*

For kG, it turns out that if the representation type is wild[3] then equality holds in the above theorem. So if $|k| = \aleph_t$ then the pure global dimension of kG is exactly $t+1$.

For $\mathbb{Z}/2 \times \mathbb{Z}/2$ in characteristic two, the pure global dimension is one for k countable and two for k uncountable. For the other tame cases, the exact value of the pure global dimension is not known. For finite representation type, the pure global dimension is zero.

Example For $G = \mathbb{Z}/2 \times \mathbb{Z}/4$ in characteristic two, the pure global dimension is $t+1$, where $|k| = \aleph_t$, but $\underline{\mathrm{Ph}}^6(M,N)$ is always zero in this case. To see this, we note that G has a subgroup H of index two and pure global dimension two. So the restriction to H of a composite of three phantom maps is projective. This implies that a composite of three phantom maps is always divisible by the degree one cohomology element $x \in H^1(G, \mathbb{F}_2)$ corresponding to H. So the composite of six phantom maps is divisible by x^2, which is zero.

Example If G has p-rank two then the Rickard idempotent E-modules are pure injective, so there are no phantom maps into them. It is not known whether this holds in higher rank.

Example Denote by $\mathcal{U}$ the collection of all proper subvarieties of V_G. Then $H^*(G, F(\mathcal{U}))$ is flat, and its projective dimension as a module over $H^*(G,k)$ is equal to the pure projective dimension of $F(\mathcal{U})$ as a kG-module.

During the conference, Henning Krause and the author found a proof that $F(\mathcal{U})$ is always pure injective. In fact, it is even endofinite, so that it can be thought of in some sense as *the* generic kG-module. It decomposes as a direct sum of indecomposables corresponding to the irreducible components of the variety V_G. In general, it is rare for Rickard's idempotent F-modules to be pure injective. This can be deduced from the known properties of its endomorphism ring.

A Spectral Sequence

One of the tools for studying composites of phantom maps is the following spectral sequence [13]. We write $M = \varinjlim_{\alpha} M_\alpha$ as a filtered colimit of finitely generated kG-modules. Then we have a spectral sequence

$$\varprojlim_{\alpha}{}^s \mathrm{Ext}^t_{kG}(M_\alpha, \Omega N) \Rightarrow \mathrm{Ext}^{s+t}_{kG}(M, \Omega N).$$

The module in the second slot can be anything, but we have chosen to make it ΩN so that the indices come out right in the five term sequence, which is as follows:

$$\begin{aligned} 0 \to \underline{\mathrm{Ph}}(M,N) \to \underline{\mathrm{Hom}}_{kG}(M,N) \to \varprojlim_{\alpha} \underline{\mathrm{Hom}}_{kG}(M_\alpha, N) \\ \to \mathrm{Pext}^2_{kG}(M, \Omega N) \to \underline{\mathrm{Ph}}^2(M, \Omega^{-1}N) \to 0. \end{aligned}$$

[3]The representation type of kG is finite if G has cyclic Sylow p-subgroups, tame if $p=2$ and G has dihedral, semidihedral or generalized quaternion Sylow 2-subgroups, and wild otherwise.

Usually, the five term sequence coming from a spectral sequence does not go to zero on the right hand end, but here we have replaced the usual version of the last term by the image of the last map.

It follows from this five term sequence that if the pure global dimension of kG is at least two, then either $\underline{\mathrm{Ph}}^2 \neq 0$ or there exists an example of a filtered system in $\mathsf{stmod}(kG)$ which does not lift to a filtered system in $\mathsf{mod}(kG)$. We know of examples where the latter occurs, but we still do not know whether the former occurs.

References

1. A. Adem and R. J. Milgram, *Cohomology of finite groups*, Grundlehren der mathematischen Wissenschaften, vol. 309, Springer-Verlag, Berlin/New York, 1994.
2. J. L. Alperin, *Local representation theory*, Santa Cruz conference on Finite Groups, Proc. Symp. Pure Math., vol. 37, 1980, pp. 369–375.
3. ———, *Weights for finite groups*, The Arcata Conference on Representations of Finite Groups, Proc. Symp. Pure Math., vol. 47, part 1, American Math. Society, 1987, pp. 369–379.
4. D. J. Benson, *Representations and Cohomology II: Cohomology of groups and modules*, Cambridge Studies in Advanced Mathematics, vol. 31, Cambridge University Press, 1991.
5. ———, *Cohomology of modules in the principal block of a finite group*, New York Journal of Mathematics **1** (1995), 196–205.
6. ———, *Cohomology of modules for a finite group*, Representation Theory of Finite Groups (R. Solomon, ed.), Ohio State University Mathematical Research Institute Publications, vol. 6, de Gruyter, 1997, pp. 11–17.
7. D. J. Benson and J. F. Carlson, *Complexity and multiple complexes*, Math. Zeit. **195** (1987), 221–238.
8. D. J. Benson, J. F. Carlson, and J. Rickard, *Complexity and varieties for infinitely generated modules, I*, Math. Proc. Camb. Phil. Soc. **118** (1995), 223–243.
9. ———, *Complexity and varieties for infinitely generated modules, II*, Math. Proc. Camb. Phil. Soc. **120** (1996), 597–615.
10. ———, *Thick subcategories of the stable module category*, Fundamenta Mathematicae **153** (1997), 59–80.
11. D. J. Benson, J. F. Carlson, and G. R. Robinson, *On the vanishing of group cohomology*, J. Algebra **131** (1990), 40–73.
12. D. J. Benson and G. Ph. Gnacadja, *Phantom maps and purity in modular representation theory, I*, Fundamenta Mathematicae **161** (1999), 37–91.
13. ———, *Phantom maps and purity in modular representation theory, II*, Algebras and Representation Theory, to appear.
14. A. K. Bousfield and D. M. Kan, *Homotopy limits, completions and localizations*, Lecture Notes in Mathematics, vol. 304, Springer-Verlag, Berlin/New York, 1972.
15. M. Broué, *Blocs, isométries parfaites, catégories dérivées*, Comptes Rendus Acad. Sci. Paris, Série I **307** (1988), 13–18.
16. ———, *Isométries de caractères et équivalences de Morita ou dérivées*, Publ. Math. Inst. Hautes Études Sci. **71** (1990), 45–63.
17. ———, *Isométries parfaites, types de blocs, catégories dérivées*, Astérisque **181–182** (1990), 61–92.
18. J. F. Carlson, *Varieties for cohomology with twisted coefficients*, Acta Math. Sin. (Engl. Ser.) **15** (1999), 81–92.
19. ———, *The varieties and cohomology ring of a module*, J. Algebra **85** (1983), 104–143.
20. ———, *Modules and Group Algebras*, Lectures in Mathematics, ETH Zürich, Birkhäuser, 1996.

21. L. Chouinard, *Projectivity and relative projectivity over group rings*, J. Pure & Applied Algebra **7** (1976), 278–302.
22. J. D. Christensen and N. P. Strickland, *Phantom maps and homology theories*, Topology **37** (1998), 339–364.
23. E. C. Dade, *Endo-permutation modules over p-groups, II*, Ann. of Math. **108** (1978), 317–346.
24. ———, *Counting characters in blocks, I*, Invent. Math. **109** (1992), 187–210.
25. ———, *Counting characters in blocks, II*, J. Reine & Angew. Math. **448** (1994), 97–190.
26. L. Evens, *The cohomology ring of a finite group*, Trans. Amer. Math. Soc. **101** (1961), 224–239.
27. G. Ph. Gnacadja, *Phantom maps in the stable module category*, J. Algebra **201** (1998), 686–702.
28. B. Gray, *Spaces of the same n-type, for all n*, Topology **5** (1966), 241–243.
29. R. M. Guralnick, *The dimension of the first cohomology group*, Representation Theory II: Groups and Orders. Springer Lecture Notes in Mathematics 1178, Springer-Verlag, Berlin/New York, 1986, pp. 94–97.
30. R. M. Guralnick and C. Hoffman, *The first cohomology group and generation of simple groups*, Groups and Geometries (Siena, 1996), Birkhäuser, Basel, 1998, pp. 81–89.
31. C. U. Jensen and H. Lenzing, *Model theoretic algebra*, Gordon and Breach, 1989.
32. R. Kessar, *Blocks and source algebras for the double covers of the symmetric and alternating groups*, J. Algebra **186** (1996), 872–933.
33. R. Knörr and G. R. Robinson, *Some remarks on Alperin's conjecture*, J. London Math. Soc. **39** (1989), 48–60.
34. B. Külshammer, *Offene Probleme in der Darstellungstheorie endlicher Gruppen*, Jahresber. Deutsch. Math.-Verein **94** (1992), 98–104.
35. M. Linckelmann, *Derived equivalence for cyclic blocks over a P-adic ring*, Math. Zeit. **207** (1991), 293–304.
36. P. A. Linnell, *Cohomology of finite soluble groups*, J. Algebra **107** (1987), 53–62.
37. P. A. Linnell and U. Stammbach, *The block structure and Ext of p-soluble groups*, J. Algebra **108** (1987), 280–282.
38. ———, *The cohomology of p-constrained groups*, J. Pure & Applied Algebra **49** (1987), 273–279.
39. C. A. McGibbon, *Phantom maps*, Handbook of algebraic topology (I. M. James, ed.), North Holland, Amsterdam, 1995, pp. 1209–1257.
40. D. G. Quillen, *The spectrum of an equivariant cohomology ring, I*, Ann. of Math. **94** (1971), 549–572.
41. ———, *The spectrum of an equivariant cohomology ring, II*, Ann. of Math. **94** (1971), 573–602.
42. J. Rickard, *Derived categories and stable equivalence*, J. Pure & Applied Algebra **61** (1989), 303–317.
43. ———, *Splendid equivalences: derived categories and permutation modules*, Proc. London Math. Soc. **72** (1996), 331–358.
44. ———, *Idempotent modules in the stable category*, J. London Math. Soc. **178** (1997), 149–170.
45. G. R. Robinson, *Alperin's conjecture, numbers of characters, and Euler characteristics of quotients of p-subgroup complexes*, J. London Math. Soc. **52** (1995), 88–96.
46. ———, *Local structure, vertices and Alperin's conjecture*, Proc. London Math. Soc. **72** (1996), 312–330.
47. G. R. Robinson and R. Staszewski, *More on Alperin's conjecture*, Astérisque **181–182** (1990), 237–255.
48. J. Scopes, *Cartan matrices and Morita equivalence for blocks of the symmetric groups*, J. Algebra **142** (1991), 441–455.

49. J. Thévenaz, *Locally determined functions and Alperin's conjecture*, J. London Math. Soc. **45** (1992), 446–468.
50. ———, *Equivariant K-theory and Alperin's conjecture*, J. Pure & Applied Algebra **85** (1993), 185–202.

DEPARTMENT OF MATHEMATICS, UNIVERSITY OF GEORGIA, ATHENS, GA 30602, USA
E-mail address: djb@byrd.math.uga.edu

Trends in Mathematics, © 2000 Birkhäuser Verlag Basel/Switzerland

BOUSFIELD LOCALIZATION FOR REPRESENTATION THEORISTS

JEREMY RICKARD

1. Introduction

Bousfield localization is a technique that has been used extensively in algebraic topology, specifically in stable homotopy theory, over the last quarter century. Its name derives from the fundamental work of Bousfield [3], although this is based on earlier work of Brown [4] and Adams [1]. In abstract terms, Bousfield localization deals with the inclusion of a thick subcategory (i.e., a triangulated subcategory closed under direct summands) into a triangulated category and the existence of adjoint functors to such an inclusion. In the original topological setting, the triangulated category was typically the stable homotopy category, and the thick subcategory was defined by the vanishing of some homology theory, but the techniques involved work much more generally. In particular, the triangulated category can be one of interest to representation theorists, such as a stable module category or the derived category of a module category: since the techniques rely heavily on limiting procedures, however, one is forced to work with infinite dimensional modules.

In this article I will look at the simplest form of Bousfield localization – so-called 'finite localization' – which is good enough for most of the applications to representation theory to date, and I will briefly describe a more general form.

For the most part, I will eschew applications, referring instead to the article by Dave Benson in these proceedings. However, in the final section I will describe a recent application of a technique very similar to Bousfield localization to the construction of equivalences of derived categories.

Except for the last section, almost nothing in this article is original. Much of what I know about Bousfield localization I learned from the papers of Neeman [2, 7, 8, 9, 10, 11], who has enthusiastically argued that these techniques are much underused outside algebraic topology and has shown that they can be used to simplify substantially the proofs of certain theorems of algebraic geometry and algebraic K-theory. This article owes a lot to this work of Neeman, though I mostly restrict my attention to the triangulated categories that are of interest in representation theory.

2. Homotopy colimits

Many of the constructions that we shall describe here depend on taking 'direct limits' of sequences of objects in some triangulated category. However, direct limits of such sequences, in the strict category-theoretic sense, rarely exist in a triangulated category, so we shall have to use a substitute.

To motivate this, let us first recall one method of constructing the direct limit of a sequence

$$X_0 \xrightarrow{\alpha} X_1 \xrightarrow{\alpha} X_2 \longrightarrow \dots$$

of objects in an abelian category, such as a module category (for convenience of notation we shall use α to denote all the maps in this sequence – it will be clear from the context which of the maps we are referring to). This method is to express the direct limit as the cokernel of the injective endomorphism $1-\alpha$ of the direct sum of the objects X_i, so we have a short exact sequence

$$0 \longrightarrow \bigoplus_{i=0}^{\infty} X_i \xrightarrow{1-\alpha} \bigoplus_{i=0}^{\infty} X_i \longrightarrow \operatorname{colim}(X_i) \longrightarrow 0.$$

In a triangulated category we can mimic this, so long as the category has countable direct sums, by substituting a distinguished triangle for the short exact sequence. Thus we define the homotopy colimit of the sequence by the distinguished triangle

$$\bigoplus_{i=0}^{\infty} X_i \xrightarrow{1-\alpha} \bigoplus_{i=0}^{\infty} X_i \longrightarrow \operatorname{hocolim}(X_i) \longrightarrow \bigoplus_{i=0}^{\infty} X_i[1].$$

The homotopy colimit shares some, but not all, of the properties of a genuine colimit. Let Y be another object of the category. If we apply the functor

$\operatorname{Hom}(?, Y)$ to the distinguished triangle above, then we get a long exact sequence including

$$\operatorname{Hom}(\operatorname{hocolim}(X_i), Y) \longrightarrow \operatorname{Hom}(\bigoplus_{i=0}^{\infty} X_i, Y) \xrightarrow{1-\alpha} \operatorname{Hom}(\bigoplus_{i=0}^{\infty} X_i, Y).$$

If we choose a collection of maps

$$\beta_i : X_i \longrightarrow Y$$

with the compatibility conditions that $\alpha\beta_i = \beta_{i-1}$, then this represents an element of the kernel of the second map in the exact sequence. By exactness, this implies that there is a homomorphism

$$\beta : \operatorname{hocolim}(X_i) \longrightarrow Y$$

such that each β_i is the composition of the natural map

$$X_i \longrightarrow \operatorname{hocolim}(X_i)$$

with β. However, for a genuine colimit we would need β to be unique, which is only the case if the first map

$$\mathrm{Hom}(\mathrm{hocolim}(X_i), Y) \longrightarrow \mathrm{Hom}(\bigoplus_{i=0}^{\infty} X_i, Y)$$

in the exact sequence is injective, which will not usually be the case.

The paper [2] of Bökstedt and Neeman deals in much greater depth with the notion of homotopy colimits in triangulated categories.

3. Finite localization

Throughout this section, $\mathcal{T}$ will denote a triangulated category with arbitrary (i.e., possibly infinite) direct sums of sets of objects. To give some representation-theoretic examples, $\mathcal{T}$ could be the stable module category $\mathsf{StMod}(\Lambda)$ of (possibly infinite-dimensional) modules for a finite-dimensional self-injective algebra Λ, or the derived category $D(\mathrm{Mod}(\Lambda))$ of (possibly unbounded) complexes of modules for a finte-dimensional algebra Λ. Note that the bounded derived category $D^b(\mathrm{Mod}(\Lambda))$ will not do, since the direct sum of an infinite number of bounded complexes may not be bounded itself.

It is automatically true that the direct sum of a set of distinguished triangles is itself a distinguished triangle: this is certainly clear in the examples above.

We shall be interested in the following finiteness condition on objects of $\mathcal{T}$.

Definition 3.1. An object C of $\mathcal{T}$ is **compact** if the functor $\mathrm{Hom}(C, ?)$ (from $\mathcal{T}$ to abelian groups) preserves arbitrary direct sums.

Another way of saying this is that any map from C to an infinite direct sum must factor through a finite subsum.

In the example $\mathsf{StMod}(\Lambda)$, it is easy to see that the compact objects are precisely those isomorphic to finitely generated modules. In the example $D(\mathrm{Mod}(\Lambda))$, the compact objects are those isomorphic to bounded complexes of finitely generated projective modules.

A related finiteness condition on the category $\mathcal{T}$, which is easily seen to be satisfied by both of the examples above, is the following.

Definition 3.2. The category $\mathcal{T}$ is **compactly generated** if there is a set $\mathcal{X}$ of compact objects of $\mathcal{T}$ such that for each object T of $\mathcal{T}$ there is an object C of $\mathcal{X}$ such that $\mathrm{Hom}(C, T) \neq 0$.

The reason for choosing the set $\mathcal{X}$ of objects rather than using *all* compact objects is that the class of compact objects may not be a set.

The notion of a compact object is related to that of a homotopy colimit by the following important but easy result.

Proposition 3.3. *If C is a compact object of $\mathcal{T}$ and*

$$X_0 \xrightarrow{\alpha} X_1 \xrightarrow{\alpha} X_2 \longrightarrow \dots$$

is a sequence of objects, then there is a natural isomorphism

$$\mathrm{Hom}(C, \mathrm{hocolim}(X_i)) \cong \mathrm{colim}\, \mathrm{Hom}(C, X_i).$$

Proof. Since C is compact, the map

$$\mathrm{Hom}(C, \bigoplus_{i=0}^{\infty} X_i[n]) \xrightarrow{1-\alpha} \mathrm{Hom}(C, \bigoplus_{i=0}^{\infty} X_i[n])$$

is isomorphic to

$$\bigoplus_{i=0}^{\infty} \mathrm{Hom}(C, X_i[n]) \xrightarrow{1-\alpha} \bigoplus_{i=0}^{\infty} \mathrm{Hom}(C, X_i[n]),$$

it is injective for all n, and the long exact sequence obtained by applying the functor $\mathrm{Hom}(C, ?)$ to the distinguished triangle that defines $\mathrm{hocolim}(X_i)$ splits into short exact sequences, one of which is

$$0 \longrightarrow \bigoplus_{i=0}^{\infty} \mathrm{Hom}(C, X_i) \xrightarrow{1-\alpha} \bigoplus_{i=0}^{\infty} \mathrm{Hom}(C, X_i) \longrightarrow \mathrm{Hom}(C, \mathrm{hocolim}(X_i)) \longrightarrow 0.$$

But since this is the short exact sequence that calculates $\mathrm{colim}\, \mathrm{Hom}(C, X_i)$, we get the required isomorphism. □

Definition 3.4. A **localizing subcategory** of $\mathcal{T}$ is a triangulated subcategory that is closed under taking direct sums of sets of objects.

The simplest version of Bousfield localization, so-called 'finite localization', involves a compactly generated localizing subcategory of $\mathcal{T}$.

Theorem 3.5. *Let $\mathcal{C}$ be a localizing subcategory of $\mathcal{T}$ and assume that $\mathcal{C}$ is compactly generated by a set $\mathcal{X}$ of objects that are compact in $\mathcal{T}$ (not just in $\mathcal{C}$). Then the inclusion functor*

$$\mathcal{C} \longrightarrow \mathcal{T}$$

has a right adjoint.

We shall describe the construction of the right adjoint and sketch the proof.

We may as well assume that the set $\mathcal{X}$ of compact objects is closed under the shift functors $C \mapsto C[n]$. Let X be an object of $\mathcal{T}$ and set $X_0 = X$. We shall construct, by induction on n, a sequence of objects X_n.

Given X_k, let $S_k \longrightarrow X_k$ be a map from a (possibly infinite) direct sum of objects of $\mathcal{X}$ with the property that *every* map from an object of $\mathcal{X}$ to X_k factors through it. This can be achieved by considering all maps $C \longrightarrow X_k$ from objects of $\mathcal{X}$ to X_k and setting S_k to be the direct sum of the objects C, one for each such map. Now form the distinguished triangle

$$S_k \longrightarrow X_k \longrightarrow X_{k+1} \longrightarrow S_k[1]$$

to define X_{k+1}. Note that, for any object C of $\mathcal{X}$, the exact sequence

$$\mathrm{Hom}(C, S_k) \longrightarrow \mathrm{Hom}(C, X_k) \longrightarrow \mathrm{Hom}(C, X_{k+1})$$

shows that any map $C \longrightarrow X_k$ composes with $X_k \longrightarrow X_{k+1}$ to give zero, since the first map in the exact sequence is surjective by construction, and hence the second map is zero.

If we set $FX = \text{hocolim}(X_i)$, then by Proposition 3.3 we have that

$$\text{Hom}(C, FX) \cong \text{colim}(\text{Hom}(C, X_k)) = 0$$

for any C in $\mathcal{X}$.

If we define an object EX by forming the distinguished triangle

$$EX \longrightarrow X \longrightarrow FX \longrightarrow EX[1],$$

then we shall show that EX is in $\mathcal{C}$. Let us first give a proof of this making the extra assumption that the quotient triangulated category $\mathcal{T}/\mathcal{C}$ exists. Then we shall describe how to sidestep this assumption.

If $\mathcal{T}/\mathcal{C}$ does exist, then proving that EX is in $\mathcal{C}$ is equivalent to proving that the map $X \longrightarrow FX$ is an isomorphism in $\mathcal{T}/\mathcal{C}$. But all the maps

$$X = X_0 \longrightarrow X_1 \longrightarrow \ldots \longrightarrow X_k \longrightarrow X_{k+1} \longrightarrow \ldots$$

are isomorphisms in $\mathcal{T}/\mathcal{C}$ by construction. The natural functor

$$\mathcal{T} \longrightarrow \mathcal{T}/\mathcal{C}$$

preserves (countable) direct sums because $\mathcal{C}$ is closed under (countable) direct sums. It follows that FX is the homotopy colimit of the sequence

$$X = X_0 \longrightarrow X_1 \longrightarrow \ldots \longrightarrow X_k \longrightarrow X_{k+1} \longrightarrow \ldots$$

in the quotient category, not just in the category $\mathcal{T}$. Hence $X \longrightarrow FX$ is an isomorphism in $\mathcal{T}/\mathcal{C}$, being the homotopy colimit of a sequence of isomorphisms.

The potential problem with the existence of the quotient $\mathcal{T}/\mathcal{C}$ is a set-theoretic one. Given objects U and V of $\mathcal{T}$, a map from U to V in $\mathcal{T}/\mathcal{C}$ would be represented by a diagram $U \longleftarrow W \longrightarrow V$ in $\mathcal{T}$, where the map from W to U occurs in a distinguished triangle

$$W \longrightarrow U \longrightarrow C \longrightarrow W[1]$$

whose third object C is in $\mathcal{C}$. Since the objects of $\mathcal{C}$ may not form a set, it is not clear that the equivalence classes of diagrams $U \longleftarrow W \longrightarrow V$ form a set. Thus it is not clear that we can form the quotient category within our original set-theoretic universe.

However, for the purpose of this proof we can sidestep these difficulties. First, if one is willing to accept that the quotient category can be constructed in some larger set-theoretic universe (which may depend on accepting some large cardinal axioms), then this will be good enough. Even if one is not, then the following argument can be used.

Let $\mathcal{C}'$ be the smallest triangulated subcategory of $\mathcal{T}$ that contains all the objects S_k used in the construction of FX and closed under *countable* direct sums. Then there is a set of isomorphism classes of objects of $\mathcal{C}'$, and hence there is no set-theoretic problem involved in the construction of the quotient $\mathcal{T}/\mathcal{C}'$. Since $\mathcal{C}'$ is closed under countable direct sums, the natural functor

$$\mathcal{T} \longrightarrow \mathcal{T}/\mathcal{C}'$$

preserves homotopy colimits of sequences, and so $X \longrightarrow FX$ is an isomorphism in $\mathcal{T}/\mathcal{C}'$. So EX is in $\mathcal{C}'$, and hence in $\mathcal{C}$.

Having shown that EX is in $\mathcal{C}$, we next want to show that $X \mapsto EX$ is the right adjoint of the inclusion functor $\mathcal{C} \longrightarrow \mathcal{T}$.

The key step that remains to prove this is to show that $\mathrm{Hom}(C, FX) = 0$ for every C in $\mathcal{C}$, not just for C in $\mathcal{X}$. It is easy to prove that $\mathrm{Hom}(C, FX) = 0$ for every C in the localizing subcategory $\mathcal{C}''$ of $\mathcal{T}$ generated by $\mathcal{X}$. Let us show that $\mathcal{C}'' = \mathcal{C}$.

To do this, we just need to note that everything we have proved so far about EX and FX works equally well if we replace $\mathcal{C}$ by $\mathcal{C}''$. In particular, if X is in $\mathcal{C}$, then EX is in $\mathcal{C}''$, but also $FX \cong 0$ and so $EX \cong X$. Hence X is in $\mathcal{C}''$.

Although he did not state it in terms of triangulated categories, this theorem essentially goes back to Brown [4].

4. Applications of finite localization in modular representation theory

As we mentioned, the stable module category $\mathsf{StMod}(\Lambda)$ of a self-injective algebra (e.g., a finite group algebra kG) is an example of a category $\mathcal{T}$ to which Theorem 3.5 applies. For a finite group algebra there are many natural choices for the localizing subcategory $\mathcal{C}$: in particular subcategories defined in terms of 'varieties for modules'. We shall not describe here the applications that have been found, but shall refer to Dave Benson's article in these proceedings for more details.

5. More general Bousfield localization

In this section we shall assume that the triangulated category $\mathcal{T}$ has direct sums and is compactly generated, but we shall not make any assumption of compact generation for the localizing subcategory $\mathcal{C}$.

There is a quite general theorem asserting the existence of a right adjoint for the inclusion functor $\mathcal{C} \longrightarrow \mathcal{T}$. Only a set-theoretic problem prevents a completely general theorem.

Theorem 5.1. *Let $\mathcal{T}$ be a compactly generated triangulated category in which direct sums of sets of objects exist. Let $\mathcal{C}$ be a localizing subcategory of $\mathcal{T}$. The inclusion functor $\mathcal{C} \longrightarrow \mathcal{T}$ has a right adjoint if and only if the quotient category $\mathcal{T}/\mathcal{C}$ (in the sense of triangulated categories) exists.*

As we remarked when dealing with finite localization, the potential problem with the existence of the quotient category is that when localizing a category, one may be forced to construct a Hom-set that is too large to be a set. To the best of my knowledge there are no examples – with $\mathcal{T}$ compactly generated – where this is *known* to be a genuine problem, although there are cases that arise in stable homotopy theory where it may well depend on the model of set theory that one uses.

We shall very briefly sketch the proof. I would like to thank Amnon Neeman for describing this to me.

In his paper [10], Neeman gives a proof of a 'Brown representability theorem' in a general compactly generated triangulated category $\mathcal{T}$ with direct sums. This states that if H is a cohomological contravariant functor from $\mathcal{T}$ to abelian groups that takes direct sums to direct products, then H is representable: i.e., there is an object T_H of $\mathcal{T}$ such that $H \cong \operatorname{Hom}(?, T_H)$. The proof of this is in many ways reminiscent of the proof of Theorem 3.5 that we sketched.

This can be applied to the construction of an adjoint as follows. Let X be an object of $\mathcal{T}$ and let $H = \operatorname{Hom}_{\mathcal{T}/\mathcal{C}}(?, X)$. This is a cohomological functor that takes direct sums to direct products, and so the representability theorem applies. Thus

$$\operatorname{Hom}_{\mathcal{T}/\mathcal{C}}(?, X) \cong \operatorname{Hom}_{\mathcal{T}}(?, FX)$$

for some object FX of $\mathcal{T}$. There is a natural map $X \longrightarrow FX$, and we can form a distinguished triangle

$$EX \longrightarrow X \longrightarrow FX \longrightarrow EX[1].$$

Then EX is in $\mathcal{C}$ and $X \mapsto EX$ is the right adjoint of the inclusion functor.

6. Constructing equivalences of derived categories

In his article in these proceedings, Dave Benson briefly describes Broué's conjecture on equivalences of derived categories between blocks of finite group algebras. A construction rather like that involved in finite Bousfield localization has recently proved useful in verifying the existence of such equivalences.

The key theorem is the following.

Theorem 6.1. *Let B be a finite-dimensional symmetric algebra over a field k, and let $X_0, \dots, X_n$ be objects of the derived category $D^b(\operatorname{mod}(B))$ such that*

(i) $\operatorname{Hom}(X_i, X_j) = 0$ *if* $i \neq j$ *and* $\operatorname{Hom}(X_i, X_i) = k$ *for all* i,

(ii) $\operatorname{Hom}(X_i, X_j[m]) = 0$ *for all* $m < 0$ *and all* i, j, *and*

(iii) $X_0, \dots, X_n$ *generate* $D^b(\operatorname{mod}(B))$ *as a triangulated category.*

Then there is another algebra C (also a finite-dimensional symmetric k-algebra) and an equivalence $D^b(\operatorname{mod}(B)) \approx D^b(\operatorname{mod}(C))$ of triangulated categories that takes $X_0, \dots, X_n$ to the simple C-modules.

Note that conditions (i) to (iii) are obviously satisfied by the simple modules: for example, condition (i) is just Schur's Lemma.

Let us sketch the proof before describing why this theorem is useful.

The idea of the proof is to mimic a construction of the injective hull of a simple module. When applied to the objects $X_1, \dots, X_n$, this will produce objects $T_0, \dots, T_n$ such that $T = T_0 \oplus \cdots \oplus T_n$ is a 'tilting complex' [13] for B. The algebra C will be the endomorphism algebra of T, and so, by the main theorem of [13], the derived categories $D^b(\operatorname{mod}(B))$ and $D^b(\operatorname{mod}(C))$ are equivalent.

One condition that would be satisfied if $X_0, \ldots, X_n$ were the simple B-modules and $T_0, \ldots, T_n$ were their injective hulls is that $\mathrm{Hom}(X_i, T_j[m]) = 0$ for $m \neq 0$. If we regard X_j as a 'first approximation' to T_j, then this Hom-space only vanishes for $m < 0$. By adjusting X_j to 'kill' all maps from $X_i[-m]$ to X_j, we shall eventually produce T_j.

Let $\mathcal{X}$ be $\{X_i[t] : 0 \leq i \leq n, t < 0\}$. Note that this is only closed under *negative* shifts.

Set $Y_0 = X_j$ for some j. We shall define a sequence $Y_0, Y_1, \ldots, Y_k, \ldots$ of objects of $D(\mathrm{Mod}(B))$ by induction on k as follows.

Given Y_k, let $S_k \longrightarrow Y_k$ be a map from a direct sum of objects of $\mathcal{X}$ to Y_k, such that *every* map from an object of $\mathcal{X}$ to Y_k factors through it. Also, choose S_k to be 'minimal': no proper direct summand of S_k has the property just described. Form a distinguished triangle

$$S_k \longrightarrow Y_k \longrightarrow Y_{k+1} \longrightarrow S_k[1]$$

to define Y_{k+1}. Finally, define $T_j = \mathrm{hocolim}(Y_k)$.

To show that $\mathrm{Hom}(X_i, T_j[m]) = 0$ for $m > 0$, we want to know that

$$\mathrm{Hom}(X_i[-m], \mathrm{hocolim}(Y_k)) = \mathrm{colim}(\mathrm{Hom}(X_i[-m], Y_k),$$

which is zero by construction (just as in the proof of Theorem 3.5), since $X_i[-m]$ is in $\mathcal{X}$. Unfortunately, we cannot deduce this from Proposition 3.3, since $X_i[-m]$ is not usually compact. However, the complexes Y_k are *uniformly* bounded below, and the functor $\mathrm{Hom}(X_i, ?)$ does preserve direct sums for such families of objects, so all is well.

The fact that B is a symmetric algebra is used in the proof (which we shall omit) that T has all the other properties of a tilting complex. The details will appear in [14].

In particular, one must show that T is isomorphic to a bounded complex of finitely generated projective modules. For a general symmetric algebra B, the intermediate objects Y_k certainly need not be bounded: however, if B is the group algebra of a finite group, it follows from the finite generation of group cohomology that it is possible to take all the objects Y_k to be bounded complexes of finitely generated modules.

Finally, let us describe the reason that this theorem is useful. This is based on an idea of Okuyama [12]. In many of the simpler situations where Broué's conjecture has yet to be verified, it is known that the blocks A and B involved are stably equivalent (i.e., $\mathsf{stmod}(A)$ and $\mathsf{stmod}(B)$ are equivalent as triangulated categories) even though it is not known that $D^b(\mathrm{mod}(A))$ and $D^b(\mathrm{mod}(B))$ are equivalent. Moreover, the stable equivalence is of 'Morita type': i.e., it is induced by an exact functor between the module categories (often this exact functor is just restriction or some mild generalization).

Usually one of the blocks involved (A, say) is much more complicated than the other one: in fact, its structure (in terms of a quiver with relations, say) is often not known. If one can construct a tilting complex T for the simpler block B, then one knows that the derived categories of B and of $C = \mathrm{End}(T)$ are equivalent, but it may be very hard to decide whether A and C are isomorphic. However, by composing the known stable equivalence $\mathsf{stmod}(A) \approx \mathsf{stmod}(B)$ with the equivalence $\mathsf{stmod}(B) \approx \mathsf{stmod}(C)$ induced by the equivalence of derived categories $D^b(\mathrm{mod}(B)) \approx D^b(\mathrm{mod}(C))$, one obtains a stable equivalence $\mathsf{stmod}(A) \approx \mathsf{stmod}(C)$ of Morita type.

Okuyama's idea was to use a theorem of Linckelmann [6] that states that any stable equivalence of Morita type (between self-injective algebras) that takes simple modules to simple modules is induced by an equivalence of module categories. If one can choose the tilting complex T so that the equivalence $\mathsf{stmod}(A) \approx \mathsf{stmod}(C)$ has this property, then $\mathrm{mod}(A) \approx \mathrm{mod}(C)$, and so one can prove that

$$D^b(\mathrm{mod}(A)) \approx D^b(\mathrm{mod}(C)) \approx D^b(\mathrm{mod}(B))$$

without needing to know anything about the structure of A except what the images of its simple modules are under the stable equivalence to $\mathsf{stmod}(B)$.

Using this idea, Okuyama verified several particular cases of Broué's conjecture.

The main problem, of course, is to construct an equivalence $D^b(\mathrm{mod}(B)) \approx D^b(\mathrm{mod}(C))$ with the required properties. Theorem 6.1 shows that one need only produce objects $X_0, \dots, X_n$ of $D^b(\mathrm{mod}(B))$ that are stably isomorphic to the images of the simple A-modules (i.e., isomorphic in $\mathsf{stmod}(B)$, considered as a quotient category of $D^b(\mathrm{mod}(B))$), and have properties (i) to (iii) of Theorem 6.1. This tends to be a lot easier than directly constructing the tilting complex T, since $X_0, \dots, X_n$ usually have a much simpler structure than T does.

Let us give an example where the coefficient field k has characteristic 3. Let G be the alternating group A_6 (which is isomorphic to $PSL_2(9)$), let $P \cong C_3 \times C_3$ be a Sylow 3-subgroup of G and let $H = N_G(P)$, which is the semidirect product of P with a cyclic group of order 4. Let A and B be the principal block algebras of kG and kH respectively. It is well-known that restriction of modules induces a stable equivalence between A and B. Okuyama [12] proved, by different methods, that A and B have equivalent derived categories, but we shall describe how Theorem 6.1 gives a simple proof.

The algebra B has four simple modules, all one-dimensional, which we shall call $k, 1, 2, 3$, the number referring to how the cyclic quotient C_4 of H acts.

Also, A has four simple modules, $k = V_0, V_1, V_2, V_3$ with dimensions $1, 3, 4$ and 3 respectively. Let Z_i be the restriction of V_i to B (for $i = 0, 1, 2, 3$). Then $Z_0 = k$ and the structure of Z_1, Z_2, Z_3 is described by the following diagrams (the rows give the Loewy layers).

$$Z_1 = \begin{array}{c} 1 \\ k \\ 3 \end{array}, \quad Z_2 = \begin{array}{ccc} & 2 & \\ 1 & & 3 \\ & 2 & \end{array}, \quad Z_3 = \begin{array}{c} 3 \\ k \\ 1 \end{array}$$

Clearly the modules $Z_0, \ldots, Z_3$, considered as objects of the derived category $D^b(\mathrm{mod}(B))$, do not satisfy the hypotheses of Theorem 6.1: there are non-zero maps between Z_1 and Z_3, and Z_2 has a two-dimensional endomorphism algebra.

But let $X_0 = Z_0 = k$, $X_1 = \Omega^2 Z_1[2]$, $X_2 = \Omega Z_2[1]$ and $X_3 = \Omega^2 Z_3[2]$. These objects are stably isomorphic to $Z_0 \ldots, Z_3$ since, for example, X_1 is isomorphic in $D^b(\mathrm{mod}(B))$ to the complex

$$\ldots 0 \longrightarrow P_1 \longrightarrow P_0 \longrightarrow Z_1 \longrightarrow 0 \longrightarrow \ldots$$

with Z_1 in degree zero, obtained by truncating a minimal projective resolution of Z_1. But this is clearly isomorphic to Z_1 in $\mathsf{stmod}(B)$, considered as the quotient of $D^b(\mathrm{mod}(B))$ by the subcategory of bounded complexes of finitely generated projective modules.

The structure of $\Omega^2 Z_1, \Omega Z_2$ and $\Omega^2 Z_3$ is described by the following diagrams.

$$\Omega^2 Z_1 = \begin{array}{c} k \\ 3 \\ 2 \end{array}, \quad \Omega Z_2 = \begin{array}{ccc} & k & \\ 1 & & 3 \\ & 2 & \end{array}, \quad \Omega^2 Z_3 = \begin{array}{c} k \\ 1 \\ 2 \end{array}$$

These modules, and hence the objects $X_0, \ldots, X_3$, have one-dimensional endomorphism algebras, so that the second part of condition (i) of Theorem 6.1 is satisfied.

The rest of condition (i) and condition (ii) follow in most cases from the fact that $\mathrm{Hom}_{D^b(\mathrm{mod}(B))}(U, V[m]) = 0$ if U and V are modules and $m < 0$. The remaining cases are:

$$\begin{gathered}
\mathrm{Hom}(k, X_j[-2]) = \mathrm{Hom}(k, \Omega^2 Z_j) = 0 \text{ (for } j = 1,3), \\
\mathrm{Hom}(k, X_j[-1]) = \mathrm{Ext}^1(k, \Omega^2 Z_j) = \underline{\mathrm{Hom}}(k, \Omega Z_j) = 0 \text{ (for } j = 1,3), \\
\mathrm{Hom}(k, X_j) = \mathrm{Ext}^2(k, \Omega^2 Z_j) = \underline{\mathrm{Hom}}(k, Z_j) = 0 \text{ (for } j = 1,3), \\
\mathrm{Hom}(k, X_2[-1]) = \mathrm{Hom}(k, \Omega Z_2) = 0, \\
\mathrm{Hom}(k, X_2) = \mathrm{Ext}^1(k, \Omega Z_2) = \underline{\mathrm{Hom}}(k, Z_2) = 0, \\
\mathrm{Hom}(X_2, X_j[-1]) = \mathrm{Hom}(\Omega Z_2, \Omega^2 Z_j) = 0 \text{ (for } j = 1,3), \\
\mathrm{Hom}(X_2, X_j) = \mathrm{Ext}^1(\Omega Z_2, \Omega^2 Z_j) = \underline{\mathrm{Hom}}(Z_2, Z_j) = 0 \text{ (for } j = 1,3),
\end{gathered}$$

and

$$\mathrm{Hom}(X_i, X_j) = \mathrm{Hom}(\Omega^2 Z_i, \Omega^2 Z_j) = 0 \text{ (for } \{i,j\} = \{1,3\}).$$

Condition (iii) is quite easy to verify. We just need to show that all the simple B-modules are in the triangulated subcategory $\mathcal{C}$ of $D^b(\mathrm{mod}(B))$ generated by $X_0, \ldots, X_3$. Since $k, \Omega^2 Z_1, \Omega Z_2$ and $\Omega^2 Z_3$ are in $\mathcal{C}$, so are the modules

$$\begin{array}{c} 3 \\ 2 \end{array}, \quad \begin{array}{ccc} 1 & & 3 \\ & 2 & \end{array}, \quad \begin{array}{c} 1 \\ 2 \end{array}$$

that are kernels of surjections from $\Omega^2 Z_1, \Omega Z_2$ and $\Omega^2 Z_3$ to direct sums of copies of k. Hence 1 and 3, which are quotients of the second module above by submodules isomorphic to the first and third, are also in $\mathcal{C}$, and finally 2 is in $\mathcal{C}$, since it is the kernel of a surjection from the first module to 3.

Theorem 6.1 has also been used to verify some new cases of Broué's conjecture:

Theorem 6.2 (Chuang [5]). *If $G = SL_2(p^2)$, H is the normalizer of a Sylow p-subgroup of G, and k is algebraically closed of characteristic p, then the principal blocks of kG and kH have equivalent derived categories.*

The case $p = 5$ of this theorem was independently proved by Holloway, who has also proved the corresponding theorem for $G = SL_2(27)$ and $p = 3$.

REFERENCES

[1] J. F. Adams, *A variant of E. H. Brown's representability theorem*, Topology **10** (1971), 185–198.

[2] M. Bökstedt, A. Neeman, *Homotopy limits in triangulated categories*, Compositio Math. **86** (1993), 209–234.

[3] A. K. Bousfield, *The localization of spectra with respect to homology*, Topology **18** (1979), 257–281.

[4] E. H. Brown, Jr., *Abstract homotopy theory*, Trans. Amer. Math. Soc. **119** (1965), 79–85.

[5] J. Chuang, *Derived equivalence in $SL_2(p^2)$*, preprint.

[6] M. Linckelmann, *Stable equivalences of Morita type for self-injective algebras and p-groups.* Math. Zeit. **223** (1996), 87–100.

[7] A. Neeman, *The Brown representability theorem and phantomless triangulated categories*, J. Algebra **151** (1992), 118–155.

[8] A. Neeman, *The chromatic tower for $D(R)$*, Topology **31** (1992), 519–532.

[9] A. Neeman, *The connection between the K-theory localisation theorem of Thomason, Trobaugh and Yao, and the smashing subcategories of Bousfield and Ravenel*, Ann. Scient. de l'ENS **25** (1992), 547–566.

[10] A. Neeman, *The Grothendieck duality theorem via Bousfield's techniques and Brown representability*, J. Amer. Math. Soc. **9** (1996), 205–236.

[11] A. Neeman, *On a theorem of Brown and Adams*, Topology **36** (1997), 619–645.

[12] T. Okuyama, *Some examples of derived equivalent blocks of finite groups*, Comm. Alg. (to appear).

[13] J. Rickard, *Morita theory for derived categories.* J. London Math. Soc. (2) **39** (1989), 436–456.

[14] J. Rickard, *Constructing derived equivalences for symmetric algebras*, in preparation.

UNIVERSITY OF BRISTOL, SCHOOL OF MATHEMATICS, UNIVERSITY WALK, BRISTOL BS8 1TW, ENGLAND

E-mail address: J.Rickard@bristol.ac.uk

Trends in Mathematics, © 2000 Birkhäuser Verlag Basel/Switzerland

THE THICK SUBCATEGORY GENERATED BY THE TRIVIAL MODULE

JON F. CARLSON*

This paper is a report of my lecture at the International Conference on the Representation Theory of Algebras which preceded the Euroconference on Infinite Length Modules. While it is true that the aim of the research is the homological algebra of finitely presented modules, a major point of the lecture was the role that infinitely generated modules could play in the investigation. The methods of idempotent modules and the theory of support varieties for infinite dimensional modules have already had a significant impact on group representation theory. It seems certain that there will be a lot more to follow. I am honored by the invitation to include the report in the conference proceedings, and I would like to thank the organizers of the conference and the workshop for the stimulating experience.

Throughout the paper G is a finite group and k is a field of characteristic $p > 0$. Ultimately the investigation is concerned with the structure and the behavior of the functors $\mathrm{Ext}^n_{kG}(M, N)$ for M and N finitely generated kG-modules. In particular we would like to understand how the properties of modules are reflected in their cohomology and in their extensions with other modules. Even more specifically, we ask for things like the following

Question 1. What are the properties of pairs of modules M and N that will produce the result that

$$\mathrm{Ext}^n_{kG}(M, N) = 0.$$

for all n.

One of the aims of this paper is to show that the above, as well as other questions, are really questions about categories of modules. For the most part we will restrict ourselves to the study of finite dimensional kG-modules. Of course the questions that we ask have analogues for all kG-modules. However for several reasons, it seems natural to look at these issues separately. Many of the techniques that we use in the study of finitely generated modules fail for infinitely generated modules. Even such a thing as the construction of the idempotent modules and functors in a later section of this paper is problematic if we do not begin with a thick subcategory of finitely generated modules.

For some notation, let **mod**(kG) be the category of finitely generated kG-modules and kG-module homomorphisms. We denote the corresponding category of all kG-modules by **Mod**(kG). Let $\phi : P \longrightarrow M$ be a projective cover of the

Partly supported by a grant from NSF.

kG-module M. Then $\Omega(M)$ is defined to be the kernel of ϕ. Thus we have an exact sequence

$$0 \longrightarrow \Omega(M) \longrightarrow P \xrightarrow{\phi} M \longrightarrow 0$$

Inductively, let $\Omega^n(M) = \Omega(\Omega^{n-1}(M))$. Similarly let $\Omega^{-1}(M)$ be the cokernel of the inclusion $M \longrightarrow Q$ of M into its injective hull. Again inductively we have $\Omega^{-n}(M) = \Omega^{-1}(\Omega^{-n+1}((M))$. So for every $n \in \mathbb{Z}$ we have an exact sequence

$$0 \longrightarrow \Omega^{n+1}(M) \longrightarrow Proj \longrightarrow \Omega^n(M) \longrightarrow 0.$$

We emphasize here that kG is a self-injective algebra. So a module is injective if and only if is projective. This holds even for infinitely generated modules.

The Natural Setting

The stable category **stmod**(kG) of finitely generated kG-modules is a natural setting for our investigation. Roughly speaking, it is the category we get by setting the projective modules equal to zero. Specifically we define the stable category as follows. The objects in **stmod**(kG) are the finite dimensional kG-modules as in **mod**(kG). For M and N objects, the morphisms from M to N are given by

$$\overline{\text{Hom}}_{kG}(M,N) = \frac{\text{Hom}_{kG}(M,N)}{\text{PHom}_{kG}(M,N)}$$

where $\text{PHom}_{kG}(M,N)$ is the set of all homomorphisms that factor through a projective module.

The operators Ω and Ω^{-1} are functors on the stable category, and they are used to create a triangulated structure on **stmod**(kG). Basically, a triangle

$$\longrightarrow L \xrightarrow{\alpha} M \xrightarrow{\beta} N \xrightarrow{\gamma} \Omega^{-1}(L) \longrightarrow$$

in **stmod**(kG) corresponds to a collection of exact sequences:

$$0 \longrightarrow L \xrightarrow{\alpha_1} M \oplus Proj \xrightarrow{\beta_1} N \longrightarrow 0,$$

$$0 \longrightarrow M \xrightarrow{\beta_2} N \oplus Proj \xrightarrow{\gamma_2} \Omega^{-1}(L) \longrightarrow 0,$$

etc. Here the class modulo $\text{PHom}_{kG}(L,M)$ of the map α_1 is α, the class of β_1 and β_2 is β and the class of γ_2 is γ. So when we pass from the module category **mod**(kG) to **stmod**(kG) we substitute "triangulated" for "abelian". There is a set of axioms that a triangulated category must satisfy. For example every morphism between two objects is contained in a unique triangle. The proof that the axioms hold for **stmod**(kG) as well as the correspondence between triangles and exact sequences is a series of exercises in homological algebra. More details are given in [8].

Likewise any exact sequence in **mod**(kG) defines a triangle in **stmod**(kG). The functor Ω^{-1} is called the translation functor for the triangulation. So a triangle consists of a triple of objects and a triple of maps and every time we go around the triangle we must apply the translation functor Ω^{-1}. There is a set of axioms that triangulated categories must satisfy. For the axioms and other details on triangulated categories see the book by Happel [13].

In a real sense, cohomology takes place in the stable category. This fact is best illustrated by the following result that is easily verified from basic principles.

Lemma 2. *For M and N kG-modules, and any $n > 0$,*

$$\overline{\mathrm{Hom}}_{kG}(\Omega^n(M), N) \cong \mathrm{Ext}^n_{kG}(M, N),$$

and

$$\overline{\mathrm{Hom}}_{kG}(M, \Omega^{-n}(N)) \cong \mathrm{Ext}^n_{kG}(M, N)$$

In fact if we substitute Tate cohomology $\widehat{\mathrm{Ext}}$ for Ext then the above lemma is true for all integers n. It follows that in the stable category the familiar long exact sequence in cohomology becomes a sequence of *Hom*'s. Namely we have the following.

Lemma 3. *Given a triangle of kG-modules*

$$\longrightarrow L \xrightarrow{\alpha} M \xrightarrow{\beta} N \xrightarrow{\gamma} \Omega^{-1}(L) \longrightarrow$$

and a kG-module S we have an exact sequence

$$\dots \xrightarrow{\gamma_*} \overline{\mathrm{Hom}}_{kG}(S, \Omega^r(L)) \xrightarrow{\alpha_*} \overline{\mathrm{Hom}}_{kG}(S, \Omega^r(M)) \xrightarrow{\beta_*}$$
$$\overline{\mathrm{Hom}}_{kG}(S, \Omega^r(N)) \xrightarrow{\gamma_*} \overline{\mathrm{Hom}}_{kG}(S, \Omega^{r-1}(L)) \xrightarrow{\alpha_*} \dots$$

Of course there is a similar sequence in the other variable.

Thick Subcategories

A full subcategory $\mathcal{M}$ of **stmod**(kG) is said to be *thick* if it is triangulated and if it is also closed under the taking of direct summands. So if $\mathcal{M}$ is a thick subcategory then it is closed under the operators Ω and Ω^{-1} and whenever we have a triangle in **stmod**(kG) (or exact sequence in **mod**(kG)) and two of the objects are in $\mathcal{M}$ then so is the third. The stable category of all modules in a particular block of kG is a thick subcategory of **stmod**(kG). Other examples include things like the collection of all periodic modules or the collection of all modules whose variety is contained in a fixed subvariety of $V_G(k)$. The subcategory of periodic modules is thick because a module is periodic if and only if its variety has dimension one (see [1]). More about this is presented later.

If M is a kG-module then the thick subcategory $\mathcal{H}(M)$ generated by M is the smallest thick subcategory of **stmod**(kG) containing M. It contains $\Omega^n(M)$ for all n. It contains any extension of one of these modules by another. It can be characterized as the full subcategory consisting of every module L that is a direct summand of a module L' that can be filtered by a sequence of submodules such that each of the factors is isomorphic to $\Omega^n(M)$ for some n. With this in mind it is not difficult to derive the following fact from the long exact sequence.

Proposition 4. *Suppose that M and N are kG-modules such that $\widehat{\mathrm{Ext}}^n_{kG}(M,N) = 0$ for all n. Then $\widehat{\mathrm{Ext}}^n_{kG}(M',N') = 0$ for all n, all M' in $\mathcal{H}(M)$ and all N' in $\mathcal{H}(N)$.*

We should note here that if $\mathrm{Ext}^n_{kG}(M,N) = 0$ for all $n > 0$ or if $\widehat{\mathrm{Ext}}^n_{kG}(M,N) = 0$ for n in a sufficiently long interval then $\widehat{\mathrm{Ext}}^n_{kG}(M,N) = 0$ for all n [4].

Proposition 5. *Suppose that M and N are kG-modules and that N is in $\mathcal{H}(M)$, the thick subcategory generated by M. If N is not a projective module, then it must be that $\mathrm{Ext}^n_{kG}(M,N) \neq 0$ for infinitely many values of n.*

The last result follows because if $\widehat{\mathrm{Ext}}^n_{kG}(M,N) = 0$ for all n then $\widehat{\mathrm{Ext}}^n_{kG}(M',N) = 0$ for all n and for all M' in $\mathcal{H}(M)$. In particular, $\widehat{\mathrm{Ext}}^0_{kG}(N,N) = 0$, and the identity homomorphism on N must factor through a projective. Thus N is a projective module.

So we see that the statement that $\mathrm{Ext}^n_{kG}(M,N) = 0$ is really a statement about the subcategories that the modules M and N generate. That is, if $\mathrm{Ext}^n_{kG}(M,N) = 0$ then the subcategories $\mathcal{H}(M)$ and $\mathcal{H}(N)$ have no nonzero object in common. As far as we know the converse of this statement is completely open.

Question 6. Suppose that M and N are kG-modules such that $\mathrm{Ext}^n_{kG}(M,N) \neq 0$ for some $n > 0$. Is it necessary that there is a nonprojective module L that is in both $\mathcal{H}(M)$ and $\mathcal{H}(N)$?

We should notice that if M and N are indecomposable kG-modules in different blocks then $\mathrm{Ext}^*_{kG}(M,N) = 0$. Also we have a standard adjointness

$$\mathrm{Ext}^n_{kG}(M,N) \cong \mathrm{Ext}^n_{kG}(k, M^* \otimes N) \cong H^n(G, M^* \otimes N)$$

for all n. Here $M^* = \mathrm{Hom}_k(M,k)$ is the k-dual and the tensor product $M^* \otimes N = M^* \otimes_k N$ ($\cong \mathrm{Hom}_k(M,N)$) is made into a G-module by letting the elements of G act diagonally. If the summand of $M^* \otimes N$ that is in the principal block is projective then there are no extensions between M and N. So it can be said that, in some sense, cohomology takes place in the principal block. Even further it measures how far the tensor product $M^* \otimes N$ is from being in the thick subcategory generated by k.

For the remainder of the paper let $\mathcal{K} = \mathcal{H}(k)$ be the thick subcategory of **stmod**(kG) generated by the trivial module. Let $\tilde{B}_0(G)$ be the thick subcategory of modules in the principal block of kG.

Varieties and the vanishing of cohomology

The cohomology ring $H^*(G,k)$ of a finite group G is a graded-commutative, finitely generated k-algebra whose maximal ideal spectrum $V_G(k)$ is a homogeneous affine variety. For a finitely generated kG-module M the variety of M is defined to be the subvariety $V_G(M) = V_G(J(M))$ of all maximal ideals that contain the annihilator $J(M)$ in $H^*(G,k)$ of the cohomology ring $\mathrm{Ext}^*_{kG}(M,M)$. The varieties of modules have been studied extensively for several years now. For details on the subject see the books [1, 12]. The properties that are most of interest to us here can be summarized in the following. Note that the zero element in $V_G(k)$ is the maximal ideal of all elements of positive degree.

Theorem 7. *Let M be a kG-module.*

(1.) *M is projective if and only if $V_G(M) = \{0\}$.*

(2.) *If*

$$\longrightarrow L \xrightarrow{\alpha} M \xrightarrow{\beta} N \xrightarrow{\gamma} \Omega^{-1}(L) \longrightarrow$$

is a triangle in **stmod**(kG) *then $V_G(M) \subseteq V_G(L) \cup V_G(N)$.*

(3.) *$V_G(M) = V_G(\Omega^n(M)) = V_G(M^*)$ for all n.*

(4.) *For M and N kG-modules*

$$V_G(M \otimes N) = V_G(M) \cap V_G(N).$$

It should be pointed out that the variety of a module can be determined directly from the structure of the module without any cohomological calculations whatsoever. It is done by computing the "rank variety" for the elementary abelian p-subgroups. See [1, 12] for details.

We note also that there is an extension of the notion of the variety of a module to infinitely generated modules [5], and the same properties are true in a suitably modified form. See the paper by Benson [3] in these proceedings for a synopsis of these results.

Several consequences of the Theorem are immediate to our investigation. First of all, if M and N have the property that $V_G(M) \cap V_G(N) = \{0\}$, then $\mathrm{Ext}^*_{kG}(M, N) = 0$. Next we can see that if V is a closed subvariety of $V_G(k)$ then the full subcategory $\mathcal{M}_V$ of all modules with $V_G(M) \subseteq V$ is a thick subcategory of **stmod**(kG). The good news is that in some cases this is close to being the whole story [6].

Theorem 8. *Suppose that the centralizers of p-elements in G are p-nilpotent. Then any thick subcategory $\mathcal{C}$ of the stable category, $\tilde{B}_0(G)$, of the principal block is determined by varieties. That is, there is a collection $\mathcal{V}$ of closed subvarieties of $V_G(k)$ that is closed under specialization and taking of finite unions such that $\mathcal{C}$ is the full subcategory $\mathcal{M}_{\mathcal{V}}$ consisting of all modules M in the principal block such that $V_G(M)$ is an element of $\mathcal{V}$.*

Of course, if G is a p-group then $\tilde{B}_0(G) =$ **stmod**(kG) and we have a complete classification of the thick subcategories of **stmod**(kG). The proof in this case uses the idempotent modules of Rickard, which we say more about later, and the extended notion of varieties for infinitely generated modules. When G is not a p-group then the proof of the Theorem requires also a deep result of Benson [2]. In any event it rests on the following facts, the first of which is proved in [4] for $p > 2$ and in [2] for the general case.

Theorem 9. *Suppose that the centralizers of p-elements in G are p-nilpotent. Then $\tilde{B}_0(G) = \mathcal{K}$.*

A thick subcategory $\mathcal{M}$ is tensor ideal closed if whenever M is in $\mathcal{M}$ and N is another module then $M \otimes N$ is in $\mathcal{M}$. The fact that we need for the proof of the theorem is that any tensor ideal closed subcategory is determined by varieties [6]. It is not difficult to verify that any thick subcategory of $\mathcal{K}$ is tensor ideal closed in $\mathcal{K}$.

All of this tells us that, if we are in the case where the centralizers of p-elements are p-nilpotent, then two modules M and N in $\tilde{B}_0(G)$ generate orthogonal subcategories ($\mathrm{Ext}^*_{kG}(M,N) = 0$) if and only if $V_G(M) \cap V_G(N) = \{0\}$. This holds more generally for any two modules in $\mathcal{K}$.

Before going further we should mention that there is another fortunate occurence in the case that $\tilde{B}_0(G) = \mathcal{K}$. Namely we have some strong control on the self-equivalences. Every stable self-equivalence $\mathcal{F} : \mathcal{K} \longrightarrow \mathcal{K}$ must take the trivial module k to an endotrivial module $\mathcal{F}(k)$. A module M is endotrivial if the k-endomorphism ring $\mathrm{Hom}_k(M,M) \cong M^* \otimes M$ is isomorphic to $k \oplus P$ where P is a projective module. That is, it is isomorphic to k in the stable category **stmod**(kG) [11]. This has some importance in the investigation of questions such as Broué's conjecture on derived equivalences of block algebras with abelian defect groups [7]. The conjecture of Broué, like Alperin's conjecture, asserts that some of the important properties of the representation theory of a group can be determined by looking at the representation theory of local subgroups, that is, subgroups that are the normalizers of nontrivial p-subgroups.

The nucleus and other thick subcategories

But what happens in other cases? What if the centralizers of p-elements are not p-nilpotent – if $\tilde{B}_0(G) \neq \mathcal{K}$? For the principal block there is a theorem of Benson that gives us a reasonably satisfactory conclusion, at least to the question of vanishing cohomology and the relation to varieties.

To state the theorem we need the following definition. Note that for H a subgroup of G we have a restriction map $res_{G,H} : H^*(G,k) \longrightarrow H^*(H,k)$ and a corresponding map on varieties $res^*_{G,H} : V_H(k) \longrightarrow V_G(k)$. We should remember Quillen's theorem that $V_G(k) = \bigcup res^*_{G,E}(V_E(k))$ where E runs through the elementary abelian p-subgroups of G [1, 12].

Definition 10. *The nucleus of G is the subvariety of $V_G(k)$ defined by*

$$\mathcal{N} = \bigcup \; res^*_{G,E}(V_E(k))$$

where E runs through the set of all elementary abelian p-subgroups such that $C_G(E)$ is not p-nilpotent.

Then Benson's Theorem can be stated as follows.

Theorem 11. [2] *Suppose that M is a module in $\tilde{B}_0(G)$ and that*

$$H^*(G,M) = \mathrm{Ext}^*_{kG}(k,M) = 0.$$

Then $V_G(M) \subseteq \mathcal{N}$.

Moreover the theorem can be extended to cover varieties defined by cohomology modules [9]. To understand this let $\mathcal{I}(M)$ denote the annihilator in $H^*(G,k)$ of the cohomology $H^*(G,M)$ for a kG-module M, and let $W_G(M) = V_G(\mathcal{I}(M))$ be the subvariety of $V_G(k)$ consisting of all maximal ideals that contain $\mathcal{I}(M)$. It is not difficult to show that $W_G(M) \subseteq V_G(M)$ because $H^*(G,M)$ is a module over $\operatorname{Ext}^*_{kG}(M,M)$. So the theorem says that for M in $\tilde{B}_0(G)$

$$V_G(M) \smallsetminus W_G(M) \subseteq \mathcal{N}.$$

For a pair of modules M and N in the principal block, we can say even more. Notice that $\mathcal{I}(M^* \otimes N)$ is the annihilator of $\operatorname{Ext}^*_{kG}(M,N) \cong H^*(G, M^* \otimes N)$. Hence we can also prove that

$$W_G(M^* \otimes N) \subseteq V_G(M^* \otimes N) = V_G(M) \cap V_G(N).$$

Furthermore it can be shown that

$$V_G(M) \cap V_G(N) \smallsetminus W_G(M^* \otimes N) \subseteq \mathcal{N}.$$

For a block B other than the principal block we can define a block nucleus $\mathcal{N}_B$ that has some of the properties of the principal block nucleus [9]. But this time the definition has some serious problems with it, as pointed out in the paper [9]. Perhaps a better definition would come from a more specialized treatment of the block algebra such as that in [14]. The approach there has the advantage of alleviating the difficulty of not having any module to play the role of the trivial module in an arbitrary block. On the other hand it has the disadvantage of not being defined in the same context as the varieties of modules in the principal block. There are a lot of open questions here.

However, even in the case of the principal block, if $\mathcal{N} \neq \{0\}$, then there is still no direct and certain way to predict the orthogonality of the subcategories generated by the modules M and N ($\operatorname{Ext}^*_{kG}(M,N) = 0$) by looking only at the module structure. The current state of affairs with regards to the thick subcategories of $\tilde{B}_0(G)$ is just as muddled. We might ask the question:

Question 12. Suppose that $\mathcal{N} \neq \{0\}$. Is it possible to get a complete classification of the thick subcategories of $\tilde{B}_0(G)$?

In [6], we looked at this and related question in some detail. In particular we considered several seemingly simple examples. In not even one case were we able to answer the above question. Indeed, the situation appears to be very chaotic. In every example that we considered there is an infinite number of (mutually orthogonal) thick subcategories of the "no cohomology" category $\mathcal{U}$. This the full subcategory of all modules M in $\tilde{B}_0(G)$ such that $H^*(G,M) = \{0\}$. It seems to be a complete mystery. Another question we ask is the following.

Question 13. Suppose that $\mathcal{N} \neq \{0\}$. Let $\mathcal{W}$ be the full subcategory of all modules M in $\tilde{B}_0(G)$ such that $\operatorname{Ext}^*_{kG}(M,N) = 0$ for all N in $\mathcal{U}$. Is $\mathcal{W} = \mathcal{K}$?

Again, in no case, not even the easiest of examples, were we able to answer this question in [6]. However some recent progress in the area is encouraging. Now we can at least say what happens in some of the very special examples discussed in [6]. The rest of this work will focus on the few new results that we have in this direction. To get started, we need a new techniques as given below.

IDEMPOTENT MODULES AND FUNCTORS

It is here that we begin to call on the methods of infinitely generated modules. A few years ago Jeremy Rickard introduced a very useful tool in the form of idempotent modules and functors corresponding to a subcategory. As we have already said, the method was used in the original solution of the theorem of Benson (Theorem 11) as well as in the classification of the thick subcategories in the case that $\mathcal{N} = \{0\}$. More details as well as other applications of the functors can be found in the survey of Benson [3]. See [8] for another leisurely treatment of the background. The statement goes as follows.

Theorem 14. [15] *Suppose that $\mathcal{C}$ is a thick subcategory of* **stmod**(kG). *Then there exist functors $\mathcal{E}_\mathcal{C}$ and $\mathcal{F}_\mathcal{C}$ such that for any module M there is a triangle*

$$\longrightarrow \mathcal{E}_\mathcal{C}(M) \longrightarrow M \longrightarrow \mathcal{F}_\mathcal{C}(M) \longrightarrow \Omega^{-1}(\mathcal{E}_\mathcal{C}(M)) \longrightarrow$$

with the two properties

(1). $\mathcal{E}_\mathcal{C}(M) \in \mathcal{C}^\oplus$ and

(2). $\mathcal{F}_\mathcal{C}(M)$ is $\mathcal{C}$-local.

Moreover the two properties characterize the triangle and the modules $\mathcal{E}_\mathcal{C}(M)$ and $\mathcal{F}_\mathcal{C}(M)$.

Here $\mathcal{C}^\oplus$ is the subcategory of the stable category **StMod**(kG) of all kG-modules consisting of all modules that are filtered colimits of objects in $\mathcal{C}$ [15]. It is the triangulated subcategory generated by the closure of $\mathcal{C}$ in **StMod**(kG) under the operation of taking arbitrary direct sums. A kG-module N in **Mod**(kG) is said to be $\mathcal{C}$-local if $\overline{\mathrm{Hom}}_{kG}(C, N) = 0$ for all $C \in \mathcal{C}$.

In the case that the subcategory $\mathcal{C}$ is tensor ideal closed, the modules $\mathcal{E}_\mathcal{C}(k)$ and $\mathcal{F}_\mathcal{C}(k)$ are idempotent in the stronger sense that $\mathcal{E}_\mathcal{C}(k) \otimes \mathcal{E}_\mathcal{C}(k) \cong \mathcal{E}_\mathcal{C}(k)$ and $\mathcal{F}_\mathcal{C}(k) \otimes \mathcal{F}_\mathcal{C}(k) \cong \mathcal{F}_\mathcal{C}(k)$ and $\mathcal{E}_\mathcal{C}(k) \otimes \mathcal{F}_\mathcal{C}(k) \cong \{0\}$ in the stable category. That is, as modules

$$\mathcal{E}_\mathcal{C}(k) \otimes \mathcal{E}_\mathcal{C}(k) \cong \mathcal{E}_\mathcal{C}(k) \oplus P$$

for P a projective module. In almost all cases the modules are infinite dimensional. Indeed, it can be shown that any finite dimensional nonprojective module M with the property that $M \otimes M \cong M$ in the stable category must have the form $M \cong k \oplus proj$ (see Thm. 3.5 of [8]).

Also for subcategories defined by varieties there is a straightforward construction of the modules $\mathcal{E}_\mathcal{C}(k)$ and $\mathcal{F}_\mathcal{C}(k)$ by direct limits (see [8] or the original paper [15]). For the subcategory $\mathcal{K}$ or any subcategory generated by a single object there is also a straightforward construction of $\mathcal{E}_\mathcal{K}(M)$ and $\mathcal{F}_\mathcal{K}(M)$ arising from the cohomology of M. In particular the modules are a direct limit of finitely generated modules and we can get some control of the situation. To illustrate what happens we elaborate on one example from the list of examples in [6].

THE EXAMPLE $C_3 \times S_3$ IN CHARACTERISTIC 3

Suppose that $G \cong C_3 \times S_3 \cong \langle z \rangle \times \langle y, x \rangle$ where $z^3 = y^3 = x^2 = 1$ and $xyx = y^2$. Let k be a field of characteristic 3. Then there are two simple kG-modules: the trivial module k and a module ϵ of dimension one on which the element x of order 2 acts by multiplication by -1. As a left kG-module

$$kG \cong P_k \oplus P_\epsilon,$$

where P_k and P_ϵ are the projective covers of k and ϵ respectively. Both of these modules have dimension 9 and their restrictions to the Sylow 3-subgroup $S = \langle z, y \rangle$ of G is a copy of kS. Let $Y = y - y^2 \in kG$, so that $xYx = -Y$. Let $Z = z - 1$. So the radical of kS is generated by Y and Z. Now notice that $L = P_k/(Y \cdot P_k)$ has dimension 3 and has a filtration

$$\{0\} \quad = \quad Z^3 \cdot L \quad \subseteq \quad Z^2 \cdot L \quad \subseteq \quad Z \cdot L \quad \subseteq \quad L$$

where $(Z^i \cdot L)/(Z^{i+1} \cdot L) \cong k$. So this module is in $\mathcal{K}$. Also $L \cong Y^2 \cdot P_k$ is in $\mathcal{K}$. So $(Y \cdot P_k)/(Y^2 \cdot P_k) \cong P_\epsilon/(Y \cdot P_\epsilon)$ is in $\mathcal{K}$.

On the other hand look at $M = P_\epsilon/(Z \cdot P_\epsilon)$. It can be checked that this is a periodic module ($\Omega(M) = Z \cdot P_\epsilon \subseteq P_\epsilon$ and $\Omega^2(M) \cong M$). Moreover there exists no nonzero homomorphism $k \longrightarrow \Omega^n(M)$. Therefore M is in the category $\mathcal{U}$ of no-cohomology modules. (Note that the principal block is the only block.) That is, $\mathrm{Ext}^*_{kG}(k, M) = \mathrm{Ext}^*_{kG}(M, k) = 0$ only because the socle of M and $\Omega(M)$ is isomorphic to ϵ and hence $\overline{\mathrm{Hom}}(k, M) = \overline{\mathrm{Hom}}(k, \Omega(M)) = 0$.

With all of this in mind we want to find the structure of the modules in the triangle

$$\longrightarrow \mathcal{E}_{\mathcal{K}}(U) \longrightarrow U \longrightarrow \mathcal{F}_{\mathcal{K}}(U) \longrightarrow \Omega^{-1}(\mathcal{E}_{\mathcal{K}}(U)) \longrightarrow$$

where $U = k \oplus \epsilon$. Now notice that $\mathcal{E}_{\mathcal{K}}(k \oplus \epsilon) \cong k \oplus \mathcal{E}_{\mathcal{K}}(\epsilon)$ and that $\mathcal{F}_{\mathcal{K}}(k \oplus \epsilon) \cong \mathcal{F}_{\mathcal{K}}(\epsilon)$ since $\mathcal{F}_{\mathcal{K}}(k) = 0$. The point of this is that U is a generator for the category **stmod**(kG) since it is the direct sum of the only two irreducible modules. Thus, **stmod**(kG) $= \mathcal{H}(U)$. Consequently for any module M, we have that $\mathrm{Ext}^*_{kG}(U, M) = 0$ or $\mathrm{Ext}^*_{kG}(M, U) = 0$ if and only if M is projective (injective).

The point of what is to follow is that the generators of the category have a lot to say about the module structure. Before going further we should remark that the converse to the above statement about the generator for the category is a completely open question. That is, we can ask the following question.

Question 15. Suppose that M is an object in a thick subcategory $\mathcal{M}$ of **stmod**(kG) and that $\mathrm{Ext}^*_{kG}(M, V) \neq 0$ for all V in the subcategory. Must it be true that M is a generator for the category in the sense that $\mathcal{H}(M) = \mathcal{M}$?

In the case that we are considering we can actually determine the module $\mathcal{F}_{\mathcal{K}}(U) \cong \mathcal{F}_{\mathcal{K}}(\epsilon)$ in terms of generators and relations. Let F be the direct sum of an infinite number of copies of P_ϵ generated by the elements $\{a_i : i = 1, 2, \dots\}$.

That is, each a_i represents the idempotent $(x-1)/2 \in kG$ that generates P_ϵ as a direct summand of kG. Thus $x \cdot a_i = -a_i$. Let R be the submodule generated by the elements

$$Z \cdot a_{2i-1} - Y^2 \cdot a_{2i} \qquad \textit{and} \qquad Z^2 \cdot a_{2i} - Y^2 \cdot a_{2i+1}$$

for $i = 1, 2, \ldots$.

We claim that $\mathcal{F} = F/R \cong \mathcal{F}_\mathcal{K}(\epsilon)$. First note that the distinguished map $\theta : \epsilon \longrightarrow \mathcal{F}$ is given by $\theta(m) = Y^2 \cdot a_1 + R$ where m is a generator for ϵ. If we let M_i be the submodule of $\mathcal{F}/\theta(\epsilon)$ generated by the classes of the elements $a_1, \ldots, a_i$ then each quotient M_i/M_{i-1} is in $\mathcal{K}$. That is, each is a quotient of $P_\epsilon/(Y^2 \cdot P_\epsilon)$ by a submodule all of whose composition factors are isomorphic to k. So $\mathcal{F}/\theta(\epsilon)$ is the direct limit of the system

$$M_1 \longrightarrow M_2 \longrightarrow M_3 \longrightarrow \ldots$$

of objects in $\mathcal{K}$. The fact that $\mathcal{F}$ is $\mathcal{K}$-local is also easy because the socle and top $(\mathcal{F}/Rad(\mathcal{F}))$ of $\mathcal{F}$ consist of direct sums of copies of ϵ and $\mathcal{F}$ is periodic of period 2 $(\Omega^2(\mathcal{F}) \cong \mathcal{F})$. Consequently, by the criteria of Theorem 14, we have our claim.

We want to look at $\overline{\text{End}}_{kG}(\mathcal{F})$, the endomorphism ring of $\mathcal{F}_\mathcal{K}(U) = \mathcal{F}_\mathcal{K}(\epsilon)$ in the stable category. It can be shown in this case that the stable endomorphism ring is isomorphic to $k[\alpha, \beta]/(\beta^2)$. Here α is the homomorphism that takes each a_i to a_{i+2} for all i, and β is represented by the homomorphism given by the images $\beta(a_{2i-1}) = Z \cdot a_{2i}$ and $\beta(a_{2i}) = a_{2i+1}$. It can be calculated that β^2 factors through a projective module and hence is zero in the stable category.

Now suppose that N is a module in the subcategory $\mathcal{U}$ of $\tilde{B}_0(G)$ consisting of all modules M with $H^*(G, M) = 0$. Then for the generator $U = k \oplus \epsilon$ of the example we have that $\text{Ext}^*_{kG}(N, \mathcal{E}_\mathcal{K}(U)) = 0$ and by the long exact sequence, $\text{Ext}^*_{kG}(N, U) = \text{Ext}^*_{kG}(N, \mathcal{F}_\mathcal{K}(U))$ which has finite dimension over k. Because U is a generator for $\tilde{B}_0(G)$ we are assured that $\text{Ext}^*_{kG}(N, U) \neq 0$. But for each n, $\text{Ext}^n_{kG}(N, \mathcal{F}_\mathcal{K}(U))$ is a module over the stable endomorphism ring $\overline{\text{End}}_{kG}(\mathcal{F}_\mathcal{K}(U))$. We emphasize that it must be a finite dimensional module. It is this fact that we need to use to answer Question 13. Even more can be proved.

THE DOUBLE PERP OF $\mathcal{K}$

Something very similar to the situation that we have seen in the above example occurs in several other cases including all of the examples in [6]. It does not seem to be universal or even generic, but it is still encouraging that we can answer some of these questions in even very special circumstance. We finish this paper by giving a brief sketch of how the information on the stable endomorphism ring $\overline{\text{End}}_{kG}(\mathcal{F}_\mathcal{K}(U))$ for a generator U of $\tilde{B}_0(G)$ can be employed effectively. Basically we define a new sort of support variety for modules in the "no cohomology" subcategory $\mathcal{U}$. The reader should be warned that we are skipping over several very sticky details. These will be dealt with fully in a future paper [10]. We hope that the account presented here will at least give the flavor of the subject.

Here is what we are assuming. We have a generator U for the stable category **stmod**(kG) (or in some cases we might want to just consider the stable category $\tilde{B}_0(G)$ of the principal block). The endomorphism ring $\overline{\mathrm{End}}_{kG}(\mathcal{F}_\mathcal{K}(U))$ has a central polynomial subgring $S = k[\alpha]$ with the property that every polynomial f in S that is not a scalar is injective and has has a finite dimensional nonzero cokernel which we call N_f. We let $\mathcal{U}$ be the category of finitely generated modules M with $H^*(G, M) = 0$. Also for the purposes of this argument we assume that the nucleus $\mathcal{N}$ has dimension one. This will mean that any module in $\mathcal{U}$ must be periodic.

As we noticed, if N is in $\mathcal{U}$ then the long exact sequence

$$\ldots \to \mathrm{Ext}^n_{kG}(N, \mathcal{E}_\mathcal{K}(U)) \to \mathrm{Ext}^n_{kG}(N, U) \to \mathrm{Ext}^n_{kG}(N, \mathcal{F}_\mathcal{K}(U)) \to \ldots$$

and the fact that $\mathrm{Ext}^n_{kG}(N, \mathcal{E}_\mathcal{K}(U)) = 0$ implies that $\mathrm{Ext}^n_{kG}(N, \mathcal{F}_\mathcal{K}(U)) \cong \mathrm{Ext}^n_{kG}(N, U)$ is a finite dimensional module over S. That is, it is finite dimensional as a vector space over the field k. Because of the periodicity of the cohomology we can show additionally that the annihilator of $\mathrm{Ext}^*_{kG}(N, \mathcal{F}_\mathcal{K}(U))$ in S is a nonzero ideal which we denote $\mathcal{J}(N)$. Hence it has a variety $\mathfrak{X}(N)$ in the maximal ideal spectrum $\mathfrak{X} \cong k$ of S. The variety $\mathfrak{X}(N)$ is the set of zeros of the polynomials in the ideal $\mathcal{J}(N)$.

The next thing that we prove is that for any triangle in $\mathcal{U}$

$$\longrightarrow N_1 \longrightarrow N_2 \longrightarrow N_3 \longrightarrow \Omega^{-1}(N_1) \longrightarrow$$

we have $\mathcal{J}(N_1) \cdot \mathcal{J}(N_3) \subseteq \mathcal{J}(N_2)$. Therefore if $\mathfrak{Y} \subseteq \mathfrak{X}$ is a subvariety and if $\mathcal{U}_\mathfrak{Y}$ is the collection of all modules N in $\mathcal{U}$ with $\mathfrak{X}(N) \subseteq \mathfrak{Y}$ then $\mathcal{U}_\mathfrak{Y}$ is a thick subcategory of $\mathcal{U}$.

By assumption the homomorphism α is injective though not surjective, and the same is true for any $f(\alpha)$ where f is a nonconstant polynomial. So for any such endomorphism we have an exact sequence in **Mod**(kG) (triangle in **StMod**(kG))

$$0 \longrightarrow \mathcal{F}_\mathcal{K}(U) \xrightarrow{f(\alpha)} \mathcal{F}_\mathcal{K}(U) \longrightarrow N_f \longrightarrow 0.$$

where N_f is finitely generated. We can prove at this point that $f^2 \in \mathcal{J}(N_f)$ so that $\mathfrak{X}(N_f) = \mathfrak{X}(f)$ the zero set of f. So the subcategory $\mathcal{U}_{\mathfrak{X}(f)}$ is not empty. Therefore we have the following.

Fact 1. The category $\mathcal{U}$ has an infinite number of thick subcategories. Moreover the modules all have the same variety.

The thick subcategories correspond to the subvarieties of $\mathfrak{X}$. If N is a module in $\mathcal{U}$ then the variety $V_G(N)$ of N is precisely the nucleus $\mathcal{N}$ which is a line in $V_G(k)$. Note also that any module M in $\mathcal{U}$ is contained in the proper subcategory $\mathcal{U}_{\mathfrak{X}(M)}$. Thus we have the following.

Fact 2. No object in $\mathcal{U}$ is a generator of the category $\mathcal{U}$.

Finally, we can get an answer to Question 13 in this particular case. We end the paper by presenting a brief sketch of the proof.

Fact 3. If M is a finitely generated kG-module with $\mathrm{Ext}^*_{kG}(M, N) = \{0\}$ for all N in $\mathcal{U}$ then M is in $\mathcal{K}$.

The first thing to notice is that, in the above notation, $\mathrm{Ext}^*_{kG}(M, N_f) = \{0\}$ for all polynomials $f(\alpha) \in S$. Thus, from the long exact sequence on cohomology we get that multiplication by $f(\alpha)$

$$f(\alpha) : \mathrm{Ext}^*_{kG}(M, \mathcal{F}_{\mathcal{K}}(U)) \longrightarrow \mathrm{Ext}^*_{kG}(M, \mathcal{F}_{\mathcal{K}}(U))$$

is an injection for all f. But we can show that $\mathrm{Ext}^*_{kG}(M, \mathcal{F}_{\mathcal{K}}(U))$ is a finitely generated module over the polynomial ring S. So the only way that this can happen is if $\mathrm{Ext}^*_{kG}(M, \mathcal{F}_{\mathcal{K}}(U)) = \{0\}$. But U is a generator for the category so we can create any module out of U by a series of extensions of translations of U by Ω^n or by the taking of direct summands. Any such process will preserve the property listed above, because $\mathcal{F}_{\mathcal{K}}$ is a functor. Hence we must have that

$$\mathrm{Ext}^*_{kG}(M, \mathcal{F}_{\mathcal{K}}(M)) = \{0\}.$$

The last statement leads to our final contradiction because in the triangle

$$\longrightarrow \mathcal{E}_{\mathcal{K}}(M) \longrightarrow M \longrightarrow \mathcal{F}_{\mathcal{K}}(M) \longrightarrow \Omega^{-1}(\mathcal{E}_{\mathcal{K}}(M)) \longrightarrow$$

the map from M to $\mathcal{F}_{\mathcal{K}}(M)$ is the zero map. Hence the identity map from M to M factors through an element of $\mathcal{K}^{\oplus}$ and M must be in $\mathcal{K}$.

References

[1] D. J. Benson, "Representations and cohomology I, II", Cambridge Univ. Press, 1991.

[2] D. J. Benson, *Cohomology of modules in the principal block of a finite group*, New York J. of Math. **1** (1995), 196–205.

[3] D. J. Benson, *Infinite dimensional modules for finite groups*, These proceedings.

[4] D. J. Benson, J. F. Carlson and G. R. Robinson, *On the vanishing of group cohomology*, J. of Algbra **131** (1990), 40–73.

[5] D. J. Benson, J. F. Carlson and J. Rickard, *Complexity and varieties for infinitely generated modules, II*, Math. Proc. Cam. Phil. Soc. **120** (1996), 597–615.

[6] D. J. Benson, J. F. Carlson and J. Rickard, *Thick subcategories of the stable module category*, Fund. Math. **153** (1997), 59–80.

[7] M. Broué, *Isométries parfaites, types de blocs, catégories dérivées*, Astérisque **181–182** (1990), 61–92.

[8] J. F. Carlson, "Modules and group algebras", ETH Lecture Notes, Birkhäuser Verlag, 1996.

[9] J. F. Carlson, *Varieties for cohomology with twisted coefficients*, Acta Acad. Sinica (to appear).

[10] J. F. Carlson, *Orthogonal subcategories of modules over group algebras*, (in preparation).

[11] J. F. Carlson and R. Rouquier, *Self-equivalences of stable module categories*, Math. Zeit. (to appear).

[12] L. Evens, "The Cohomology of Groups", Oxford University Press, 1991.

[13] D. Happel, "Triangulated categories in the Representation Theory of Finite Dimensional Algebras", London Math. Soc. Lecture Notes No. 119, Cambridge Univ. Press, 1988.

[14] M. Linckelmann, *Varieties in block theory*, (preprint).

[15] J. Rickard, *Derived equivalences as derived functors*, J. London Math. Soc. **43** (1991), 37–48.

Department of Mathematics, University of Georgia, Athens, Georgia 30602, USA
E-mail address: jfc@sloth.math.uga.edu

Trends in Mathematics, © 2000 Birkhäuser Verlag Basel/Switzerland

BIRATIONAL CLASSIFICATION OF MODULI SPACES

AIDAN SCHOFIELD

1. Introduction

In this paper, I want to introduce a new way to study moduli spaces of objects in certain kinds of abelian categories. I have applied this method to the study of moduli spaces of representations of quivers, of vector bundles over smooth projective curves and of vector bundles over P^2. In the first and third case, this allows the rationality problem to be reduced to the case of a suitable number of suitably sized matrices up to simultaneous conjugacy; in the second case, one can reduce the rationality problem to that of vector bundles of degree 0, which leads to a proof that the moduli space of vector bundles of rank r and determinant line bundle L is rational provided that the rank r and the degree of the line bundle are co-prime. I shall give a general idea of the method largely without proof and present in incomplete detail how the method is applied in the case of representations of quivers where the preparatory work is of interest in its own right since it gives an efficient algorithm to compute the canonical decomposition of a dimension vector. Here the new idea is the notion of *a rigid sub-dimension vector* of a dimension vector. If α is a dimension vector then β is said to be a rigid sub-dimension vector of α if and only if a general representation of dimension vector α has a unique subrepresentation of dimension vector β. A dimension vector α is said to be *uniform* if and only if it is a multiple of a Schur root in which case it is either a Schur root or else the Schur root β in question is real or isotropic and the canonical decomposition of α is $\alpha = n\beta$. A major result proved in this paper is that if α is a Schur root then α has a rigid sub-dimension vector β such that both β and $\alpha - \beta$ are uniform. This implies that Schur roots are built up from smaller Schur roots in an inductive way.

For the purposes of this introduction to these ideas I shall assume that we are working over an algebraically closed field of characteristic 0, k; many of these results carry over to arbitrary characteristic and to fields that are not algebraically closed. For an introduction to moduli problems the reader should consult [10] for general information and [7] for the special case of quivers.

In the paper [9], the author and L. leBruyn showed that a moduli space of representations of dimension vector α where α is a Schur root is stably birational to a suitable number of matrices up to simultaneous conjugacy. Here we say that the algebraic varieties V and W are stably birational if there exist integers a and b such that $V \times A^a$ is birational to $W \times A^b$ where A^c is c-dimensional affine space. The size of the matrices is $\gcd_v \alpha(v)$ where v runs through the vertices of the quiver. Of course there are many special cases where such a moduli space is known to be birational to a suitable number of matrices up to simultaneous conjugacy. However, even in the case where this greatest common divisor is 1 it was not known in general that birationality holds which implies that the moduli space is rational. The status of the rationality problem for matrices up to simultaneous conjugacy at present is unsatisfactory; thus we know that the space of n by n matrices up to simultaneous conjugacy is rational provided $n \leq 4$ and is stably rational for 5 and 7 and hence is stably rational for any n dividing 420 by the results of [13] and [1]. The case of vector bundles over a smooth projective curve of fixed rank and determinant bundle has been much studied at least in the case where the rank and degree are co-prime. With this restriction, many cases have been shown to be rational varieties. The methods in this paper lead to a direct proof that these moduli spaces are rational when the rank and degree are co-prime. In the paper [6], we apply these methods to show that the moduli space of vector bundles of rank r and degree d over a smooth projective curve is birationally linear over the moduli space of vector bundles of rank h and degree 0 where $h = \gcd(r, d)$. The reader may consult [10] for a general introduction to this problem and [5] for a survey of results on the moduli space of vector bundles over P^2.

Section 2 describes without proof the geometric ideas used to study the moduli spaces in these three different contexts. Section 3 presents formally the results on uniform rigid sub-dimension vectors that lead into this geometric territory for the case of moduli spaces of representations of quivers and section 4 presents the inductive arguments that are needed for representations of quivers in an informal style. The last part of section 3 also describes the algorithm for computing the canonical decomposition of an arbitrary dimension vector. Throughout the paper we shall use the results and notation of [12] where definitions not given here may be found.

2. Geometric methods

This section is not rigorous in any way. My intention is to give a general idea of how the geometric method works in a fairly general context that covers the known cases where the method has applied. I shall be interested in describing moduli spaces birationally. Typically I shall be constructing rational morphisms between these moduli spaces and other varieties closely connected to them. I need a name for the best case; I shall say that a rational dominant morphism $\phi : V \to W$ between irreducible algebraic varieties V and W is *birationally linear* if the function field of V is purely transcendental over the image of the function field of W. I shall also say that V is birationally linear over W if such a morphism exists.

Let $\underline{A}$ be an abelian category where for all objects $B, C \in \underline{A}$, $\mathrm{Hom}(B, C)$ is a finite dimensional vector space over a field k. I shall assume that there is a function from the objects of $\underline{A}$ to a discrete abelian group which is additive on short exact sequences; call this the *type* of an object. In the case of quivers this is just the dimension vector and for a vector bundle E over a smooth projective curve it is the pair $(r,d) \in \mathbb{Z}^2$ where r is the rank of E and d is its degree. I shall assume that there is an irreducible algebraic variety V_α that carries a family of objects of $\underline{A}$ of type α; if $p \in V_\alpha$, then B_p is the corresponding object in $\underline{A}$. The set of objects $\{B_p : p \in V_\alpha\}$ should be morally most of the objects of this type. There is also a semisimple algebraic group G_α acting on V_α such that p and q are in the same orbit if and only if B_p and B_q are isomorphic. In these circumstances, there is a notion of stability and semistability for objects of $\underline{A}$, usually several, depending on choices of linearisation of the action of G_α on V_α. Define a type α to be *stable* if there are stable objects of type α. For representations of quivers, stable types are just Schur roots and for vector bundles over smooth projective curves every type of a vector bundle is stable. If α is a stable type then there is an open non-empty subset of V_α, $V_{\alpha,s}$, the set of stable points with respect to some linearisation of the action of G_α on which G_α acts freely and $M_\alpha = V_{\alpha,s}/G_\alpha$ is called *the moduli space of objects of* $\underline{A}$ *of type* α. I am justified here in ignoring the choice of linearisation since I shall be interested in describing the moduli space birationally and the moduli spaces for different choices of linearisation are typically birational.

I shall say that a *general* object of type α has property P if and only if there is a non-empty open subset U of V_α such that for all $p \in U, B_p$ has property P. Now suppose that a general object of type α has a unique subobject isomorphic to a particular object $B \cong C^t$ where C is stable and therefore has endomorphism ring k. Assume also that the factor is stable; let $\beta = \alpha-\text{type}(B)$. This induces a rational morphism from M_α to M_β. Usually it is clear that this morphism is dominant, that is, its image contains an open subset of M_β. I can describe this morphism in the following way. Let s be the minimal value of $\text{ext}(B_q, C)$ where $\text{ext}(E,D)$ is the dimension of $\text{Ext}(E,D)$, as q varies over V_β, then a general object of type β satisfies $\text{ext}(B_q, C) = s$. This defines a vector bundle of rank s, E, over a suitable open subset of V_β such that the fibre above a point q is $\text{Ext}(B_q, C)$. The problem that arises is that the action of G_β does not usually extend to an action on E; however, a central extension $\tilde{G}_\beta$ of G_β does act on E compatibly with the action of G_β on V_β where the central subgroup acts via a character on the fibres of the vector bundle. Associated to E is a Grassmannian bundle $\mathfrak{Gr}\binom{E}{t}$ such that the fibre above a point q is the Grassmannian of t-dimensional subspaces of $\text{Ext}(B_q, C)$ which inherits an action of $\tilde{G}_\beta$ on which the central subgroup acts trivially; that is, it is an action of G_β compatible with the action of G_β on V_β. Now the variety $\mathfrak{Gr}\binom{E}{t}/G_\beta$ is a moduli space for objects of type β together with a subspace of dimension t of $\text{Ext}(\ ,C)$. This determines an object of type α by considering the corresponding extension of B_q on C^t and since a general object of type α has a unique subobject isomorphic to C^t, I can usually identify birationally the spaces M_α and $\mathfrak{Gr}\binom{E}{t}/G_\beta$. Thus the rational morphism from M_α to M_β has fibres that are birational to Grassmannians over a closed point. The problem is that this Grassmannian bundle is usually not locally trivial in the Zariski topology and this morphism is usually not birationally linear. However, it turns out that birationally these morphisms are controlled by certain central simple algebras over the function field of M_β. The reader may wish to consult [3] for a detailed discussion of central simple algebras and the Brauer group; there is no obvious reference for the Grassmannian bundles; however, their relationship to ordinary Grassmannian bundles locally trivial in the Zariski topology is the same as the relationship between Brauer-Severi bundles and projective space bundles. The techniques used to study these Grassmannian bundles is outlined below.

I can control this morphism from M_α to M_β in the following way. Let F_β be the function field of M_β. Since the centre of $\tilde{G}_\beta$ acts via a character on E, G_β acts on the sheaf of endomorphisms of E, $E^\vee \otimes E$, and therefore there is a bundle of central simple algebras $E^\vee \otimes E/G_\beta$ defined on a suitable open subset of M_β. The fibre above the generic point gives a central simple algebra S_E of dimension s^2 over F_β. I can describe the generic fibre of the morphism from $\mathfrak{Gr}\binom{E}{t}/G_\beta$ to M_β in terms of the central simple algebra S_E. The algebraic group $PGl_s(F_\beta)$ is the automorphism group of $M_s(F_\beta)$ and consequently the central simple algebra S_E which is a twisted form of $M_s(F_\beta)$ is given by an element $\kappa(S_E) \in H^1(Gal(\overline{F_\beta}/F_\beta), PGl_s(\overline{F_\beta}))$ where $\overline{F_\beta}$ is the separable closure of F_β. Since $PGl_s(F_\beta)$ acts on $\mathfrak{Gr}\binom{s}{t}$, the image of $\kappa(S_E)$ in $H^1(Gal(\overline{F_\beta}/F_\beta), Aut(\mathfrak{Gr}\binom{s}{t})(\overline{F_\beta}))$ determines a twisted form $\mathfrak{Gr}\binom{s}{t}_{S_E}$ of $\mathfrak{Gr}\binom{s}{t}$ defined over F_β and this can be shown to be the generic fibre of the morphism from $\mathfrak{Gr}\binom{E}{t}/G_\beta$ to M_β. This is a variety about which we know a little. $\mathfrak{Gr}\binom{s}{t}$ represents the functor of t-dimensional subspaces of ${F_\beta}^s$ and passing by Morita equivalence to $M_s(F_\beta)$, this becomes the functor of right ideals in $M_s(F_\beta)$ of dimension st over F_β. On passing to the twisted form $\mathfrak{Gr}\binom{s}{t}_{S_E}$, this represents the functor of right ideals of dimension st in S_E. It is a fairly simple matter to see that this has a point in F_β if and only if it is a rational variety and this is the case if and only if $S_E \cong M_u(S)$ where $u = s/\gcd(s,t)$ and S is some central simple algebra of dimension $\gcd(s,t)^2$.

Now let χ be the character on the central subgroup of $\tilde{G}_\beta$ by which it acts on E and let E' be another vector bundle of rank s' over some non-empty open subset of V_β on which $\tilde{G}_\beta$ acts compatibly with the action of G_β on V_β where the centre acts via the character χ. Then on a suitable open subset G_β acts on $E^\vee \otimes E'$ and the generic fibre of $E^\vee \otimes E'/G_\beta$ is a left S_E, right $S_{E'}$ bimodule of dimension ss' and consequently the Brauer class of S_E and $S_{E'}$ are equal. If $s > s'$, one can show that $\mathfrak{Gr}\binom{s}{t}_{S_E}$ is birationally linear over $\mathfrak{Gr}\binom{s'}{t}_{S_{E'}}$ because of this equality of Brauer classes. This freedom I now have to change the vector bundle allows me to proceed by induction on the dimension of the moduli space.

There are various modifications to this general plan that are needed in different contexts. It may be the case that a general object of type α is a factor of $B = C^t$ in a unique way. Now let $\beta = \text{type}(B) - \alpha$. Then there is a rational morphism from M_α to M_β which we may know to dominant. This time I consider a vector bundle over a non-empty open subset of V_β where the fibres are of the form $\text{Hom}(B_q, C)$ and the Grassmannian of t-dimensional subspaces of this. This is in fact the procedure used to deal with vector bundles over a smooth projective curve. A further modification I need to consider is that a general object of type α may contain a unique object of type β and the corresponding morphism from M_α to $M_\beta \times M_{\alpha-\beta}$ is dominant. This is approximately what I need to deal with the case of moduli spaces of representations of quivers. Again this reduces to at worst twisted forms of Grassmannians which I can handle for a reduction process.

3. Rigid subrepresentations

In this section, I return to a more rigorous presentation of the material. The main aim is to find information on the internal structure of a general representation of dimension vector α. This takes the form of an algorithm that computes the canonical decomposition if α is not a Schur root or else finds a rigid sub-dimension vector β of α such that both β and $\alpha - \beta$ are uniform dimension vectors.

I use the notation $R(Q, \alpha)$ for the vector space that parametrises representations of dimension vector α on which the algebraic group PGl_α acts. Recall that the canonical decomposition of α is a sum $\alpha = \sum_i \beta_i$ where each β_i is a Schur root and a general representation of dimension vector α is isomorphic to a direct sum $\oplus_i R_i$ where for each i, R_i is a representation of dimension vector β_i such that $\mathrm{End}(R_i) = k$, the ground field. Call each β_i a *canonical summand* of α.

For the moment I make the assumption that Q is a quiver without directed loops. Recall that a dimension vector β is said to be a rigid sub-dimension vector of the dimension vector α if and only if a general representation of dimension vector α has a unique subrepresentation of dimension vector β. In [12], the author showed in characteristic 0 that every representation of dimension vector α contains a subrepresentation of dimension vector β if and only if $\mathrm{ext}(\beta, \alpha - \beta) = 0$. We shall use this result repeatedly in this section. Recall also that a dimension vector α is said to be uniform if it is a multiple of a Schur root or equivalently its canonical decomposition is of the form $\alpha = n\beta$; call β the *root* of the uniform dimension vector. I shall be interested in finding uniform rigid sub-dimension vectors of a particular dimension vector α; I may as well assume that the support of α is Q. When Q is a quiver without directed loops then there are obviously such dimension vectors since if v is a sink vertex, then $\alpha(v)e_v$ is a uniform rigid sub-dimension vector of α where e_v is the dimension vector such that $e_v(v) = 1$ and $e_v(w) = 0$ when $v \neq w$.

Firstly recall some facts about the canonical decomposition. The sum $\alpha = \sum_i \beta_i$ is the canonical decomposition if and only if each β_i is a Schur root and $\mathrm{ext}(\beta_i, \beta_j) = 0$ if $i \neq j$. If $\beta_i = \beta_j$ and $i \neq j$ then either β_i is a real Schur root, that is, $\hom(\beta_i, \beta_i) = 1$, or else $\hom(\beta_i, \beta_i) = 0$ that is, β_i is an isotropic Schur root. If $\beta_i \neq \beta_j$ then one of $\hom(\beta_i, \beta_j)$ and $\hom(\beta_j, \beta_i)$ must be 0. In fact the following lemma contains a stronger statement.

Lemma 3.1. *Let $\{\beta_i : i = 1 \to m\}$ be pairwise distinct canonical summands of the dimension vector α. Then one of $\hom(\beta_i, \beta_{i+1})$ for $i = 1 \to m - 1$ and $\hom(\beta_m, \beta_1)$ must be 0.*

Proof. Let $\{R_i : i = 1 \to m\}$ be representations of dimension vector β_i respectively such that $\mathrm{Ext}(R_i, R_j) = 0$ for $i \neq j$. Then by lemma 4.1 of [4], any homomorphism from R_i to R_j must be either injective or surjective. Assume our conclusion is false; then there is a non-zero homomorphism from R_i to R_{i+1} for each $i < m$ and a non-zero homomorphism from R_m to R_1. Each homomorphism in this chain must be either surjective or injective but not both and no surjective homomorphism may be followed by an injective homomorphism since their composition would then be neither injective nor surjective; hence these homomorphisms must be either all surjective or all injective which is absurd. □

Lemma 3.2. *Let $\alpha = \sum_i n_i\beta_i$ be the canonical decomposition of α where $\beta_i = \beta_j$ if and only if $i = j$. Then it is possible to choose the indexing so that $i < j \Rightarrow$* $\hom(\beta_i, \beta_j) = 0$.

Proof. Define a transitive relation $<$ on the dimension vectors $\{\beta_i\}$ by $i < j$ if $\hom(\beta_j, \beta_i) > 0$. Then the preceding lemma shows that $<$ is a partial order and can therefore be extended to a total order which is the conclusion of the lemma. □

Thus one can assume that the canonical decomposition of α is $\alpha = \sum_i n_i\beta_i$ where $\beta_i = \beta_j$ if and only if $i = j$ and $\hom(\beta_1, \beta_i) = 0$ for $i > 1$.

Lemma 3.3. *Assume the notation of the preceding paragraph. The dimension vector $n_1\beta_1$ is a uniform rigid sub-dimension vector of α.*

Proof. To begin with, note that $\hom(\beta_1, \beta_i) = 0 = \mathrm{ext}(\beta_1, \beta_i)$ for $i > 1$ and hence $\hom(n_1\beta_1, \alpha - n_1\beta_1) = 0 = \mathrm{ext}(n_1\beta_1, \alpha - n_1\beta_1)$. Recall from section 3 of [12] the algebraic variety $R(Q, n_1\beta_1 \subset \alpha)$ which parametrises representations of dimension vector α with a distinguished subrepresentation of dimension vector $n_1\beta_1$. Then from section 3 of [12] it follows that the morphism from $R(Q, n_1\beta_1 \subset \alpha)$ to $R(Q, \alpha)$ is surjective since the fibre above a point p is bijective with the set of subrepresentations of dimension vector $n_1\beta_1$ in R_p. Moreover, the conditions above mean that the dimension of $R(Q, n_1\beta_1 \subset \alpha)$ equals the dimension of $R(Q, \alpha)$ so that the fibre above a general point is finite. Both varieties are irreducible. Moreover, there is a rational section, a morphism defined on an open subset of $R(Q, \alpha)$ to $R(Q, n_1\beta_1 \subset \alpha)$ which sends the point p to the point in the fibre above p corresponding to the subrepresentation that is the direct summand of dimension vector $n_1\beta_1$. But a morphism between irreducible algebraic varieties that is generically finite and has a rational section must be generically bijective (consider the effect of the rational section on the function fields of the two varieties). This implies that the fibre above a general point of $R(Q, \alpha)$ consists of 1 point which means that $n_1\beta_1$ is a rigid sub-dimension vector of α and it is uniform by assumption. □

Let $\alpha = \sum_i n_i\gamma_i$ be the canonical decomposition of α where $\gamma_i = \gamma_j$ if and only if $i = j$. If j is an index such that $\hom(\gamma_j, \gamma_i) = 0$ for $i \neq j$ then call the dimension vector $n_j\gamma_j$ a *uniform rigid summand* of α.

The next two lemmas give ways to find new rigid sub-dimension vectors of a dimension vector which will form the basis of an algorithm to compute the canonical decomposition of a dimension vector α or else to find a rigid sub-dimension vector β of α when it is a Schur root such that both β and $\alpha - \beta$ are uniform.

Lemma 3.4. *Let $\alpha = \beta + \gamma + \delta$ where $\beta + \gamma$ is a rigid sub-dimension vector of α and β is a uniform rigid summand of $\beta + \gamma$. Then β is a rigid sub-dimension vector of α.*

Proof. Let R be a general representation of dimension vector α; in particular, it has a unique subrepresentation R' of dimension vector $\beta + \gamma$ which in turn may be taken to have a unique subrepresentation R'' of dimension vector β. Since β is a uniform rigid summand of $\beta + \gamma$, $\operatorname{ext}(\beta, \gamma) = 0$ and also $\operatorname{ext}(\beta, \delta) = 0$ since $\operatorname{ext}(\beta + \gamma, \delta) = 0$ and β is a canonical summand of $\beta + \gamma$. So, $\operatorname{ext}(\beta, \gamma + \delta) = 0$ and hence every representation of dimension vector α has a subrepresentation of dimension vector β. Also since γ is a canonical summand of $\beta+\gamma$ and $\operatorname{ext}(\beta+\gamma, \delta) = 0$ then $\operatorname{ext}(\gamma, \delta) = 0$ so a general representation of dimension vector $\gamma + \delta$ has a subrepresentation of dimension vector γ. So let S be a subrepresentation of R of dimension vector β; then R/S has a subrepresentation of dimension vector γ, T/S where T has to be R' since it is a subrepresentation of dimension vector $\beta + \gamma$ of R. So, S is a subrepresentation of dimension vector β in R' and must be R''. Thus β is a rigid sub-dimension vector of α. □

Lemma 3.5. *Let $\alpha = \beta + \gamma + \delta$ where β is a rigid sub-dimension vector of α and γ is a uniform rigid summand of $\gamma + \delta$. Then $\beta + \gamma$ is a rigid sub-dimension vector of α.*

Proof. Since $\operatorname{ext}(\beta, \gamma + \delta) = 0$ and γ is a uniform rigid summand of $\gamma + \delta$, $\operatorname{ext}(\beta, \gamma) = 0 = \operatorname{ext}(\beta, \delta)$. Also $\operatorname{ext}(\gamma, \delta) = 0$ and so $\operatorname{ext}(\beta + \gamma, \delta) = 0$. Hence a general representation of dimension vector α has a subrepresentation of dimension vector $\beta+\gamma$. In turn this has a subrepresentation of dimension vector β which must be the unique one and the factor must be the unique subrepresentation of dimension vector γ. So there is a unique subrepresentation of dimension vector $\beta + \gamma$. □

These results are the basis for the following lemma which contains most of the work for our understanding of the internal structure of Schur roots and representations and is also the basis for an algorithm for constructing the canonical decomposition of a dimension vector.

Lemma 3.6. *Let $\alpha = \beta + \gamma + \delta$ be a dimension vector for the quiver Q without directed loops where β is a uniform rigid sub-dimension vector of α and γ is a uniform rigid summand of $\gamma + \delta$. Let $\beta = m\beta'$ and $\gamma = n\gamma'$ be the canonical decompositions of β and γ so that β' is the root of β. Then either γ' is a canonical summand of α or else any uniform rigid summand of $\beta + \gamma$ is a uniform rigid sub-dimension vector of α whose root is larger than β'.*

Proof. Our assumptions imply that $\hom(\beta, \gamma+\delta) = 0 = \operatorname{ext}(\beta, \gamma+\delta)$ and hence $\hom(\beta, \gamma) = 0 = \operatorname{ext}(\beta, \gamma)$ and $\hom(\beta, \delta) = 0 = \operatorname{ext}(\beta, \delta)$ since γ and δ are canonical summands of $\gamma + \delta$. For the same reason, $\operatorname{ext}(\gamma, \delta) = 0 = \operatorname{ext}(\delta, \gamma)$. If $\operatorname{ext}(\gamma, \beta) = 0$ then γ is a summand of α because $\alpha - \gamma = \beta + \delta$ and so $\operatorname{ext}(\gamma, \alpha-\gamma) = 0 = \operatorname{ext}(\alpha - \gamma, \gamma)$. Thus if γ' is not a canonical summand of α then $\operatorname{ext}(\gamma, \beta) > 0$ and so $\operatorname{ext}(\gamma', \beta') > 0$. It follows that $\hom(\gamma', \beta') = 0$ since $\operatorname{ext}(\beta', \gamma') = 0$ and so one of $\hom(\gamma', \beta')$ and $\operatorname{ext}(\gamma', \beta')$ must be 0 by theorem 4.1 of [12].

By lemma 3.5, $\beta + \gamma$ is a rigid sub-dimension vector of α. So consider its canonical decomposition. Firstly $\hom(\beta', \gamma') = 0 = \operatorname{ext}(\beta', \gamma')$ and $\hom(\gamma', \beta') = 0$. On the other hand, $\operatorname{ext}(\gamma', \beta') \neq 0$. Let S be a general representation of dimension vector $\beta + \gamma$. Then S has a subrepresentation R of dimension vector β such that both R and S/R are general representations. So $R \cong \oplus_{j=1}^m R_j$ and $S/R \cong \oplus_{i=1}^n S_i$ where $\mathbf{dim}\, R_j = \beta'$, $\mathbf{dim}\, S_i = \gamma'$, each R_j and S_i is a Schur representation, $\operatorname{Hom}(R_j, S_i) = 0 = \operatorname{Hom}(S_i, R_j)$, $\operatorname{Hom}(R_j, R_l) = 0$ for $j \neq l$ unless β' is a real Schur root in which case they are isomorphic, and similarly, $\operatorname{Hom}(S_i, S_l) = 0$ for $i \neq l$ unless γ' is a real Schur root in which case they are isomorphic. By Ringel's simplification process [11] one knows that any summand of S must have a filtration by subrepresentations such that the factors are isomorphic to either an S_i or an R_j. No S_i can be a summand of S since then its dimension vector γ' would be a summand in the canonical decomposition of α. Thus any summand in the canonical decomposition of $\beta + \gamma$ must be of the form $a\beta' + b\gamma'$ where $a > 0$. If β' is not a summand in the canonical decomposition of $\beta + \gamma$, then every canonical summand of $\beta + \gamma$ is larger than β'. If β' is a summand in the canonical decomposition of $\beta + \gamma$ then $\langle (m-1)\beta' + n_1\gamma_1, \beta' \rangle \geq 0$ and so $\langle \beta', \beta' \rangle > 0$ which implies that β' is a real Schur root and so each R_j is isomorphic to the real Schur representation $G(\beta')$ of dimension vector β'. Therefore, there is a summand of S that has a proper subrepresentation isomorphic to $G(\beta')$ which is assumed to be a summand of S. It follows that a uniform rigid summand of $\beta + \gamma$ cannot be a multiple of β' and therefore the root of a uniform rigid summand κ of $\beta + \gamma$ must be larger than β'. Therefore, in either case there is a uniform rigid summand κ of $\beta + \gamma$ whose root is larger than β'. However, by lemma 3.4, κ is a uniform rigid sub-dimension vector of α which completes our proof. □

The main result on the internal structure of Schur roots is now a simple induction.

Theorem 3.7. *Let α be a Schur root for the quiver Q without directed loops. Then there exists a rigid sub-dimension vector β of α such that both β and $\alpha - \beta$ are uniform dimension vectors.*

Proof. Assume that the support of α is Q. Let v be a sink vertex; then as discussed before $\alpha(v)e_v$ is a uniform rigid sub-dimension vector of α where e_v is the dimension vector of the simple representation at the vertex v. If the factor is uniform nothing remains to be done. Otherwise, assume that β is a uniform rigid sub-dimension vector of α such that $\alpha - \beta$ is not uniform and proceed by induction on $\alpha - \beta'$ where β' is the root of β. Since $\alpha - \beta$ is not uniform, $\alpha - \beta = \gamma + \delta$ where γ is a uniform rigid summand of $\alpha - \beta$. Since the root of γ cannot be a summand of α one concludes by lemma 3.6 that there is a uniform rigid sub-dimension vector of α whose root is larger than β'. By induction, the result follows. □

Lemma 3.6 also gives an algorithm to compute the canonical decomposition of a dimension vector α in the following way. I begin by dealing with a quiver without directed loops. First of all, it is a simple matter to compute whether a dimension vector is a Schur root on a quiver with 2 vertices v and w with no arrows from w to v though all other possible arrows are allowed, since all roots are Schur. Now assume that Q is a quiver without directed loops. As at the beginning of the proof of theorem 3.7, one may assume that the support of α is the quiver Q and v is a sink vertex so that $\alpha(v)e_v$ is a uniform rigid sub-dimension vector of α and $\alpha - \alpha(v)e_v$ is a smaller dimension vector. One can compute the canonical decomposition of $\alpha - \alpha(v)e_v$ by induction. If $\alpha - \alpha(v)e_v$ is not uniform then lemma 3.6 either gives a canonical summand of α and we may proceed by induction or else it gives a new uniform rigid sub-dimension vector of α which has a larger root. Thus the only time a problem occurs is when there is a uniform rigid sub-dimension vector β of α such that $\gamma = \alpha - \beta$ is also uniform. Let $\beta = m\beta'$ and $\gamma = n\gamma'$ be their canonical decompositions. If both β' and γ' are not real then it is a simple matter to see that α must be a Schur root and β is a uniform rigid sub-dimension vector of α such that $\alpha - \beta$ is also uniform. Without loss of generality one may assume that β' is real (for example by reversing all the arrows of the quiver). Let $t = \mathrm{ext}(\gamma', \beta')$. If γ' is also real, one considers the canonical decomposition of (n, m) for the 2-vertex quiver with t arrows from the first vertex to the second and no loops. If this is $(n, m) = a(b, c) + d(e, f)$ then the canonical decomposition of α is $\alpha = a(b\gamma + c\beta) + d(e\gamma + f\beta)$. Otherwise, α is a Schur root. If γ' is isotropic, one again considers the dimension vector (n, m) for the 2-vertex quiver with t arrows from the first to the second vertex but with 1 loop at the first vertex. Again its canonical decomposition determines the canonical decomposition of α using the same formula. Finally, if γ' is neither real nor isotropic then $\gamma = \gamma'$ and α is a Schur root if and only if $m \leq t$; otherwise its canonical decomposition is $\alpha = (\gamma + t\beta') + (m - t)\beta'$. Note that when α is a Schur root we have also calculated a dimension vector β that is a rigid sub-dimension vector of α such that both β and $\alpha - \beta$ are uniform.

In [12], there is a different algorithm for computing the canonical decomposition of a dimension vector. One knows that β is a canonical summand of the dimension vector α if and only if it is a Schur root and $\mathrm{ext}(\beta, \alpha - \beta) = 0 = \mathrm{ext}(\alpha - \beta, \beta)$ or equivalently a representation of dimension vector α has subrepresentations of dimension vector β and $\alpha - \beta$. On the other hand, $\mathrm{ext}(\beta, \alpha - \beta)$ and $\mathrm{ext}(\alpha - \beta, \beta)$ may be calculated by knowing the dimension vectors of subrepresentations of β. Since $\beta < \alpha$ inductively we know all of these dimension vectors. This algorithm is substantially more complicated than the one in this paper and is correspondingly much slower.

One should also note that this extends to computing the canonical decomposition of a dimension vector over an arbitrary quiver. If Q is a quiver with vertex set V and arrow set A and α is a dimension vector for the quiver Q then one constructs a new quiver Q' with vertex set $V \times \{0,1\}$, and arrow set $V \cup (A \times \{0\})$ where $iv = (v,0)$, $tv = (v,1)$, $i(a,0) = (ia,0)$ and $t(a,0) = (ta,1)$. Then αp_1 is a dimension vector on Q' and if $\alpha = \sum_i \beta_i$ is the canonical decomposition of α then the canonical decomposition of αp_1 is $\alpha p_1 = \sum_i \beta_i p_i$ since for a general representation of dimension vector αp_1, the arrows in V are invertible.

I implemented this algorithm as an `EmacsLisp` program which is probably not the swiftest language in which to do it. It seems to handle arbitrary dimension vectors instantaneously and took 10 minutes to compute the canonical decompositions of all 5-subspace problems where the total space has dimension at most 13. Once I have tidied it up I shall make it available at
`http://www.maths.bris.ac.uk/ pure/staff/maas/maas`.
It runs inside `Emacs` and I shall explain how to use it in the file itself.

4. Moduli spaces of representations

This section will be less rigorous than the last one. I wish to give an outline of the following theorem.

Theorem 4.1. *Let α be a Schur root for the quiver Q. Let $h = \gcd_v(\alpha_v)$ and $p = 1 - \langle \alpha/h, \alpha/h \rangle$. Then a moduli space of representations of dimension vector α is birational to p h by h matrices up to simultaneous conjugacy.*

I shall explain extremely telegraphically at the end of this section how to deal with 2-vertex quivers without directed loops since this involves entirely different techniques; however, I shall explain how to reduce to this case in much greater detail.

In the last section, I showed that every Schur root α has a uniform rigid subdimension vector β such that $\gamma = \alpha - \beta$ is also uniform. Let $m = \gcd_v(\beta_v)$ and $n = \gcd_v(\gamma_v)$ and let $\beta = m\beta'$ and $\gamma = n\gamma'$. Let $t = \operatorname{ext}(\gamma', \beta')$, let $p = 1 - \langle \beta', \beta' \rangle$ and let $q = 1 - \langle \gamma', \gamma' \rangle$. If β' is real, then a general representation of dimension vector β is isomorphic to $G(\beta')^m$ where $G(\beta')$ is the unique real Schur representation of dimension vector β'; if β' is isotropic then a general representation of dimension vector β is isomorphic to $\oplus_{i=1}^m R_i$ where each R_i is a Schur representation of dimension vector β' such that $\operatorname{Hom}(R_i, R_j) = 0 = \operatorname{Ext}(R_i, R_j)$ for $i \neq j$; and if β' is neither real nor isotropic then a general representation of dimension vector β is already Schur. The same comments apply to γ and γ'. By induction on the dimension vector, assume that the moduli space of representations of dimension vector β' in the first two cases is birational to a point and P^1 respectively and in the third case, the moduli space of representations of dimension vector β is birational to p m by m matrices up to simultaneous conjugacy. The same comments apply to γ' and γ.

Now it is possible to show using the methods outlined in section 2 that the moduli space of representations of dimension vector α is birational to the moduli space of representations of dimension vector (n, m) for the 2-vertex quiver with q loops at the first vertex, t arrows from the first to the second vertex and p loops at the second vertex. We shall be looking at a number of these so we call this quiver $KQ(q,t,p)$. I shall use induction on the 5 numbers n, m, q, t, p where we know that $n, m, t > 0$. The case where $q = p = 0$ is the case I have already said requires entirely different methods and will not be discussed here. One needs to know that $\gcd_v(\alpha(v)) = \gcd(m, n)$ and again this can be shown to be true.

If the dimension vector (n, m) is a Schur root for the quiver $KQ(q', t', p')$ where $(q', t', p') < (q, t, p)$ then the result follows by induction since the moduli space of representations of dimension vector (n, m) for the quiver $KQ(q, t, p)$ is birationally linear over the moduli space of representations of dimension vector (n, m) for the quiver $KQ(q', t', p')$; see for example [9] . This means that we have certain minimal cases to look at. These are when (q, t, p) is one of $(1, 1, 1)$, $(0, t, p)$, and $(q, t, 0)$ since if both p and q are both positive then it is clear that (n, m) is a Schur root. The first of these in fact reduces to one of the other two cases since if $n \geq m$ then (n, m) is a Schur root because (for example) the dimension vector (n, n, m) for the quiver with vertices u, v and w with 2 arrows from u to v and 1 arrow from v to w lies in the fundamental region for the action of the Weyl group. The second and third case are essentially equivalent by reversing all the arrows of the quiver so we shall deal with the case $(q, t, 0)$. If (n, m) is a Schur root for $KQ(q, t, 0)$ then $m \leq nt$. We need to show that these are all Schur roots and determine their moduli spaces. If $m = nt$ then the reflection functor at the second vertex shows that the moduli space of representations of dimension vector (n, m) is birational to the moduli space of representations of dimension vector $(0, n)$ for $KQ(0, t, q)$ which is just q n by n matrices up to simultaneous conjugacy. If $m < nt$ we may still apply the reflection functor at the second vertex which preserves the value of $\gcd(m, n)$ followed by duality to ensure that $0 < m \leq \frac{nt}{2}$. If $t > 2$ or if $t = 2$ and $m < n$, then $m < n(t-1)$ and we may conclude by induction that (n, m) is a Schur root for $KQ(q, t-1, 0)$. Thus we may reduce either to the case (n, n) for $KQ(q, 2, 0)$ which is clearly birational to $q+1$ n by n matrices up to simultaneous conjugacy and thus a Schur root with the correct moduli space; or else we reduce to (n, m) where $\frac{n}{2} \geq m > 0$ for the quiver $KQ(q, 1, 0)$ and it is enough to deal with the case where $q = 1$ or equivalently by duality the dimension vector (m, n) for the quiver $KQ(0, 1, 1)$ with $0 < m \leq \frac{n}{2}$.

Consider the quiver Q' with 3 vertices u, v and w with 1 arrow from u to v and 2 arrows from v to w and the dimension vector (m, n, n); this is also clearly a Schur root since it lies in the fundamental region for the action of the Weyl group. Also a general representation of this dimension vector inverts the first arrow from v to w and hence the moduli space of representations of this dimension vector is birational to the moduli space of representations of dimension vector (m, n) for $KQ(0, 1, 1)$. Now $(m, m, 2m)$ is a uniform rigid sub-dimension vector of (m, n, n) and its root is $(1, 1, 2)$ which is a real Schur root. Let (a, b, c) be the rigid sub-dimension vector of (m, n, n) that is constructed from this one such that both it and $(m-a, n-b, n-c)$ are uniform by the method of lemma 3.6 in theorem 3.7. By induction on the pair (n, m) it follows that the moduli spaces of representations of dimension vector the root of (a, b, c) and the root of $(m-a, n-b, n-c)$ satisfy the theorem since their construction from smaller Schur roots must involve a smaller pair than (n, m) and if $m' = \gcd(a, b, c)$ and $n' = \gcd(m-a, n-b, n-c)$ then $(m', n') < (m, n)$ so again the proof is complete by induction.

At this stage, the only part of the proof of this theorem about which I have said nothing is the case of 2-vertex quivers with no directed loops in the quiver. This case appears to require substantially different arguments. I do not intend to discuss this in great detail; however, if $\alpha = (m, n)$ is a dimension vector for the quiver $KQ(0, t, 0)$, $h = \gcd(m, n)$, and $p = 1 - \langle \alpha/h, \alpha/h \rangle$ then it is possible to construct representations R and S such that $\hom(R, S) = 1 + p$ and if T is a general representation of dimension vector α then $\hom(R, T) = \hom(S, T) = h$. Thus $\mathrm{Hom}(R \oplus S, \ )$ defines a functor to representations of the quiver $KQ(0, 1+p, 0)$ that takes a general representation of dimension vector (m, n) to a representation of dimension vector (h, h); moreover it can be shown that this functor defines a birational morphism from the moduli space of representations of dimension vector (m, n) for the quiver $KQ(0, t, 0)$ to the moduli space of representations of dimension vector (h, h) for the quiver $KQ(0, 1+p, 0)$ and this is clearly birational to p h by h matrices up to simultaneous conjugacy.

I hope that this summary gives an impression of the strategy of the proof of theorem 4.1 and the ideas involved.

References

[1] Ch. Bessenrodt, L. Le Bruyn: *Stable rationality of certain* PGL_n *quotients,* Invent. Math. **104** (1991) 179-199.

[2] W.W. Crawley-Boevey: *Subrepresentations of general representations of quivers,* Bull. London Math. Soc. **28** (1996), 363–366.

[3] P. K. Draxl, *Skew fields*, London Mathematical Society Lecture Note Series **81** (1981), Cambridge University Press, Cambridge.

[4] D. Happel, C.-M. Ringel, Tilted algebras, Trans. Amer. Math. Soc. 274 (1982) 399–443.

[5] P.L Katsylo, *Birational geometry of moduli varieties of vector bundles over* $\mathbb{P}^2$, Izv. Akad. Nauk SSSR Ser. Mat. **55** (1991), 429–438, translation in: Math. USSR-Izv. **38** (1992), 419–428.

[6] Alastair King, Aidan Schofield, *Rationality of moduli of vector bundles on curves*, to appear in Indagationes Mathematicae (1999).

[7] A.D. King, *Moduli of representations of finite dimensional algebras,* Quart. J. Math. Oxford (2), **45** (1994), 515–530.

[8] L. Le Bruyn: *Simultaneous equivalence of square matrices,* Séminaire d' Algèbre Paul Dubreil et Marie-Paul Malliavin, 39ème Année (Paris, 19887/1988), 127-136, Lect. Notes in Math., 1404, Springer, Berlin-New York, 1989.

[9] L. Le Bruyn, A. Schofield: *Rational invariants of quivers and the ring of matrix invariants,* Perspectives in ring theory (Antwerp, 1987), 21–29, NATO Adv. Sci. Inst. Sec. C: Math. Phys. Sci., 233, Kluwer Acad. Publ., Dodrecht, 1988.

[10] P.E. Newstead: *Introduction to Moduli Problems and Orbit Spaces,* T.I.F.R. Lecture Notes, Springer-Verlag, 1978.

[11] C.-M. Ringel, Representations of K-species and bimodules, J. Algebra 41 (1976) 269–302.

[12] Aidan Schofield, General representations of quivers, Proc. London Math. Soc. (3) 65 (1992) 46–64.

[13] Aidan Schofield, *Matrix invariants of composite size,* J. Algebra **147**m (1992), 345–349.

UNIVERSITY OF BRISTOL, SCHOOL OF MATHEMATICS, UNIVERSITY WALK, BRISTOL BS8 1TW, ENGLAND

Trends in Mathematics, © 2000 Birkhäuser Verlag Basel/Switzerland

TAME ALGEBRAS AND DEGENERATIONS OF MODULES

GRZEGORZ ZWARA

In representation theory of finite dimensional algebras one studies first of all their finite dimensional modules. But sometimes when studying finite dimensional modules it appears that we have to use infinite dimensional modules. One of the main problems in representation theory is to find a classification of the isomorphism classes of all finite dimensional modules. This reduces to a classification of the indecomposable modules, by the Krull-Remak-Schmidt Theorem. The aim of this paper is to show how infinite dimensional modules arise in two problems: classifications of indecomposable finite dimensional modules and characterizations of degenerations of modules.

1. Module varieties.

Throughout the paper k denotes a fixed algebraically closed field of arbitrary characteristic. All considered algebras will be associative k-algebras with 1. Let Λ be a finitely generated algebra. Hence, Λ can be presented as a quotient of a free algebra $k\langle X_1, \dots, X_t\rangle$ by a two-sided ideal I. We denote by $mod\Lambda$ the category of finite dimensional over k left Λ-modules.

Now, for any positive integer d we define the affine variety $mod_\Lambda^d(k)$ of Λ-module structures on k^d consisting of t-tuples $m = (m_1, \dots, m_t)$ of $d \times d$-matrices with coefficients in k satisfying the relation ρ for any (non-commutative) polynomial ρ in I. Any such t-tuple m corresponds to a d-dimensional Λ-module M in the obvious way. The general linear group $Gl_d(k)$ acts on $mod_\Lambda^d(k)$ by conjugation:

$$g \star m = (gm_1g^{-1}, \dots, gm_tg^{-1})$$

and the orbits correspond to the isomorphism classes of d-dimensional Λ-modules. The variety $mod_\Lambda^d(k)$ depends on the choice of the presentation of Λ only up to $Gl_d(k)$-equivariant isomorphism. In Section 2 we will use module varieties to give geometric interpretations of the representation type of an algebra. It turns out that this concept has in fact a geometric nature. On the other hand, the $Gl_d(k)$-variety $mod_\Lambda^d(k)$ is an interesting object in itself. In Section 3 we will be interested in an algebraic description of (Zariski) closures of $Gl_d(k)$-orbits in $mod_\Lambda^d(k)$.

We shall need not only the affine variety $mod_\Lambda^d(k)$, but more general - the affine scheme mod_Λ^d (see [3]). In fact, we will use a functorial interpretation of mod_Λ^d, that is for any algebra R, we have the set $mod_\Lambda^d(R)$ defined in a similar way as $mod_\Lambda^d(k)$. The set $mod_\Lambda^d(R)$ consists of t-tuples $m = (m_1, \dots, m_t)$ of $d \times d$-matrices with coefficients in R satisfying the relation ρ for any ρ in I. Any such

t-tuple m corresponds to an Λ-R-bimodule M, which as an R-module is free of rank d. (Observe that we obtain Λ-modules of infinite length provided $\dim_k R = \infty$.) The sets $mod_A^d(R)$ have some nice properties. For example, let $\mathcal{C}$ be an affine variety and by $k[\mathcal{C}]$ and $k(\mathcal{C})$ we denote the coordinate ring of $\mathcal{C}$ and the field of rational functions on $\mathcal{C}$, respectively. Then any point of $mod_A^d(k[\mathcal{C}])$ corresponds to a regular morphism $\mu : \mathcal{C} \to mod_A^d(k)$ and any point of $mod_A^d(k(\mathcal{C}))$ corresponds to a rational morphism $\mu : \mathcal{C} \to mod_A^d(k)$. For basic background on regular and rational morphisms and algebraic group actions we refer to [4].

2. Representation type of algebras.

Henceforth A is a fixed finite dimensional algebra. One of the most fundamental problems in the representation theory of algebras is to find a classification of the isomorphism classes of the indecomposable modules in $modA$. Because of this classification problem there were introduced the following definitions.

Definition 2.1. *One says that the algebra A is of:*

- **finite type**, *if there is only finitely many isomorphism classes of indecomposable modules in $modA$;*
- **tame type**, *if for each natural number d there is a finite number of A-$k[X]$-bimodules $M_1, \dots, M_n$ which are free of rank d as right $k[X]$-modules, such that every indecomposable A-module of dimension d is isomorphic to $M_i \otimes_{k[X]} k[X]/(X-\lambda)$ for some $1 \le i \le n$ and $\lambda \in k$;*
- **wild type**, *if there is a finitely generated $A - k\langle X,Y\rangle$-bimodule M which is free as a right $k\langle X,Y\rangle$-module, such that the functor*

$$M \otimes (-) : mod\, k\langle X,Y\rangle \longrightarrow modA$$

preserves indecomposability and isomorphism classes.

There is the well-known Tame and Wild Dichotomy Theorem which is due to Drozd.

Theorem 2.2. *The algebra A is either of tame or wild type, and not both.*

For any finitely generated algebra Λ there is a full exact embedding $\mathcal{F} : mod\Lambda \to modk\langle X,Y\rangle$. This implies that the classification problem for any wild algebra "contains" the classification problem for all finitely generated algebras. Hence, we can realistically hope to describe all modules only for tame algebras.

The above definitions of representation types have the following geometric inerpretations.

Corollary 2.3. *The algebra A is of*

- *finite type if and only if $mod_A^d(k)$ has only finitely many $Gl_d(k)$-orbits, for any $d \ge 1$;*
- *tame type if and only if for each $d \ge 1$ there is a finite number of regular morphisms*

$$\mu_1, \dots, \mu_n : \mathbb{A}^1 \to mod_A^d(k)$$

such that every indecomposable A-module of dimension d is isomorphic to the A-module corresponding to $\mu_i(s)$, for some $i \leq n$ and $s \in \mathbb{A}^1$;

- *wild type if and only if for all $n \geq 1$ there is $d = d(n) \geq 1$ and a regular morphism $\mu : \mathbb{A}^n \to mod_A^d(k)$ such that for every $s \in \mathbb{A}^n$ the A-module corresponding to $\mu(s)$ is indecomposable and two points $\mu(s)$, $\mu(s')$ belong to different orbits provided $s \neq s'$.*

The last part follows from the fact that for any $n \geq 1$, there is an A-$k[X_1, X_2, \ldots, X_n]$-bimodule M_n which is free of finite rank as a right $k[X_1, X_2, \ldots, X_n]$-module and such that the functor

$$M_n \otimes (-) : mod\, k[X_1, X_2, \ldots, X_n] \longrightarrow modA$$

preserves indecomposability and isomorphism classes.

Now we introduce some class of indecomposable A-modules of infinite length, the so called generic modules. In case A is of tame type, the generic modules will be used to describe one-parameter families of indecomposable modules in $modA$.

Definition 2.4. *If G is an A-module, then we call its length as an $End_A(G)$-module the* **endolength** *of G. An indecomposable module G is called* **generic** *if it is of infinite length over A, but finite endolength.*

Example 2.5. *Let A be the Kronecker algebra, that is, the path algebra kQ of the quiver*

$$Q: \quad 1 \rightrightarrows 2$$

Let G be the following representation of Q:

$$k(X) \underset{\cdot X}{\overset{\cdot 1}{\rightrightarrows}} k(X)$$

Then $End_A(G) \simeq k(X)$ and G is a generic A-module of endolength 2. (In fact this is the only generic A-module, up to isomorphism).

Then we have an alternative definition of representation types, which seems to be more forthright then Definition 2.1 (see [2])

Theorem 2.6.

1. *The algebra A is of finite type if and only if there is no generic left A-module.*
2. *The algebra A is of tame type if and only if for each $d \geq 1$ there are only finitely many isomorphism classes of generic left A-modules of endolength d.*

In the case A is of tame type, generic modules control indecomposable modules in $modA$ as follows (see [2]).

Theorem 2.7. *Assume that A is of tame type. Then for any generic module G, the ring $End_A(G)/radEnd_A(G)$ is a rational function field in one variable over k and we can choose an A-R_G-bimodule M_G, such that the following hold:*

1. *R_G is a finitely generated localization of $k[X]$. As a right R_G-module, M_G is free of rank equal to the endolength of G. If K_G is the quotient field of R_G, then*
   $M_G \otimes_{R_G} K_G \simeq G$.
2. *The functor $M_G \otimes_{R_G} (-) : modR_G \longrightarrow modA$ preserves isomorphism classes, indecomposability and Auslander-Reiten sequences.*
3. *For each $d \geq 1$, all but finitely many isomorphism classes of d-dimensional A-modules arise as $M_G \otimes_{R_G} R_G/(r)$ for some G and $0 \neq r \in R_G$.*

Look at a geometric consequences of the above theorem. Let A be of tame type. Then every generic A-module is isomorphic to the A-module corresponding to some point $m \in mod_A^d(k(X))$. Moreover, for every $d \geq 1$, there are a finite subset $C_0 \subset \mathbb{A}^1$, cofinite subsets $C_1, \ldots, C_l \subseteq \mathbb{A}^1$ and regular morphisms $\nu_i : C_i \to mod_A^d(k)$, $0 \leq i \leq l$, such that the A-module corresponding to $\nu_i(s)$ is indecomposable and $\nu_i(s)$, $\nu_j(s')$ belong to the same $Gl_d(k)$-orbit if and only if $i = j$ and $s = s'$.

Now for tame algebras, we want to compare the number of one-parameter families of indecomposable modules and the number of generic modules. Let $\mu_A(d)$ be the smallest integer l such that there are A-$k[X]$-bimodules $M_1, \ldots, M_l$ which are free of rank d as right $k[X]$-modules, and such that the set of modules

$$\{M_i \otimes_{k[X]} k[X]/(X - \lambda) | \lambda \in k, 1 \leq i \leq l\}$$

meets all but finitely many isomorphism classes of indecomposable d-dimensional A-modules. Let $g_A(d)$ be the number of isomorphism classes of generic A-modules of endolength d. Then the numbers $\mu_A(d)$ and $g_A(d)$ are related as follows (see [2]).

Theorem 2.8. *The equality $\mu_A(n) = \Sigma_{d|n} g_A(d)$ holds.*

The last theorem leads to a simpler definition of domestic algebras. Recall that A is of **domestic type**, if there is $N \geq 1$ such that $\mu_A(d) \leq N$, for all $d \geq 1$. As a direct consequence we get the following characterization of algebras of domestic type (see [2]).

Corollary 2.9. *The algebra A is of domestic type if and only if there are only finitely many isomorphism classes of generic left A-modules.*

Now, we are interested in a construction of the generic modules which reflect the correspondence between families of finite dimensional modules and generic ones. First we note that homogeneous tubes are "typical" components of Auslander-Reiten quivers of tame algebras (see [1]):

Theorem 2.10. *If A is of tame type, then for any $d \geq 1$, all but finitely many isomorphism classes of indecomposable A-modules of dimension d lie in homogeneous tubes.*

Let $\mathcal{T} = (T_i)_{i \geq 1}$ be a homogeneous tube.

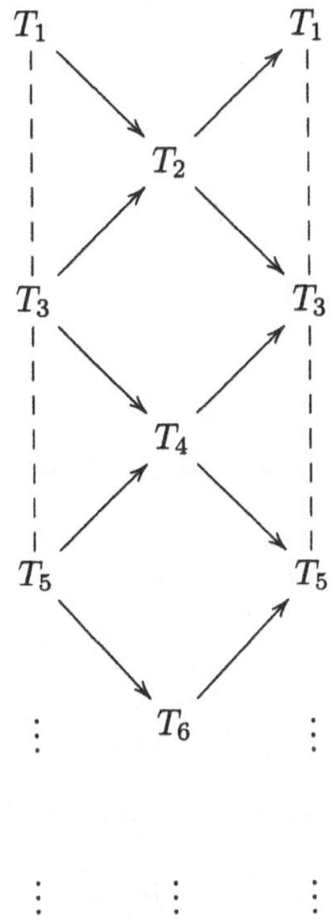

Observe that we have the following direct system of A-modules and monomorphisms:

$$T_1 \longrightarrow T_2 \longrightarrow T_3 \longrightarrow \cdots .$$

Let $T = \varinjlim T_i$. For any module M and any set I, let M^I denote the product of copies of M, indexed by I.

The following theorem shows a relation between homogeneous tubes and generic modules (see [5]). This theorem uses the concept of the Ziegler closure. However, in this particular situation, there is an elementary description of the points in the Ziegler closure. For more information about the Ziegler closure, see the contributions of Krause and Prest in this volume.

Theorem 2.11. *The module T is Σ-pure-injective, indecomposable, but not generic. The set of generic modules which belong to the Ziegler closure of T is non-empty and finite. These modules are direct summands of a product T^I of T, for some set I.*

Observe that there is a direct system of exact sequences which is indexed by $i \geq 1$:

$$\begin{array}{ccccccccc} 0 & \longrightarrow & T_1 & \longrightarrow & T_i & \longrightarrow & T_{i-1} & \longrightarrow & 0 \\ & & \downarrow{\scriptstyle =} & & \downarrow & & \downarrow & & \\ 0 & \longrightarrow & T_1 & \longrightarrow & T_{i+1} & \longrightarrow & T_i & \longrightarrow & 0. \end{array}$$

Taking direct limits gives an exact sequence

$$0 \to T_1 \to T \xrightarrow{\varphi} T \to 0.$$

Then the endomorphism φ is locally nilpotent, that is

$$T = \bigcup_{i \geq 1} Ker\varphi^i.$$

The next theorem describes more precisely the structure of the module T^I (see [7]).

Theorem 2.12. *Let X be an infinite dimensional A-module, let φ be a locally nilpotent endomorphism of X with finite dimensional kernel. Let I be some infinite set. Then the product X^I is the direct sum of a non-zero module $\overline{P}$ of finite endolength and copies of X. The endolength of $\overline{P}$ is bounded by the minimum of the dimensions of the vector spaces $Ker\varphi^t / Ker\varphi^{t-1}$, with $t \geq 1$.*

The proof of the above theorem uses only modules (of infinite length) and their elements. We present the idea of the proof.

Let $P = X^I$ and define a submodule P' of P:

$$P' = \bigcup_{t \geq 1} (Ker\varphi^t)^I.$$

The module P' is a direct sum of copies of X. The endomorphism φ is locally nilpotent, and therefore X is Σ-algebraically compact and P' is algebraically compact. The module P' is also a pure submodule of P, hence $P \simeq P' \oplus \overline{P}$, where $\overline{P} = P/P'$.

For any map $\gamma : I \to I$ we define an endomorphism $\overline{\gamma} : P \to P$ as follows: if $y = (y_i)_{i \in I} \in P$, then

$$\overline{\gamma}(y) = (\varphi(y_{\gamma(i)}))_{i \in I}.$$

Let C be the k-subalgebra of $End_A(P)$ generated by such endomorphisms. Then $\overline{P}$ is a C-module of length n, where n is the minimum of the dimension of the vector spaces $Ker\varphi^t / Ker\varphi^{t-1}$, with $t \geq 1$. Hence, $\overline{P}$ has length at most n as an $End_A(P)$-module.

3. Degenerations of modules.

We have seen in Section 2 that module varieties appear naturally in fundamental problems in representation theory of algebras. Now we want to analyse the module varieties with its additional structure given by the action of the general linear group. First we intoduce the relation of degeneration.

Definition 3.1. *Let M and N be two d-dimensional A-modules corresponding to points m and n of $mod_A^d(k)$. One says that N is a* **degeneration** *of M if n belongs to the (Zariski) closure $\overline{Gl_d(k)(m)}$ in $mod_A^d(k)$, and we denote this fact by $M \leq_{deg} N$.*

Observe that $\leq_{deg}$ is a partial order on the set of isomorphism classes in $modA$. There is the following algebraic description of the partial order $\leq_{deg}$ (see [8]).

Theorem 3.2. *Let $M, N \in modA$. Then the following conditions are equivalent:*

1. *$M \leq_{deg} N$.*
2. *There is a short exact sequence $0 \to Z \to Z \oplus M \to N \to 0$ in modA, for some module Z.*
3. *There is a short exact sequence $0 \to N \to M \oplus Z \to Z \to 0$ in modA, for some module Z.*

The implications (2) $\Rightarrow$ (1) and (3) $\Rightarrow$ (1) follow by Proposition 3.4 in [6]. The reason is that "generically" the quotient of a monomorphism $Z \to Z \oplus M$ is M. More precisely, if $n \in \overline{Gl_d(k)m}$, then there is a cofinite subset $\mathcal{C}$ of $\mathbb{A}^1$, a point c in $\mathcal{C}$ and a morphism $\mu : \mathcal{C} \to \overline{Gl_d(k)m}$, such that $\mu(c) = n$ and $\mu(c') \in Gl_d(k)m$ for all $c' \neq c$.

The proof of the implication (1) $\Rightarrow$ (3) is an example for the use of modules of infinite length to show some statements about finite dimensional modules. Now we present the main steps in the proof of the implication (1) $\Rightarrow$ (3). In fact, the main steps reduce the statement to some properties of discrete valuation algebras which are formulated in Proposition 3.3.

Assume that $M \leq_{deg} N$ ($n \in \overline{Gl_d(k)m}$). There are a nonsingular affine curve $\mathcal{C}$, a point c in $\mathcal{C}$ and a regular morphism $\mu : \mathcal{C} \to \overline{Gl_d(k)m}$, such that $\mu(c) = n$ and $\mu(c') \in Gl_d(k)m$ for $c' \neq c$.

Let $k[\mathcal{C}]$ be the coordinate ring of $\mathcal{C}$. Let R be the local ring of the point c in $\mathcal{C}$ with maximal ideal $\mathfrak{m}$. Then R is a discrete valuation k-algebra. Let K be the field of rational functions on $\mathcal{C}$. Hence, $k[\mathcal{C}] \subset R \subset K$ and the morphism μ corresponds to a point of $mod_A^d(k[\mathcal{C}]) \subseteq mod_A^d(R)$. In this way we get an A-R bimodule Y such that Y as an R-module is free of rank d and the A-module $Y/Y\mathfrak{m}$ is isomorphic to N. Since the image of μ "generically" belongs to the orbit of m, there is an isomorphism of A-K-bimodules

$$\psi : M \otimes_k K \to Y \otimes_R K.$$

There is $s \in K \setminus \{0\}$ such that $(s \cdot \psi)(M \otimes_k R) \subseteq Y$. Then

$$\varphi = s \cdot \psi|_{M \otimes_k R} : M \otimes_k R \to Y$$

is a monomorphism of A-R-bimodules. The following exact sequences of R-modules:

$$0 \to R/\mathfrak{m} \to R/\mathfrak{m}^{i+1} \to R/\mathfrak{m}^i \to 0, \quad i \geq 1,$$

give the exact sequences

$$0 \to Y/Y\mathfrak{m} \to Y/Y\mathfrak{m}^{i+1} \to Y/Y\mathfrak{m}^i \to 0, \quad i \geq 1,$$

in *modA*. Since the A-modules $Y/Y\mathfrak{m}$ and N are isomorphic, our statement follows from the proposition below:

Proposition 3.3. *Assume that R is a discrete valuation k-algebra with the maximal ideal $\mathfrak{m}$ and residue field $R/\mathfrak{m} = k$. Let Y be an A-R-bimodule which is, as an R-module, free of rank d. Assume that there is a monomorphism $\varphi : M \otimes_k R \to Y$ of A-R-bimodules, for some d-dimensional A-module M. Then, for a sufficiently large h, there is an A-module isomorphism*

$$Y/Y\mathfrak{m}^{h+1} \simeq Y/Y\mathfrak{m}^h \oplus M.$$

As a consequence of the proof of Riedtmann's theorem and the above characterization of degenerations we get:

Corollary 3.4. *Let $m, n \in mod_A^d(k)$ such that $n \in \overline{Gl_d(k)m}$. Then there is a regular morphism $\mu : \mathbb{A}^1 \to \overline{Gl_d(k)m}$, such that $\mu(0) = n$ and $\mu(t) \in Gl_d(k)m$ for t in a cofinite subset of $\mathbb{A}^1$.*

We end the paper with an example illustrating how we obtain from a degeneration $M \leq_{deg} N$ an exact sequence of the form $0 \to N \to M \oplus Z \to Z \to 0$.

Example 3.5. *Let $A = kQ/I$, where*

$$Q: \qquad 1 \xrightarrow{\ \beta\ } 2 \circlearrowleft \alpha \qquad , \qquad I = \langle \alpha^2 \rangle.$$

Let M and N be two A-modules given by the following representations:

$$M: \quad k \xrightarrow{\begin{bmatrix}1\\0\end{bmatrix}} k^2 \circlearrowleft \begin{bmatrix}0&0\\1&0\end{bmatrix} \qquad N: \quad k \xrightarrow{\begin{bmatrix}0\\1\end{bmatrix}} k^2 \circlearrowleft \begin{bmatrix}0&0\\1&0\end{bmatrix}$$

We define a regular morphism $\mu : \mathbb{A}^1 \to mod_A^3(k)$, such that the point $\mu(t)$ corresponds to the following representation:

$$k \xrightarrow{\begin{bmatrix}t\\1\end{bmatrix}} k^2 \circlearrowleft \begin{bmatrix}0&0\\1&0\end{bmatrix}$$

Then the point $\mu(0)$ corresponds to N and the points $\mu(t)$ correspond to modules isomorphic to M, for all $t \neq 0$. Hence, $M \leq_{deg} N$.

Let $R = k[T]_{(T)}$ and $\mathfrak{m} = T \cdot R$. Then the A-module Y is given by the following representation:

$$R \xrightarrow{\begin{bmatrix}T\\1\end{bmatrix}} R \oplus R \circlearrowleft \begin{bmatrix}0&0\\1&0\end{bmatrix}$$

Observe that the A-modules $Y/Y\mathfrak{m}$ and N are isomorphic to each other, and for any $h \geq 1$, the A-module $Y/Y\mathfrak{m}^h$ is given by the following representation:

$$
{}_AY/Y\mathfrak{m}^h : \quad k^h \xrightarrow{\begin{bmatrix} J \\ 1_h \end{bmatrix}} k^h \oplus k^h \circlearrowleft \begin{bmatrix} 0 & 0 \\ 1_h & 0 \end{bmatrix}
$$

where J denotes an $h \times h$ Jordan block with eigenvalue 0. One can show that $Y/Y\mathfrak{m}^{h+1} \simeq Y/Y\mathfrak{m}^h \oplus M$, for all $h \geq 2$. Hence, the sequence

$$0 \to Y/Y\mathfrak{m} \to Y/Y\mathfrak{m}^{h+1} \to Y/Y\mathfrak{m}^h \to 0$$

has the form $0 \to N \to M \oplus Z \to Z \to 0$, for $Z = Y/Y\mathfrak{m}^{\mathfrak{h}}$ and $h \geq 2$.

Acknowledgement. The author thanks H. Krause for many stimulating remarks concerning the preparation of the paper.

References

[1] **W. Crawley-Boevey**, "On tame algebras and bocses", Proc. London Math. Soc. **56** (1988), 451–483.

[2] **W. Crawley-Boevey**, "Tame algebras and generic modules", Proc. London Math. Soc. **63** (1991), 241–265.

[3] **P. Gabriel**, "Finite representation type is open", in: Lecture Notes in Math. **488** (1975), 132–155.

[4] **H. Kraft**, "Geometrische Methoden in der Invariantentheorie", Vieweg (1985).

[5] **H. Krause**, "Generic modules over artin algebras", Proc. London Math. Soc. **76** (1998), 276–306.

[6] **C. Riedtmann**, "Degenerations for representations of quivers with relations", Ann. Sci. École Normale Sup. **4** (1986), 275–301.

[7] **C.M. Ringel**, "A construction of endofinite modules", in: Advances in Algebra and Model Theory, Gordon - Breach, (ed. M. Droste and R. Göbel), London (1997), 387–399.

[8] **G. Zwara**, "Degenerations of finite dimensional modules are given by extensions", Compositio Math. (in press)

Faculty of Mathematics and Informatics, Nicholas Copernicus University, Chopina 12/18, 87–100 Toruń, Poland

E-mail address: `gzwara@mat.uni.torun.pl`

Trends in Mathematics, © 2000 Birkhäuser Verlag Basel/Switzerland

ON SOME TAME AND DISCRETE FAMILIES OF MODULES

RAYMUNDO BAUTISTA

1. Introduction

Here we consider algebras Λ over an algebraically closed field k, which are k-finite dimensional.

One of the main purposes of the theory of representations of finite dimensional algebras is the study of the category of k-finite dimensional left (or right) Λ-modules. This category is denoted by *mod* Λ.

It is known that the Krull-Schmidt theorem holds for *mod* Λ, so any object can be decomposed into indecomposables modules, esentially in a unique way. Thus, in some sense, the properties of *mod*Λ can be reduced to those of *ind* Λ, the full subcategory of *mod* Λ whose objects are the indecomposable modules.

An algebra Λ is said to be of finite representation type if *ind* Λ has only a finite number of isomorphism classes. On the other hand Λ is said to be of tame representation type if for each dimension d, there is a finite number of one parameter families of indecomposable modules having dimension d, such that with the exception of only a finite number isomorphism classes in

$$ind\,\Lambda_d = \{X \in ind\Lambda \mid dim\,X = d\}$$

a X in $ind\,\Lambda_d$ is isomorphic to some module in some of the families.

Algebras of finite and tame representation type have been extensively studied. Here we consider families $\mathcal{F}$ of modules, closed under isomorphisms, which for any d, all but a finite number of isomorphism classes in

$$\mathcal{F}_d = \{\, X \in \mathcal{F} \mid dimX = d \,\}$$

have a representative in the union of a finite number of one-parameter families. These families will be called *tame families.* We will study this kind of families for Λ of tame representation type.

As it is well known [6], the tame representation type of an algebra is characterized in terms of generic modules. We will see that some properties of generic modules are related with the polynomial growth type of the algebra. In particular we introduce the concept of *curly generic module*, which apparently appears only in the non polynomial growth algebras. For string algebras, a strong version of curly generic module has been recently discovered by C.M. Ringel.

2. Definitions and general results

In the following, a *rational algebra* Γ is an algebra of the form $\Gamma = k[x]_{f(x)}$, $S(\Gamma) = \{\lambda \in k | f(\lambda) \neq 0\}$. We recall that the simple Γ-modules are of the form $S_\lambda = k[x]/(x-\lambda)$ with $\lambda \in S(\Gamma)$.

Definition 2.1. *A subset $\mathcal{T}$ of $ind\,\Lambda_d$ is called a one parameter family if there is a Λ-Γ bimodule M, Γ a rational algebra, M free finitely generated over Γ, such that*

$$\mathcal{T} = \{M \otimes_\Gamma S_\lambda | \, \lambda \in S(\Gamma)\,\}$$

Definition 2.2. *If $\mathcal{U} \subset ind\Lambda_d$ is a family closed under isomorphisms, we will say that the one parameter families $\mathcal{T}_1, \mathcal{T}_2, \ldots, \mathcal{T}_s$ cover $\mathcal{U}$ if each $\mathcal{T}_i \subset \mathcal{U}$ and for almost all isomorphism classes $\{X\}$, X in $\mathcal{U}$, $X \cong X' \in \mathcal{T}_i$ for some i, $1 \leq i \leq s$. The minimal possible s is denoted by $\mu(\mathcal{U})$.*

An algebra is of tame representation type if, for each natural number d, $ind\Lambda_d$ is covered by a finite number of one parameter families. The number $\mu(ind\Lambda_d)$ will be denoted by μ_d. The study of μ_d in terms of d is of great interest in the theory of representations of tame algebras.

We recall that the Auslander-Reiten quiver of $mod\Lambda$ is an oriented quiver Γ_Λ whose vertices are the isomorphism classes of $ind\Lambda$ $\{X\}$. There is an arrow from $\{X\}$ to $\{Y\}$ iff there is an irreducible map $u : X \to Y$. By a component of Γ_Λ we mean a connected component. A vertex is called projective (respectively injective) if the corresponding module is projective (respectively injective). In Γ_Λ there is a function τ, called translation,

$$\tau : \text{ non projective vertices of } \Gamma_\Lambda \to \text{ non injective vertices of } \Gamma_\Lambda,$$

given by $\tau\{X\} = \{DtrX\}$.

The geometrical shapes of most of the components of the Auslander-Reiten quiver of $mod\Lambda$, for Λ of tame representation type, are tubes:

A tube is an oriented graph of the form:

$$\begin{array}{ccccccc}
\cdots & x_{1,i} & & x_{1,i+1} & & x_{1,i+2} & \cdots \\
& \downarrow & \nearrow & \downarrow & \nearrow & \downarrow & \\
\cdots & x_{2,i} & & x_{2,i+1} & & x_{2,i+2} & \cdots \\
& \downarrow & \nearrow & \downarrow & \nearrow & \downarrow & \\
\cdots & x_{3,i} & & x_{3,i+1} & & x_{3,i+2} & \cdots \\
& \downarrow & \nearrow & \downarrow & \nearrow & \downarrow & \\
& \vdots & & \vdots & & \vdots &
\end{array}$$

where the i-th column coincides with the $i+n$ column. The translation function is given by $\tau(x_{i,v}) = x_{i,(v+1)}$. The number n is the rank of the tube. If $n = 1$ the tube is called *homogeneous*.

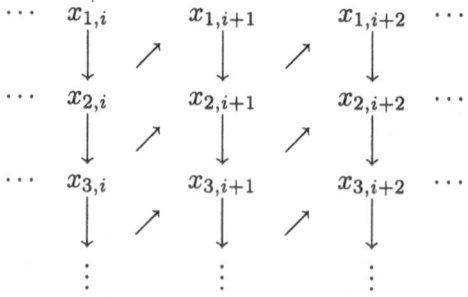

A homogeneous tube has the form:

$$\begin{array}{c} x_1 \\ \uparrow\downarrow \\ x_2 \\ \uparrow\downarrow \\ x_3 \\ \uparrow\downarrow \\ \vdots \end{array}$$

with $\tau(x_i) = x_i$. If $\{X\} = x_i$, i is the quasilength of X.

In some cases there is a nice distribution of the homogeneous tubes in Γ_Λ. For instance in the book of Ringel [11] is considered the case of the algebra $B = End_A(T)$, where A is a tame hereditary algebra and T is a tilting A-module, which is sum of preprojective modules. In this case it is proved that

$$\Gamma_\Lambda = \mathcal{P} \vee \mathcal{T} \vee \mathcal{I}$$

where $\mathcal{P}$ is the preprojective component, $\mathcal{I}$ the preinjective component, and $\mathcal{T}$ is a union of tubes almost all of them homogeneous. Moreover there are only maps from $\mathcal{P}$ to $\mathcal{T}$ and to $\mathcal{I}$ and from $\mathcal{T}$ to $\mathcal{I}$. In addition any map from one object in $\mathcal{P}$ to an object in $\mathcal{I}$ is factorized through each component in $\mathcal{T}$. The algebra B is called *tame concealed.* Other important class of algebras which also have a nice distribution of the homogeneous tubes in Γ_Λ are the *tubular algebras* . These algebras are also studied in [11]. If Λ is a concealed algebra there exists a constant b such that $\mu_d \leq d$ for all d. Algebras for which such constant exists are called *domestic.* If for some algebra Λ there are a constants t and u such that for all d, $\mu_d \leq d^t + u$ the algebra is said to be of *polynomial growth* . For instance, tubular algebras are of polynomial growth. Recent results on tubular algebras and algebras of polynomial growth can found in [8] and in [14].

In this survey we will consider some families of indecomposable modules which in some cases determine if the algebra is of polynomial growth.

Definition 2.3. *Take $\mathcal{F}$ a full subcategory of $ind\Lambda$, closed under isomorphisms. Then*

1. *$\mathcal{F}$ is discrete if, for each $d \in \mathbb{N}$, $\mathcal{F}_d = \{X \in \mathcal{F} \mid dimX = d\}$ has only a finite number of isomorphism classes.*
2. *$\mathcal{F}$ is tame if, for each $d \in \mathbb{N}$, $\mathcal{F}_d$ is covered by a finite number of one parameter families. We will put $\mu_d(\mathcal{F}) = \mu(\mathcal{F}_d)$.*

For instance if Λ is of tame representation type

$$\mathcal{D} = \{X \in ind\Lambda \mid DtrX \ncong X\}$$

is a discrete family by [5]. We call X in $mod\Lambda$ *homogeneous* if $DtrX \cong X$.

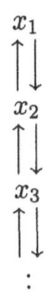

For M in $mod\Lambda$ we will consider the following number introduced by J. Boza [4],

$$\delta(M) = dim_k Hom_\Lambda(M, DtrM) - dim_k \mathcal{S}(M, DtrM)$$

where $\mathcal{S}(X,Y) = \{f : X \to Y | f$ factorizes through some semisimple module$\}$. Having this number we can introduce the following partition of $ind\Lambda$:

For $i \in \{0,1,2,...\}$, $\mathcal{C}(i) = \{X \in ind\Lambda | \delta(X) = i\}$. Each $\mathcal{C}(i)$ is closed under isomorphisms. Our first result is the following:

Theorem 2.1 (R. Bautista, J. Boza [2]). *For any algebra Λ, we have that $\mathcal{C}(0)$ is discrete and $\mathcal{C}(1)$ is tame.*

The behaviour of $\mu_d(\mathcal{C}(1))$ can be understood more easily if instead
of using the dimension of M, we use the coordinates $\underline{c}(M)$ of M. For introducing this vector, take a minimal projective presentation of M:

$$\coprod_{i=1}^{s} m_i P_i \to \coprod_{i=1}^{s} n_i P_i \to M \to 0$$

where $P_1,...,P_s$ is a representative system of the isomorphism classes of indecomposable projectives. Then

$$\underline{c}(M) = (m_1,...,m_s; n_1,...,n_s) \in \mathbb{N}^{2s}$$

Lemma 2.1. *If $\mathcal{T}$ is a one parameter family with $\mathcal{T} \subset \mathcal{C}(1)_d$ then there exists a $\underline{c} \in \mathbb{N}^{2s}$ such that for all $X \in \mathcal{T}$, $\underline{c}(X) = \underline{c}$. We will put $\underline{c}(\mathcal{T}) = \underline{c}$.*

Proposition 2.1 (R. Bautista, J. Boza [2]).
1. *If $\mathcal{T}$ and $\mathcal{T}'$ are one parameter families in $\mathcal{C}(1)$ and $\underline{c}(\mathcal{T}) = \underline{c}(\mathcal{T}')$, then for almost all isoclasses $\{X\}$ with X in $\mathcal{T}$, there is a X' in $\mathcal{T}'$ with $X \cong X'$.*
2. *If X and X' are in $\mathcal{C}(0)$ and $\underline{c}(X) = \underline{c}(X')$ then $X \cong X'$.*

From the above one can see that the family

$$\mathcal{C}(1)_{\underline{c}} = \{X \in \mathcal{C}(1) | \underline{c}(X) = \underline{c}\}$$

is covered by at most one one parameter family.

3. The families $\mathcal{C}(u)$ for Λ tame

If $\mathcal{F}$ is any family closed under isomorphisms in $mod\Lambda$, Λ of tame representation type, $\mathcal{F}$ need not to be of tame type. It can happen, for instance that for infinitely many isoclasses $\{X\}$, $X \in \mathcal{F}_d$ is in a one parameter family $\mathcal{T}$ but also an infinite number of isoclasses $\{X\}$, $X \in \mathcal{F}_d$ is not in $\mathcal{T}$. In this case $\mathcal{F}_d$ can not be covered by a finite number of one parameter families.

We have the following.

Proposition 3.1. *Let Λ be a tame algebra and $\mathcal{T}$ a one parameter family with infinitely many non isomorphic modules. Then there is a number u such that for almost all X in $\mathcal{T}$*

$$u = dim_k Hom_\Lambda(X, DtrX)/\mathcal{S}(X, DtrX)$$

We will put $u(\mathcal{T}) = u$.

¿From 3.1 we obtain the following result.

Proposition 3.2. *If Λ is an algebra of tame representation type, then all the families $\mathcal{C}(u)$ are tame.*

Proof. Take a $d \in \mathbb{N}$, then because Λ is tame, $ind\Lambda_d$ is covered by a finite number of one parameter families $\mathcal{T}_1, ..., \mathcal{T}_l$. Consider now

$$J = \{i \in \{1, ..., l\} | u(\mathcal{T}_i) = u\}.$$

Clearly the families $\mathcal{T}_j$ with j in J cover $\mathcal{C}(u)_d$.

We recall that in a homogeneous tube of the Auslander-Reiten quiver of $mod\Lambda$ all modules are homogeneous and there is a module C for which his almost split sequence has indecomposable middle term, this module is called *quasisimple.*

Proposition 3.3. *Let Λ be of tame representation type . Then if $\mathcal{T}$ is an infinite one parameter family there are integer numbers $q(\mathcal{T})$ and $e(\mathcal{T})$ such that*

1. *for almost all X in $\mathcal{T}$, X lie in a homogeneous tube and $quasilength X = q(\mathcal{T})$;*
2. *for almost all X in $\mathcal{T}$, $dim_k End_\Lambda(X) = e(\mathcal{T})$.*

As a consequence we have:

Corollary 3.1. $\mathcal{C}(u, q) = \{X \in \mathcal{C}(u) | X$ *is in a tube and has quasilength* $= q\}$ *is a tame family.*

Proposition 3.4. *Let Λ be an algebra of tame representation type, and $\mathcal{T}$ a one parameter family. Then there is a $\underline{c} \in \mathbb{N}^{2s}$, s=number of isomorphism classes of simples in modΛ, such that for almost all $X \in \mathcal{T}$, $\underline{c}(X) = \underline{c}$. We will put $\underline{c}(\mathcal{T}) = \underline{c}$.*

From the above it is clear that if $\mathcal{F}$ is a tame family, $\mathcal{F}_{\underline{c}} = \{X \in \mathcal{F} | \underline{c}(X) = \underline{c}\}$ is covered by a finite number of one parameter families, the minimal possible number of such families is denoted by $\nu_{\underline{c}}(\mathcal{F})$. In particular, $\nu_{\underline{c}} = \nu_{\underline{c}}(ind\Lambda)$.

In the following if $\underline{x}$ and $\underline{y}$ are non zero vectors in $\mathbb{N}^{2s}$, we will say that $\underline{x}$ is a divisor of $\underline{y}$ if there is an integer m such that $\underline{y} = m\underline{x}$. In this case we put $\underline{x}|\underline{y}$. Finally one can prove the following inequality.

Proposition 3.5. *For Λ tame and for each $\underline{c} \in \mathbb{N}^{2s}$, $s =$ number of isoclasses of simples in modΛ,*

$$\nu_{\underline{c}}(\mathcal{C}(u)) \leq \sum_{v \leq u, \underline{l} | \underline{c}} \nu_{\underline{l}}(\mathcal{C}(v, 1)$$

4. Relations between $\nu_{\underline{c}}$ and μ_d and polynomial growth

In the literature on tame algebras, the numbers μ_d are used, we will see the relation between these numbers and $\nu_{\underline{c}}$. Then we will give a sufficient condition for polynomial growth representation type.

For simplicity we will assume Λ basic. Take $P_1,...,P_s$ a representative system of the isomorphism classes of indecomposable projective modules. We have $dim_k P_i/radP_i = 1$. Take $t = max\{dim_k P_i\}_{i=1,...s}$

We put

$$I(t) = \{\underline{c} \in \mathbb{N}^{2s}, \underline{c} = (m_1,...m_s; n_1,...n_s) | \sum n_i \leq t \, and \sum m_i \leq t(\sum n_i)\}$$

Then if $\mathcal{U}$ is a tame family in $ind\Lambda$ we have:

$$\mu_d(\mathcal{U}) \leq \sum_{\underline{c} \in I(t)} \nu_{\underline{c}}(\mathcal{U})$$

$$\nu_{\underline{c}}(\mathcal{U}) \leq \sum_{d \leq |c|} \mu_d(\mathcal{U})$$

where $|\underline{c}| = \sum m_i + \sum n_i$ for $\underline{c} = (m_1, ..., m_s; n_1, ..., n_s)$.

Definition 4.1. *A tame family $\mathcal{U} \subset ind\Lambda$ is called domestic, if there is a number L such that for each d in $\mathbb{N}$, $\mu_d(\mathcal{U}) \leq L$. The family $\mathcal{U}$ is called c-domestic if there is a number L' such that for all $\underline{c}$ in $\mathbb{N}^{2s}$, $\nu_{\underline{c}}(\mathcal{U}) \leq L'$.*

The family $\mathcal{U}$ is called of polynomial growth if there are numbers m and v in $\mathbb{N}$ such that for each natural number d, $\mu_d(\mathcal{U}) \leq d^m + v$.

Similarly $\mathcal{U}$ is of c-polynomial growth if there are numbers m and v in $\mathbb{N}$ such that for each $\underline{c} \in \mathbb{N}^{2s}$, $\nu_{\underline{c}}(\mathcal{U}) \leq |\underline{c}|^m + v$

From the previous inequalities one easily obtains the next lemma.

Lemma 4.1. *$\mathcal{U}$ is of polynomial growth if and only if $\mathcal{U}$ is of c-polynomial growth.*

Recently the polynomial growth strongly simply connected algebras have been intensively investigated (see [8], [9], [13], [14],[15]).

In a nice characterization of strongly simply connected algebras of polynomial growth, the family $\mathcal{D}$ of non homogeneous indecomposable modules plays an important role. In this characterization appears the vector dimension of a module M,

$$\underline{dim}M = (dim_k Hom_\Lambda(P_i, M)) \in \mathbb{N}^s$$

For $\underline{d} \in \mathbb{N}^s$ we put $\mathcal{D}_{\underline{d}} = \{X \in \mathcal{D} | \underline{dim}X = \underline{d}\}$

Proposition 4.1 (A. Skowronski, G. Zwara [16]). . *Let Λ be a strongly simply connected algebra, then Λ is of polynomial growth iff there is a number b such that for all $\underline{d} \in \mathbb{N}^s$, the number of isoclasses in $\mathcal{D}_{\underline{d}} \leq b$.*

We recall the following conjecture [[13] problem 15] due to A. Skowronski.

If for any $M \in ind\Lambda$, $rad^\omega(M, M) = \bigcap_{i \geq 1} rad^i(M, M) = 0$, then Λ is of polynomial growth.

In next section we prove the above conjecture. In section 6 we will give a result similar to the above proposition.

5. Generic modules and one parameter families

Generic modules are infinite k-dimensional Λ-modules which play an important role in the properties of the modules which belong to one parameter families. Here we will see some relations between the properties of generic modules and the representation type of Λ.

Definition 5.1. *A generic Λ-module G is an indecomposable Λ-module, infinite dimensional over k, having finite length as left module over $End_\Lambda(G)$.*

We recall that ([6]) if Λ is tame and G a generic Λ-module, then $G/radG \cong k(x)$.

It is known from [6] that for a generic module G over a tame algebra Λ, $G \cong M \otimes_\Gamma k(x)$ for some $\Lambda - \Gamma$ bimodule M; Γ rational and M free, finitely generated over Γ. Moreover the family $\{M \otimes_\Gamma S_\lambda, \lambda \in S(\Gamma)\}$ is in $ind_d(\Lambda)$ for some d. This one parameter family is denoted by $\mathcal{T}_G$.

One can see that if $\mathcal{T}'$ is another one parameter family associated to G as before, then there is only a finite number of modules in $\mathcal{T}_G$ (respectively $\mathcal{T}'$) which are not isomorphic to some module in $\mathcal{T}'$ (respectively $\mathcal{T}_G$). Moreover we can assume that all X in $\mathcal{T}_G$ lie in a homogeneous tube and have quasilength 1. By propositions 3.1 and 3.3 we have the numbers $\delta(\mathcal{T}_G)$, $e(\mathcal{T}_G)$ and $\underline{c}(\mathcal{T}_G) \in \mathbb{N}^{2s}$.

We will put $\delta(G) = \delta(\mathcal{T}_G)$, $e(G) = e(\mathcal{T}_G)$, $\underline{c}(G) = \underline{c}(\mathcal{T}_G)$.

On the other hand it is known from [6] that if $\mathcal{T}$ is a one parameter family then there is a generic module G such that $\mathcal{T}_G$ covers $\mathcal{T}$.

Definition 5.2 ([7]). *An algebra Λ is called generically directed if there are no cycles $G_0 \to G_1 \to \ldots \to G_0$ of non zero nonisomorphisms between generic Λ modules G_i.*

For instance, tubular algebras are generically directed.

Now we introduce the following.

Definition 5.3. *A tame algebra Λ is called generically Schurian if for all generic Λ- modules G we have $e(G) = 1$.*

Lemma 5.1. *Suppose the algebra Λ is tame, and G is a generic Λ-module. Then if $End_\Lambda(G)$ is a division ring, $e(\mathcal{T}_G) = 1$.*

Corollary 5.1. *Any generically directed algebra is generically Schurian.*

We recall (see[1]) that an algebra Λ is said to be cycle-finite if there are not cycles of the form

$$X_1 \xrightarrow{f_1} X_2 \to \cdots \to X_n \xrightarrow{f_n} X_{n+1} = X_1$$

with some $f_i \in rad^\omega(X_i, X_{i+1})$.

Proposition 5.1. *If Λ is a cycle-finite algebra, then Λ is generically Schurian.*

Proof. Take G a generic Λ-module. For almost all X in $\mathcal{T}_G$, X is quasisimple in a homogeneous tube, then

$$1 = dim_k End_\Lambda(X)/rad^\omega(X,X) = dim_k End_\Lambda(X)$$

(here $rad^\omega(X,X) = 0$), therefore, $e(G) = 1$.

Theorem 5.1. *If Λ is a generically Schurian algebra, then Λ is of polynomial growth.*

Proof. Take G a generic Λ-module. Let $\mathcal{T}_G$ be the corresponding one parameter family. Then for almost all X in $\mathcal{T}_G$, $dim_k End_\Lambda(X) = 1$. Therefore:

$$\delta(X) = dim_k Hom_\Lambda(X,X)/\mathcal{S}(X,X) = 1$$

consequently $X \in \mathcal{C}(1)$. Thus $\mathcal{C}(u,1)$ is discrete for $u \geq 1$. On the other hand $\mathcal{C}(1,1) = \mathcal{C}(1)$.

Therefore $\nu_{\underline{c}}(\mathcal{C}(1,1)) = \nu_{\underline{c}}(\mathcal{C}(1)) \leq 1$. From proposition 3.4 we obtain

$$\nu_{\underline{c}} = \sum_u \nu_{\underline{c}}(\mathcal{C}(u)) \leq \sum_{v,\underline{l}|\underline{c}} \nu_{\underline{l}}(\mathcal{C}(v,1))$$

$$= \sum_{\underline{l}|\underline{c}} \nu_{\underline{l}}(\mathcal{C}(1)) \leq \tau(\underline{c})$$

where $\tau(\underline{c})$ is the number of divisors of $\underline{c}$. Then Λ is of polynomial growth.

6. Polynomial growth of $\mathcal{C}(u)$ and curly generic modules

In this section we will see some conditions for the polynomial growth of the families $\mathcal{C}(u)$. These conditions are in terms of the non homogeneous indecomposable modules and some especial kind of generic modules.

Definition 6.1. *Let Λ be a tame algebra, then G a generic module is called curly if there is an infinite sequence of generic modules $G_1 = G$, G_2,...G_n,... such that*

1. *for any $i \neq j$ there are non zero maps which are not invertible from G_i to G_j;*
2. *$e(G_i) \geq i$.*

For the following result we recall that $\mathcal{D} = \{X \in ind(\Lambda) | DtrX \not\cong X\}$.

Proposition 6.1 (R. Bautista, R. Zuazua, [3]). *Let Λ be a tame algebra and assume there is a number b such that for each $\underline{c} \in \mathbb{N}^{2s}$, number of isoclasses in $\mathcal{D}_{\underline{c}}$ + number of isoclasses of curly generic modules with $\underline{c}(G) = \underline{c}$ is less than b, then $\mathcal{C}(u)$ is of polynomial growth for $u = 1, 2, \ldots$.*

Definition 6.2. *A tame algebra Λ is called generically e-bounded if there is a number L such that for all generic Λ-module G, $e(G) \leq L$.*

Now we can deduce a result similar to proposition 4.4.

Theorem 6.1. *Let Λ be a tame generically e-bounded algebra. Suppose there is a number b such that for all $\underline{c} \in \mathbb{N}^{2s}$ the number of isoclasses of $\mathcal{D}_{\underline{c}} \leq b$. Then Λ is of polynomial growth.*

Proof. Here Λ is generically e-bounded, then there are not curly generic modules. Therefore all $\mathcal{C}(u)$ are of polynomial growth by proposition 6.1. On the other hand, if G is a generic module and $\mathcal{T}_G$ is a one parameter family associated to G, for almost all $X \in \mathcal{T}_G$, X is quasisimple homogeneous. Moreover:

$$\delta(X) = dim_k Hom_\Lambda(X,Y)/\mathcal{S}(X,Y) \leq dim_k Hom_\Lambda(X,X) = e(G) \leq L$$

Consequently all $\mathcal{C}(v,1)$ with $v \geq L$ are discrete families. Using proposition 3.5 we obtain:

$$\nu_{\underline{c}} = \sum_u \nu_{\underline{c}}(\mathcal{C}(u)) \leq \sum_{u \leq L, \underline{l} | \underline{c}} \nu_{\underline{l}}(\mathcal{C}(u,1)) \leq \sum_{u \leq L, \underline{l} | \underline{c}} \nu_{\underline{l}}(\mathcal{C}(u))$$

therefore Λ is of polynomial growth.

We can mention that all known examples of polynomial growth algebras are generically e-bounded. On the other hand, for string algebras of non polynomial growth there are always curly generic modules, according with a recent result of C.M. Ringel. In fact a stronger version occurs in this case.

Definition 6.3. *Let Λ be a tame algebra, a generic Λ-module G is called strongly curly if:*

1. *there is a chain of generic modules G_1, G_2,..,G_s,... such that for $i < j$ there is a non splittable monomorphism $f_{ij} : G_i \to G_j$ and a non splittable epimorphism $f_{ji} : G_j \to G_i$;*
2. *$e(G_i) \geq i$.*

The proofs of the main results of this survey are done (see [3]) using the techniques developed for bocses (see [12] and [10]). Results similar to the ones mentioned here are also true for those matrix problems which are equivalent to the category of representations of some finite dimensional triangular free bocs.

References

[1] I. Assem, A. Skowronski. Indecomposable modules over multicoil algebras. Math. Scand. 71 (1992), 31-61.

[2] R. Bautista, J. Boza. Reduction Algorithms and Quadratic Forms. Preprint 492. Instituto de Matemáticas UNAM.

[3] R. Bautista, R. Zuazua. On one parameter families of modules for tame algebras and bocses. In preparation.

[4] J. Boza,Algoritmos de Reducción en la Teoria de Representaciones de Algebras. Tesis Doctoral, UNAM 1996.

[5] W.W. Crawley- Boevey. On tame algebras and Bocses. Proc. London Math. Soc. 56 (1988), 451-483.

[6] W.W. Crawley- Boevey, Tame Algebras and generic modules. Proc. London Math. Soc. 63(1991),241-265

[7] W.W. Crawley- Boevey. Modules of finite length over their endomorphism ring. Representations of Algebras and Related Topics. London Math. Soc. Lecture Note Series 168(Cambridge University Press 1992) 127-184

[8] J.A. de la Peña. The Tits form of a tame algebra. Canadian Mathematical Society Conference Proceedings. Vol.19,1996, 159-183

[9] J.A. de la Peña and A. Skowronski. Geometric and homological characterizations of polynomial growth strongly simply connected algebras. Invent. Math. 126,1996, 287-296

[10] Y.A. Drozd. Tame and wild matrix problems. Amer.Math.Soc. Transl.(2) Vol128, 1986, 31-55

[11] C. M. Ringel. Tame algebras and integral quadratic forms. Lecture Notes in Math. 1099 (Springer, 1984)

[12] A.V. Roiter, Matrix problems and representations of bocss. Lecture Notes in Mathematics 831, Springer 1980, 288-324

[13] A. Skowronski. Cycles in module categories. Finite Dimensional Algebras and Related Topics.NATO ASI, series C, vol.424(Kruwer Acad. Publ. 1994),309-345.

[14] A. Skowronski. Module Categories over Tame Algebras. Canadian Mathematical Society Conference Proceedings.Vol.19,1996,281-348.

[15] A. Skowronski. Simply connected algebras of polynomial growth. Compositio Math. 109 (1977), 99-133.

[16] A. Skowronski and G. Zwara. On the number of discrete indecomposable modules over tame algebras. Colloquium Mathematicum, vol.73(1997) No.1,93-114.

INSTITUTO DE MATEMATICAS, UNAM, MEXICO
E-mail address: raymundo@matem.unam.mx

Trends in Mathematics, © 2000 Birkhäuser Verlag Basel/Switzerland

PURITY, ALGEBRAIC COMPACTNESS, DIRECT SUM DECOMPOSITIONS, AND REPRESENTATION TYPE

BIRGE HUISGEN-ZIMMERMANN

Helmut Lenzing zu seinem 60. Geburtstag gewidmet

INTRODUCTION

The idea of purity pervades human culture, from religion through sexual and moral codes, cleansing rituals and dietary rules, to stratifications of societies into castes. It does so to an extent which is hardly backed up by the following definition found in the Oxford English Dictionary: *freedom from admixture of any foreign substance or matter*. In fact, in a process of blending and fusing of ideas and ideals (which stands in stark contrast to the dictionary meaning of the term itself), the concept has acquired dozens of additional connotations. They have proved notoriously hard to pin down, to judge by the incongruence of the emotional and poetic reactions they have elicited. Here is a small sample:

Purity is obscurity. Ogden Nash.

One cannot be precise and still be pure. Marc Chagall.

Unto the pure all things are pure. New Testament.

To the pure all things are impure. Mark Twain.

To the pure all things are indecent. Oscar Wilde.

Blessed are the pure in heart for they have so much more to talk about. Edith Wharton.

Be thou as chaste as ice, as pure as snow, thou shalt not escape calumny. Get thee to a nunnery, go. W. Shakespeare.

Necessary, for ever necessary, to burn out false shames and smelt the heaviest ore of the body into purity. D. H. Lawrence.

Mathematics possesses not only truth, but supreme beauty – a beauty cold and austere, like that of a sculpture, without appeal to any part of our weaker nature, sublimely pure, and capable of a stern perfection such as only the greatest art can show. Bertrand Russell.

Purity strikes me as the most mysterious of the virtues and the more I think about it the less I know about it. Flannery O'Connor.

Even in the 'sublimely pure' subject of mathematics, several notions of purity were competing with each other a few decades ago. However, while the state of

This work was partially supported by a grant from the National Science Foundation (USA).

affairs is still rather muddled in contemporary society at large, at least mathematicians have come to an agreement in the meantime. Section 1 traces the development of the concept within algebra in rough strokes. Due to the large number of contributions to the subject, we can only highlight a selection.

The survey to follow should be read in conjunction with those of M. Prest and G. Zwara ([**59**] and [**84**]). The first emphasizes the model-theoretic aspect of the topic, while the second describes the impact of generic modules on representation theory. The generic modules form a distinguished subclass of the class of pure injective (= algebraically compact) modules, the importance of which became apparent through the seminal work of Crawley-Boevey [**16**]. Since we wanted to minimize the overlap with an overview article on endofinite modules by the latter author, the article [**17**] should be consulted as another supplement to our survey. For the sake of orientation, we offer the following diagram displaying a hierarchy of noteworthy classes of algebraically compact modules.

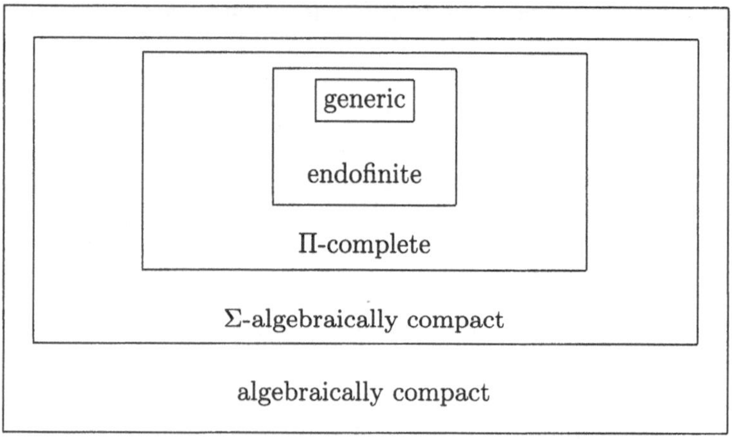

In the present overview, we mainly address the outer layers of the stack. We conclude with a discussion of product-complete modules, the study of which provides a natural bridge to the class of endofinite modules. The interest of the latter class also becomes clear as one approaches it from alternate angles, e.g., via the duality theory developed in [**17**]. Zwara's survey picks up the story at this point and zeroes in on the generic modules, i.e., the non-finitely generated indecomposable endofinite modules. The extensive spread of applications of these concepts is further illustrated by the contributions of D. Benson and H. Krause to this volume ([**10**] and [**50**]): Benson encounters them in his study of phantom maps between modules over group algebras, while Krause uses them towards new characterizations of tame representation type.

What follows below is almost exclusively expository. However, a few results rounding off the picture (e.g., the characterization of purity in terms of matrix groups in Proposition 4) appear to be new, as are several of our arguments. Moreover, in some instances, we streamline results which can essentially be found in the literature by taking the present state of the art into account.

The red thread that will lead us through this discussion is provided in Section 2, in the form of a number of global decomposition problems going back to work of Koethe (1935) and Cohen-Kaplansky (1951). Long before the impact of the subject on the representation theory of tame finite dimensional algebras surfaced, these problems had put a spotlight on (Σ-)algebraic compactness and linked it to finite representation type. Section 3 contains the most important characterizations of the algebraically compact and the Σ-algebraically compact modules, as well as the functorial underpinnings on which they are based. Section 4 takes us back to the decomposition problems stated at the outset, and Section 5 evaluates the outcome in representation-theoretic terms. The topic of product-completeness, addressed in Section 6, supplements both the decomposition theory of direct powers of modules as described in Theorem 10 of Section 4, and the impact of 'globally nice' decompositions of products on the representation type of the underlying ring discussed in Section 5. The very short final section should be seen as an appendix, meant to trigger further research in a direction that has been somewhat neglected in the recent past.

1. Purity and algebraic compactness – definitions and a brief history

It was already in the first half of this century that the concept of a pure subgroup of an abelian group proved pivotal in accessing the structure, first of p-groups, then also of torsionfree and mixed abelian groups, as well as of modules over PID's. Given a PID R, a submodule A of an R-module B is called *pure* if, for all $r \in R$, the intersection $A \cap rB$ equals rA. As is well-known, the submodule $T(B)$ consisting of the torsion elements of B is always pure, and in exploring the p-primary components of $T(B)$, major headway is gained by studying a tell-tale class of pure submodules of the simplest possible structure, namely the 'basic' ones. They are determined up to isomorphism by the following requirements: Given a prime $p \in R$, a submodule B_0 of a p-primary R-module B is called *basic* if it is a direct sum of cyclic groups which is pure in B and has the additional property of making B/B_0 divisible. In fact, a natural extension of this concept to arbitrary modules over discrete valuation domains provides us with one of the reference points in the classification of the algebraically compact abelian groups sketched below.

In the 1950's, it became apparent that suitable variations of the original notion of purity should yield important generalizations of split embeddings in far more general contexts, and several extensions of the concept, naturally all somewhat akin in spirit, appeared in the literature. Finally, in 1959, Cohn's definition [**14**] won the 'contest'. Given left modules A and B over an arbitrary associative ring R, Cohn called a monomorphism $f : A \to B$ *pure* in case tensoring with any right R-module X preserves injectivity in the induced map $id \otimes f : X \otimes_R A \to X \otimes_R B$; by extension, a short exact sequence $0 \to A \to B \to C \to 0$ is labeled *pure exact* in case the monomorphism from A to B is pure. When, in 1960,

Maranda [**54**] undertook a study of the modules M for which the contravariant Hom-functor $\text{Hom}_R(-, M)$ takes maps from certain restricted classes of monomorphisms in R-Mod to epimorphisms in the category of abelian groups, the class which quickly became the most popular in this game was that of pure monomorphisms. Accordingly, modules M such that $\text{Hom}_R(-, M)$ preserves exactness in pure exact sequences were labelled *pure injective*.

One of the fundamental demands on a good concept of purity is this: It should yield convenient criteria for recognizing direct summands – the philosophy being that purity is a first step toward splitting. One hopes for readily verifiable conditions which take a pure inclusion the rest of the way to a split one. Here are two sample statements of this flavor for Cohn's purity; we will re-encounter them in Section 3.

Observations 0.

• *If R is a Dedekind domain and M a pure submodule of an R-module N such that $Ann_R(M) \neq 0$, then M is a direct summand of N.*

• *If R is a left perfect ring and Q a pure submodule of a projective left R-module P, then Q is a direct summand of P.*

Proof. The first statement will arise as an immediate consequence of Theorem 6 in Section 3. To establish the second we will show that, given a pure exact sequence

$$0 \to Q \to P \to P/Q \to 0$$

of left modules over a ring R, such that P is flat, the quotient P/Q is flat as well. This will yield our claim since a perfect base ring will make flat modules projective.

To check flatness of P/Q, let $0 \to A \to B \to C \to 0$ be any short exact sequence of right R-modules, and consider the following diagram which has exact rows and columns due to our setup.

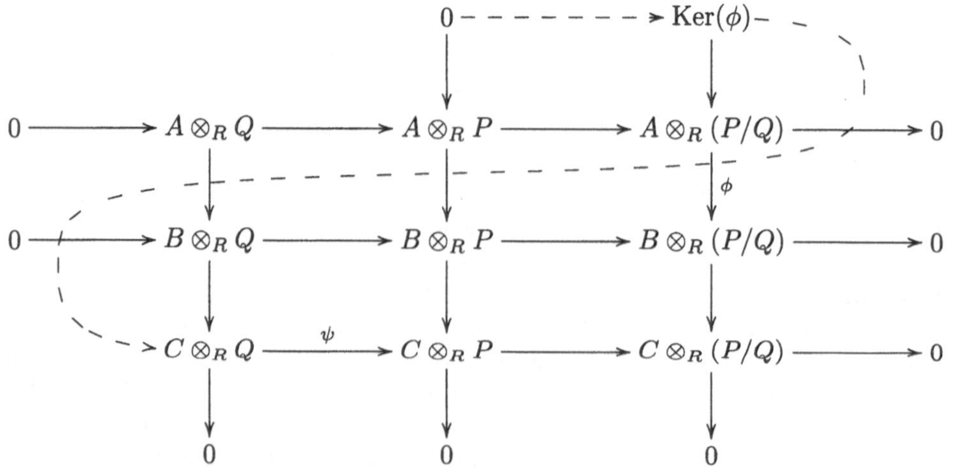

Applying the Snake Lemma and using the fact that the map ψ is injective by hypothesis, we deduce that $\mathrm{Ker}(\phi) = 0$ as required. □

The theory of purity and pure injectivity could also be baptized *the theory of systems of linear equations for modules*. A typical such system has the following format: Starting with a left R-module M, an element $(m_i)_{i \in I} \in M^I$, and a row-finite matrix $(r_{ij})_{i \in I, j \in J}$ of elements of R, one considers the system

$$\sum_{j \in J} r_{ij} X_j = m_i \qquad (i \in I), \tag{†}$$

and calls each element in M^J which satisfies all equations of (†) a solution in M. A first indication of a connection between such systems and the theory of purity and pure injectivity was exhibited by Fieldhouse (see [**22**]), who observed that purity of a submodule A of an R-module B is equivalent to the following equational condition: *Every finite system of linear equations with right-hand sides in A, which is solvable in B, has a solution in A.* This result makes one anticipate a bridge between linear systems and pure injectivity as well. Such a bridge does in fact exist, but relies on systems that need not be finite. Here it is:

Theorem 1. (Warfield, 1969 [**72**]) *For $M \in R$-Mod, the following statements are equivalent:*

(1) *M is pure injective.*

(2) *Any system of the form (†) which is finitely solvable in M (i.e., has the property that, for any finite subset $I' \subseteq I$, the finite subsystem of equations labeled by $i \in I'$ is solvable in M) has a global solution in M.*

(3) *M is a direct summand of a compact Hausdorff R-module N (the latter is to mean that N is a compact Hausdorff abelian group such that all multiplications by elements of R are continuous).*

In this instance, Warfield acted primarily as a coordinator of results and ideas from various parts of the literature, most of the implications having been previously known in a variety of specialized contexts, some in full generality. It is quite enlightening to follow the line of successive modifications of the pertinent concepts. From work of Kaplansky [**45**], Łoś [**53**], and Balcerzyk [**9**] done in the fifties, through papers of Mycielski [**55**], Butler-Horrocks [**11**], Kiełpiński [**46**], Weglorz [**75**], Fuchs [**23**], and Stenstrøm [**71**] scattered over the sixties, it incrementally leads to the present theory.

Warfield's coordination was very much called for, as the obvious interest of the topological condition (3) of Theorem 1 had triggered various exploratory trips in its own right. In 1954, Kaplansky published a characterization of those abelian groups which arise as algebraic direct summands of compact Hausdorff groups, referring to them as 'algebraically compact' [**45**]. Originally, he had set out to describe the compact abelian groups, but realized that, from an algebraic viewpoint, the former class allowed for a cleaner description and appeared more natural. In particular, he provided a complete classification of the algebraically compact abelian groups: They are precisely the direct sums of divisible groups and groups which are Hausdorff and complete in their $\mathbb{Z}$-adic topologies. The latter can be written uniquely as direct products of factors A_p which are complete and Hausdorff in their p-adic topologies, respectively, with p tracing the primes. Each A_p, in turn, can be canonically viewed as a module over the ring of p-adic integers and pinned down up to isomorphism in terms of its basic submodule, a direct sum of cyclic p-goups and copies of the p-adic integers, thus leading to a convenient full set of invariants. Subsequently, equivalent descriptions of the class of groups exhibited by Kaplansky were given by Mycielski, Łoś, Weglorz, and Fuchs (in the setting of noetherian rings). Most notable were the descriptions in terms of 'equational compactness' conditions. The first general concept of algebraic compactness was introduced by Mycielski in 1964, for arbitrary algebraic systems in fact. Restricted to modules, it just amounts to condition (2) of Theorem 1. For the sake of emphasis, we repeat:

Definition. Given an associative ring R with identity, a left R-module M is called *algebraically compact* if any system

$$\sum_{j \in J} r_{ij} X_j = m_i \qquad (i \in I),$$

based on a row-finite matrix (r_{ij}) with entries in R and $m_i \in M$, which is finitely solvable in M, has a global solution in M.

Moreover, M is said to be Σ-*algebraically compact* in case all direct sums of copies of M are algebraically compact.

So, in particular, Theorem 1 tells us that the pure injective and the algebraically compact modules coincide. We will give the latter terminology preference in the sequel.

Remark. It is of course natural to also wonder about characterizations of pure projective modules, i.e., of those modules which are projective relative to pure exact sequences. Such characterizations are much more readily obtained than useful descriptions of their pure injective counterparts. According to Warfield [**72**] and Fieldhouse [**22**], the pure projective modules are precisely the direct summands of direct sums of finitely presented modules.

With little effort, one obtains the following first list of examples of algebraically compact (alias pure injective) modules that goes beyond the most obvious ones, the injectives; a second installment of examples can be found in part B of Section 3.

1. [**72**] Any module M can be purely embedded into a pure injective module, namely its 'Bohr compactification', as follows:

$$M \to \mathrm{Hom}_{\mathbb{Z}}(\mathrm{Hom}_{\mathbb{Z}}(M, \mathbb{R}/\mathbb{Z}), \mathbb{R}/\mathbb{Z}),$$

the assignment being evaluation. If one equips $\mathrm{Hom}_{\mathbb{Z}}(M, \mathbb{R}/\mathbb{Z})$ with the discrete topology and the 'double dual' of M with the compact-open topology, one thus arrives at a compact Hausdorff R-module. As a consequence, one can construct pure injective resolutions of a module M and measure the deviation of M from pure injectivity by means of its 'pure injective dimension'.

2. [**72**] If R is a commutative local noetherian domain with maximal ideal $\mathfrak{m}$, complete in its $\mathfrak{m}$-adic topology, then R is algebraically compact (as an R-module). In particular, this is true for the ring of p-adic integers. Note, however, that the ring of p-adic integers fails to be Σ-algebraically compact (both as an abelian group and as a module over itself).

3. [**24**] If R is a countable ring, A any left R-module, and $\mathcal{F}$ a non-principal ultrafilter on $\mathbb{N}$, then the ultrapower $A^{\mathbb{N}}/\mathcal{F}$ is algebraically compact.

4. Every artinian module over a commutative ring is Σ-algebraically compact (see the examples following Theorem 6 below).

5. If R is a commutative artinian principal ideal ring, then all R-modules are direct sums of cyclic submodules. The latter being finite in number, up to isomorphism, Example 4 shows all objects of R-Mod to be algebraically compact in that case.

We will sketch a proof for the fifth remark (apparently folklore), since it is part of the red thread that will lead us through this survey. By the Chinese Remainder Theorem, R is a finite direct product of local rings, and hence it is harmless to assume that R is a local artinian principal ideal ring with maximal ideal $\mathfrak{m}$ say. Now all ideals of R are powers of $\mathfrak{m}$, and each homomorphism from a power $\mathfrak{m}^n$ to R sends $\mathfrak{m}^n$ back to $\mathfrak{m}^n$ and can therefore be extended to R – just use the fact that $\mathfrak{m}$ is principal. This shows that R is self-injective; in fact, R being an artinian ring, R is Σ-injective as an R-module. Now let M be any nonzero module. We provide a decomposition of the desired ilk by induction on the least natural number N such that $\mathfrak{m}^{N+1}M = 0$. Clearly M is a module over the ring $R/\mathfrak{m}^{N+1}$ which, as we just saw, is Σ-injective over itself. Let $(x_i)_{i \in I}$ be a maximal family of elements of M such that the sum of the Rx_i is direct, with each of the Rx_i isomorphic to R. Due to injectivity, the sum $\sum_{i \in I} Rx_i$ is then a direct summand of M, say $M = \sum_{i \in I} Rx_i \oplus M'$, and due to the maximal choice of our family, M' does not contain a copy of $R/\mathfrak{m}^{N+1}$, i.e., M' is annihilated by $\mathfrak{m}^N$. Our claim follows by induction.

Point 5 shows, in particular, that all artinian principal ideal rings have finite representation type in the sense to follow. In fact, it has long been known that, among the commutative artinian rings, the principal ideal rings are precisely the ones having this property.

Definition. A ring R is said to have *finite representation type* if it is left artinian and if, up to isomorphism, there are only finitely many indecomposable finitely generated left R-modules.

Finite representation type is actually left-right symmetric, as was shown by Eisenbud and Griffith in [**18**].

2. A global decomposition problem

For better focus, we interject two problems which, on the face of it, are only loosely connected with our main theme. The connection turns out to be much closer than anticipated at first sight. In fact, these problems have motivated a major portion of the subsequent work on algebraic compactness.

Global Problems. (Koethe [**48**], Cohen-Kaplansky [**13**]) *For which rings R is every right R-module*

(a) *a direct sum of finitely generated modules?*

(b) *a direct sum of indecomposable modules?*

Work on the commutative case was initiated by Koethe in 1935, continued by Cohen-Kaplansky in 1951, and completed by Griffith in 1970 for part (a), and by Warfield in 1972 for part (b).

Theorem 2. ([**48, 13, 28, 74**]) *For any commutative ring R, conditions* (a) *and* (b) *above are equivalent and satisfied if and only if R is an artinian principal ideal ring.*

As we noted at the end of Section 1, among the commutative rings, the artinian principal ideal rings are precisely the ones having finite representation type. Moreover, we observed that these rings enjoy the property that all their modules are algebraically compact. As we will see in Theorem 13 of Section 4, this property in turn characterizes the artinian principal ideal rings, which rounds off the 'commutative solution' to our problems. The reasons for the most interesting implications, namely that either of the two conditions (a), (b) forces R to be an artinian principal ideal ring, can be roughly summarized as follows: In general, large direct products of modules exhibit a high resistance to infinite direct sum decompositions; more precisely, in most cases, infinite direct sum decompositions of direct products can be traced back to infinite decompositions of finite subproducts.

In the noncommutative situation, the problems become far more challenging. In his 1972 paper, Warfield stated: "For non-commutative rings, the questions raised in this paper seem to be much more difficult. All that seems to be known is that any ring satisying [the above conditions (a), (b)] is necessarily [left] artinian." At present, there is still at least one link missing to a truly satisfactory resolution of these problems. The main key to what is known is an equivalent characterization of Σ-algebraic compactness, which will be presented in the next section.

3. Characterizations of (Σ-)algebraically compact modules

We begin with a subsection dedicated to the main technical resource of the subject, introduced independently by Gruson-Jensen [**30**] and W. Zimmermann [**80**]. It is the more general functorial framework of [**80**] and [**37**] which we will describe below, since the extra generality adds to the transparency of the arguments.

A. Product-compatible functors and matrix functors.

Definition. (1) A *p-functor* on R-Mod is a subfunctor P of the forgetful functor $R\text{-Mod} \to \mathbf{Ab}$ which commutes with direct products, i.e., P assigns to each left R-module M a subgroup PM such that $f(PM) \subseteq PN$ for any homomorphism $f : M \to N$, and $P(\prod_{i\in I} M_i) = \prod_{i\in I}(PM_i)$ for any direct product $\prod_{i\in I} M_i$ in R-Mod.

Note that any *p*-functor automatically commutes with direct sums (this being actually true for any subfunctor of the forgetful functor).

(2) A *pointed matrix* over R is a row-finite matrix $\mathcal{A} = (a_{ij})_{i\in I, j\in J}$ of elements from R, paired with a column index $\alpha \in J$. Given a pointed matrix $(\mathcal{A}, \alpha)$, we call the following *p*-functor $[\mathcal{A}, \alpha]$ on R-Mod a *matrix-functor*: For any R-module M, the subgroup $[\mathcal{A}, \alpha]M$ is defined to be the α-th projection of the solution set in M of the homogeneous system

$$\sum_{j\in J} a_{ij} X_j = 0 \qquad \text{for all } \ i \in I;$$

in other words,

$$[\mathcal{A}, \alpha]M = \{m \in M \mid \exists \text{ a solution } (m_j) \in M^J \text{ of the above system with } m_\alpha = m\}.$$

Further, we call $[\mathcal{A}, \alpha]$ a *finite matrix functor* in case the matrix $\mathcal{A}$ is finite.

Given a (finite) matrix functor $[\mathcal{A}, \alpha]$ on R-Mod and a left R-module M, we will call the subgroup $[\mathcal{A}, \alpha]M$ a *(finite) matrix subgroup* of M. The finite matrix subgroups were labeled "sousgroupes de définition fini" by Gruson and Jensen, and "*pp*-definable subgroups" by the model theorists, Prest, Herzog, Rothmaler, and others.

Observation 3. Basic properties of matrix functors. *The first two give alternate descriptions of matrix functors.*

(1) $[\mathcal{A}, \alpha] = \mathrm{Hom}_R(Z, -)(z)$ *for a suitable left R-module Z and $z \in Z$; conversely, every functor of the form* $\mathrm{Hom}_R(Z, -)(z)$ *is a matrix functor for a suitable matrix $\mathcal{A}$. The finite matrix functors are precisely the functors* $\mathrm{Hom}_R(Z, -)(z)$ *with finitely presented Z.*

(2) *The finite matrix subgroups of M moreover coincide with the kernels of the $\mathbb{Z}$-linear maps $M \to Z \bigotimes_R M$, $m \mapsto z \otimes m$, where Z is a finitely presented right R-module and $z \in Z$.*

(3) *The class of matrix functors is closed under arbitrary intersections and finite sums. The finite matrix functors are closed under* finite *intersections and finite sums.*

(4) *If M is an R-S-bimodule, then every matrix subgroup of ${}_RM$ is an S-submodule of M.*

In the context of part (4), the most important candidates for S will be the opposite of the endomorphism ring of M, as well as subrings of the center of R. How easily matrix functors can be manipulated is evidenced by the easy proofs of the above observations; we include one sample argument to make our point.

Proof of part of (3). We will show that finite sums of matrix functors are again of that ilk. So let $(\mathcal{A} = (a_{ij})_{i\in I, j\in J}, \alpha)$ and $(\mathcal{B} = (b_{kl})_{k\in K, l\in L}, \beta)$ be two pointed matrices with entries in R. It is clearly harmless to assume that the occurring index sets are all disjoint. Let $(\mathcal{C} = (c_{uv})_{u\in U, v\in V}, \gamma)$ be defined as follows: Assuming that γ belongs to none of the sets I, J, K, L, we set $U = \{\gamma\} \cup I \cup K$ and $V = \{\gamma\} \cup J \cup L$, and define c_{uv} via $c_{\gamma\gamma} = 1$, $c_{\gamma\alpha} = c_{\gamma\beta} = -1$, $c_{ij} = a_{ij}$ whenever $(i,j) \in I \times J$, $c_{kl} = b_{kl}$ whenever $(k,l) \in K \times L$, and $c_{uv} = 0$ in all other cases. It is straightforward to check that, for any left R-module M, we have $[\mathcal{A}, \alpha]M + [\mathcal{B}, \beta]M = [\mathcal{C}, \gamma]M$. Moreover, we remark that $\mathcal{C}$ is finite if $\mathcal{A}$ and $\mathcal{B}$ are. □

Important instances of matrix functors.

- *Whenever $\mathfrak{a}$ is a finitely generated right ideal of R, the assignment $M \mapsto \mathfrak{a}M$ defines a finite matrix functor on R-*Mod.
- *For every subset T of R, the assignment $M \mapsto \mathrm{Ann}_M(T)$ is a matrix functor; it is finite in case T is.*

More generally: Whenever $\mathfrak{a}$ is a finitely generated right ideal and T a subset of R, the conductor $(\mathfrak{a}M : T)$ is a matrix subgroup of M.

- *All finitely generated $\mathrm{End}_R(M)$-submodules of M are matrix subgroups.*

Pinning down matrices for the first three types of examples is straightforward. To verify the last: By Observation 3(1), each cyclic $\mathrm{End}_R(M)$-submodule of M is a matrix subgroup; now use Observation 3(3) to move to finite sums.

As a first application of matrix functors to our present objects of interest, we will give a characterization of purity in terms of finite matrix subgroups. Ironically, this description provides a perfect formal parallel to that of purity over PID's, while all of the initial attempts to generalize the concept of purity, which were based on formal analogy (with requirements such as $\mathfrak{a}N \cap M = \mathfrak{a}M$ for all cyclic or finitely generated right ideals of the base ring), were eventually discarded as not quite strong enough to best serve their purpose.

Proposition 4. *A submodule M of a left R-module N is pure if and only if $[\mathcal{A}, \alpha]N \cap M = [\mathcal{A}, \alpha]M$ for all finite matrix functors $[\mathcal{A}, \alpha]$ on R-*Mod.

Proof. The proof for pure left exactness of finite matrix functors is straightforward (cf. [**79**]).

Now assume that $M \subseteq N$ satisfies the above intersection property. To verify purity of the inclusion, let

$$\sum_{j=1}^{s} a_{ij}X_j = m_i \qquad (1 \le i \le r) \tag{$\dagger$}$$

be a finite linear system with $m_i \in M$, which is solvable in N. We prove its solvability in M by induction on r. For $r = 1$, let $\mathcal{A} = (-1, a_{11}, \dots, a_{1s})$ be a single row with the first column labeled α; then $m_1 \in [\mathcal{A}, \alpha]N \cap M = [\mathcal{A}, \alpha]M$ by construction and hypothesis, which provides a solution of (†) in M. Next suppose that $r \ge 2$, pick a solution (v_j) of (†) in N, and use the induction hypothesis to procure a solution (u_j) in M of the first $r-1$ equations. Define $m'_r := \sum_{j=1}^{s} a_{rj}(v_j - u_j) = m_r - \sum_{j=1}^{s} a_{rj}u_j \in M$, and observe that $\sum_{j=1}^{s} a_{ij}(v_j - u_j) = 0$ for $i = 1, \dots, r-1$. We infer that m'_r belongs to $[\mathcal{B}, \beta]N \cap M$, where $\mathcal{B}$ is the matrix

$$\begin{pmatrix} 0 & a_{11} & a_{12} & \cdots & a_{1s} \\ \vdots & \vdots & \vdots & & \vdots \\ 0 & a_{r-1,1} & a_{r-1,2} & \cdots & a_{r-1,s} \\ -1 & a_{r1} & a_{r2} & \cdots & a_{rs} \end{pmatrix},$$

and β labels the first column of $\mathcal{B}$. By hypothesis, we thus have $m'_r \in [\mathcal{B}, \beta]M$, which provides us with a family (u'_j) in M satisfying $\sum_{j=1}^{s} a_{ij}u'_j = 0$ for $1 \le i \le r-1$ and $\sum_{j=1}^{s} a_{rj}u'_j = m'_r$. This yields a solution $(u_j + u'_j)$ of (†) in M as required. □

Moreover, numerous properties of a ring R linked to the behavior of direct products in R-Mod can be conveniently recast in terms of matrix functors. We give two examples taken from [**79**, **80**]:

• A left R-module M is flat if and only if $[\mathcal{A}, \alpha]M = [\mathcal{A}, \alpha]R \cdot M$ for arbitrary finite matrix functors $[\mathcal{A}, \alpha]$. One deduces that all direct sums of copies of a flat module M are flat if and only if, for each finite matrix functor $[\mathcal{A}, \alpha]$, there exists a finitely generated right ideal $\mathfrak{a} \subseteq [\mathcal{A}, \alpha]R$ with $[\mathcal{A}, \alpha]M = \mathfrak{a}M$.

• Thus: flatness is inherited by arbitrary direct products of flat left R-modules (i.e., R is right coherent) if and only if all finite matrix subgroups of ${}_RR$ are finitely generated right ideals.

B. p-functors and algebraic compactness.

The following equivalent descriptions of algebraically compact and Σ-algebraically compact modules will prove extremely useful in the sequel. In each of the two theorems, the equivalence of (1) and (3) was independently established by Gruson-Jensen and Zimmermann ([**31**, **80**]); condition (2) – a significant strengthening of the necessary side, in light of the third of the above instances of matrix subgroups – was added by the latter.

Theorem 5. *The following statements are equivalent for any left R-module M:*

(1) *M is algebraically compact.*

(2) *Every family of residue classes $(m_l+P_lM)_{l\in L}$ with $m_l \in M$ and p-functors P_l, which has the finite intersection property, has non-empty intersection.*

(3) *Same as (2) with P_l replaced by finite matrix functors.*

In the special situation of a Prüfer domain R (i.e., a commutative semihereditary integral domain), a specialized version of this result had already been obtained by Warfield [**72**]: In that case, the matrix subgroups of the form $(\mathfrak{a}M : T)$, where $\mathfrak{a}$ is a finitely generated ideal and T a finite subset of R, are representative.

Since Zimmermann's elementary proof of this theorem [**80**, Satz 2.1] does not exist in English translation, we include a sketch below. We begin with a remark which will come in handy in other contexts as well. Namely, locally – i.e., on the closure of a given module M under direct sums and products in R-Mod – any p-functor acts like a matrix functor. Indeed, given a p-functor P on R-Mod, we write $PM = \{m_i \mid i \in I\}$, let M_i be a copy of M for each i, and set $\overline{m} = (m_i)_{i\in I} \in \prod_{i\in I} M_i$. Then $PM = \mathrm{Hom}_R(\prod_{i\in I} M_i, M)(\overline{m})$, and Observation 3(1) yields our claim.

Proof of Theorem 5. '(1) $\Longrightarrow$ (2)': Assuming (1), we start with a family of residue classes $(m_l + P_lM)_{l\in L}$ having the finite intersection property. By the preceding remark, we may assume that the P_l are matrix functors, say $P_l = [\mathcal{A}_l, \alpha_l]$ with $\mathcal{A}_l = (a^l_{ij})_{i\in I_l, j\in J_l}$. We construct a system of equations which is finitely solvable in M, and a global solution to which will provide us with an element in the intersection of our family. It is clearly harmless to assume that all of the index sets I_l and J_l are pairwise disjoint. Choose an element α contained in none of these. We set $I = \bigcup_{l\in L} I_l$, $J = \{\alpha\} \cup \bigcup_{l\in L} J_l \setminus \{\alpha_l\}$, and define a row-finite matrix $\mathcal{A} = (a_{ij})_{i\in I, j\in J}$ as follows: $a_{i\alpha} = a^l_{i\alpha_l}$ if $i \in I_l$; $a_{ij} = a^l_{ij}$ if $i \in I_l$ and $j \in J_l \setminus \{\alpha_l\}$; finally we set $a_{ij} = 0$ in all other cases. It is immediate that $[\mathcal{A}, \alpha]M = \bigcap_{l\in L}[\mathcal{A}_l, \alpha_l]M$. The matrix $\mathcal{A}$ will serve as coefficient matrix of our system. As for its right-hand side, we define $m \in M^I$ by stringing up the elements $m_i = a_{i\alpha}m_l$ for $i \in I_l$ in the only plausible fashion. It is then straightforward to check that the system $\sum_{j\in J} a_{ij}X_j = m_i$ for $i \in I$ is finitely solvable by construction. Hence it has a global solution, say $(z_i)_{i\in I}$, and one readily verifies that z_α belongs to the intersection of the family of residue classes with which we started out.

'(3) $\Longrightarrow$ (1)': This time, we begin with a system

$$\sum_{j\in J} a_{ij}X_j = m_i \qquad \text{for} \qquad i \in I, \tag{$\dagger$}$$

which is finitely solvable in M. The crucial step of our argument is the following easy consequence of (3) and finite solvability of ($\dagger$): Namely, for each index $\alpha \in J$, there exists an element $y_\alpha \in M$ such the the system

$$\sum_{j\in J\setminus\{\alpha\}} a_{ij}X_j = m_i - a_{i\alpha}y_\alpha$$

is again finitely solvable, as follows. Given any finite subset $I' \subseteq I$, let $y_\alpha(I')$ be the α-th component of a solution of the finite system $\sum_{j\in J} a_{ij}X_j = m_i$ for $i \in I'$, and let $\mathcal{A}(I')$ be the matrix consisting of the rows of $\mathcal{A}$ labelled by I'. Then the family $\big(y_\alpha(I') + [\mathcal{A}(I'), \alpha]M\big)$, where I' runs through the finite subsets of I, has the finite intersection property. Therefore its intersection is nonempty by (3), and any element y_α in this intersection satisfies our requirement.

Next we consider the set $\mathcal{K}$ of all pairs (K, y), where K is a subset of J and $y = (y_k) \in M^K$ is such that the system

$$\sum_{j\in J\setminus K} a_{ij}X_j \quad = \quad m_i - \sum_{k\in K} a_{ik}y_k \qquad \text{for} \qquad i \in I$$

is in turn finitely solvable in M. We equip this set of pairs with the standard order, namely $(K, y) \leq (K', y')$ if $K \subseteq K'$ and $y_k = y'_k$ whenever $k \in K$. Clearly, $\mathcal{K} \neq \varnothing$. One checks that the set $\mathcal{K}$ is inductively ordered, and denotes by (K_0, z) a maximal element of $\mathcal{K}$. Our initial statement, applied to the finitely solvable system $\sum_{j\in J\setminus K_0} a_{ij}X_j = m_i - \sum_{k\in K_0} a_{ik}z_k$, now yields $J = K_0$, which makes z a global solution of (†). □

Theorem 6. *The following statements are equivalent for any left R-module M:*

(1) *M is Σ-algebraically compact.*

(2) *Every countable descending chain $P_1 \supseteq P_2 \supseteq P_3 \supseteq \cdots$ of p-functors becomes stationary on M.*

(3) *Same as* (2) *with P_l replaced by finite matrix functors.*

We point out that conditions (2) and (3) could just as well have been phrased as follows: 'M has the descending chain condition for p-functorial subgroups' (or, equivalently, 'M has the descending chain condition for finite matrix subgroups'), since the classes of p-functors and finite matrix functors are both closed under finite intersections. Indeed, given a descending chain $P_1M \supseteq P_2M \supseteq \ldots$ of p-functorial subgroups of M, we obtain $P_iM = Q_iM$ for the descending chain $Q_i = P_1 \cap \cdots \cap P_i$ of p-functors.

Proof of Theorem 6. In view of Theorem 5, the implication '(3) $\implies$ (1)' is just an analogue of the much older result that artinian modules are linearly compact [**77**]; it is left as an exercise. In the arguments given by Zimmermann and Gruson-Jensen, the novel implication '(1) $\implies$ (2)' is obtained via a detour through direct products, which we sketch because it pinpoints the importance of product-compatibility of the functors we are considering: Clearly, Σ-algebraic compactness of M forces the natural (pure) embedding of the direct sum $M^{(\mathbb{N})}$ in the direct product $M^{\mathbb{N}}$ to split, say $\prod_{n\in\mathbb{N}} M_n = C \oplus \bigoplus_{n\in\mathbb{N}} M_n$, where each M_n is a copy of M. Let $\pi : \prod_{n\in\mathbb{N}} M_n \to C$ be the corresponding projection, and $\pi_n : \prod_{i\in\mathbb{N}} M_i \to M_n$ the canonical maps. Moreover, assume we have a descending chain $P_1 \supseteq P_2 \supseteq P_3 \supseteq \ldots$ of p-functors with $m_n \in P_nM_n \setminus P_{n+1}M_n$ for $n \in \mathbb{N}$. Set $x = (m_n)$, $y_n = (m_1, \ldots, m_n, 0, \ldots)$, $x_n = x - y_n$, and decompose x and x_n in the form

$x = s + c$ and $x_n = s_n + c_n$ with $s, s_n \in \bigoplus_{n\in\mathbb{N}} M_n$ and $c, c_n \in C$; then, clearly $s = y_n + s_n$. Since x_n belongs to $\prod_{n\in\mathbb{N}} P_{n+1}M_n = P_{n+1}(\prod_{n\in\mathbb{N}} M_n)$ by construction, s_n belongs to $P_{n+1}(\bigoplus_{n\in\mathbb{N}} M_n)$. So if we choose an index $N \in \mathbb{N}$ with $\pi_N(s) = 0$, we infer that $m_N = \pi_N(y_N) = \pi_N(s - s_N) = -\pi_N(s_N)$ lies in $P_{N+1}M_N$, which contradicts our choice of m_N. $\square$

In Section 4, these concepts and results will turn out tailored to measure for tackling the problems advertized in Section 2. However, the two preceding theorems lend themselves to a large number of further applications. We present a small selection, starting with a few additional examples; most of them are obvious in light of the above theory, the others we tag with references.

Further examples of (Σ-)algebraically compact modules.

- Whenever M is an R-S-bimodule which is artinian over S, M is Σ-algebraically compact as an R-module. This is, for instance, true if M is an R-module which is artinian over its endomorphism ring or over the center of R. In particular, for any Artin algebra R, the category R-mod consists entirely of Σ-algebraically compact modules.
- Suppose that M is a (Σ)-algebraically compact left R-module, P a p-functor on R-Mod and S a subring of R such that PM is an S-submodule of M. Then both PM and M/PM are (Σ-)algebraically compact over S. In particular: If $\mathfrak{a}$ is an ideal of R which is finitely generated on the right, then $\mathfrak{a}M$ and $M/\mathfrak{a}M$ are (Σ-)algebraically compact (see [**37**]).
- If M is a module over a Dedekind domain R, then M is Σ-algebraically compact if and only if $M \doteq M_1 \oplus M_2$ where M_1 is divisible and M_2 has nonzero R-annihilator.
- If M is finitely generated over a commutative noetherian ring, then M is (Σ)-algebraically compact precisely when M is artinian. (For the nontrivial implication, see [**80**].)
- Suppose S is a left Σ-algebraically compact ring (e.g. a field), $(X_i)_{i\in I}$ a family of independent indeterminates over S, and k any positive integer. Then the truncated polynomial ring

$$R = S[X_i \mid i \in i]/(X_i \mid i \in i)^k$$

is in turn Σ-algebraically compact as a left module over itself (see [**37**]).

- If S is left algebraically compact, then so is every power series ring $R = S[[X_i \mid i \in I]]$ ([loc. cit.]). Note, however, that R fails to be Σ-algebraically compact in case $I \neq \varnothing$.
- If S is any ring and G a group, then the group ring SG is left algebraically compact if and only if S has this property and G is finite (see [**81**]).
- Suppose M is an R-S-bimodule. If A is an algebraically compact left R-module, then the left S-module $\mathrm{Hom}_R(M, A)$ is algebraically compact as well – see below. (This 'compactification by passage to a suitable dual' is akin to the Pontrjagin dual.) In particular, the endomorphism ring S of any algebraically

compact left module is algebraically compact on the right, the change of side being due to the convention that S acts on the left of M.

Since the last of the above remarks, concerning the passage of algebraic compactness from ${}_RA$ to ${}_S\operatorname{Hom}_R(M,A)$, will be relevant in the sequel, we include a justification based on the adjointness of Hom and tensor product: Start with a pure monomorphism $U \to V$ of left S-modules, and observe that the pure injectivity of A forces the upper row (and hence also the lower one) of the following commutative diagram to be an epimorphism:

$$\begin{array}{ccc}
\operatorname{Hom}_R(M \otimes_S V, A) & \longrightarrow & \operatorname{Hom}_R(M \otimes_S U, A) \\
\downarrow \cong & & \downarrow \cong \\
\operatorname{Hom}_S(V, \operatorname{Hom}_R(M,A)) & \longrightarrow & \operatorname{Hom}_S(U, \operatorname{Hom}_R(M,A))
\end{array}$$

The following equivalence is due to Faith [**21**].

Corollary 7. *For any injective left R-module M, the following statements are equivalent:*

(1) *M is Σ-injective (meaning that every direct sum of copies of M is injective).*

(2) *R has the ascending chain condition for M-annihilators, i.e., for left ideals of the form $\operatorname{Ann}_R(U)$, where U is a subset of M.*

Proof. Note that condition (1) is tantamount to Σ-algebraic compactness of M under our hypotheses, since $M^{(I)}$ is pure in M^I for any set I. The matrix subgroups of an injective module M are precisely the annihilators in M of subsets of R (prove $[\mathcal{A},\alpha]M = \operatorname{Ann}_M \operatorname{Ann}_R[\mathcal{A},\alpha]M$ or consult [**80**, Beispiel 1.1]). Thus Theorem 6 shows (1) to be equivalent to the descending chain condition for R-annihilators in M. But the latter descending chain condition has an equivalent flip side, namely the ascending chain conditon for M-annihilators in R. □

Corollary 8. *Any pure submodule of a Σ-algebraically compact module is a direct summand.*

Proof. Let M be Σ-algebraically compact and $U \subseteq M$ a pure submodule. By Proposition 4, $[\mathcal{A},\alpha]M \cap U = [\mathcal{A},\alpha]U$ for every finite matrix functor $[\mathcal{A},\alpha]$, and hence U inherits the descending chain condition for finite matrix subgroups from M. Therefore, U is in turn Σ-algebraically compact and consequently a direct summand of M. □

The next result is part of the bridge taking us back to our global decomposition problems. We will obtain the first part as another consequence of Theorem 6, while the second can be most elegantly derived from the following category equivalence due to Gruson-Jensen [**32**] which provides a very general tool for extending properties of injective modules to algebraically compact ones: Let fin.pres-R be the full subcategory of mod-R consisting of the finitely presented right R-modules, and let (fin.pres-R, $\mathbf{Ab}$) be the category of all additive covariant functors from fin.pres-R to the abelian goups. This functor category is a Grothendieck category, and the functor

$$R\text{-Mod} \to (\text{fin.pres-}R, \mathbf{Ab})$$

defined by $M \mapsto - \otimes_R M$ induces a category equivalence from the full subcategory of algebraically compact objects of R-Mod to the full subcategory of injective objects of (fin.pres-R, $\mathbf{Ab}$).

Proposition 9. (Further assets of (Σ-)algebraically compact modules)

(1) *Every Σ-algebraically compact module is a direct sum of indecomposable summands with local endomorphism rings.*

(2) *If M is an algebraically compact left module with endomorphism ring S, then $S/\operatorname{rad}(S)$ is von Neumann regular and right self-injective; moreover, idempotents can be lifted modulo* $\operatorname{rad}(S)$.

In particular, S is local in case M is indecomposable.

(3) *Each strongly invariant submodule M of an algebraically compact module has the exchange property, i.e., can be shifted inside direct sum grids as follows: Given any equality of left R-modules $M \oplus X = \bigoplus_{i \in I} Y_i$, there exist submodules $Y_i' \subseteq Y_i$ such that $\bigoplus_{i \in I} Y_i = M \oplus \bigoplus_{i \in I} Y_i'$.*

(Here a submodule M of a module N is said to be strongly invariant in case each homomorphism ϕ from M to N satisfies $\phi(M) \subseteq M$.)

Proof. Part (1) [**37**]: Suppose that M is Σ-algebraically compact, and let $(U_i)_{i \in I}$ be a maximal family of independent indecomposable submodules of M such that $U = \bigoplus_{i \in I} U_i$ is pure in M. By Corollary 8, U is a direct summand of M, say $M = U \oplus V$. If V were nonzero, we could pick a nonzero element $v \in V$ and choose a pure submodule $W \subseteq V$ which is maximal with the property of excluding v. Again we would obtain splitting, $V = W \oplus Z$, which would provide us with an indecomposable summand Z of V. But the existence of such a Z is incompatible with the maximal choice of the family (U_i). That the U_i even have local endomorphisms rings, will follow from part (2).

For part (2), we use the above category equivalence and the well-known fact that endomorphism rings of injectives have the listed properties; indeed, the argument given by Osofsky in [**57**] for modules carries over to Grothendieck categories. (Note however that the direct argument given in [**37**] is a bit slicker than the original proof for injective modules, one of the reasons being that algebraic compactness is passed on to endomorphism rings by the last of the examples following Theorem 6.)

As for part (3), we will only sketch a proof for the finite exchange property and refer the reader to [**38**] for a complete argument. Let M be a strongly invariant submodule of an algebraically compact left R-module N, and let S be the endomorphism ring of M. From the last of the remarks following Theorem 6, we know that $S = \operatorname{Hom}_R(M, N)$ is an algebraically compact right S-module. Therefore S

is an exchange ring in the sense of Warfield by part (2), which easily implies the finite exchange property of M (see [**73**]). □

To conclude the section, we point to two examples to mark what we consider potentially dangerous curves: Namely, while any right artinian ring is necessarily Σ-algebraically compact on the left, on the right it need not even be algebraically compact. An example demonstrating this was given by Zimmermann in [**81**]. Moreover, while the descending chain condition for finite matrix subgroups of a module M entails the descending chain condition for arbitrary matrix subgroups, the analogous implication fails for the ascending chain condition as the following example shows: One of the consequences we derived from Theorem 6 says that the algebra $R = K[X_i \mid i \in \mathbb{N}]/(X_i \mid i \in \mathbb{N})^2$ over a field K is Σ-algebraically compact. Hence, if E denotes the minimal injective cogenerator for R, then the R-module $E = \mathrm{Hom}_R(R, E)$ has the ascending chain condition for finite matrix subgroups (use Tool 17 below to see this). Note that the ascending chain condition for *arbitrary* matrix subgroups would amount to E being noetherian over its endomorphism ring S, since all finitely generated S-submodules of E are in fact matrix subgroups. But the S-module E fails to be noetherian: Indeed, if it were, then the equality $A = \mathrm{Ann}_R \mathrm{Ann}_E(A)$ for any ideal A of R would make R artinian, an obvious absurdity. On the other hand:

Remark. Suppose that M is a direct sum of finitely presented objects in R-mod. Then the R-module M satisfies the ascending chain condition for finite matrix subgroups if and only if it is noetherian over its endomorphism ring.

To see this, let $M = \bigoplus_{i\in I} M_i$ with finitely presented summands M_i, and let S be the endomorphism ring of M. For the nontrivial implication, suppose that M has the ascending chain condition for finite matrix subgroups. Clearly, M is noetherian over S in case all finitely generated S-submodules are finite matrix subgroups. By Observation 3(3), it thus suffices to prove that each cyclic S-submodule Sm of M is a finite matrix subgroup. Choose a finite subset $I' \subseteq I$ such that $m \in \bigoplus_{i\in I'} M_i$. Then $Sm = \mathrm{Hom}_R(\bigoplus_{i\in I'} M_i, M)(m)$ is indeed a finite matrix subgroup of M by Observation 3(1). □

To learn about the module-theoretic impact of the maximum condition for finite matrix subgroups, consult [**82**], for generalizations of classical results to modules satisfying this condition, see [**83**].

4. Return to our global decomposition problems

As we mentioned in Section 2, the arguments settling the commutative case rest on the fact that certain large direct products tend to resist nontrivial direct sum decompositions. It is therefore not surprising that, also in the noncommutative situation, the decomposition properties of such products are crucial.

Theorem 10. *For a left R-module M, the following conditions are equivalent:*

(1) *There exists a cardinal number $\aleph$ such that every direct product of copies of M is a direct sum of $\aleph$-generated modules.*

(2) *Every direct product of copies of M is a direct sum of submodules with local endomorphism rings.*

(3) *M is Σ-algebraically compact.*

The implication '(1) $\Longrightarrow$ (3)' is due to Gruson-Jensen [**31**], '(2) $\Longrightarrow$ (3)' to the author of this article [**35**], whereas '(3) $\Longrightarrow$ (1)' can already be found in work of Kiełpiński going back to 1967 [**46**]. The implication '(3) $\Longrightarrow$ (2)' is a consequence of Proposition 9. Credit should also go to Chase, since the ideas he developed in [**12**] play an essential role in the proofs. These ideas can be extracted and upgraded to the following format:

Lemma 11. (Chase) *Let*

$$f : \prod_{i \in \mathbb{N}} U_i \longrightarrow \bigoplus_{j \in J} V_j$$

be a homomorphism of R-modules, and $P_1 \supseteq P_2 \supseteq P_3 \supseteq \cdots$ a descending chain of p-functors. Then there exists a natural number n_0 such that

$$f\Big(P_{n_0} \prod_{i \geq n_0} U_i\Big) \quad \subseteq \quad \bigoplus_{\text{finite}} V_j \; + \; \bigcap_{n \in \mathbb{N}} P_n\Big(\bigoplus_J V_j\Big).$$

(Here $\bigoplus_{\text{finite}}$ stands for a direct sum extending over some finite subset of J.)

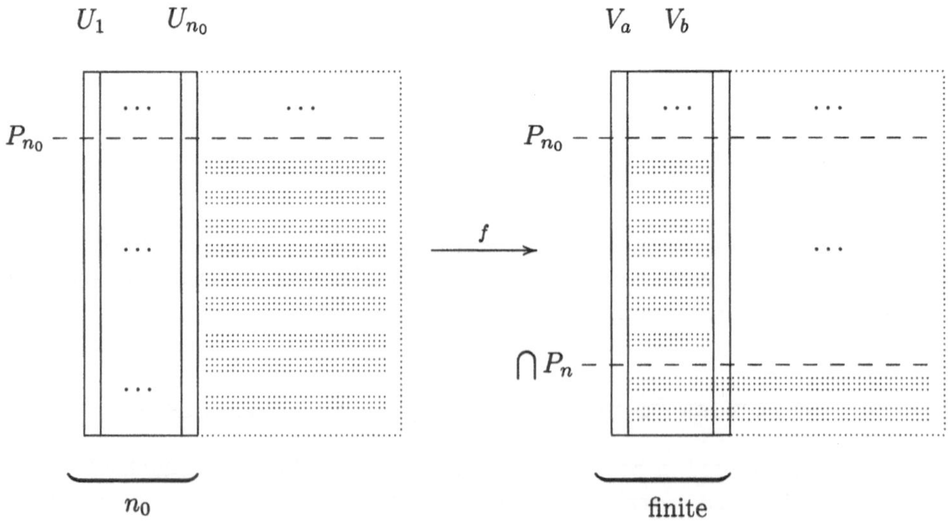

Since the summand $\bigcap_{n\in\mathbb{N}} P_n(\bigoplus_J V_j)$ on the far right of the pivotal inclusion is a correction term which can be suppressed in most applications, the lemma says that, after a bit of trimming on both sides, a cofinite subproduct of the U_i's maps to a finite subsum of the V_j's.

Before we give some of the arguments to illustrate the techniques, we derive the following consequence of Theorem 10.

Corollary 12. (Chase [**12**])

• *If there exists a cardinal number $\aleph$ such that all of the left R-modules R^I are direct sums of $\aleph$-generated submodules (this being, e.g., the case if the projectives in R-*Mod *are closed under direct products), then R is left perfect.*

• *If there is a cardinal number $\aleph$ such that all left R-modules are direct sums of $\aleph$-generated modules, then R is left artinian.*

Proof of Corollary 12. From the hypothesis of the first assertion we deduce that R has the descending chain condition for finitely generated right ideals, those being among the matrix subgroups of the right R-module R by Section 3.

Now suppose that we have the global decomposition property of the second assertion. Then R is left perfect by the first part. Moreover, all left R-modules have the descending chain condition for annihilators of subsets of R, and hence R has the ascending chain condition for annihilators of subsets of arbitrary left R-modules; in other words, R is left noetherian. This implies that R is indeed left artinian. □

We will sketch a proof of Chase's Lemma, as the argument is clarified by the use of p-functors.

Proof of Lemma 11. The natural projection $\bigoplus_{k\in J} V_k \to V_j$ will be denoted by q_j. Assume the conclusion to be false. Then a standard induction yields a sequence $(n_k)_{k\in\mathbb{N}}$ of natural numbers with $n_{k+1} > n_k$, together with sequences of pairwise different elements $j_k \in J$, resp. $x_k \in P_{n_k}(\prod_{i\geq n_k} U_i)$, such that

$$q_{j_k} f(x_k) \notin P_{n_{k+1}} V_{j_k} \qquad \text{and} \qquad q_{j_k} f(x_l) = 0 \quad \text{for} \quad l < k.$$

Note that the definition $x = \sum_{k\in\mathbb{N}} x_k \in \prod_{i\in\mathbb{N}} U_i$ makes sense (indeed, in view of $x_k \in \prod_{i\geq n_k} U_i$, the sum of the x_k reduces to a finite sum in each U_i-component), and that for all $k \in \mathbb{N}$ we have

$$q_{j_k} f(x) = q_{j_k} f(x_k) + q_{j_k} f\big(\sum_{l>k} x_l\big) \neq 0,$$

because the first summand does not lie in $P_{n_{k+1}} V_{j_k}$, whereas the second does. But this contradicts the fact that $f(x)$ belongs to a finite subsum of V_j's. □

Note that the implication '(3) $\implies$ (2)' of Theorem 10 is an immediate consequence of Proposition 9. For the converse, we refer the reader to [**35**]. We include a proof for the equivalence of (1) and (3) however.

Proof of '(1) $\iff$ (3)' of Theorem 10. First suppose that (1) is satisfied, and let $P_1 \supseteq P_2 \supseteq P_3 \supseteq \cdots$ be a chain of p-functors. Abbreviate the intersection $\bigcap_{n\in\mathbb{N}} P_n$ by P. We want to prove the existence of an index n_0 such that $P_{n_0}M = PM$. For that purpose, it is clearly harmless to assume that $PM = 0$. This implies that $PV = 0$ for arbitrary direct summands V of direct powers of M, since P is in turn a p-functor and, as such, commutes with direct sums and direct products.

Set $\tau = \max\{\aleph_0, \aleph, |R|\}$, and choose a set I of cardinality at least τ. By hypothesis, $M^I \cong \prod_{n\in\mathbb{N}} M^I$ has a direct decomposition into $\aleph$-generated summands V_j. Applying Chase's Lemma to

$$\prod_{n\in\mathbb{N}} M^I \xrightarrow{\text{id}} \bigoplus_{j\in J} V_j,$$

we obtain a natural number n_0 with

$$\prod_{n\geq n_0} (P_{n_0}M)^I \hookrightarrow \bigoplus_{\text{finite}} V_j \;+\; P(\bigoplus_J V_j),$$

the final summand on the right being zero by our assumption on M. But this shows the right-hand side to have cardinality at most τ, while the left-hand side has cardinality $> \tau$ if $P_{n_0}M \neq 0$. To make the left-hand side small enough to fit into the right, we thus need to have $P_{n_0}M = 0$, which shows that our chain of p-functors becomes stationary on M. Now apply Theorem 6 to obtain (3).

For the converse, assume M to be Σ-algebraically compact, and set $\aleph = \max(\aleph_0, |R|)$. It is straightforward to see that each $\aleph$-generated submodule U of M can be embedded into a pure $\aleph$-generated submodule (just close U under solutions of finite linear systems with right-hand sides in U, repeat the process $\aleph_0$ times, and take the union of the successive closures), the latter being a direct summand of M by Corollary 8. Another application of this corollary thus shows that any maximal family of independent $\aleph$-generated submodules, summing up to a pure submodule of M, must generate all of M. □

As a matter of course, Theorem 10 leads us to the following answer to the questions concerning global decompositions of modules posed in Section 2.

Theorem 13. (Gruson-Jensen, Huisgen-Zimmermann, [loc. cit.], Zimmermann [**78**]) *For a ring R, the following statements are equivalent:*

(1) *Every left R-module is a direct sum of finitely generated submodules.*

(1′) *There exists a cardinal number $\aleph$ such that every left R-module is a direct sum of $\aleph$-generated submodules.*

(2) *Every left R-module is a direct sum of indecomposable submodules.*

(3) *Every left R-module is algebraically compact.*

Proof. Clearly '(1) $\Longrightarrow$ (1′) $\Longrightarrow$ (3)' by Theorem 10. The implication '(3) $\Longrightarrow$ (2)' follows from Proposition 9.

To derive '(2) $\Longrightarrow$ (3)' from Theorem 10, keep in mind that each left R-module M can be embedded as a pure submodule into an algebraically compact module N. By (2) all direct products N^I are direct sums of indecomposable modules, all of which have local endomorphism rings by Proposition 9. Thus N is Σ-algebraically compact by Theorem 10, and so is M by Corollary 8.

'(3) $\Longrightarrow$ (1)'. By (3), all pure inclusions of left R-modules split, and therefore all left R-modules are pure projective. As we mentioned before, due to Warfield [**72**], this means that every module M is a direct summand of a direct sum of finitely presented modules U_i, say $M \oplus M' = \bigoplus_{i \in I} U_i$. On the other hand, we already know that (3) implies decomposability of all left R-modules into submodules with local endomorphism rings, and hence the Krull-Remak-Schmidt-Azumaya Theorem, applied to $M \oplus M'$, yields the required decomposition property for M. □

In particular, we retrieve the following result of Fuller [**25**]: Condition (1) of Theorem 13 is equivalent to

(4) Every left R-module M has a direct sum decomposition $M = \bigoplus_{i \in I} M_i$ which complements direct summands (i.e., given any direct summand N of M, there exists a subset $I' \subseteq I$ such that $M = N \oplus \bigoplus_{i \in I'} M_i$).

Due to the fact that every module can be purely embedded into an algebraically compact (= pure injective) one – go back to the first of the examples in Section 1 – one can build a homology theory based on pure injective resolutions, in analogy with the traditional homology theories. (Alternatively, one can use the fact that every module is an epimorphic image of a pure projective module under an epimorphism with pure kernel and consider pure projective resolutions.) In particular, it makes sense to speak of the pure injective and pure projective dimensions of a module, and to define the left pure global dimension of a ring R to be the supremum of the pure injective dimensions of its left modules. By playing off the two arguments of the Hom-functor against each other as in the case of the traditional global dimensions, one observes that the left pure global dimension of a ring equals the supremum of the pure projective dimensions of its left modules.

Thus the rings pushed into the limelight by Theorem 10 are precisely the ones for which the left pure global dimension is zero. Since they can be equivalently described by the requirement that all pure inclusions of left modules split, they are also referred to as the *left pure semisimple* rings. We already know from Corollary 12 that they are necessarily left artinian, but this of course does not tell much of the story.

5. Rings of vanishing left pure global dimension

On the negative side, the rings of the title are still not completely understood. Ironically, however, this fact also has an upside: Namely, as a host of inconclusive arguments on this theme appeared in circulation, numerous interesting insights resulted. The purpose of this section is to describe a representative selection of such insights and to delineate the *status quo* for further work on the subject.

The first milestone along the way was the recognition that vanishing of the pure global dimension on both sides takes us to a class of thoroughly studied rings. In fact, the solution to our problems in this left-right symmetric situation closely parallels the outcome in the commutative case.

Theorem 14. (Auslander [**2**], Ringel-Tachikawa [**62**], Fuller-Reiten [**26**]) *A ring R has finite representation type if and only if the left* and *right pure global dimensions of R are zero.*

The implication 'only if' was shown independently by Auslander and Ringel-Tachikawa in [**2**] and [**26**]; the following easy argument is due to Zimmermann [unpublished]. The converse was first established by Fuller-Reiten; we will re-obtain it as a consequence of the versatile Tool 17 at the end of this section.

Proof of 'only if'. Suppose that R has finite representation type, and set $M = \bigoplus_{1\le i\le n} M_i$, where $M_1, \dots, M_n$ represent the isomorphism types of the indecomposable finitely generated left R-modules. The ring R being left artinian, the module M has finite length, whence its endomorphism ring is semiprimary and so, in particular, left perfect. We denote the opposite of this endomorphism ring by S, thus turning M into a right S-module. To prove that the left pure global dimension of R is zero, it suffices to check that every left R-module A is a direct sum of copies of the M_i. For that purpose, we start by choosing a pure projective presentation of A. Since in our present situation the pure projective modules are precisely the objects in $\text{Add}(M)$ (see the remark following the definition of algebraic compactness in Section 1), this amounts to the existence of a short exact sequence

$$0 \to K \to M^{(I)} \to A \to 0, \qquad (\dagger)$$

such that K is pure in $M^{(I)}$. Our goal is to show that this sequence splits. The module $M^{(I)}$ being a direct sum of finitely generated modules with local endomorphism rings, the Crawley-Jónsson-Warfield theorem (see, e.g.,[**1**]) will then tell us that A is in turn a direct sum of copies of the M_i.

To show splitness of ($\dagger$), we observe that the pure projectivity of M guarantees the following sequence of left S-modules to be exact:

$$0 \to \text{Hom}_R(M, K) \to \text{Hom}_R(M, M^I) \to \text{Hom}_R(M, A) \to 0 \qquad (\ddagger)$$

In fact, this sequence is even pure exact. We will deduce this from the elementary fact that the pure exact sequences are precisely those short exact sequences whose exactness is preserved by all functors $\text{Hom}_S(B, -)$ with finitely presented first argument B. So let B be a finitely presented left S-module, and consider the following commutative diagram, the columns of which reflect the adjointness of Hom and tensor product; for compactness, Hom-groups are denoted by square brackets:

$$\begin{array}{ccccccccc}
0 & \longrightarrow & [B,[M,K]] & \longrightarrow & [B,[M,M^{(I)}]] & \longrightarrow & [B,[M,A]] & \longrightarrow & 0 \\
 & & \downarrow \cong & & \downarrow \cong & & \downarrow \cong & & \\
0 & \longrightarrow & [M\otimes_S B,\, K] & \longrightarrow & [M\otimes_S B,\, M^{(I)}] & \longrightarrow & [M\otimes_S B,\, A] & \longrightarrow & 0
\end{array}$$

The lower row is exact, since $M \otimes_S B$ is a finitely presented R-module, and pure projective as such. Consequently, the upper row is exact as well.

The sequence (‡) thus provides us with a pure inclusion of left S-modules,

$$0 \to \operatorname{Hom}_R(M, K) \to \operatorname{Hom}_R(M, M^{(I)}) \cong S^{(I)}.$$

Since S is left perfect, this sequence actually splits by Observations 0 of Section 1. The splitness of (†) can now be gleaned from the following commutative diagram which results from the fact that M is a generator for R-Mod and hence provides us with a functorial isomorphism $M \otimes_S \operatorname{Hom}_R(M, -) \to \operatorname{id}_{R\text{-Mod}}$:

$$\begin{array}{ccccccccc}
0 & \longrightarrow & M \otimes_S [M, K] & \longrightarrow & M \otimes_S [M, M^{(I)}] & \longrightarrow & M \otimes_S [M, A] & \longrightarrow & 0 \\
 & & \cong \downarrow & & \cong \downarrow & & \downarrow \cong & & \\
0 & \longrightarrow & K & \longrightarrow & M^{(I)} & \longrightarrow & A & \longrightarrow & 0
\end{array}$$

Indeed, splitness of the upper row yields splitness of the lower. □

Still open: The pure semisimplicity problem. *Is every ring with vanishing one-sided pure global dimension of finite representation type?*

In view of the preceding theorem, the question can be rephrased as to whether pure semisimplicity is a left-right symmetric property. In the sequel, we list a number of partial results which resolve the problem in the positive for classes of rings exhibiting some – even if faint – 'commutativity symptoms'. Subsequently, we will see that in general the left pure semisimple rings at least come very close to having finite representation type, in a sense to be made precise.

The first statement of the next theorem is due to Auslander [**3**], while the second strengthened version was proved by Herzog [**34**]. The third assertion was first obtained by Herzog and later derived by Schmidmeier from a more general duality principle (see [**34**] and [**63**]).

Theorem 15. *The answer to the pure semisimplicity question is 'yes' within the following classes of rings:*

- *Artin algebras.*
- *More generally, rings with self-duality.*
- *P.I. rings.*

We will sketch an elementary proof for a slight generalization of the first assertion. It relates left-right symmetry of pure semisimplicity directly to the existence of almost split maps. Namely, as the author showed in [**36**], the following is true:

- If R-mod has left almost split maps, then vanishing of the left pure global dimension of R implies finite representation type.

Proof of the preceding statement. The following argument is inspired by [**4**], where the notions of a preprojective/preinjective partition are introduced. Suppose that R has left pure global dimension zero. By Corollary 12, this forces R to be left artinian. Moreover, suppose that each indecomposable object A in R-mod is the source of a left almost split map, i.e., of a nonsplit monomorphism $\phi : A \to B$ such that each homomorphism from A to another object in R-mod, which is not a split monomorphism, factors through ϕ. Our proof for representation-finiteness of R uses Tool 17 below, as well as the following well-known duality discovered by Hullinger [**41**] and Simson [**65**]: If E is the minimal injective cogenerator for R-Mod and T the opposite of the endomorphism ring of E, then T is twosided artinian, again has left pure global dimension zero, and the functor $\mathrm{Hom}_R(-, E)$ induces a Morita duality R-mod $\to$ mod-T. As in the proof of Corollary 18, we exploit the descending chain condition for finite matrix subgroups, satisfied globally in T-mod, to obtain the ascending chain condition for finite matrix subgroups in arbitrary right T-modules. The remark at the end of Section 3 now shows that each direct sum of finitely generated right T-modules is in fact noetherian over its endomorphism ring.

Let $\mathfrak{D}$ be a transversal of the finitely generated indecomposable right T-modules. In a first step we will establish a *strong preprojective partition* on $\mathfrak{D}$, namely an ordinal-indexed partition $\mathfrak{D} = \bigcup_\alpha \mathfrak{D}_\alpha$ of $\mathfrak{D}$ into pairwise disjoint finite subsets $\mathfrak{D}_\alpha$ such that each $\mathfrak{D}_\alpha$ is a minimal generating set for $\mathfrak{D} \setminus \bigcup_{\beta<\alpha} \mathfrak{D}_\beta$ and such that, for each α, $\mathfrak{D}_\alpha$ consists precisely of those objects in $\mathfrak{D} \setminus \bigcup_{\beta<\alpha} \mathfrak{D}_\beta$ for which every epimorphism $\bigoplus_{finite} D_i \to D$ with $D_i \in \mathfrak{D} \setminus \bigcup_{\beta<\alpha} \mathfrak{D}_\beta$ splits. This claim will follow from an obvious transfinite induction if we can show that every nonempty subset $\mathfrak{D}' \subseteq \mathfrak{D}$ contains a finite generating set, and that any minimal such generating set $\mathfrak{D}_0$ consists precisely of those $D \in \mathfrak{D}'$ which have the property that all epimorphisms from $\mathrm{add}\,\mathfrak{D}'$ onto D split. For simplicity, we denote the subset $\mathfrak{D}'$ again by $\mathfrak{D}$. Let $M = \bigoplus_{i \in I} M_i$ be the direct sum of the objects in $\mathfrak{D}$, and S the endomorphism ring of M. Since M is noetherian over S, we obtain a finite subset $I' \subseteq I$ such that $M = S\big(\bigoplus_{i \in I'} M_i\big)$; this shows that the set $\{M_i \mid i \in I'\}$ generates $\mathfrak{D}$. Let $\mathfrak{D}_0 \subseteq \mathfrak{D}$ be a minimal finite generating set for $\mathfrak{D}$. Then, clearly, each object $D \in \mathfrak{D}$, with the property that arbitrary epimorphisms from $\mathrm{add}\,\mathfrak{D}$ onto D split, belongs to $\mathfrak{D}_0$. For the converse, let $D \in \mathfrak{D}_0$ and $f : X \to D$ be an epimorphism with $X \in \mathrm{add}\,\mathfrak{D}$. Moreover choose an epimorphism $g : D^l \oplus \bigoplus_{1 \le i \le m} D_i \to X$, where the D_i are objects in $\mathfrak{D}_0 \setminus \{D\}$. Note that splitting of $h = fg$ will imply splitting of f. In order to see that h splits, denote the endomorphism ring of D by $S(D)$, and let $h_1, \dots, h_l$ be the restrictions of h to the various copies of D occurring as summands of the domain of h. Assume that all of the h_i are non-isomorphisms. The ring $S(D)$ being local, this means that the h_i belong to the radical $J(D)$ of $S(D)$. Let $D' \subseteq D$ be the trace of $\bigoplus_{1 \le i \le m} D_i$ in D. The fact that $J(D)D$ is superfluous in D (keep in mind that D is noetherian over $S(D)$), applied to the equality $D = h_1 D + \cdots + h_l D + D'$ yields $D = D'$. But this means that $\mathfrak{D}_0 \setminus \{D\}$ is still a generating set for $\mathfrak{D}$, a contradiction to our minimal choice of $\mathfrak{D}_0$. Hence one of the h_i is an isomorphism, and consequently h splits.

Next we apply the duality $\mathrm{Hom}_R(-, E)$ to $\mathfrak{D}$, to obtain a *strong preinjective partition* of the transversal $\mathfrak{C} = \mathrm{Hom}_R(\mathfrak{D}, E)$ of the indecomposable objects in R-mod. In other words, we obtain a partition $\mathfrak{C} = \bigcup_\alpha \mathfrak{C}_\alpha$, such that each $\mathfrak{C}_\alpha$ is a minimal finite cogenerating set for $\mathfrak{C} \setminus \bigcup_{\beta<\alpha} \mathfrak{C}_\beta$, where the $\mathfrak{C}_\alpha$ are pinned down by the following additional property (†): Namely, each $\mathfrak{C}_\alpha$ consists precisely of those objects C in $\mathfrak{C}\setminus\bigcup_{\beta<\alpha} \mathfrak{C}_\beta$ for which all monomorphisms from C to $\mathrm{add}\big(\mathfrak{C}\setminus\bigcup_{\beta<\alpha} \mathfrak{C}_\beta\big)$ split.

The smallest ordinal τ such that $\mathfrak{C}_\tau = \varnothing$ is called the length of the partition; clearly $\mathfrak{C}_\beta = \varnothing$ for all $\beta > \tau$. Observe that τ is a successor ordinal. Indeed, if $S_1, \ldots, S_n$ are the simple modules in $\mathfrak{C}$ – say $S_i \in \mathfrak{C}_{\beta_i}$ – then $\beta = 1 + \max(\beta_1, \ldots, \beta_n)$ is the length of the partition. This is due to the fact that each object $C \in \mathfrak{C}_\beta$ is non-simple indecomposable and hence gives rise to a non-split embedding of one of the S_i into C; but such a nonsplit monomorphism is incompatible with property (†) of $\mathfrak{C}_{\beta_i}$.

Finally, we show that τ is bounded above by the first infinite ordinal number ω. In light of the preceding paragraph, this will yield finiteness of τ and thus finiteness of $\mathfrak{C}$, i.e., finitenenss of the representation type of R. Suppose to the contrary that there exists an object $A \in \mathfrak{C}_\omega$, and let $\phi : A \to B$ be a left almost split monomorphism with $B \in R$-mod. Moreover, for $\alpha < \omega$, let B^α be the reject of $\mathfrak{C}_\alpha$ in B, and observe that $B^\beta \subseteq B^\alpha$ whenever $\beta < \alpha < \omega$. Due to the finite length of B, the resulting chain of rejects becomes stationary, say at $\gamma < \omega$. This shows that B/B^γ is cogenerated by $\mathfrak{C}_\alpha$ for all ordinal numbers α between γ and ω, which places a copy of B/B^γ into $\mathrm{add}\big(\bigcup_{\beta\geq\omega} \mathfrak{C}_\beta\big)$. On the other hand, since each of the $\mathfrak{C}_\alpha$ for $\alpha < \omega$ cogenerates A – recall that $A \in \mathfrak{C}_\omega$ – we obtain a family of non-split monomorphisms $f_\alpha : A \to C_\alpha$, where $C_\alpha \in \mathrm{add}\,\mathfrak{C}_\alpha$ and α again runs through the ordinals between γ and ω. Now left almost splitness of ϕ permits us to factor all of these maps through ϕ, say $f_\alpha = g_\alpha \phi$ for a suitable homomorphism $g_\alpha : B \to C_\alpha$. By the choice of γ, the reject B^γ is contained in the kernel of each g_α, which shows that the homomorphism $\overline{\phi} : A \to B/B^\gamma$ obtained by composing ϕ with the canonical map $B \to B/B^\gamma$ is still a monomorphism. It is in turn nonsplit because ϕ does not split. But in view of the above placement of B/B^γ relative to our preprojective partition, this contradicts property (†) of $\mathfrak{C}_\omega$. Hence the assumption that $\mathfrak{C}_\omega$ be nonempty is absurd and our argument is complete. □

In the meantime, Simson has replaced the pure semisimplicity problem by a conjecture, to the effect that the answer is negative (see [**68**, **69**]). His key to a potential class of counterexamples is the following:

Connection with a strong Artin problem for division rings. *The answer to the pure semisimplicity question is positive if and only if the following is true:*

For every simple D-E-bimodule M, where D and E are division rings, such that $\dim({}_D M) < \infty$ *and* $\dim(M_E) = \infty$, *there exists a non-finitely generated* *indecomposable left module over the triangular matrix ring*

$$\begin{pmatrix} D & M \\ 0 & E \end{pmatrix}.$$

For a proof see [**68**].

On the other hand, the next theorem guarantees the rings of vanishing left pure global dimension to at least have a very sparse supply of indecomposable finitely generated modules. The result was independently proved by Prest [**58**] and Zimmermann and the author [**39**]. In particular, it relates the pure semisimplicity conjecture to the quest for a better understanding of those left artinian rings of infinite representation type which fail to satisfy the conclusion of the second Brauer-Thrall Conjecture. Recall that for a left artinian ring R of infinite cardinality, the latter postulates the following: If R has infinite representation type, then there exist infinitely many distinct positive integers d_n such that, for each n, there are infinitely many isomorphism classes of indecomposable left R-modules of composition length d_n. While this conjecture has long been confirmed for finite dimensional algebras over algebraically closed fields, in [**60**] Ringel constructed a class of artinian rings with infinite center which violate this implication.

Theorem 16. *Suppose that R has left pure global dimension zero. Then:*

- *For each $d \in \mathbb{N}$, there exist only finitely many left R-modules of length d, up to isomorphism.*
- *For each $d \in \mathbb{N}$, there exist only finitely many length-d modules in* fin.pres-R, *up to isomorphism.*

The obstacle one meets in trying to strengthen the latter assertion for <u>right</u> R-modules to the level of that for the left lies in the fact that it is not known whether left pure semisimple rings are necessarily right artinian. In fact, it is known that a positive answer to this question would resolve the pure semisimplicity problem in the positive.

The following duality for finite matrix subgroups, proved in [**39**] and – in model-theoretic terms – also in [**58**], is one of the pivotal tools in proving Theorem 16. We include it here because it yields some by-products of independent interest.

Tool 17. *If M is an R-S-bimodule and C an injective cogenerator for Mod-S, then the following lattices are anti-isomorphic:*

- *the lattice of finite matrix subgroups of the left R-module M*

and

- *the lattice of finite matrix subgroups of the right R-module $\mathrm{Hom}_S(M,C)_R$.*

Sketch of proof. Keep in mind that the finite matrix subgroups of ${}_RM$ are precisely the kernels of the maps $M \to Q \otimes_R M$, $m \mapsto q \otimes m$ with Q finitely presented and $q \in Q$, and that they can alternately be given the form $\mathrm{Hom}_R(P,M)(p)$ with P finitely presented and $p \in P$ (see Observations 3(1),(2) – a proof can be found in [**39**, Lemma 1]).

First one checks that, given a finite matrix subgroup $U = \mathrm{Hom}_R(P, M)(p)$ of $_RM$ with P and p as above, the C-dual $\mathrm{Hom}_S(M/U, C)$ is a finite matrix subgroup of the right R-module $M^+ = \mathrm{Hom}_S(M, C)$. From the exactness of the sequence

$$\mathrm{Hom}_R(P, M) \xrightarrow{\psi} M \xrightarrow{\text{canon}} M/U \to 0,$$

where ψ is evaluation at p, one obtains exactness of the sequence of induced maps

$$0 \to \mathrm{Hom}_S(M/U, C) \to \mathrm{Hom}_S(M, C) \xrightarrow{\psi^*} \mathrm{Hom}_S(\mathrm{Hom}_R(P, M), C).$$

Using the canonical isomorphism $\tau : \mathrm{Hom}_S(M, C) \otimes_R P \to \mathrm{Hom}_S(\mathrm{Hom}_R(P, M), C)$ and the fact that the map $\tau^{-1}\psi^*$ sends $f \in \mathrm{Hom}_S(M, C)$ to $f \otimes p$, one deduces the first claim. Clearly, this finite matrix subgroup of M^+ coincides with the annihilator $\mathrm{Ann}_{M^+}(U)$.

Similarly, one verifies that, for any finite matrix subgroup V of the right R-module M^+, the annihilator $\mathrm{Ann}_M(V)$ is a finite matrix subgroup of the left R-module M.

It is now a question of routine to check that the lattice anti-homomorphisms $U \mapsto \mathrm{Ann}_{M^+}(U)$ and $V \mapsto \mathrm{Ann}_M(V)$ are inverse to each other. □

It was shown by Crawley-Boevey that, among the finite dimensional algebras over an algebraically closed field, the ones of finite representation type are characterized by the nonexistence of generic modules, i.e., of non-finitely generated indecomposable endofinite modules. (As the term suggests, a module is *endofinite* if it has finite length over its endomorphism ring.) The following consequence of the above duality shows that the absence of generic objects is offset by the richest possible supply of endofinite modules in the representation-finite case.

Corollary 18. ([**58**, **39**]) *For any ring R, the following statements are equivalent:*
(1) *All objects in R-*Mod *are endofinite.*
(2) *Same as* (1) *for* Mod-R.
(3) *The left and right pure global dimensions of R vanish.*

Proof. It clearly suffices to show the equivalence of (1) and (3). If (1) holds, Tool 17, applied to R and the opposites of the endomorphism rings of the modules considered, yields the descending chain condition for finite matrix subgroups in all *right* R-modules. That the left modules satisfy this chain condition is immediate from our hypothesis. By Theorem 6, this shows that all R-modules, left and right, are algebraically compact, i.e. (3) holds.

Now assume (3). In view of the descending chain condition for matrix subgroups, satisfied globally for left and right R-modules, Tool 17 provides us with the ascending chain condition for finite matrix subgroups as well. Let M be any left R-module and S its endomorphism ring. We know that all finitely generated S-submodules of M are matrix subgroups, so if we can show that all matrix subgroups of $_RM$ arise from finite matrices, we are done. But this finiteness condition follows easily from our hypothesis as follows: Indeed, let $[\mathcal{A}, \alpha]$ be any matrix functor on R-Mod, where $\mathcal{A} = (a_{ij})_{i\in I, j\in J}$. Moreover, for any finite subset I' of I, let $\mathcal{A}(I')$ be the matrix consisting of those rows of $\mathcal{A}$ which are indexed by I'. Since $\mathcal{A}$ is row-finite, the matrix $\mathcal{A}(I')$ is actually finite in effect. Now let I_1 be any finite subset of I and note that the finite matrix subgroup $[\mathcal{A}(I_1), \alpha]M$ of M contains $[\mathcal{A}, \alpha]M$. If the inclusion is proper, there exists a finite subset $I_2 \subseteq I$ containing I_1 such that $[\mathcal{A}(I_2), \alpha]M$ is properly contained in $[\mathcal{A}(I_1), \alpha]M$. If $[\mathcal{A}, \alpha]M$ is still strictly contained in $[\mathcal{A}(I_2), \alpha]M$, we repeat the process. Eventually, it has to terminate by our chain condition, which shows that $[\mathcal{A}, \alpha]M$ equals some finite matrix subgroup $[\mathcal{A}(I_m), \alpha]M$, and our argument is complete. □

We conclude this section by looping back to its beginning and filling in the proof of the as yet unjustified implication of Theorem 14. We prepare with the following proposition (not needed in full strength here) which generalizes Crawley-Boevey's observation that an endofinite direct sum of modules with local endomorphism rings involves only finitely many isomorphism types of summands [**17**, Proposition 4.5].

Proposition 19. *Suppose that $(M_i)_{i\in I}$ is a family of left R-modules with local endomorphism rings. Let $M = \bigoplus_{i\in I} M_i$, and denote by S the endomorphism ring of M, by $J(S)$ the Jacobson radical of S.*

If there exists a natural number N with $J(S)^N M = 0$ such that, moreover, $J(S)^k M$ is finitely generated over S for all $0 \le k \le N$, then the family $(M_i)_{i\in I}$ involves only finitely many isomorphism classes.

Proof. Suppose N is as in the claim, and start by noting that $\mathrm{Hom}_R(M_i, M_j) \subseteq J(S)$, whenever $M_i \not\cong M_j$; this is due to the locality of the endomorphism rings of the M_i. Next observe that, for any finite sequence $x_1, \dots, x_r$ in M, we have

$$\sum_{i=1}^{r} Sx_i \subseteq \bigoplus_{i\in I'} M_i + \sum_{i=1}^{r} J(S)x_i,$$

where $I' \subseteq I$ is the 'closure of the supports of the x_l, $1 \le l \le r$, under isomorphism'; by this we mean that

$$I' = \{i \in I \mid M_i \cong M_j \text{ for some } j \in \bigcup_{1\le l\le r} \mathrm{supp}(x_l)\}.$$

Now pick generators $x_{k1}, \dots, x_{kr_k}$ for each of the left S-modules $J(S)^k M$, where k runs from 0 to $N-1$. Moreover, for each k, let I_k be the closure of $\bigcup_{1\le l\le r_k} \mathrm{supp}(x_{kl})$ under isomorphism. Then

$$\begin{aligned} M = \sum_{l=1}^{r_0} Sx_{0l} &\subseteq \bigoplus_{i\in I_0} M_i + \sum_{l=1}^{r_0} J(S)x_{0l} \subseteq \bigoplus_{i\in I_0} M_i + \sum_{l=1}^{r_1} Sx_{1l} \\ &\subseteq \bigoplus_{i\in I_0\cup I_1} M_i + \sum_{l=1}^{r_1} J(S)x_{1l} \subseteq \bigoplus_{i\in I_0\cup I_1} M_i + \sum_{l=1}^{r_2} Sx_{2l}, \end{aligned}$$

and since $J(S)x_{N-1,l} = 0$ for all l by hypothesis, an obvious induction yields

$$M = \bigoplus_{i \in I_0 \cup \cdots \cup I_{N-1}} M_i.$$

But the modules M_i with $i \in I_0 \cup \cdots \cup I_{N-1}$ fall into finitely many isomorphism classes by construction, and our argument is complete. □

On the side, we point out that, in Proposition 19, neither the hypothesis that $J(S)^N M = 0$ for some N, nor the condition that $J(S)^k M$ be finitely generated over S for all k, suffices to yield the conclusion: If $R = \mathbb{Z}$ and $M = \bigoplus_{p \text{ prime}} \mathbb{Z}/(p)$, then clearly $J(S) = 0$. An example to justify the second remark is as follows: Let R be the Kronecker algebra and M_n the preprojective string module of dimension $2n+1$. If we set $M = \bigoplus_{n \in \mathbb{N}} M_n$, then $J(S)^k M = \bigoplus_{n \geq k+1} M_n$ is finitely generated over S for each k.

Proof of the implication 'if' in Theorem 14. Our hypothesis being condition (3) of Corollary 18, we see that all left R-modules are endofinite. Since we already know R to be artinian (Corollary 12), we only need to show that any transversal $(M_i)_{i \in I}$ of the indecomposable objects in R-mod is finite. But this is an immediate consequence of Proposition 19. □

For the reader interested in picking up the threads that are still dangling, we list a number of additional references addressing the pure semisimplicity problem: [**65**], [**66**], [**41**], [**67**], [**42**], [**76**], [**6**], [**19**], [**43**], [**44**], [**36**], [**5**], [**33**], [**52**], [**70**].

6. Vanishing pure global dimension and product-completeness

From Theorem 10 we know that the Σ-algebraically compact modules M are precisely those with the property that all direct products of copies of M are direct sums of indecomposable components with local endomorphism rings. This naturally raises the question as to the structure of the indecomposable summands in such decompositions.

More generally, we ask: *What are the indecomposable summands of large Σ-algebraically compact direct products $\prod_{i \in I} M_i$?*

This problem is further motivated by the following points: (a) The question is intimately related to the pure semisimplicity problem. The connection first surfaced in a theorem of Auslander (Theorem 20 below), and is reinforced by results of Gruson, Garavaglia, and Krause-Saorín (also compare with Corollary 18). And (b), direct powers of non-generic Σ-algebraically compact modules provide prime hunting ground for generic objects, as evidenced by Theorems 2.11 and 2.12 (due to Krause and Ringel, respectively) in Zwara's contribution to this volume.

Slightly extending the terminology of Krause-Saorín [**51**], we call a family $(M_i)_{i \in I}$ *product-complete*, if the direct product $\prod_{i \in I} M_i$ belongs to $\mathrm{Add}(\bigoplus_{i \in I} M_i)$, the closure of $\{M_i \mid i \in I\}$ in R-Mod under formation of direct summands and direct sums. As in [**51**], a module M will be labeled product-complete if all families consisting of copies of M have this property.

According to Gruson [**29**] and Garavaglia [**27**], an indecomposable module M is product-complete if and only if it is endofinite. This equivalence was generalized by Krause-Saorín as follows: For an arbitrary module M, finite endolength is equivalent to the postulate that all direct summands of M be product-complete [**51**, Theorem 4.1].

This yields the following hierarchy within the class of algebraically compact modules:

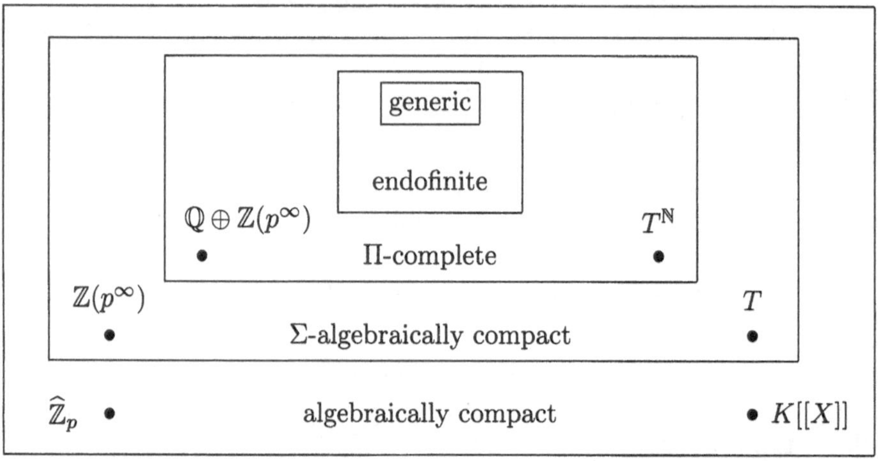

Each of the inclusions is proper: The algebra of power series $K[[X]]$ is algebraically compact without being Σ-algebraically compact; and the same is true for the ring $\widehat{\mathbb{Z}}_p$ of p-adic integers, viewed either as $\mathbb{Z}$- or $\widehat{\mathbb{Z}}_p$-module. The ring $T = K[X_i \mid i \in \mathbb{N}]/(X_i \mid i \in \mathbb{N})^m$, where K is a field and m some integer ≥ 2, is Σ-algebraically compact (c.f. examples following Theorem 6), but not product-complete; indeed, $T^{\mathbb{N}}$ is not projective, as T fails to be coherent. On the other hand, Schulz proved the non-projective summands in arbitrary direct products of copies of T to be all isomorphic to the unique simple T-module (see [**64**]), whence $T^{\mathbb{N}}$ is product-complete without being endofinite. Furthermore, the direct sum of the group of rational numbers and the Prüfer group $\mathbb{Z}(p^\infty)$ for some prime p is a product-complete abelian group, but fails to be endofinite. Over any Artin algebra R, finally, all objects in R-mod are endofinite without being generic.

As is to be expected in light of our previous discussion, global product-completeness of families of R-modules occurs only rarely. More precisely, we have:

Theorem 20. (Auslander [**3**]) *For any ring R, the following statements are equivalent:*

(1) *R has finite representation type.*

(2) *All families of finitely generated indecomposable left R-modules are product-complete.*

We give a proof relying only on the results established in the previous sections. Note that the first part of our argument is very similar to the reasoning of Krause-Saorín ([**51**, 3.8]).

Proof. '(1) $\Longrightarrow$ (2)'. Given (1), we know from Theorem 14 that the left pure global dimension of R is zero. Let $(M_i)_{i\in I}$ be a family of indecomposable objects in R-mod. To see that each indecomposable direct summand of $\prod_{i\in I} M_i$ is isomorphic to one of the M_j, it suffices to show this for the case where all M_i are pairwise isomorphic since, by hypothesis, our family contains only finitely many isomorphism types. So suppose $M_i \cong M$ for all i. Denote the endomorphism ring of M by S. From Corollary 18, we know that M has finite length over S, and hence S is left artinian: Indeed, letting $m_1, \ldots, m_n$ be a generating set for M over R, we see that S embeds into M^n as a left S-module, via $s \mapsto (sm_k)$. In particular, this implies that S^I is projective as a right S-module. The ring S being local, this means that S^I is free, say $S^I \cong S^{(J)}$. Using the fact that the tensor functor $- \otimes_S M$ commutes with direct products in our setting, we see that $M^I \cong S^I \otimes_S M \cong S^{(J)} \otimes M \cong M^{(J)}$, which yields (2).

For the converse, assume that (2) is true, and let $(M_i)_{i\in I}$ be a transversal of the isomorphism types of the indecomposable objects in R-mod. We will obtain finiteness of I by showing that the direct product $\prod_{i\in I} M_i$ equals the direct sum $\bigoplus_{i\in I} M_i$. We start by observing that $\bigoplus_{i\in I} M_i$ is algebraically compact: Indeed, all direct powers of $\prod_{i\in I} M_i$ are direct sums of copies of the M_i by hypothesis, whence $\prod_{i\in I} M_i$ is Σ-algebraically compact by Theorem 10; but by Theorem 6, this is tantamount to Σ-algebraic compactness of $\bigoplus_{i\in I} M_i$. In particular, the pure inclusion $\bigoplus_{i\in I} M_i \subseteq \prod_{i\in I} M_i$ splits, say $\prod_{i\in I} M_i = \bigoplus_{i\in I} M_i \oplus N$, and N is in turn a direct sum of copies of M_i's. If N were nonzero, we could thus find a direct summand isomorphic to some M_k in N. On the other hand, we may cancel M_k from the above product-sum equality to obtain $\prod_{i\in I\setminus\{k\}} M_i \cong \bigoplus_{i\in I\setminus\{k\}} M_i \oplus N$ (keep in mind that $\mathrm{End}(M_k)$ is local). This makes M_k a direct summand of $\prod_{i\in I\setminus\{k\}} M_i$, thus contradicting our hypothesis. We conclude $N = 0$, which forces I to be finite as required. $\square$

The following result, due to Krause and Saorín [**51**, Proposition 4.2], characterizes the product-complete modules in terms of their matrix subgroups. It continues the line of Theorem 10, where we related direct sum decompositions of direct products M^I to finiteness conditions on the lattice of matrix subgroups of M.

Proposition 21. *An object $M \in R$-*Mod *is product-complete if and only if M has the descending chain condition for (finite) matrix subgroups and all (finite) matrix subgroups of M are finitely generated over the endomorphism ring of M.*

We include an elementary proof for one implication.

Proof of 'only if'. Suppose M is product-complete. Then, clearly, M is Σ-algebraically compact by Theorem 10, which is tantamount to M satisfying the descending chain condition for matrix subgroups by Theorem 6. To see that any matrix subgroup $[\mathcal{A},\alpha]M$ is finitely generated over the endomorphism ring S of M, recall that

$$[\mathcal{A},\alpha]M = \mathrm{Hom}_R(M^I, M)(\underline{m})$$

for some set I and some element $\underline{m} \in M^I$ (this was explained after the statement of Theorem 5). By hypothesis, M^I is a direct summand of a suitable direct sum $M^{(J)}$, and hence $[\mathcal{A},\alpha]M = \mathrm{Hom}_R(M^{(J)}, M)(m)$ for some $m = (m_j) \in M^{(J')}$, where J' is a finite subset of J. This gives $[\mathcal{A},\alpha]M = \sum_{j\in J'} Sm_j$ as required. □

In view of the final remark of Section 3A, a ring R is right coherent if and only if all finite matrix subgroups of the regular left module R are finitely generated as right ideals. We can thus supplement Corollary 12 to retrieve Chase's characterization of the rings whose left projective modules are closed under direct products.

Corollary 22. (Chase [**12**]) *R has the property that all direct products of projective left R-modules are again projective if and only if R is left perfect and right coherent.*

The following question appears of significantly lower importance than the one with which we opened the section. However, the fact that it is not yet answered shows the lacunary state of our present understanding of direct sum decompositions of large direct products.

Given a Σ-algebraically compact module M, is $M^{\mathbb{N}}$ product-complete?

As was already observed by Krause and Saorín [**51**], there is *some* power M^I which is product-complete; indeed, the fact that all powers of M split into summands of cardinalities bounded above by $\max(|R|, \aleph_0)$ guarantees that, eventually, saturation with respect to the appearance of new direct summands is reached.

7. Concluding remarks on pure global dimension

From the previous section we know that a ring R has finite representation type if and only if its left and right pure global dimensions are zero. Moreover, by Theorem 15, the latter condition is left-right symmetric for Artin algebras. What can one say about the pure global dimensions of R when they do not vanish?

The question of how these invariants relate to other properties of the ring and its module categories is largely open. Some general facts of interest are available, however, as well as some classes of algebras where the connection is understood. We include only a few results to trigger interest in further investigation of the problem.

A very rough, but not unreasonable, answer to the above question is this: "That depends on the cardinality of R." As a first step in justifying this response, we present an insight due to Gruson and Jensen [**32**]; for an alternate approach, see [**47**].

Theorem 23. *The following implications hold for any ring R.*

(1) *If R is countable, but not of finite representation type, then the left (right) pure global dimension of R equals 1.*

(2) *If the cardinality of R is bounded from above by $\aleph_t$ for some integer $t \geq 0$, the pure global dimensions of R are at most $t+1$.*

One can do far better for specialized classes of rings, however. For example, as was observed by Kiełpiński and Simson [**47**], if $R = S[X]$ is a polynomial ring over a commutative ring S of cardinality $\aleph_s$ in a set X of indeterminates which has cardinality $\aleph_t$ (with $s, t \geq 0$), then the pure injective dimension of R equals $\max(s,t)+1$. Our final result, due to D. Baer, Brune, and Lenzing [**7**], presents a smooth picture of pure homology for hereditary algebras over algebraically closed base fields subject to certain cardinality restrictions. In that case, the pure global dimension mirrors the representation type of R as follows:

Theorem 24. *Suppose that R is a hereditary finite dimensional algebra over an algebraically closed field of cardinality $\aleph_t$, where $t \geq 2$ is an integer. Then:*

(1) *The left (right) pure global dimension of R equals 2 if and only if R has tame, but infinite, representation type.*

(2) *The left (right) pure global dimension of R equals $t+1$ if and only if R has wild representation type.*

Finally, we refer the reader to [**56**], [**8**] and [**7**] for further instances in which the pure global dimension is understood. Motivated by the lectures of Benson at the 1998 conference in Bielefeld, we conclude with the following

Problem. *Given a finite group G and a field K of suitable cardinality, how does the pure global dimension of the group algebra KG reflect the representation type?*

References

1. F.W. Anderson and K.R. Fuller, *Rings and Categories of Modules*, Graduate Texts in Math. 13, Springer-Verlag, New York-Heidelberg, 1974; Second Edition, Springer-Verlag, New York, 1992.
2. M. Auslander, *Representation theory of Artin Algebras II*, Communic. in Algebra **1** (1974), 293-310.
3. ———, *Large modules over Artin algebras*, in Algebra, topology and categories, Academic Press, New York, 1976, pp. 1-17.
4. M.Auslander and S.O. Smalø, *Preprojective modules over Artin algebras*, J. Algebra **66** (1980), 61-122.
5. G. Azumaya, *Countable generatedness version of rings of pure global dimension zero*, in: London Math. Soc. Lecture Notes Series, vol. 168, Cambridge University Press, 1992, pp. 43-79.
6. G. Azumaya and A. Facchini, *Rings of pure global dimension zero and Mittag-Leffler modules*, J. Pure Appl. Algebra **62** (1989), 109-122.
7. D. Baer, H. Brune and H. Lenzing, *A homological approach to representations of algebraas II: tame hereditary algebras*, J. Pure and Appl. Algebra **26** (1982), 141-153.
8. D. Baer and H. Lenzing, *A homological approach to representations of algebras I: the wild case*, J. Pure and Appl. Algebra **24** (1982), 227-233.
9. S. Balcerzyk, *On the algebraically compact groups of I. Kaplansky*, Fund. Math. **44** (1957), 91-93.
10. D.J. Benson, *Infinite dimensional modules for finite groups*, this volume.
11. M.C.R. Butler and G. Horrocks, *Classes of extensions and resolutions*, Philos. Trans. Roy. Soc., London **264** (1961), 155-222.
12. S.U. Chase, *Direct products of modules*, Trans. Amer. Math. Soc. **97** (1960), 457-473.
13. I.S. Cohen and I. Kaplansky, *Rings for which every module is a direct sum of cyclic modules*, Math. Zeitschr. **54** (1951), 97-101.
14. P.M. Cohn, *On the free product of associative rings*, Math. Zeitschr. **71** (1959), 380-398.
15. P. Crawley and B. Jónnson, *Refinements for infinite direct decompositions of algebraic systems*, Pacific J. Math. **14** (1964), 797-855.
16. W. W. Crawley-Boevey, *Tame algebras and generic modules*, Proc. London Math. Soc. **63** (1991), 241-264.
17. ———, *Modules of finite length over their endomorphism ring*, in Representations of Algebras and Related Topics (S. Brenner and H. Tachikawa, Eds., eds.), London Math. Soc. Lec. Note Series 168, Cambridge Univ. Press, Cambridge, 1992, pp. 127-184.
18. D. Eisenbud and P. Griffith, *The structure of serial rings*, Pacific J. Math. **36** (1971), 109-121.
19. A. Facchini, *Anelli di tipo di rappresentazione finito, di dimensione pura globale zero, e moduli di Mittag-Leffler*, Rend. Sem. Mat. Fis. Milano **59** (1989), 65-80.
20. A. Facchini, *Mittag-Leffler modules, reduced products and direct products*, Rend. Sem. Mat. Univ. Padova **85** (1991), 119-132.
21. C. Faith, *Rings with ascending condition on annihilators*, Nagoya Math. J. **27** (1966), 179-191.
22. D. J. Fieldhouse, *Aspects of purity*, in Ring Theory, Proc. Conf. Univ. Oklahoma 1973, Lecture Notes in Pure and Applied Math. 7, Dekker, New York, 1974, pp. 185-196.

23. L. Fuchs, *Algebraically compact modules over Noetherian rings*, Indian J. Math. **9** (1967), 357-374.

24. L. Fuchs, *Infinite Abelian Groups*, Academic Press, New York and London, 1970.

25. K.R. Fuller, *On rings whose left modules are direct sums of finitely generated modules*, Proc. Amer. Math. Soc. **54** (1976), 39-44.

26. K.R. Fuller and I. Reiten, *Note on rings of finite representation type and decompositions of modules*, Proc. Amer. Math. Soc. **50** (1975), 92-94.

27. S. Garavaglia, *Dimension and rank in the theory of modules*, Preprint, 1979.

28. P.A. Griffith, *On the decomposition of modules and generalized left uniserial rings*, Math. Ann. **184** (1970), 300-308.

29. L. Gruson, *Simple coherent functors*, in Representations of Algebras, Lecture Notes in Math. 488, Springer-Verlag, 1975, pp. 156-159.

30. L. Gruson and C.U. Jensen, *Modules algébriquement compacts et foncteurs* $\varprojlim^{(i)}$, C. R. Acad. Sci. Paris, Sér. A **276** (1973), 1651-1653.

31. ———, *Deux applications de la notion de L-dimension*, C. R. Acad. Sci. Paris, Sér. A **282** (1976), 23-24.

32. ———Dimensions cohomologiques reliées aux foncteurs $\varprojlim^{(i)}$, in Sém. d'Algèbre P. Dubreil et M.-P. Malliavin, Lecture Notes in Math. 867, Springer-Verlag, Berlin, 1981, pp. 234-249.

33. I. Herzog, *Elementary duality for modules*, Trans. Amer. Math. Soc. **340** (1993), 37-69.

34. ———, *A test for finite representation type*, J. Pure Appl. Algebra **95** (1994), 151-182.

35. B. Huisgen-Zimmermann, *Rings whose right modules are direct sums of indecomposable modules*, Proc. Amer. Math. Soc. **77** (1979), 191-197.

36. ———, *Strong preinjective partitions and representation type of artinian rings*, Proc. Amer. Math. Soc. **109** (1990), 309-322.

37. B. Huisgen-Zimmermann and W. Zimmermann, *Algebraically compact rings and modules*, Math. Zeitschr. **161** (1978), 81-93.

38. ———, *Classes of modules with the exchange property*, J. Algebra **88** (1984), 416-434.

39. ———, *On the sparsity of representations of rings of pure global dimension zero*, Trans. Amer. Math. Soc. **320** (1990), 695-711.

40. ———, *On the abundance of* $\aleph_1$*-separable modules*, in Abelian Groups and Noncommutative Rings, A Collection of Papers in Memory of Robert B. Warfield, Jr. (L. Fuchs, K.R. Goodearl, J.T. Stafford, and C. Vinsonhaler, Eds.), Contemp. Math. **130** (1992), 167-180.

41. H. Hullinger, *Stable equivalence and rings whose modules are a direct sum of finitely generated modules*, J. Pure Appl. Algebra **16** (1980), 265-273.

42. C.U. Jensen and H. Lenzing, *Algebraic compactness of reduced products and applications to pure global dimension*, Comm. Algebra **11** (1983), 305-325.

43. ———, *Model theoretic algebra with particular emphasis on fields, rings, modules*, Algebra, Logic and Applications, vol. 2, Gordon & Breach Science Publishers, 1989.

44. C.U. Jensen and B. Zimmermann-Huisgen, *Algebraic compactness of ultrapowers and representation type*, Pac. J. Math. **139** (1989), 251-265.

45. I. Kaplansky, *Infinite Abelian Groups*, Univ. of Michigan Press, Ann Arbor, 1954.

46. R. Kiełpiński, *On* Γ*-pure injective modules*, Bulletin de L'Académie Polonaise des Sciences, Sér. des sciences math., astr. et phys. **15** (1967), 127-131.

47. R. Kiełpiński and D. Simson, *On pure homological dimension*, Bulletin de L'Académie Polonaise des Sciences, Sér. des sciences math., astr. et phys. **23** (1975), 1-6.

48. G. Koethe, *Verallgemeinerte abelsche Gruppen mit hyperkomplexem Operatorenring*, Math. Zeitschr. **39** (1935), 31-44.

49. H. Krause, *Generic modules over Artin algebras*, Proc. London Math. Soc. **76** (1998), 276-306.

50. ———, *Finite versus infinite dimensional representations – a new definition of tameness*, this volume.

51. H. Krause and M. Saorín, *On minimal approximations of modules*, in Trends in the Representation Theory of Finite Dimensional Algebras (E.L. Green and B. Huisgen-Zimmermann, Eds.), Contemp. Math. **229** (1998), 227-236.

52. A. Laradji, *On duo rings, pure semisimplicity and finite representation type*, Communic. in Algebra **25** (1997), 3947-3952.

53. J. Łoś, *Abelian groups that are direct summands of every abelian group which contains them as pure subgroups*, Fund. Math. **44** (1957), 84-90.

54. J. M. Maranda, *On pure subgroups of abelian groups*, Arch. Math. **11** (1960), 1-13.

55. J. Mycielski, *Some compactifications of general algebras*, Colloq. Math. **13** (1964), 1-9.

56. F. Okoh, *Hereditary algebras that are not pure hereditary*, in Representation theory II, Proc. 2nd ICRA, Ottawa 1979, Lecture Notes in Math. 832, Springer-Verlag, Berlin-New York, 1980, pp. 432-437.

57. B. L. Osofsky, *Endomorphism rings of quasi-injective modules*, Canad. J. Math. **20** (1968), 895-903.

58. M. Prest, *Duality and pure semisimple rings*, J. London Math. Soc. **38** (1988), 403-409.

59. ———, *Topological and geometric aspects of the Ziegler spectrum*, this volume.

60. C.M. Ringel, *Representations of K-species and bimodules*, J. Algebra **41** (1976), 269-302.

61. ———, *A construction of endofinite modules*, Preprint.

62. C.M. Ringel and H. Tachikawa, *QF-3 rings*, J. reine angew. Math. **272** (1975), 49-72.

63. M. Schmidmeier, *The local duality for homomorphisms and an application to pure semisimple PI-rings*, Colloq. Math. **77** (1998), 121-132.

64. R. Schulz, *Reflexive modules over perfect rings*, J. Algebra **61** (1979), 527-537.

65. D. Simson, *Pure semisimple categories and rings of finite representation type*, J. Algebra **48** (1977), 290-296; *Corrigendum*, J. Algebra **67** (1980), 254-256.

66. ———, *On pure global dimension of locally finitely presented Grothendieck categories*, Fund. Math. **96** (1977), 91-116.

67. ———, *Partial Coxeter functors and right pure semisimple hereditary rings*, J. Algebra **71** (1981), 195-218.

68. ———, *On right pure semisimple hereditary rings and an Artin problem*, J. Pure Appl. Algebra **104** (1995), 313-332.

69. ———, *A class of potential counterexamples to the pure semisimplicity conjecture*, in Advances in Algebra and Model Theory (M. Droste and R. Göbel, eds.), Algebra Logic and Applications Series 9, Gordon and Breach, Amsterdam, 1997, pp. 345-373.

70. ———, *Dualities and pure semisimple rings*, in Abelian Groups, Module Theory, and Topology (Padova 1997) (D. Dikranjan and L. Salce, eds.), Lecture Notes in Pure and Appl. Math., vol. 201, Dekker, New York, 1998, pp. 381-388.

71. B. Stenström, *Pure submodules*, Arkiv för Mat. **7** (1967), 159-171.

72. R.B. Warfield, Jr., *Purity and algebraic compactness for modules*, Pac. J. Math. **28** (1969), 699-719.

73. ———, *Exchange rings and decompositions of modules*, Math. Annalen **199** (1972), 31-36.

74. ———, *Rings whose modules have nice decompositions*, Math. Zeitschr. **125** (1972), 187-192.

75. B. Weglorz, *Equationally compact algebras, I*, Fund. Math. **59** (1966), 289-298.

76. M. Zayed, *Indecomposable modules over right pure semisimple rings*, Monatsh. Math. **105** (1988), 165-170.

77. D. Zelinsky, *Linearly compact modules and rings*, Amer. J. Math. **75** (1953), 79-90.

78. W. Zimmermann, *Einige Charakterisierungen der Ringe über denen reine Untermoduln direkte Summanden sind*, Bayer. Akad. Wiss. Math.-Natur. Kl. S.-B. 1972, Abt. II (1973), 77-79.

79. ———, *Rein-injektive direkte Summen von Moduln*, Habilitationsschrift, Universität München, 1975.

80. ———, *Rein-injektive direkte Summen von Moduln*, Communic. in Algebra **5** (1977), 1083-1117.

81. ———, *(Σ-) algebraic compactness of rings*, J. Pure Appl. Algebra **23** (1982), 319-328.

82. ———, *Modules with chain conditions for finite matrix subgroups*, J. Algebra **190**, 68-87.

83. ———, *Extensions of three classical theorems to modules with maximum condition for finite matrix subgroups*, forum Math. **10** (1998), 377-392.

84. G. Zwara, *Tame algebras and degenerations of modules*, this volume.

Department of Mathematics, University of California, Santa Barbara, CA 93106, USA

E-mail address: birge@math.ucsb.edu

Trends in Mathematics, © 2000 Birkhäuser Verlag Basel/Switzerland

TOPOLOGICAL AND GEOMETRIC ASPECTS OF THE ZIEGLER SPECTRUM

MIKE PREST

ABSTRACT. The aim here is to emphasise the topological and geometric structure that the Ziegler spectrum carries and to illustrate how this structure may be used in the analysis of particular examples. There is not space here for me to give a survey of what is known about the Ziegler spectrum so there are a number of topics that I will just mention in order to give some indication of what lies beyond what is discussed here.

1. The Ziegler spectrum
2. Various dimensions
3. These dimensions for artin algebras
4. These dimensions in general
5. Duality
6. The complexity of morphisms in mod-R
7. The Gabriel-Zariski topology
8. The sheaf of locally definable scalars

1. THE ZIEGLER SPECTRUM

1.1. **A reminder on purity and pure-injectives.** Suppose that M is a submodule of N. Consider a finite system $\Sigma_{i=1}^n x_i r_{ij} = a_j$ $(j = 1, ...m)$ of R-linear equations over M: that is, the r_{ij} are in R, the a_j are in M and the x_i are indeterminates. Suppose that there is a solution $b_1, ..., b_n$ to this system in N (that is, the b_i are in N and for each i we have $\Sigma_{i=1}^n b_i r_{ij} = a_j$). If it happens that M is a direct summand of N then, projecting the b_i from N to M, we obtain a solution in M to this system. More generally, we say that M is **purely embedded** in N if every such system with a solution in N also has a solution in M.

A module N is **pure-injective** if every pure embedding of N into any module is a split embedding. An equivalent definition is that N is pure-injective if every infinite system of R-linear equations over N (allowing infinitely many indeterminates) has a solution in N provided that every finite sub-system has a solution in N.

There are many equivalent ways of defining pure embeddings and pure-injectivity (see [29] in this volume). One which, in Section 1.5, we shall recall and use is the characterisation in terms of the functor category (R-mod, **Ab**).

1.2. **A little history.** The subject of this article is a topological space, whose underlying set consists of the isomorphism classes of indecomposable pure-injective modules, which was introduced by Ziegler in his pivotal paper [73] on the model theory of modules.

The model theoretic investigation of modules began in the early 1970s with papers of Eklof and Sabbagh [63], [64], [65], [15], [66] in particular ([70] was a precursor). Already at this time the importance of pure-injectives in this project was apparent since Sabbagh had shown [64] that every module is elementarily equivalent to (has the same first-order logical properties as) a pure-injective module (in fact, to its pure-injective hull). In the later 1970s Garavaglia wrote an influential paper [18] centred around the question of when a module has what he called elementary Krull dimension and on the consequences of a module having this dimension. He defined this dimension in terms of the lattice of pp formulas but one can re-phrase this and say that he was looking at the Krull dimension (in the sense of [58], see [36]) of the lattice of finitely generated subfunctors of the forgetful functor in (mod-R, **Ab**).

Garavaglia never published his paper but it was a major inspiration for those working in the area at the time and, in particular, for Ziegler's work [73] which, especially in his introduction of this topological space, transformed the subject.

1.3. **A little model theory.** Here I will give just a brief indication of why the Ziegler spectrum plays a key role in the model theory of modules. One of the important concepts of model theory is elementary equivalence: two structures are elementarily equivalent if they satisfy the same first-order sentences. This is a much coarser equivalence relation than isomorphism and the model-theoretic approach to classification is usually to identify the elementary equivalence classes first and then to go on to develop structure theory within a given elementary equivalence class (in the best case, obtaining invariants which give a classification to isomorphism).

Ziegler showed that there is a bijection between the closed subsets of this space and those elementary equivalence classes of modules which are closed under direct sum (and hence with definable categories of modules in the sense of [14]). If R is, for example, an algebra over an infinite field then every elementary equivalence class is closed under direct sum.

It follows that if one has a complete description of the Ziegler spectrum (points and topology) of a ring then one has a great deal of information about the model theory of modules over that ring: many model-theoretic questions are thereby reduced to some kind of check using, say, a list of points and an explicitly described basis of open sets.

1.4. **Definition of the Ziegler spectrum: first formulation.** Here is the definition as given by Ziegler.

Given a ring R we define what we now call its **right Ziegler spectrum**, Zg_R, to be the topological space whose points are the (isomorphism classes of) indecomposable pure-injective R-modules and whose topology is defined by taking all the sets of the following form for a basis.

$(\phi/\psi) = \{N \in \mathrm{Zg}_R : \phi(N)/\psi(N) \neq 0\}$.

Here ϕ and ψ are pp (=positive primitive) formulas with $\phi \geq \psi$.

I will not say what is meant by a pp formula (but you can find out by looking at just about any paper on the model theory of modules e.g. [17], [40], [62], [73], [43], [23]) since we can immediately re-phrase this definition of the open sets in terms of concepts which have been discussed in other articles in this volume ([29], [35]). Namely, to the "pp formula" ϕ is associated a subgroup, $\phi(M)$, of M (or of a finite power of M) for any module M and this assignment $M \mapsto \phi(M)$ defines a functor, a p-functor in the terminology of [74] (see [29]), from the category of R-modules to the category of abelian groups (other terminology that has been used for the "pp-definable subgroup" $\phi(M)$ is "subgroup of finite definition" [20], and "finitely matrizable subgroup" [75]). Let us denote this functor F_ϕ. The condition "$\phi \geq \psi$" appearing above is simply the condition that F_ψ is a subfunctor of F_ϕ. So then the condition "$\phi(N)/\psi(N) \neq 0$" appearing above is just the condition that the quotient functor F_ϕ/F_ψ evaluated on N is non-zero: $F_\phi/F_\psi(N) \neq 0$. This functor F_ϕ/F_ψ is, in the terminology of [14], see [35], a coherent functor (in fact every coherent functor is isomorphic to one of this form).

Remark. Notice that Zg_R is an object which is associated to the category Mod-R of right R-modules (it is easy to see that it can be defined purely in terms of Mod-R and hence that any Morita equivalence between rings induces a homeomorphism of Ziegler spectra). Similarly one has the left Ziegler spectrum $_R\mathrm{Zg}$ associated to the category, R-Mod, of left R-modules. In many cases these spaces are homeomorphic, with the homeomorphism being given by "elementary duality". I will say a little about duality in Section 5 of this article but for more you should consult [23] or, for a summary, [44]. Herzog showed that for every countable ring the spaces become homeomorphic after topologically indistinguishable points have been identified (5.2). It is not known whether or not this is true for rings of arbitrary size.

1.5. **Definition of the Ziegler spectrum: second formulation.** Recall [21], see [29], that there is a full and faithful embedding of Mod-R into the functor category $(R\text{-mod}, \mathbf{Ab})$ (where R-mod is the category of finitely presented left R-modules) which is given on objects by taking the R-module M_R to the tensor functor $M \otimes_R - : R\text{-mod} \longrightarrow \mathbf{Ab}$. Recall also that the functor $M \otimes -$ is an injective object of $(R\text{-mod}, \mathbf{Ab})$ if and only if M is a pure-injective module and that every injective functor has this form. In particular, the points of the Ziegler spectrum can be identified with the (isomorphism classes of) indecomposable injective functors.

If one, so to speak, applies this functor to the definition of the basis of open sets, then the following equivalent definition of the Ziegler spectrum is the result.

The points of Zg_R are the (isomorphism classes of) indecomposable injective functors in $(R\text{-mod}, \mathbf{Ab})$.

A basis of open sets consists of the sets of the form

$(F) = \{E \in \mathrm{Zg}_R : (F, E) \neq 0\}$

as F ranges over $(R\text{-mod}, \mathbf{Ab})^{\text{fp}}$ - the full subcategory whose objects are the finitely presented functors in $(R\text{-mod}, \mathbf{Ab})$ and where (F, E) denotes the group of morphisms (natural transformations) from F to E.

Let me make the precise translation between the two definitions of the basic open sets. In order to do so I need to say a little more about coherent functors and duality. First we recall [1], [21] (see Section 5) that there is a duality, D, between the categories of finitely presented functors $(R\text{-mod}, \mathbf{Ab})^{\text{fp}}$ and $(\text{mod-}R, \mathbf{Ab})^{\text{fp}}$. I should remark that these functors are sometimes referred to as the "coherent" functors but in [14] and [35] the term is used to mean the unique extension of such a finitely presented functor F to a functor $\tilde{F}$ from the category of all R-modules to $\mathbf{Ab}$ which commutes with direct limits. In any case it is rather harmless to conflate a finitely presented functor F with its extension $\tilde{F}$.

Now, there is a natural isomorphism $(F, M \otimes -) \simeq (\tilde{DF})M$. Here M is any right R-module and F is any finitely presented functor in $(R - \text{mod}, \mathbf{Ab})^{\text{fp}}$. An indecomposable injective functor such as E above has (up to isomorphism) the form $N\otimes$ - for some indecomposable pure-injective right R-module N. Then the precise connection between the above definitions of Zg_R, and using the notations of those definitions, is that $(F) = (\phi/\psi)$ exactly when $\tilde{DF} \simeq F_\phi/F_\psi$.

Ziegler showed that these basic open sets are compact.

2. Various dimensions

2.1. **Cantor-Bendixson analysis of a space.** The Cantor-Bendixson analyis of a space layers the space according to the "degree of isolation" of a point. This analysis has turned out to be very useful in understanding the structure of the Ziegler spectrum.

Recall that a point p of a topological space T is **isolated** if $\{p\}$ is an open set, that is, if p is not in the closure of its complement in T. Denote by T^0 the set of isolated points of T and by T' the set of non-isolated points. Points p in T^0 are assigned **Cantor-Bendixson rank** 0, $\text{CB}(p) = 0$. Now we may consider the closed subset T' with its induced topology: there may well be isolated points of this space and so we continue the process. That is, we set $T^{(0)} = T$, $T^{(1)} = T'$ and inductively define $T^{(\alpha+1)} = T^{(\alpha)'}$. Set $\text{CB}(p) = \alpha$ if $p \in T^{(\alpha)} \setminus T^{(\alpha+1)}$. There is no reason to restrict this process to finite steps so we set $T^{(\omega)} = \bigcap_n T^{(n)}$ and, more generally, at limit ordinals λ, set $T^{(\lambda)} = \bigcap_{\alpha<\lambda} T^{(\alpha)}$. Since an intersection of closed sets is closed the sets $T^{(\alpha)}$ are all closed subsets of T. In particular, if T is **compact** (that is, if every open cover of T has a finite subcover) then each $T^{(\alpha)}$ will be compact.

It may happen that at some stage we reach the empty set, $T^{(\alpha)} = \emptyset$. If so, choose the minimal such α (recall that every set of ordinals has a least member). If T, and hence all the $T^{(\beta)}$, are compact and if λ is a limit ordinal then we cannot have $\bigcap_{\beta<\lambda} T^{(\beta)} = \emptyset$ unless one of the $T^{(\beta)}$ with $\beta < \lambda$ already is empty. Hence α is not a limit ordinal and so $\alpha - 1$ exists. Then we say that T has **Cantor-Bendixson rank** $\alpha - 1$, $\text{CB}(T) = \alpha - 1$ and so $\text{CB}(T)$ is the maximum Cantor-Bendixson rank of any point of T. For instance $\mathrm{CB}(T) = 0$ exactly if every point of T is isolated. Notice also that the penultimate **Cantor-Bendixson derivative** $T^{(\alpha-1)}$ is a finite set (still assuming T to be compact) since otherwise, $T^{(\alpha-1)}$ being compact and infinite, there would be a non-isolated point in $T^{(\alpha-1)}$.

The other possibility is that at some stage we reach a Cantor-Bendixson derivative $T^{(\alpha)}$ of T which is non-empty but which contains no isolated points. In this case we assign the Cantor-Bendixson rank ∞ to the points of $T^{(\alpha)}$ and write $\mathrm{CB}(T) = \infty$, saying that the Cantor-Bendixson rank of T is ∞ or "undefined".

2.2. **Krull-Gabriel dimension and m-dimension.** Recall ([19], see [69]) that the Krull-Gabriel dimension of the functor category (R-mod, **Ab**), denoted

$$\mathrm{KGdim}(R\text{-mod}, \mathbf{Ab})$$

or simply $\mathrm{KG}(R)$, is defined inductively by the following process. Let $\mathcal{C}$ be a locally finitely presented Grothendieck abelian category (such as (R-mod, **Ab**) or any of the localisations of this category which will appear in a moment). Let $\mathcal{S}_0$ be the Serre subcategory with objects the *finitely presented* finite length objects and let $\mathcal{T}_0$ be the localising subcategory generated by $\mathcal{S}_0$. Set $\mathcal{C}' = \mathcal{C}/\mathcal{T}_0$ to be the localisation of $\mathcal{C}'$ at $\mathcal{T}_0$. Then $\mathcal{C}'$ also is locally finitely presented Grothendieck and we can continue the process, transfinitely, to obtain a succession of localisations (which can be represented as $\mathcal{C}/\mathcal{T}_\alpha$ for suitable localising subcategories $\mathcal{T}_\alpha$ of $\mathcal{C}$) of $\mathcal{C}$. If the zero category is reached at some stage $\mathcal{C}/\mathcal{T}_\alpha$ then the ordinal index α of the first such stage is the **Krull-Gabriel dimension**, $\mathrm{KGdim}(\mathcal{C})$, of $\mathcal{C}$. If eventually we reach a category with no finitely presented simple objects (so the process cannot continue) then we set $\mathrm{KGdim}(\mathcal{C}) = \infty$.

The duality between $(R\text{-mod}, \mathbf{Ab})^{\mathrm{fp}}$ and $(\text{mod-}R, \mathbf{Ab})^{\mathrm{fp}}$ (see Section 5) and induced dualities between corresponding localisations of these categories give the following.

Proposition 2.1. $\mathrm{KGdim}(R\text{-mod}, \mathbf{Ab}) = \mathrm{KGdim}(\text{mod-}R, \mathbf{Ab})$.

Hence the notation $\mathrm{KG}(R)$ is unambiguous.

Comment. One may compare this dimension with the Gabriel dimension, $\mathrm{Gdim}(\mathcal{C})$, of $\mathcal{C}$ which is defined similarly but with, at each stage, $\mathcal{S}_0$ replaced by the Serre subcategory of all finite length objects, finitely presented or not. Therefore $\mathrm{Gdim}(\mathcal{C}) \leq \mathrm{KGdim}(\mathcal{C})$ and, indeed, the values may well be different. But it is shown in [7, 5.1] that if $\mathrm{Gdim}(\mathcal{C}) < \infty$ then so is $\mathrm{KGdim}(\mathcal{C}) < \infty$ (note that the use of the term "Krull-Gabriel dimension" in [7] conflicts with that here).

As mentioned earlier, Garavaglia investigated what he called **elementary Krull dimension**. He defined this to be the Krull dimension (see [36]) of the lattice of pp formulas but, since this lattice may be identified with the lattice, $\mathrm{Latt}^{\mathrm{fg}}(R, -)$, of finitely generated subfunctors of the forgetful functor $(R, -)$ in (mod-R, **Ab**), we may alternatively define it to be the Krull dimension of $\mathrm{Latt}^{\mathrm{fg}}(R, -)$. One has the following connection (see [43, 10.26] where it is formulated in terms of elementary Krull dimension and m-dimension, for which see just below).

Proposition 2.2. $\mathrm{KG}(R) < \infty$ *if and only if* $\mathrm{Kdim}(\mathrm{Latt}^{\mathrm{fg}}(R,-)) < \infty$. *Always one has* $\mathrm{KG}(R) \geq \mathrm{Kdim}(\mathrm{Latt}^{\mathrm{fg}}(R,-))$.

Comment. The condition $\mathrm{KG}(R) = 0$ is equivalent to the condition that R be of finite representation type. The condition $\mathrm{Kdim}(\mathrm{Latt}^{\mathrm{fg}}(R_R,-)) = 0$ is equivalent to the condition that R be right pure-semisimple. Therefore the pure-semisimplicity conjecture (see [29]) is the conjecture that if $\mathrm{Kdim}(\mathrm{Latt}^{\mathrm{fg}}(R_R,-)) = 0$ then $\mathrm{KG}(R) = 0$.

A dimension which tracks Krull-Gabriel dimension precisely is one that was introduced in [73] and which was called **m-dimension** and denoted m- dim in [43]. Briefly, this is defined for modular lattices as is Krull dimension but, instead of, at each stage, "collapsing" intervals which are artinian, we collapse only the intervals of finite length. Thus m-dimension grows more slowly than Krull dimension but the dimensions co-exist: a modular poset has Krull dimension if and only if it has m-dimension if and only if there is no embedding of the rationals (regarded as a poset) in it. Then 2.2 and 2.3 below follow easily from the observation that every functor in $(\text{mod-}R, \mathbf{Ab})^{\mathrm{fp}}$ is a subquotient of $(R,-)^n$ for some n. The connection between m-dimension and Krull-Gabriel dimension was "folklore" for some time but can be found explicitly in [6, 4.2.8] (see [7, 4.5]) and [33, 1.1].

Proposition 2.3. $\mathrm{KG}(R) = \text{m-dim}(\mathrm{Latt}^{\mathrm{fg}}(R,-))$.

2.3. Krull-Gabriel dimension and Cantor-Bendixson rank.

Theorem 2.4. *([73, 8.4(2),8.8(1)]) If* $\mathrm{KG}(R) = \alpha < \infty$ *then* $\mathrm{CB}(\mathrm{Zg}_R) = \alpha$.

Of course one does not see Krull-Gabriel dimension nor functors in [73], rather conditions on the lattice of pp formulas, but this lattice is isomorphic (see [43, Chapter 12]) to $\mathrm{Latt}^{\mathrm{fg}}(R,-)$ and so one obtains this formulation of Ziegler's result (and others, such as that just below) by applying 2.3.

In fact, the Krull-Gabriel and Cantor-Bendixson analyses move in step through the ordinals. To make this precise and to give a local version of these results, we discuss the connection between closed subsets and localisations in the next subsection.

It is not known whether or not the converse of the above result is true in general but it is known under various conditions, for example if R is countable and, more generally, under the "isolation condition ($\wedge$)" which is discussed after 3.13.

Theorem 2.5. *([73, 8.6] also see [43, 10.19]) If R is countable (or just if R satisfies the isolation condition) and if* $\mathrm{CB}(\mathrm{Zg}_R) = \alpha < \infty$ *then* $\mathrm{KG}(R) = \alpha$.

2.4. Localisation and Ziegler-closed sets.

Theorem 2.6. *([26, 3.8], [32, 4.3], [45, Section A3]) There are natural bijections between the following objects:*

(i) closed subsets of Zg_R*;*

(i)^{op} *closed subsets of* ${}_R\mathrm{Zg}$;
(ii) Serre subcategories of $(R\text{-mod}, \mathbf{Ab})^{\mathrm{fp}}$;
(ii)^{op} *Serre subcategories of* $(\text{mod-}R, \mathbf{Ab})^{\mathrm{fp}}$;
(iii) hereditary torsion theories on $(R\text{-mod}, \mathbf{Ab})$ *of finite type;*
(iii)^{op} *hereditary torsion theories on* $(\text{mod-}R, \mathbf{Ab})$ *of finite type.*

In particular, to every closed subset of the Ziegler spectrum there corresponds a locally finitely presented (as may be checked) quotient category of $(R\text{−mod}, \mathbf{Ab})$.

We describe these bijections. Let M be any module. The Serre subcategory, $\mathcal{S}_M$, of $(\text{mod-}R, \mathbf{Ab})^{\mathrm{fp}}$ corresponding to M is the set of restrictions to mod-R of coherent functors F with $F(M) = 0$. If we take the closure of $\mathcal{S}_M$ in $(\text{mod-}R, \mathbf{Ab})$ under direct limits then we obtain a localising subcategory $\mathcal{T}_M$ and so a corresponding torsion theory, τ_M, of finite type, on $(\text{mod-}R, \mathbf{Ab})$. We denote by $(-)_M$ the corresponding localisation functor from $(\text{mod-}R, \mathbf{Ab})$ to $(\text{mod-}R, \mathbf{Ab})_M = (\text{mod-}R, \mathbf{Ab})/\mathcal{T}_M$.

We may alternatively work with the dual Serre subcategory, $\{G \in (R\text{-mod}, \mathbf{Ab})^{\mathrm{fp}} : (G, M \otimes -) = 0\} = \{DF : F \in \mathcal{S}_M\}$ and the torsion theory of finite type that it generates. We use similar notation here.

By the **support** of M, $\mathrm{supp}(M)$, we mean the following closed subset of Zg_R: $\{N \in \mathrm{Zg}_R :$ for every $F \in \mathcal{S}_M$ we have $\tilde{F}(N) = 0\}$. Similarly for the dual Serre subcategory (the condition being $(G, N \otimes -) = 0$). Every Ziegler-closed set C has the form $\mathrm{supp}(M)$ for some module M and we use notation such as $\mathcal{S}_C$ for $\mathcal{S}_M$ if $C = \mathrm{supp}(M)$. One may check, for instance, that, given a Ziegler-closed set C, the images of the functors $N \otimes -$ for $N \in C$ in the localised category $(R\text{-mod}, \mathbf{Ab})_C$ form a complete (up to isomorphism) set of indecomposable injective objects of $(R\text{-mod}, \mathbf{Ab})_C$.

We define the **Krull-Gabriel dimension** of a module M by $\mathrm{KGdim}(M) = \mathrm{KGdim}((\text{mod-}R, \mathbf{Ab})_M)$. Observe that $\mathrm{KG}(R)$ denotes the Krull-Gabriel dimension of the whole functor category but $\mathrm{KGdim}(R)$, which denotes the Krull-Gabriel dimension of the localisation of this functor category at the torsion theory corresponding to R, is in general strictly less than $\mathrm{KG}(R)$. There will be some module M such that $\mathrm{KGdim}(M) = \mathrm{KG}(R)$ (for instance take M to be a direct sum of one copy of each point of Zg_R). We also extend this notation to closed subsets C of Zg_R by setting $\mathrm{KGdim}(C) = \mathrm{KGdim}((\text{mod-}R, \mathbf{Ab})_C)$. Then we have the following relative versions of 2.2, 2.3 and 2.4 (they specialise to those results if one takes M to be a module with support the whole of Zg_R).

Proposition 2.7. *(see [43, 10.26])* $\mathrm{KGdim}(M) < \infty$ *if and only if* $\mathrm{Kdim}(\mathrm{Latt}^{\mathrm{fg}}(R, -)_M) < \infty$ *where* $\mathrm{Latt}^{\mathrm{fg}}(R, -)_M$ *denotes the lattice of finitely generated subfunctors of the localisation of the forgetful functor,* $(R, -)$*, at the torsion theory corresponding to* M*. Always* $\mathrm{KGdim}(M) \geq \mathrm{Kdim}(\mathrm{Latt}^{\mathrm{fg}}(R, -)_M)$.

Proposition 2.8. *For any module* M *we have* $\mathrm{KGdim}(M) = \text{m-dim}(\mathrm{Latt}^{\mathrm{fg}}(R, -)_M)$.

Theorem 2.9. *([73, 8.6], see [43, 10.19, 10.16]) If* $\mathrm{KGdim}(M) = \alpha < \infty$ *then* $\mathrm{CB}(\mathrm{supp}(M)) = \alpha$.

Again, one has the converse to 2.9 in the presence of the isolation condition and so, in particular, in the case that R is countable [73, 8.1, 8.4].

We remark that $\mathrm{Latt}^{\mathrm{fg}}(R,-)_M$ may be identified with the lattice of pp-definable subgroups (=subgroups of finite definition in the terminology of [29]) of M. So, for instance, M is **of finite endolength** (that is, of finite length over its endomorphism ring) if and only if $\mathrm{KGdim}(M) = 0$ (or $M = 0$).

An indecomposable module M is **generic** if it is not finitely presented but is of finite endolength.

Theorem 2.10. [73, Section 9] *If* $\mathrm{KGdim}(M) < \infty$ *then there is a point of finite endolength in* $\mathrm{supp}(M)$. *If this point is not a direct summand of M then it is not isolated in* $\mathrm{supp}(M)$.

In fact, if $\mathrm{KGdim}(M) < \infty$ then each point of $\mathrm{supp}(M)$ with maximum CB-rank will be of finite endolength.

Since, for an artin algebra R, any non-isolated point of finite endolength is generic (3.9) one has the following corollary.

Corollary 2.11. *If R is an artin algebra and if M is an R-module with* $\mathrm{KGdim}(M) < \infty$ *and with no finite-dimensional direct summand then there is a generic point in the Ziegler closure of M.*

See [25, Sections 2,3] and [33, Sections 4,8] for extensions of this.

3. KG dimension and CB rank for artin algebras

Question 3.1. *What are the possible values for* $\mathrm{KG}(R)$ *and* $\mathrm{CB}(\mathrm{Zg}_R)$ *for R a finite-dimensional algebra over a field (or, more generally, for R an artin algebra)?*

Theorem 3.2. *([2, 3.14], [30], [43, 6.28,11.29]) Let R be any ring. Then* $\mathrm{KGdim}(R) = 0$ *if and only if R has finite representation and in this case* $\mathrm{CB}(\mathrm{Zg}_R) = 0$. *If R satisfies the isolation condition (e.g. if R is countable) then this last condition is equivalent to the other two.*

Theorem 3.3. *([33],[27]) Let R be an artin algebra. Then* $\mathrm{KG}(R) \neq 1$.

Corollary 3.4. *If R is a countable artin algebra then* $\mathrm{CB}(\mathrm{Zg}_R) \neq 1$.

I see no reason for the corollary above to depend on the countability restriction but there is no proof of this.

Theorem 3.5. *([19, 4.3]) Let R be a tame hereditary finite-dimensional algebra. Then* $\mathrm{KG}(R) = 2$.

Corollary 3.6. *Let R be a tame hereditary finite-dimensional algebra. Then* $\mathrm{CB}(\mathrm{Zg}_R) = 2$.

In fact, the complete description of the points and topology of the Ziegler spectrum of a tame hereditary algebra is known and may be found in [51, Section 2], [61]. There is a sketch of the Ziegler spectrum of $\tilde{A}_1$ below.

Theorem 3.7. ([3, 4.2]) *Let R be a finite-dimensional algebra of wild representation type. Then* $\mathrm{KG}(R) = \infty$.

Corollary 3.8. *Let R be a countable finite-dimensional algebra of wild representation type. Then* $\mathrm{CB}(\mathrm{Zg}_R) = \infty$.

It had seemed to me reasonable to conjecture [43, p.350] that, for a finite-dimensional algebra, only the values 1, 2, ∞ occur for $\mathrm{CB}(\mathrm{Zg}_R)$, with rank 2 corresponding exactly to domestic representation type (this is the case where there is at least one but only finitely many generic modules - see [76] and [35]). It was known [41, 3.4], for instance, that for some non-domestic but tame algebras the CB-rank is ∞. This conjecture turned out to be false, and I will describe some algebras which show this. It is still, however, consistent with the known cases that the line between $\mathrm{CB}(\mathrm{Zg}_R) < \infty$ and $= \infty$ coincides with that between domestic and non-domestic representation type.

3.1. **The Ziegler spectrum of Λ_2 : part 1.** Let Λ_2 be the quiver with relations shown

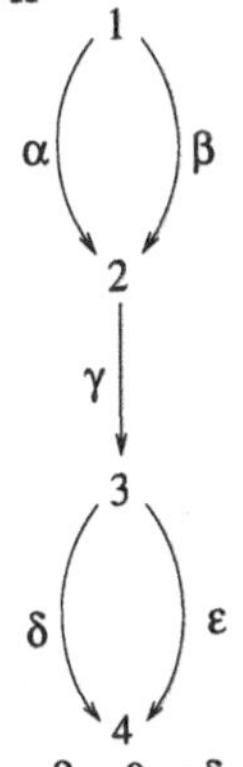

$\beta\gamma=0=\gamma\delta$

and let R denote the path algebra, over an algebraically closed field k, of this quiver, modulo the ideal generated by these relations. Then R is a string algebra and so (see [10]) the indecomposable finite-dimensional modules are string modules and band modules. This algebra is domestic and there are just two bands, corresponding to the two copies of the Kronecker quiver $\tilde{A}_1$

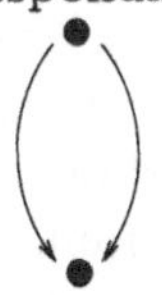

which are subquivers of Λ_2. We will describe the Ziegler spectrum of R. First we state a result which applies to all artin algebras.

Theorem 3.9. ([43, 13.1,13.3]) *Let R be an artin algebra. Then the finite length indecomposable R-modules are exactly the isolated points of* Zg_R. *The set of finite length points is dense in* Zg_R.

Since R is not of finite representation type, we know that $\mathrm{CB}(\mathrm{Zg}_R) \geq 1$ (therefore R has at least one infinitely generated indecomposable pure-injective module) and, at least if R is countable, we have by 3.4 that $\mathrm{CB}(\mathrm{Zg}_R) \geq 2$.

We can say rather more than this by using the obvious representation embeddings from the Kronecker algebra $k\tilde{A}_1$ to R. Namely, let $F_{12} : \text{Mod-}k\tilde{A}_1 \longrightarrow \text{Mod-}R$, respectively, $F_{34} : \text{Mod-}k\tilde{A}_1 \longrightarrow \text{Mod-}R$, be the functor which takes a representation

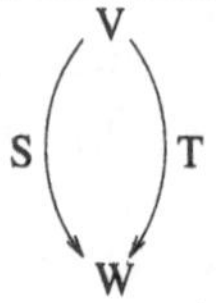

of the Kronecker quiver to the representation

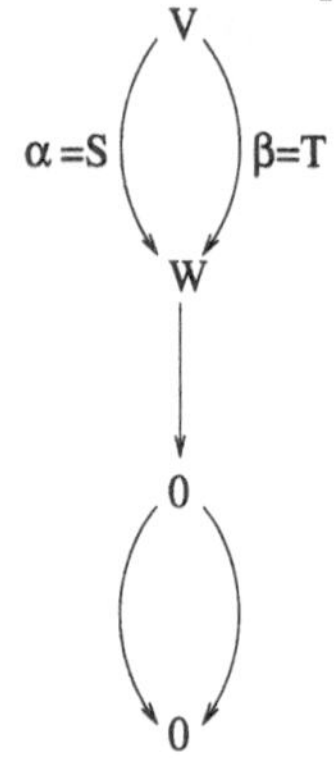

respectively

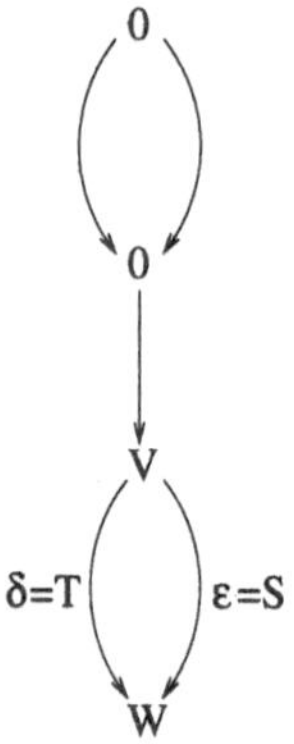

of R. Clearly F_{12} and F_{34} are representation embeddings. Now we apply the following result:

Theorem 3.10. ([47, Theorem 7]) *Let $F : \text{Mod-}S \longrightarrow \text{Mod-}R$ be a representation embedding. Then F induces a homeomorphic embedding of Zg_S as a closed subspace of Zg_R.*

Therefore, if we let $C_{12} = \mathrm{im}F_{12}$ and $C_{34} = \mathrm{im}F_{34}$ (note that these sets are disjoint) then each of C_{12}, C_{34} is a homeomorphic copy of $\mathrm{Zg}_{\tilde{A}_1}$.

3.2. **The Ziegler spectrum of $\tilde{A}_1$.** The Ziegler spectrum of $\tilde{A}_1$ is shown below. The finite-dimensional points are all isolated (and are split into the classes of preprojective, regular and preinjective modules) and, together, are dense in the space. In fact, every infinite-dimensional point is in the closure of the set of regular modules ([43, 13.6]). There are two types of point of Cantor-Bendixson rank 1: the adic points (which are analogous to the modules over the polynomial ring $k[X]$ which are completions of $k[X]$ at some prime) and the prüfer points (which are analogous to the injective hulls of simple $k[X]$-modules). Then there is the sole point of rank 2 - the generic point (analogous to the field of quotients $k(X)$). In fact all these infinite-dimensional modules can be obtained as images of the corresponding kind of module over $k[X]$ under suitable representation embeddings.

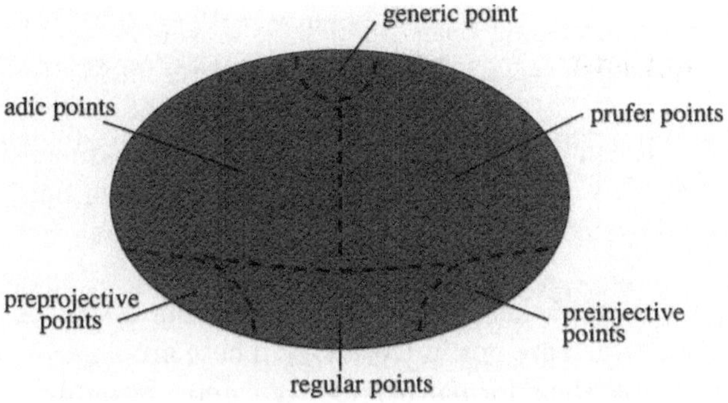

The Ziegler spectrum of any finite-dimensional tame hereditary algebra has the same shape: for details, and for an account of this work, see [51, Section 2], [61]. I will say a little about the method used in [51] to describe the spectrum.

Let S be a tame hereditary finite-dimensional algebra. Let U be a set of quasisimple regular S-modules (these being the modules which occur at the mouths of tubes in the Auslander-Reiten quiver of S). Using universal localisation in the sense of [11], [67] one may "remove" the modules in U and the result of doing so is a ring epimorphism $S \longrightarrow S'$ where S' is either a finite-dimensional tame hereditary algebra with fewer simple modules or a hereditary order in a central simple algebra -see [12, 4.2].

Theorem 3.11. *([46, Corollary 9]) Let $S \longrightarrow S'$ be an epimorphism of rings. Then the induced full embedding* Mod-S' $\longrightarrow$ Mod-S *induces a homeomorphic embedding of* $\mathrm{Zg}_{S'}$ *as a closed subspace of* Zg_S.

In fact, by choosing appropriate sets U, one obtains a finite number of universal localisations $S \longrightarrow S_i$ where the S_i are maximal orders and where, if $D = \bigcup_i \mathrm{Zg}_{S_i}$ (by the theorem above we may regard Zg_{S_i} as a subset of Zg_S) then D contains every infinite-dimensional point of Zg_S. A maximal order is a Dedekind prime ring, hence (see [38, 5.2.12]) is Morita equivalent to a Dedekind domain which may not be commutative but which will satisfy a polynomial identity (see [38, 5.3.10]). As remarked already, the Ziegler spectrum may be defined purely in terms of the category Mod-R and hence is Morita-invariant. The Ziegler spectra of PI Dedekind domains are known (see [51, 1.6]) and hence the spectra of the S_i are known. Then we patch together the descriptions of the spectra of the S_i to obtain a description of D and hence, with some more work, of Zg_S.

3.3. **The Ziegler spectrum of Λ_2 : part 2.** By 3.10 we know that each of C_{12}, C_{34} is a copy of $\mathrm{Zg}_{\tilde{A}_1}$. Let $V = \mathrm{Zg}_R \setminus (C_{12} \cup C_{34})$. Then V is a compact open set: this is not immediate from the (obvious) fact that the $\mathrm{Zg}_R \setminus C_i$ are compact because the intersection of compact Ziegler-open sets is not, in general, compact. So we show compactness of V directly. It is easy to see that any indecomposable module M which satisfies $M\gamma = 0$ is of the form $M = Me_1 \oplus Me_2$ or $M = Me_3 \oplus Me_4$ (where e_i is the idempotent of the path algebra corresponding to vertex i) and hence is in the image of one of the functors F_{12}, F_{34}. Therefore we have, using an informal but convenient way of describing basic open sets, $V = (\text{“}N\gamma \neq 0\text{”})$. In terms of coherent functors, we have $V = (F)$ where F is the representable functor $(R/\gamma R, -)$.

Note that V is not empty: it contains all the indecomposable finite-dimensional string modules N which satisfy $N\gamma \neq 0$. There are infinitely many such points in V, they are all isolated and V is compact: hence V must contain at least one infinitely generated module.

In fact, looking in [60], we see that there are infinitely many infinite-dimensional indecomposable pure-injectives which are not in $C_{12} \cup C_{34}$. These are described by infinite words in α, β, γ, δ, ϵ and their formal inverses α^{-1}, etc. Roughly, these (infinite) words correspond to (infinite) walks in the quiver. We give an example to illustrate this.

Consider the two-sided-infinite string $...\alpha\beta^{-1}\alpha\beta^{-1}\alpha\gamma\epsilon\delta^{-1}\epsilon\delta^{-1}...$ which we abbreviate as ${}^{\infty}(\beta^{-1}\alpha)\gamma(\epsilon\delta^{-1})^{\infty}$. Corresponding to this string is the module $M = M({}^{\infty}(\beta^{-1}\alpha)\gamma(\epsilon\delta^{-1})^{\infty})$ which has basis $\{z_i\}_i$ and the structure indicated by the diagram.

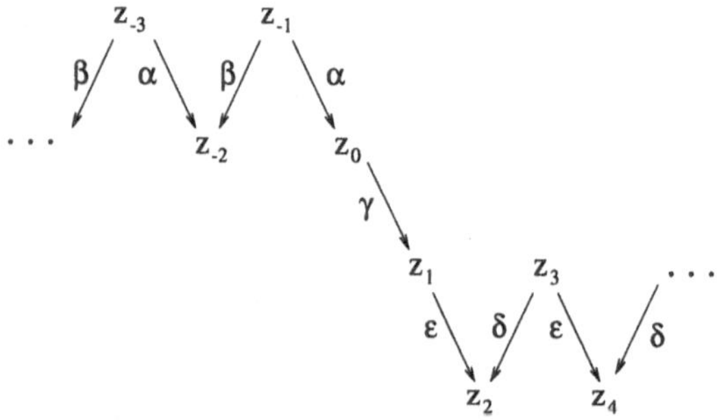

This module has, among its endomorphisms, an “expanding” endomorphism which sends $Me_3 \oplus Me_4$ to 0 and which “shifts $Me_1 \oplus Me_2$ one place to the left”, and a “contracting” endomorphism which sends $Me_1 \oplus Me_2$ to 0 and which “shifts $Me_3 \oplus Me_4$ one place to the left”. Notice that the first endomorphism acts somewhat like multiplication by p in the adic module $\mathbb{Z}_{(p)}$ (the localisation of $\mathbb{Z}$ at a prime p) and the second endomorphism acts (dually) somewhat like multiplication by p in the prufer module $\mathbb{Z}_{p^\infty}$. Because of the “expanding” part ${}^{\infty}(\beta^{-1}\alpha)$ of the word ${}^{\infty}(\beta^{-1}\alpha)\gamma(\epsilon\delta^{-1})^{\infty}$ the module M is not pure-injective but the following module is ([60, Section 5]). Let ${}^{\infty}\Pi\gamma\Sigma^{\infty}$ denote the submodule of the product $\Pi_n z_n k$ consisting of all elements $(z_n\nu_n)_n$ with $\nu_n \in k$, $\nu_n = 0$ for all $n >> 0$ (in the notation of [60] this is $\bar{M}({}^{\infty}(\beta^{-1}\alpha)\gamma(\epsilon\delta^{-1})^{\infty})$). Then ${}^{\infty}\Pi\gamma\Sigma^{\infty}$ is an indecomposable pure-injective module which clearly lies in the open set V.

Furthermore, we obtain infinitely many more points of V by taking the various “truncations” of the word ${}^{\infty}(\beta^{-1}\alpha)\gamma(\epsilon\delta^{-1})^{\infty}$ and the corresponding submodules and quotient modules of ${}^{\infty}\Pi\gamma\Sigma^{\infty}$. Here are two examples.

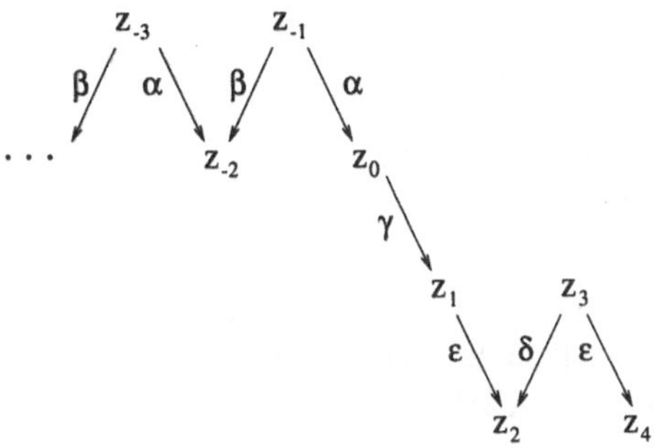

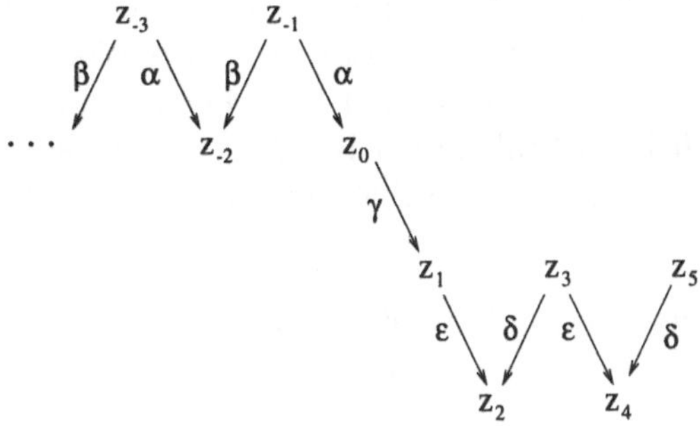

By [60, Section 5] these truncations all are indecomposable pure-injective and so we obtain infinitely many non-isolated points of V and hence the Cantor-Bendixson rank of the compact open set V is at least 2. In fact ${}^\infty\Pi\gamma\Sigma^\infty$ is a point of rank at least 2 since it is the direct limit of the natural embeddings between the various ${}^\infty\Pi\gamma\Sigma^n$, from which, [14, Section 2.3], it follows that ${}^\infty\Pi\gamma\Sigma^\infty$ is in the Ziegler-closure of the set of these modules and hence has CB-rank strictly greater than theirs.

One can check that each 1-sided infinite word N has CB-rank 1 - that is, there is an open neighbourhood of N which contains only finite-dimensional points apart from N. For instance, if we take N to be the first of the two one-sided infinite modules above then one such neighbourhood is that which is informally described by the conditions: "δ is not epi, $\ker(\delta\epsilon^{-1}) > \ker(\delta\epsilon^{-1}) \cap \operatorname{im}(\delta\epsilon^{-1})^2$ and $e_4M > \ker(\delta\epsilon^{-1})$". It is straightforward to write down a formal definition of the neighbourhood in terms of either functors or pp formulas from this informal description.

In fact, with these modules, we have now found all the points of Zg_R.

Theorem 3.12. *([9, 2.2]) Any infinite-dimensional indecomposable pure-injective module over Λ_2 is either a band module (that is, in $C_{12} \cup C_{34}$), the two-sided infinite string module above or one of its one-sided infinite truncations.*

Therefore, since V is compact, we conclude that, in the relative topology on V, the Cantor-Bendixson rank of $({}^\infty\Pi\gamma\Sigma^\infty)$ is 2 and hence, since V is an open subset of Zg_R, that $\mathrm{CB}({}^\infty\Pi\gamma\Sigma^\infty) = 2$. Next we see ${}^\infty\Pi\gamma\Sigma^\infty$ is not Ziegler-closed. Recall [29, Section 5], [76] that a module is of finite endolength if it is of finite length over its endomorphism ring.

Proposition 3.13. *(essentially [73, 8.3], see [49, 2.11]) Let R be any countable ring. If $N \in \mathrm{Zg}_R$ is a closed point then N is of finite endolength.*

Again, this is a result where one may suspect that the countability condition on the ring is not necessary but where there is no proof for arbitrary rings nor any counterexample to the general statement.

In fact a weaker condition under which 3.13 and, for example, the converse of 2.9 hold is a condition denoted $(\wedge)$ in [43, Chapter 10] and re-christened with the more informative name the **isolation condition** in [71]. This is the condition that if N is an isolated point of a closed subset C of Zg_R then there is a finitely presented functor F in $(\text{mod-}R, \mathbf{Ab})$ such that the image of F in the localisation $(\text{mod-}R, \mathbf{Ab})_C$ is simple and such that $(F) \cap C = \{N\}$. (In [43] the condition is said in terms of pp formulas.) If R is countable or if $\mathrm{KG}(R) < \infty$ (or even just if every non-zero pure-injective has an indecomposable summand [73, 8.11]) then this condition holds.

Question 3.14. *Does the isolation condition $(\wedge)$ hold for all rings? for all finite-dimensional algebras?*

The shift endomorphisms of ${}^{\infty}\Pi\gamma\Sigma^{\infty}$ defined above show that this module is not of finite endolength and so we conclude (at least, if R is countable, but we will remove this restriction in a moment) that the Ziegler-closure of ${}^{\infty}\Pi\gamma\Sigma^{\infty}$ contains at least one other module. In fact it is easily computed (independently of whether or not R is countable) that the Ziegler-closure of ${}^{\infty}\Pi\gamma\Sigma^{\infty}$ contains two other modules - the generic modules G_{12} and G_{34} which are the images of the unique generic $\tilde{A}_1$-module under the functors F_{12} and F_{34} respectively. We conclude that $\mathrm{CB}(\mathrm{Zg}_R) \geq 3$. It follows that $\mathrm{CB}(\mathrm{Zg}_R) \leq 5$ (since, once we remove the points of V, the remaining space has rank at most 2). At this point we cannot say that $\mathrm{CB}(\mathrm{Zg}_R) = 3$ because the Cantor-Bendixson ranks of non-isolated points of $C_{12} \cup C_{34}$ may be greater in Zg_R than in $C_{12} \cup C_{34}$ (since the latter is not an open set). This does in fact happen but, as it turns out, it does not affect the ranks of the highest-ranked points G_{12}, G_{34}. One may find the complete description of the topology of Zg_R in [9, Section 2] together with the following conclusion, which also follows from Schröer's computation of the Krull-Gabriel dimension.

Theorem 3.15. *([9, 2.2], [68, Theorem 14]) Let R be the path algebra of Λ_2. Then* $\mathrm{CB}(\mathrm{Zg}_R) = \text{m-dim}R = \mathrm{KGdim}(R) = 3$.

(I comment that the argument above shows that $3 \leq \mathrm{CB}(\mathrm{Zg}_R) \leq 5$ but, as remarked already, one cannot proceed from this to the same conclusion for Krull-Gabriel dimension (unless, say, the ring is countable). To reach the latter conclusion, we need the further observation that ${}^{\infty}\Pi\gamma\Sigma^{\infty}$, regarded as a module over its endomorphism ring, has Krull dimension (in the usual sense) and hence $\mathrm{m} - \dim({}^{\infty}\Pi\gamma\Sigma^{\infty}) < \infty$. Then we appeal to 2.4.)

3.4. **The Ziegler spectrum of Λ_n.** More generally, define the quivers Λ_n.

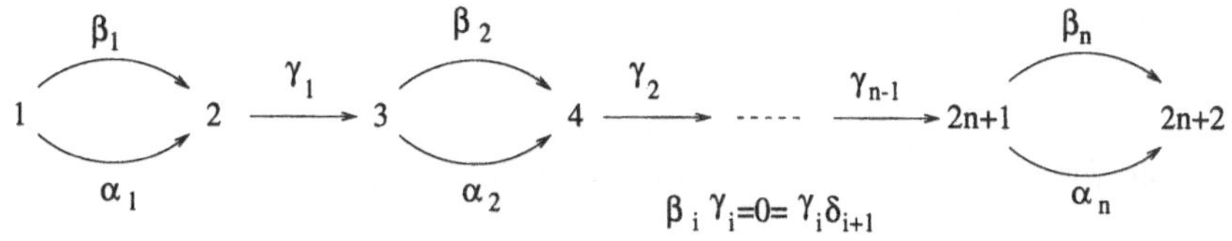

Theorem 3.16. *([9, 4.1], [68, Theorem 14]) Let R be the path algebra of Λ_n. Then* $\mathrm{CB}(\mathrm{Zg}_R) = \text{m-dim}R = \mathrm{KG}(R) = n+1$.

4. These dimensions in general

Here I give some examples of what is known about the Ziegler spectra of rings other than artin algebras, in order to illustrate the range of diversity of this space.

1. Let R be a commutative von Neumann regular ring. Then all embeddings between R-modules are pure embeddings, hence pure-injectivity reduces to injectivity and so the Ziegler spectrum coincides with the Pierce spectrum [39]. The lattice of finitely generated subfunctors of the forgetful functor can be identified with the boolean algebra of idempotents of R and the connection between the Cantor-Bendixson analysis of the Pierce spectrum and (though not with this terminology) the m-dimension of this lattice is well-known (e.g. [39], [31]). In particular $\mathrm{CB}(\mathrm{Zg}_R) = \mathrm{KG}(R)$ without restriction. Furthermore (see, e.g. [31]) all values (ordinal or ∞) for $\mathrm{CB}(\mathrm{Zg}_R)$ are obtained as R varies (in fact even as R varies over boolean algebras - that is rings satisfying the polynomial identities $x^2 = x$, $x+x=0$). The boolean algebras R for which $\mathrm{CB}(\mathrm{Zg}_R) < \infty$ are the **superatomic** boolean algebras. This case was thoroughly treated, in terms of "elementary Krull dimension" by Garavaglia [18] (see [43, 16.25]).

2. If R is a commutative serial ring of Krull dimension α, then [56] $\mathrm{KG}(R) = \mathrm{CB}(\mathrm{Zg}_R) = 2\alpha$. In the case of a non-commutative serial ring with Krull dimension α, one has 2α as an upper bound for these dimensions [59].

3. Let R be a differential polynomial ring over a universal field with derivation. Then, [55, 5.3], Zg_R has exactly three points: the unique simple (injective) R-module S, the dual, DS' of the unique simple left R-module and the injective hull, $E(R)$, of R. The first two modules are isolated and the last has CB-rank 1. The module S is Σ-injective (see [29]) with $\mathrm{KGdim}(S) = 1$, $\mathrm{KGdim}(DS') = 1$ and $E(R)$ is Σ-injective of finite endolength. So, in contrast with the case of artin algebras, this is an example of a ring with $\mathrm{CB}(\mathrm{Zg}_R) = 1 = \mathrm{KG}(R)$. Also see [57] for more on this.

4. Let $R = k\langle X, Y : YX - XY = 1\rangle$ be the first Weyl algebra over a field k of characteristic 0. The indecomposable torsion modules M are not pure-injective but each has indecomposable pure-injective hull $\bar{M}$ (and is uniquely determined by its

pure-injective hull). Locally, one has $\text{KGdim}(\bar{M}) = 1$, but none of these modules $\bar{M}$ is isolated. Since the set of such pure-injective hulls of indecomposable torsion modules is dense in Zg_R one concludes that Zg_R has no isolated points. Similar results hold for generalised Weyl algebras in the sense of [5]: for all this, see [55].

5. A ring R is **indiscrete** if the Ziegler spectrum is infinite but has the trivial topology (that is, the only open sets are $\emptyset$ and Zg_R). Any non-artinian simple von Neumann regular ring is indiscrete [72, Theorem 1]. It was an open question whether there were any other indiscrete rings until, in [54, Section 2.2], it was shown how to construct, from any non-regular ring of finite representation type, a non-regular indiscrete ring.

5. Duality

Functors which link the categories of right and left modules over a given ring often prove to be particularly useful (for a recent example see [22]). One such functor is the duality between $(R\text{-mod}, \mathbf{Ab})^{\text{fp}}$ and $(\text{mod-}R, \mathbf{Ab})^{\text{fp}}$ which may be found in [1] and [21]. This is the functor $D : (\text{mod-}R, \mathbf{Ab})^{\text{fp}} \simeq ((R\text{-mod}, \mathbf{Ab})^{\text{fp}})^{\text{op}}$ defined by $DF(L) = (F, - \otimes L)$ for $F \in (\text{mod-}R, \mathbf{Ab})^{\text{fp}}$ and $L \in R\text{-mod}$. For instance $D(M, -) \simeq M \otimes -$ and $D(- \otimes L) \simeq (L, -)$ for $M \in \text{mod-}R$ and $L \in R\text{-mod}$. One has that D^2 is naturally equivalent to the identity functor.

As examples of immediate consequences of the existence of this duality one has, taking examples from this paper, right/left symmetry of: $\text{KG}(R)$ (2.1); finite representation type (3.2); CB-rank of the Ziegler spectrum when R is countable (2.4, 2.5); existence of a factorisable system of morphisms between finitely presented modules (6.1). Furthermore, this functor is a key ingredient in the proofs of many results.

5.1. **Elementary duality.** A model-theoretic expression of the duality

$$(\text{mod-}R, \mathbf{Ab})^{\text{fp}} \simeq ((R\text{-mod}, \mathbf{Ab})^{\text{fp}})^{\text{op}}$$

was introduced in [42]. It was given in terms of pp formulas but, expressed in terms of functors, it is the fact that the lattice, $\text{Latt}^{\text{fp}}(R_R, -)$, of finitely generated subfunctors of the functor $(R_R, -) \in (\text{mod-}R, \mathbf{Ab})^{\text{fp}}$ is naturally isomorphic to the lattice of finitely generated subfunctors of the functor $({}_RR, -) \in (R\text{-mod}, \mathbf{Ab})^{\text{fp}}$. This "elementary duality" is given by: if $F \in \text{Latt}^{\text{fp}}(R_R, -)$ and i is the inclusion of F in $(R_R, -)$ then define $dF = \ker Di$. Elementary duality was extended by Herzog [23] to obtain a bijection between complete theories of right and left modules and "homeomorphism at the level of topology" between the right and left Ziegler spectra of a ring. In particular he obtained a bijection between the closed subsets of the right and left Ziegler spectra of any ring, as follows. If C is any closed subset of Zg_R and if $\mathcal{S}$ is the Serre subcategory of $(\text{mod-}R, \mathbf{Ab})^{\text{fp}}$ associated to C then the closed subset of ${}_R\text{Zg}$ dual to C is that associated to the Serre subcategory $D\mathcal{S} = \{DF : F \in \mathcal{S}\}$ of $(R\text{-mod}, \mathbf{Ab})^{\text{fp}}$.

Theorem 5.1. *([23, 4.4]) Let R be any ring. Then there is an inclusion-preserving bijection between the open subsets of* Zg_R *and* ${}_R\mathrm{Zg}$*. This bijection commutes with finite intersection and arbitrary unions.*

This "homeomorphism at the level of open sets" (more precisely, an isomorphism between the Heyting algebras of open sets) is, in some cases, literally a homeomorphism. If T is a topological space, let $\approx$ denote the equivalence relation on T which identifies topologically indistinguishable points (that is, points which belong to exactly the same open sets) and let $T/\approx$ denote the quotient space by this relation.

Theorem 5.2. *([23, 4.9]) Let R be any countable ring. Then* $\mathrm{Zg}_R/\approx$ *and* ${}_R\mathrm{Zg}/\approx$ *are homeomorphic.*

Theorem 5.3. *([23, 4.10], also see [24, 3.2]) Suppose that R is a ring with* $\mathrm{KG}(R) < \infty$*. Then there is a natural homeomorphism between* Zg_R *and* ${}_R\mathrm{Zg}$.

For instance, in the description [9] of the topology on the Ziegler spectrum of Λ_2, we use this duality as follows. By direct means, we compute neighbourhood bases of all infinite-dimensional "band" modules in the image of F_{12}. Now notice that Λ_2^{op} is isomorphic to Λ_2 and hence this description applies to Λ_2^{op}, yielding descriptions of neighbourhood bases of the "34"-points of Λ_2^{op}. Then we use elementary duality to transfer these to descriptions of neighbourhood bases of the "band" modules over Λ_2 which are in the image of F_{34}.

6. Complexity of morphisms in mod-R

We have indicated already that one may discuss these matters using the language of model theory or of functor category theory. It is also the case that one may, alternatively, phrase just about everything in terms of the category, mod-R, of finitely presented modules (e.g. see [14, 2.1]).

I will quote some results which link the complexity of the category mod-R with global properties of pure-injectivity in Mod-R.

A collection $\{M_\alpha, a_\alpha, f_{\alpha\beta} : \alpha < \beta, \alpha, \beta \in \mathbb{Q} \cap [0,1]\}$ of finitely presented R-modules M_α, elements $a_\alpha \in M_\alpha$, and morphisms $f_{\alpha\beta} : M_\alpha \longrightarrow M_\beta$ between them is a **factorisable system** if:
$f_{\beta\gamma} f_{\alpha\beta} = f_{\alpha\gamma}$ whenever $\alpha < \beta < \gamma$;
$f_{0\alpha}(a_0) = a_\alpha$ for all $\alpha > 0$;
for each $\alpha < \beta$ there is no $g : M_\beta \longrightarrow M_\alpha$ such that $g f_{\alpha\beta}(a_\alpha) = a_\alpha$

The next result is what one obtains if one translates Proposition 2.3 into this third language.

Theorem 6.1. *([50, 0.3]) Let R be any ring. Then there is a factorisable system of morphisms in mod-R if and only if* $\mathrm{KG}(R) = \infty$.

Corollary 6.2. *(see [50, 0.1,0.3]) Let R be any countable ring. Then the following are equivalent:*

(i) there is no factorisable system of morphisms in mod-R;

(ii) Zg_R *is countable;*

(iii) there are fewer than continuum many points in Zg_R.

Examples 6.3. *The (non-)existence of a factorisable system of morphisms is a condition that lends itself to being checked in particular cases. Rings which have such a system in mod-R include the tubular and wild canonical algebras* [50, p.450] *and non-domestic string algebras (*[50, p.450] *for some cases,* [68, Proposition 2] *in general).*

Let me also mention that Ziegler defines another "dimension" on the lattice of finitely generated subfunctors of the forgetful functor, called "width" (of course he used the language of model theory and so defined it on the lattice of pp formulas). He showed that if there exists a non-zero pure-injective R-module without indecomposable summands (that is a **superdecomposable** or **continuous** pure-injective module) then the width of this lattice is "∞" (that is, undefined) and hence if the width of this lattice is defined then every pure-injective module is the pure-injective hull of a direct sum of indecomposable pure-injectives. For countable rings he showed the converse: if the width of this lattice is ∞ then there is a non-zero continuous pure-injective module. This is another result which is known in the countable case but which is open in general. Notice that, by elementary duality, it is immediate that, at least for a countable ring, the existence of a continuous pure-injective module is therefore a right/left symmetric condition (since the property of having width $=\infty$ is trivially so). The condition of having width $=\infty$ can also be re-phrased in terms of systems of morphisms in mod-R [48].

6.1. **The transfinite radical of mod-R.** The above considerations are related to the transfinite radical of the category mod-R.

We may define a radical series for the category mod-R where R is an artin algebra, as follows. Let $\mathbf{R} = \mathbf{R}^1$ denote the radical of the category mod-R (it is generated as a two-sided ideal of mod-R by the non-invertible morphisms between indecomposable finitely presented R-modules). For Σ any ideal of morphisms in a category denote by Σ^n the ideal of morphisms generated by all compositions of n morphisms from Σ (so $\Sigma^n.\Sigma^m = \Sigma^{n+m}$). If λ is a limit ordinal, define $\mathbf{R}^\lambda = \bigcap\{\mathbf{R}^\nu : \nu < \lambda\}$ and if ν is a non-limit infinite ordinal - so uniquely of the form $\nu = \lambda + n$ for some limit ordinal λ and natural number $n \geq 1$ - set $\mathbf{R}^\nu = (\mathbf{R}^\lambda)^{n+1}$. Finally set $\mathbf{R}^\infty = \bigcap\{\mathbf{R}^\nu : \nu \text{ an ordinal}\}$ and $\mathbf{R}^0 = \text{mod-}R$.

Theorem 6.4. *(*[50, 0.6]*) Let R be an artin algebra. Then the following are equivalent:*

(i) there is a factorisable system of morphisms in $\mathbf{R}$*;*

(ii) the transfinite radical, $\mathbf{R}^\infty$*, of* mod-R *is non-zero;*

(iii) there is a factorisable system of morphisms in $\mathbf{R}^\infty$.

Comparing 6.1 with 6.4 we see that if $\mathbf{R}^\infty \neq 0$ then $\mathrm{KG}(R) = \infty$. More precisely ([34, 8.14]), if $\mathrm{KG}(R) = \alpha$ then $\mathbf{R}^{\omega\alpha+n} = 0$ for some natural number n. The converse is open.

Question 6.5. *Let R be an artin algebra and suppose that $\mathbf{R}^{\alpha} = 0$ for some ordinal α. Is $\mathrm{KG}(R) < \infty$?*

The question is open even for ordinals of the form $\omega + n$.

7. The Gabriel-Zariski topology

Now we say a little about the geometric aspects of the Ziegler spectrum. In fact, to see the geometric face we must dualise the topology as follows. Define a new topology on the underlying set of Zg_R by taking, for the basic open sets, the complements of the compact Ziegler-open sets. That is, a basis for this new topology consists of the sets of the form:
$[F] = \{N \in \mathrm{Zg}_R : (F, N \otimes -) = 0\}$ as F ranges over the finitely presented functors in (R-mod, $\mathbf{Ab}$).

In contrast with the case of spectral spaces [28] on which this is modelled, one cannot dualise again to recover the Ziegler topology (in fact by taking R to be the path algebra of Λ_2 one obtains an example which demonstrates this [9, 3.2]). We call this dual-Ziegler topology the (right) **Gabriel-Zariski topology** of R. The reason for the name is as follows.

Let R be a commutative noetherian ring. Then the Zariski spectrum, Spec$R$, of R is defined to be the set of all prime ideals of R, together with a topology which has, as a basis of open sets, those of the form $D(r) = \{P \in \mathrm{Spec}R : r \notin P\}$ as r varies over R.

Following Gabriel [16] one may reformulate this in terms of the category Mod-R by replacing each prime P with the (indecomposable injective) hull E_P of R/P and rewriting the condition which defines $D(r)$ as $\mathrm{Hom}(R/rR, E_P) = 0$ and then noting that one obtains the same topology if one replaces the cyclic modules R/rR by arbitrary finitely presented modules.

Thus, in any locally finitely presented Grothendieck category $\mathcal{C}$ one may reasonably define the "Zariski spectrum" of $\mathcal{C}$ to be the space which has, for its points, the (isomorphism classes of) indecomposable injective objects E of $\mathcal{C}$ and which has, for a basis of open sets for the topology, those sets of the form $\{E : (F, E) = 0\}$ as F ranges over (a generating set of) the finitely presented objects of $\mathcal{C}$.

The connection [44, p.202] (or, for more detail, [49, 1.8]) is that, if we let $\mathcal{C} = (R\text{-mod}, \mathbf{Ab})$ and if we regard the Ziegler spectrum of R as consisting of the indecomposable injective objects of $\mathcal{C}$ (the second formulation of the definition of Zg_R) then the dual-Ziegler topology defined above coincides with this topology that we have just defined on $\mathcal{C}$. Denote this topological space Zar_R.

We remark that modules which are generic in the sense of Crawley-Boevey [13] are "generic" in this space in the sense that they are not in the Zariski-closure of any other point. Cantor-Bendixson rank does not seem to be a useful tool in analysing this space, which has a much more "geometric" character.

For artin algebras every finite length point is both open and closed in the Zariski topology, just as for the Ziegler topology.

Proposition 7.1. *([43, 13.1, 11.15]) Let R be an artin algebra. Then every finite length point of Zar_R is both open and closed.*

For more on this topology, see [49].

8. The sheaf of locally definable scalars

8.1. **Rings of definable scalars.** Let C be a closed subset of Zg_R. Associated to C is the **ring**, R_C or, more precisely, the R-algebra $R \longrightarrow R_C$, **of definable scalars** of C [8]. One way to define this is to consider the localisation, $(R,-)_C$, of the forgetful functor $(R,-) \in (R\text{-mod}, \mathbf{Ab})$ at the torsion theory on $(R\text{-mod}, \mathbf{Ab})$ corresponding to C and then to define R_C to be the endomorphism ring of this localised functor. The natural isomorphism $R \simeq \mathrm{End}(R,-)^{\mathrm{op}}$ composed with the natural map $\mathrm{End}(R,-)^{\mathrm{op}} \longrightarrow \mathrm{End}((R,-)_C)^{\mathrm{op}} = R_C$ gives the R-algebra structure on R_C. The ring gets its name from the equivalent definition of it as the ring of all pp-definable maps on any module M with $\mathrm{supp}(M) = C$ (such a map is one whose graph is a subgroup of M^2 defined by a 2-place pp formula). This ring can also be obtained as the biendomorphism ring of a suitably "large" module with support equal to C. For all this, see [8] where the following is also shown where, if M is any module, we write R_M for $R_{\mathrm{supp}(M)}$.

Theorem 8.1. *([8, 3.6]) Let M be a module of finite endolength (in fact, M being Σ-pure-injective and finitely generated over its endomorphism ring is enough). Then the canonical morphisms $R \longrightarrow R_M$ and $R \longrightarrow \mathrm{Biend}(M)$ coincide.*

One also has, for example.

Theorem 8.2. *([46, Theorem 1(c)]) Let $R \longrightarrow S$ be any epimorphism of rings and let C be the image of the induced embedding of Zg_S as a closed subset of Zg_R. Then the canonical morphism $R \longrightarrow R_C$ may be identified with the original morphism $R \longrightarrow S$.*

It is not, however, the case that for every closed subset C one has $R \longrightarrow R_C$ an epimorphism (see [8, 2.7]).

8.2. **The sheaf of locally definable scalars.** In particular, for each basic Zariski-open set $[F]$, we have the R-algebra $R_{[F]}$. If $D \subseteq C$ are closed sets of Zg_R then there is a natural restriction map $R_C \longrightarrow R_D$ of R-algebras. Thus we have the data of a presheaf defined on an open basis of Zar_R: we call this (and its extension [52, Section 4] to a presheaf defined on Zar_R) the **presheaf of definable scalars** of R.

In general this presheaf is not a sheaf. For instance, if R is a ring of finite representation type then Zg_R is discrete. This means that there are no restrictions on "patching" together ring elements from the various stalks over the elements of Zg_R. Hence the ring of global sections of the sheafification of this presheaf is the direct product of the biendomorphism rings of the various indecomposable R-modules, whereas the ring of definable scalars of Zg_R itself is, as for any ring, just R. We therefore define the **sheaf of locally definable scalars**, LocDef_R, to be the sheafification of the presheaf of definable scalars.

One can check that, as for the topology, this generalises the usual Zariski structure sheaf of a commutative noetherian ring.

Proposition 8.3. *([45, C1.1], see [52, 3.1]) Let N be an indecomposable pure-injective R-module. Then the stalk of* LocDef_R *at N is R_N.*

In general, R_N need not be even Morita equivalent to a local ring [45, Section A1] but one does have the following.

Proposition 8.4. *([52, 6.1]) Let R be any ring. Then the centre of* LocDef_R *is a sheaf of commutative local rings.*

For the case of R a finite-dimensional algebra Def_R is almost never a sheaf, because the finite-dimensional points, being open and closed, are connected components of the space and hence give rise to many elements in the sheaf of locally definable scalars which are not in the presheaf of definable scalars. It seems reasonable, therefore, to remove these points in order to get a clear picture of the essential structure of the sheaf of locally definable scalars.

Theorem 8.5. *([34, 14.5],[53, 3.2,4.5]) Let R be a tame hereditary finite-dimensional algebra or a hereditary order in a central simple algebra. Let* $\mathrm{Zar}_R^{(1)}$ *denote the set of infinite-length points of* Zar_R. *Then the presheaf of definable scalars, restricted to* $\mathrm{Zar}_R^{(1)}$ *is already a sheaf.*

For example, if we take R to be the path algebra of the Kronecker quiver and if, furthermore, we identify adic and prüfer points which correspond to the same quasisimple regular module (see [51], [61]) then the ringed space that we obtain is exactly the structure sheaf of the projective line over the underlying field k.

The sheaf obtained in the result above is very closely related [34, Section 14], [53, Section 4] to those introduced in [4], [37], [12].

References

[1] M. Auslander, Isolated singularities and existence of almost split sequences (Notes by Louise Unger), pp. 194-242 in Representation Theory II, Groups and Orders, Proceedings, Ottawa, 1984, Lecture Notes in Mathematics, Vol. 1178, Springer-Verlag, 1986.

[2] M. Auslander, A functorial approach to representation theory, pp. 105-179 in Representations of Algebras, Lecture Notes in Mathematics, Vol. 944, Springer, (1980).

[3] D. Baer, Homological properties of wild hereditary artin algebras, pp. 1-12 in Representation Theory I: Finite-dimensional Algebras, Lecture Notes in Math., Vol. 1177, Springer-Verlag, 1986.

[4] D. Baer, W. Geigle and H. Lenzing, The preprojective algebra of a tame hereditary Artin algebra, Comm. Algebra, 15 (1987), 425-457.

[5] V. Bavula, Generalized Weyl algebras and their representations, Algebra and Analiz, 4(1) (1992), 75–97; English transl. in St. Petersburg Math. J., 4(1) (1993), 71–92.

[6] K. Burke, Some Model-Theoretic Properties of Functor Categories for Modules, Doctoral Dissertation, University of Manchester, 1994.

[7] K. Burke, Co-existence of Krull filtrations, J. Pure Appl. Algebra, to appear.

[8] K. Burke and M. Prest, Rings of definable scalars and biendomorphism rings, pp. 188-201 in Model Theory of Groups and Automorphism Groups, London Math. Soc. Lect. Note Ser., Vol. 244, Cambridge University Press, 1997.

[9] K. Burke and M. Prest, The Ziegler and Zariski spectra of some domestic string algebras, preprint, 1998

[10] M. C. R. Butler and C. M. Ringel, Auslander-Reiten sequences with few middle terms and applications to string algebras, Comm. Algebra, 15 (1987), 145-179.

[11] P. M. Cohn, Free Rings and Their Relations, Academic Press, New York, (Second Edn.) 1985.

[12] W. Crawley-Boevey, Regular modules for tame hereditary algebras, Proc. London Math. Soc., 62 (1991), 490-508.

[13] W.Crawley-Boevey, Tame algebras and generic modules, Proc. London Math. Soc., 63 (1991), 241-265.

[14] W. Crawley-Boevey, Infinite-dimensional modules in the representation theory of finite-dimensional algebras, pp. 29-54 in Algebras and Modules I, Canadian Math. Soc. Conf. Proc., Vol 23, American Math. Soc., 1998.

[15] P. Eklof and G. Sabbagh, Model-completions and modules, Ann. Math. Logic, 2 (1971), 251-295.

[16] P. Gabriel, Des catégories abéliennes, Bull. Soc. Math. France, 90 (1962), 323-448.

[17] S. Garavaglia, Decomposition of totally transcendental modules, J. Symbolic Logic, 45 (1980), 155-164.

[18] Garavaglia, S., Dimension and rank in the model theory of modules, University of Michigan, East Lansing, preprint, 1979.

[19] W. Geigle, The Krull-Gabriel dimension of the representation theory of a tame hereditary Artin algebra and applications to the structure of exact sequences, Manuscripta Math., 54 (1985), 83-106.

[20] L. Gruson and C. U. Jensen, Modules algébriquement compacts et foncteurs $\varprojlim^{(i)}$, C. R. Acad. Sci. Paris, Sér. A, 276 (1973), 1651-1653.

[21] L. Gruson and C. U. Jensen, Dimensions cohomologiques reliées aux foncteurs $\varprojlim^{(i)}$, pp. 243-294 in Lecture Notes in Mathematics, Vol. 867, Springer-Verlag, 1981.

[22] R. Hartshorne, Coherent functors, Adv. in Math., 140 (1998), 44-94.

[23] I. Herzog, Elementary duality of modules, Trans. Amer. Math Soc., 340 (1993), 37-69.

[24] I. Herzog, The Auslander-Reiten translate, Contemp. Math., 130 (1992), 153-165.

[25] I. Herzog, A test for finite representation type, J. Pure and Applied Algebra, 95 (1994), 151-182.

[26] I. Herzog, The Ziegler spectrum of a locally coherent Grothendieck category, Proc. London Math. Soc., 74 (1997), 503-558.

[27] I. Herzog, The endomorphism ring of a localised coherent functor, J. Algebra, 191 (1997), 416-426.

[28] M. Hochster, Prime ideal structure in commutative rings, Trans. Amer. Math. Soc., 142 (1969), 43-60.

[29] B. Huisgen-Zimmermann, Purity, algebraic compactness, direct sum decompositions and representation type, this volume.

[30] C.U. Jensen and H. Lenzing, Model Theoretic Algebra, Gordon and Breach, 1989.

[31] J. Ketonen, The structure of countable boolean algebras, Ann. Math., 108 (1978), 41-89.

[32] H. Krause, The spectrum of a locally coherent category, J. Pure Appl. Algebra, 114 (1997), 259-271.

[33] H. Krause, Generic modules over artin algebras, Proc. London Math. Soc., 76 (1998), 276-306.

[34] H. Krause, The Spectrum of a Module Category, Habilitationsschrift, Universität Bielefeld, 1997, to appear in Mem. Amer. Math. Soc.

[35] H. Krause, Finite versus infinite dimensional representations - a new definition of tameness, this volume.
[36] T. Lenagan, Dimension theory of Noetherian rings, this volume.
[37] H. Lenzing, Curve singularities arising from the representation theory of tame hereditary algebras, pp. 199-231 in Representation Theory I: Finite Dimensional Algebras, Lecture Notes in Mathematics, Vol. 1177, Springer-Verlag, 1986.
[38] J. C. McConnell and J. C. Robson, Noncommutative Noetherian Rings, Wiley, 1987.
[39] R. S. Pierce, Modules over commutative regular rings, Mem. Amer. Math. Soc., Vol. 70, 1967.
[40] A. Pillay and M. Prest, Forking and pushouts for modules, Proc. London Math. Soc., 46 (1983), 365-384.
[41] M. Prest, Model theory and representation type of algebras, pp. 219-260 in Logic Colloquium '86 (Proceedings of the European meeting of the Association for Symbolic Logic, Hull, 1986), North-Holland, Amsterdam, 1988.
[42] M. Prest, Duality and pure-semisimple rings, J. London Math. Soc., 38 (1988), 403-409.
[43] M. Prest, Model Theory and Modules, London Math. Soc. Lecture Notes Ser., Vol. 130., Cambridge University Press, 1988.
[44] M. Prest, Remarks on elementary duality, Ann. Pure Appl. Logic, 62 (1993), 183-205.
[45] M. Prest, The (pre)sheaf of definable scalars, University of Manchester, preprint, 1995.
[46] M. Prest, Epimorphisms of rings, interpretations of modules and strictly wild algebras, Comm. Algebra, 24 (1996), 517-531.
[47] M. Prest, Representation embeddings and the Ziegler spectrum, J. Pure Appl. Algebra, 113 (1996), 315-323.
[48] M. Prest, On the existence of many indecomposable pure-injectives, University of Manchester, preprint, 1995.
[49] M. Prest, The Zariski spectrum of the category of finitely presented modules, University of Manchester, preprint, 1997.
[50] M. Prest, Morphisms between finitely presented modules and infinite-dimensional representations, Canad. Math. Soc. Conf. Proc. Ser., Vol. 24 (1998), 447-455.
[51] M. Prest, Ziegler spectra of tame hereditary algebras, J. Algebra, 207 (1998), 146-164.
[52] M. Prest, The sheaf of locally definable scalars over a ring, pp. 339-351 in Models and Computability, London Math. Soc. Lecture Note Ser., Vol. 259, Cambridge University Press, 1999.
[53] M. Prest, Sheaves of definable scalars over tame hereditary algebras, University of Manchester, preprint, 1998.
[54] M. Prest, P. Rothmaler and M. Ziegler, Absolutely pure modules and indiscrete rings, J. Algebra, 174 (1995), 349-372.
[55] M. Prest and G. Puninski, Some model theory over hereditary noetherian domains, J. Algebra, 211 (1999), 268-297.
[56] G. Puninski, Cantor-Bendixson rank of the Ziegler spectrum over a commutative valuation domain, J. Symbolic Logic, to appear.
[57] G. Puninski, Pure-injective and finite length modules over certain rings of differential polynomials, Proc. Moscow-Tainan Conf., to appear.
[58] R. Rentschler and P. Gabriel, Sur la dimension des anneaux et ensembles ordonnès, C. R. Acad. Sci. Paris, Ser. A, 265 (1967), 712-715.
[59] G. Reynders, Ziegler spectra of serial rings with Krull dimension, Comm. Algebra, to appear.
[60] C. M. Ringel, Some algebraically compact modules I, pp. 419-439 in Abelian Groups and Modules, eds. A. Facchini and C. Menini, Kluwer, 1995.
[61] C. M. Ringel, The Ziegler spectrum of a tame hereditary algebra, Colloq. Math., 76 (1998), 105-115.

[62] Ph. Rothmaler, Some model theory of modules, II: on stability and categoricity of flat modules, J. Symbolic Logic, 48 (1983), 970-985.
[63] G. Sabbagh, Sur la pureté dans les modules, C. R. Acad. Sci. Paris, 271 (1970), 865-867.
[64] G. Sabbagh, Aspects logiques de la pureté dans les modules, C. R. Acad. Sci. Paris, 271 (1970), 909-912.
[65] G. Sabbagh, Sous-modules purs, existentiellement clos et élémentaires, C. R. Acad. Sci. Paris, 272 (1970), 1289-1292.
[66] G. Sabbagh and P. Eklof, Definability problems for rings and modules, J. Symbolic Logic, 36 (1971), 623-649.
[67] A. H. Schofield, Representations of Rings over Skew Fields, London Math. Soc. Lecture Note Ser., Vol. 92, Cambridge University Press, 1985.
[68] J. Schröer, On the Krull-Gabriel dimension of an algebra, Math. Zeit., 233 (2000), 287-303.
[69] J. Schröer, The Krull-Gabriel dimension of an algebra - open problems and conjectures, This volume.
[70] W. Szmielew, Elementary properties of abelian groups, Fund. Math., 41 (1954), 203-271.
[71] J. Trlifaj, Two problems of Ziegler and uniform modules over regular rings, pp. 373-383 in Abelian Groups and Modules, M. Dekker, New York, 1996.
[72] L. V. Tyukavkin, Model completeness of certain theories of modules, Algebra i Logika, 21 (1982), 73-83, translated in Algebra and Logic, 21 (1982), 50-57.
[73] M. Ziegler, Model theory of modules, Ann. Pure Appl. Logic, 26 (1984), 149–213.
[74] B. Zimmermann-Huisgen and W. Zimmermann, Algebraically compact rings and modules, Math. Zeit., 161 (1978), 81-93.
[75] W. Zimmermann, Rein injektive direkte Summen von Moduln, Comm. Algebra, 5 (1977), 1083-1117.
[76] G. Zwara, Tame algebras and degenerations of modules, this volume.

Department of Mathematics, University of Manchester, Manchester M13 9PL, UK

E-mail address: mprest@ma.man.ac.uk

Trends in Mathematics, © 2000 Birkhäuser Verlag Basel/Switzerland

FINITE VERSUS INFINITE DIMENSIONAL REPRESENTATIONS – A NEW DEFINITION OF TAMENESS

HENNING KRAUSE

This article is dedicated to Professor Herbert Kupisch on the occasion of his 70th birthday.

Let Λ be a finite dimensional algebra over some algebraically closed field k. In this note I discuss the relationship between finite and infinite dimensional modules over Λ. This discussion is based on the following three fundamental concepts:

- fp-idempotent ideals in the category $\operatorname{mod}\Lambda$ of finite dimensional Λ-modules
- endofinite modules in the category $\operatorname{Mod}\Lambda$ of all Λ-modules
- coherent functors $\operatorname{Mod}\Gamma \to \operatorname{Mod}\Lambda$ between two module categories

In order to illustrate the use of these concepts I present a new definition of tame representation type for Λ which only involves the category of finite dimensional Λ-modules.

This note follows closely the talk that I gave in Bielefeld in Summer 1998. I tried to cover a number of new results from my Habilitationsschrift [12]. In this thesis I introduced fp-idempotent ideals and used them to give a new tameness definition. My motivation was the following question raised by Ringel [14, p.144]: *Is there a definition of tameness which only involves finite dimensional modules, and avoids any reference to algebraic geometry?* The approach presented here which leads to a positive answer is based on an analysis of certain infinite dimensional modules. These are the endofinite modules which Crawley-Boevey used for his definition of generic tameness [2]. The main ingredients of my analysis can be summarized as follows:
• the Ziegler spectrum, which contains all indecomposable endofinite modules and puts some structure on the set of these modules;
• the fp-idempotent ideals, which represent the infinite dimensional endofinite modules inside the category of finite dimensional modules;
• the coherent functors, which transfer the structure of the collection of all endofinite modules from one algebra to another.
In the first three sections of this paper I will cover these aspects. Section 4 is devoted to giving two new characaterizations of tameness, and Section 5 illustrates the relationship between finite and infinite dimensional modules for tame algebras.

A final word about proofs: There are a number of proofs given in the first three sections, with the exception of some deeper results where I need to refer to the literature. The results in Section 4 are presented with complete proofs, whereas no proofs are given in the final Section 5.

1. Fp-idempotent ideals

We fix a finite dimensional algebra Λ over some algebraically closed field k. The category of (right) Λ-modules is denoted by $\operatorname{Mod}\Lambda$, and $\operatorname{mod}\Lambda$ denotes the category of finite dimensional Λ-modules. In this section we introduce fp-idempotent ideals in the category $\operatorname{mod}\Lambda$ and discuss their relationship with a number of concepts which are defined in the category $\operatorname{Mod}\Lambda$ of all Λ-modules.

A basic tool is the category $(\operatorname{mod}\Lambda, \mathrm{Ab})$ of additive functors $\operatorname{mod}\Lambda \to \mathrm{Ab}$ into the category Ab of abelian groups. Note that $(\operatorname{mod}\Lambda, \mathrm{Ab})$ is an abelian category. The natural transformations between functors form the morphisms in this category, and (co)kernels, (co)products, etc. are defined pointwise. A functor $F\colon \operatorname{mod}\Lambda \to \mathrm{Ab}$ is *finitely presented* if there exists an exact sequence

$$(*)\qquad \operatorname{Hom}_\Lambda(Y,-) \longrightarrow \operatorname{Hom}_\Lambda(X,-) \longrightarrow F \longrightarrow 0,$$

and $\operatorname{fp}(\operatorname{mod}\Lambda, \mathrm{Ab})$ denotes the full subcategory of finitely presented functors. Using the fact that $\operatorname{mod}\Lambda$ has cokernels, it is not difficult to show that $\operatorname{fp}(\operatorname{mod}\Lambda, \mathrm{Ab})$ is an abelian category. A functor $F\colon \operatorname{Mod}\Lambda \to \mathrm{Ab}$ is called *coherent* if there exists an exact sequence of the form $(*)$ with X and Y in $\operatorname{mod}\Lambda$. It is clear that $F \mapsto F|_{\operatorname{mod}\Lambda}$ defines an equivalence between the category $\operatorname{Coh}\Lambda$ of coherent functors and the category $\operatorname{fp}(\operatorname{mod}\Lambda, \mathrm{Ab})$.

Viewing $\operatorname{mod}\Lambda$ as a ring with several objects, we recall that an *ideal* $\mathfrak{I}$ in $\operatorname{mod}\Lambda$ consists of subgroups $\mathfrak{I}(X,Y)$ in $\operatorname{Hom}_\Lambda(X,Y)$ for every pair of objects X, Y in $\operatorname{mod}\Lambda$ such that for all ϕ in $\mathfrak{I}(X,Y)$ and all maps $\alpha\colon X' \to X$ and $\beta\colon Y \to Y'$ in $\operatorname{mod}\Lambda$ the composition $\beta\circ\phi\circ\alpha$ belongs to $\mathfrak{I}(X',Y')$. Note that an ideal $\mathfrak{I}$ in $\operatorname{mod}\Lambda$ is idempotent, i.e. $\mathfrak{I}^2 = \mathfrak{I}$, if and only if the class of functors in $(\operatorname{mod}\Lambda, \mathrm{Ab})$ vanishing on $\mathfrak{I}$ is closed under extensions. This motivates the following definition.

Definition 1.1. An ideal $\mathfrak{I}$ in $\operatorname{mod}\Lambda$ is called *fp-idempotent* if the class of finitely presented functors in $(\operatorname{mod}\Lambda, \mathrm{Ab})$ vanishing on $\mathfrak{I}$ is closed under extensions.

Observe that any idempotent ideal is fp-idempotent; however the converse is usually not true. There are three other concepts equivalent to fp-idempotent ideals:

- *Definable subcategories* of $\operatorname{Mod}\Lambda$. These are full subcategories of $\operatorname{Mod}\Lambda$ of the form $\{M \in \operatorname{Mod}\Lambda \mid F_i(M) = 0 \text{ for all } i \in I\}$ for a family $(F_i)_{i\in I}$ of functors in $\operatorname{Coh}\Lambda$.
- *Closed subsets* of the set $\operatorname{Ind}\Lambda$ of isomorphism classes of indecomposable pure-injective Λ-modules. These are subsets of the form $\mathcal{X} \cap \operatorname{Ind}\Lambda$ for some definable subcategory $\mathcal{X}$ of $\operatorname{Mod}\Lambda$.
- *Serre subcategories* of $\operatorname{Coh}\Lambda$. These are full subcategories of $\operatorname{Coh}\Lambda$ which are closed under forming subobjects, quotient objects, and extensions.

In order to formulate the correspondence between these concepts we introduce the following notation. Let $\mathcal{X}$ be a class of Λ-modules. We denote by $[\mathcal{X}]$ the ideal of maps in $\operatorname{mod}\Lambda$ which factor through a finite coproduct of modules in $\mathcal{X}$. The class of Λ-modules which are products of modules in $\mathcal{X}$ is denoted by $\prod \mathcal{X}$.

Fundamental correspondence. *There are bijections between*

- *the set of definable subcategories* $\mathcal{X}$ *of* $\operatorname{Mod}\Lambda$,
- *the set of closed subsets* $\mathbf{U}$ *of* $\operatorname{Ind}\Lambda$,
- *the set of Serre subcategories* $\mathcal{S}$ *of* $\operatorname{Coh}\Lambda$,
- *the set of fp-idempotent ideals* $\mathfrak{I}$ *in* $\operatorname{mod}\Lambda$.

These bijections are defined as follows:

$$\mathcal{X} \mapsto \begin{cases} \mathbf{U} = \mathcal{X} \cap \operatorname{Ind}\Lambda \\ \mathcal{S} = \{F \in \operatorname{Coh}\Lambda \mid F(M) = 0 \textit{ for all } M \in \mathcal{X}\} \\ \mathfrak{I} = [\mathcal{X}] \end{cases}$$

$$\mathbf{U} \mapsto \begin{cases} \mathcal{X} = \{M \in \operatorname{Mod}\Lambda \mid M \textit{ is a pure submodule of some } N \in \prod \mathbf{U}\} \\ \mathcal{S} = \{F \in \operatorname{Coh}\Lambda \mid F(M) = 0 \textit{ for all } M \in \mathbf{U}\} \\ \mathfrak{I} = [\prod \mathbf{U}] \end{cases}$$

$$\mathcal{S} \mapsto \begin{cases} \mathcal{X} = \{M \in \operatorname{Mod}\Lambda \mid F(M) = 0 \textit{ for all } F \in \mathcal{S}\} \\ \mathbf{U} = \{M \in \operatorname{Ind}\Lambda \mid F(M) = 0 \textit{ for all } F \in \mathcal{S}\} \\ \mathfrak{I} = \{\phi \in \operatorname{mod}\Lambda \mid F(\phi) = 0 \textit{ for all } F \in \mathcal{S}\} \end{cases}$$

$$\mathfrak{I} \mapsto \begin{cases} \mathcal{X} = \{M \in \operatorname{Mod}\Lambda \mid F(M) = 0 \textit{ for all } F \in \operatorname{Coh}\Lambda \textit{ with } F(\mathfrak{I}) = 0\} \\ \mathbf{U} = \{M \in \operatorname{Ind}\Lambda \mid F(M) = 0 \textit{ for all } F \in \operatorname{Coh}\Lambda \textit{ with } F(\mathfrak{I}) = 0\} \\ \mathcal{S} = \{F \in \operatorname{Coh}\Lambda \mid F(\phi) = 0 \textit{ for all } \phi \in \mathfrak{I}\} \end{cases}$$

The correspondence is based on the work of several mathematicians. Ziegler introduced the closed subsets of $\operatorname{Ind}\Lambda$ in model-theoretic terms and noticed that they form the closed sets of a quasi-compact space [15]; see also [8] for an algebraic argument. The correspondence between closed subsets of $\operatorname{Ind}\Lambda$ and Serre subcategories of $\operatorname{Coh}\Lambda$ has been established by Herzog in [6]; see also [8]. The definable subcategories were introduced by Crawley-Boevey [4]. Finally, fp-idempotent ideals were introduced in [12]. There one also finds a complete proof for the above correspondence.

The set $\operatorname{Ind}\Lambda$ together with Ziegler's topology is often called the *Ziegler spectrum* of Λ. We shall denote for every subset $\mathbf{U}$ of $\operatorname{Ind}\Lambda$ by

$$\overline{\mathbf{U}} = \bigcap_{\mathbf{U} \subseteq \mathbf{V} \text{ closed}} \mathbf{V}$$

the *closure* of $\mathbf{U}$ which is the smallest closed subset of $\operatorname{Ind}\Lambda$ containing $\mathbf{U}$. Note that $\mathbf{U}$ is closed with respect to Ziegler's topology if and only if $\mathbf{U} = \overline{\mathbf{U}}$.

2. Endofinite modules

The *endolength* endol(M) of a Λ-module M is the length of M viewed as a module over its endomorphism ring $\mathrm{End}_\Lambda(M)$. Following Crawley-Boevey, a Λ-module M is said to be *endofinite* provided that endol(M) is finite. It follows a list of basic properties of endofinite modules.

(E1) *If M is finite dimensional, then* $\mathrm{endol}(M) \leq \dim_k(M)$, *and equality holds if M is indecomposable.*

Proof. The endomorphism ring $\mathrm{End}_\Lambda(M)$ is a k-algebra and therefore the length of M is bounded by $\dim_k(M)$. If M is indecomposable then

$$\mathrm{endol}(M) = \dim_k(M),$$

since $\mathrm{End}_\Lambda(M)$ is local and k is algebraically closed. □

(E2) *An endofinite module M is pure-injective and has a decomposition $M = \coprod_{i\in I} M_i$ into indecomposable endofinite modules with local endomorphism rings. Conversely, such a coproduct is endofinite if and only if there are only finitely many isomorphism classes involved.*

Proof. See Proposition 4.3 in [3]. □

We denote the set of isomorphism classes of indecomposable endofinite modules as follows:

$$\mathrm{Ind}_n\,\Lambda = \{M \in \mathrm{Ind}\,\Lambda \mid \mathrm{endol}(M) = n\},$$

$$\mathrm{ind}_n\,\Lambda = \{M \in \mathrm{ind}\,\Lambda \mid \mathrm{endol}(M) = n\},$$

where $\mathrm{ind}\,\Lambda = \mathrm{Ind}\,\Lambda \cap \mathrm{mod}\,\Lambda$. Given a Λ-module M, we denote by $\mathrm{Add}\,M$ the full subcategory of Λ-modules which are direct factors of coproducts of copies of M.

(E3) *Let M be an endofinite Λ-module and denote by $M_1, \dots, M_n$ the isomorphism classes of indecomposable modules which occur in a decomposition $M = \coprod_{i\in I} M_i$ into indecomposables. Then* $\mathrm{Add}\,M$ *is a definable subcategory of* $\mathrm{Mod}\,\Lambda$ *and* $\{M_1, \dots, M_n\} = \mathrm{Add}\,M \cap \mathrm{Ind}\,\Lambda$ *is the corresponding closed subset of* $\mathrm{Ind}\,\Lambda$.

Proof. A full subcategory of $\mathrm{Mod}\,\Lambda$ is definable if and only if it is closed under products, direct limits, and pure submodules [4, 3.2]; see also [12, Theorem 2.1]. Using this result, it is proved in [12, Theorem 6.16] that $\mathrm{Add}\,M$ is a definable subcategory if M is endofinite. It follows from (E2) that $\mathrm{Add}\,M \cap \mathrm{Ind}\,\Lambda = \{M_1, \dots, M_n\}$. □

(E4) *Let M be an endofinite module. Then $[M]$ is the fp-idempotent ideal corresponding to the definable subcategory* $\mathrm{Add}\,M$. *Moreover, the ideal $[M]$ is nilpotent if and only if M has no finite dimensional indecomposable direct factor. In this case $[M]^{n+1} = 0$ where $n =$* $\mathrm{endol}(M)$.

Proof. Let $\phi\colon X \to Y$ be a map which factors through a coproduct $\coprod_{i\in I} M_i$. If X is finitely generated, then ϕ factors through $\coprod_{i\in J} M_i$ for some finite subset $J \subseteq I$. Therefore $[\operatorname{Add} M] = [M]$. The second part of the assertion follows directly from the fact that $(\operatorname{rad}\operatorname{End}_\Lambda(M))^n = 0$ by Nakayama's lemma, where $n = \operatorname{endol}(M)$. □

(E5) *The set* $\{M \in \operatorname{Ind}\Lambda \mid \operatorname{endol}(M) \leq n\}$ *is closed for all* $n \in \mathbb{N}$.

Proof. See [6, p.554]; see also [9, Proposition 5.4]. □

(E6) *Let* $M \in \operatorname{Ind}\Lambda$. *Then* $\{M\}$ *is open if and only if* $M \in \operatorname{ind}\Lambda$.

Proof. This is proved, using Auslander-Reiten theory, in [13, Proposition 13.1]. □

(E7) *Let* $U \subseteq \operatorname{ind}\Lambda$. *Then the closure* $\overline{\mathbf{U}}$ *of* $\mathbf{U}$ *corresponds to the fp-idempotent ideal* $[\mathbf{U}]$, *and* $\overline{\mathbf{U}} \setminus \mathbf{U} \subseteq \operatorname{Ind}\Lambda \setminus \operatorname{ind}\Lambda$. *Moreover,* $\overline{\mathbf{U}} = \mathbf{U}$ *if and only if* $\mathbf{U}$ *is finite.*

Proof. Clearly, $[\mathbf{U}]$ is an idempotent ideal and therefore fp-idempotent. The corresponding closed subset of $\operatorname{Ind}\Lambda$ is certainly the smallest containing $\mathbf{U}$ which is $\overline{\mathbf{U}}$. Any module in $\overline{\mathbf{U}} \setminus \mathbf{U}$ is infinite dimensional by (E6). If $\mathbf{U}$ is finite, then $\overline{\mathbf{U}} = \mathbf{U}$ since $\{M\}$ is closed for every $M \in \operatorname{ind}\Lambda$ by (E1) and (E3). Conversely, $\overline{\mathbf{U}} = \mathbf{U}$ implies that $\bigcup_{M\in\mathbf{U}}\{M\}$ is an open covering of a quasi-compact space by (E6) since $\operatorname{Ind}\Lambda$ is quasi-compact. Therefore $\mathbf{U}$ needs to be finite. □

3. Coherent functors

In this section we study functors $\operatorname{Mod}\Gamma \to \operatorname{Mod}\Lambda$ between two module categories. We assume that Λ and Γ are arbitrary rings. This is essential for the applications which will follow. Note that all concepts which have been defined in the previous sections still make sense if one replaces finite dimensional by finitely presented modules. A functor $F\colon \operatorname{Mod}\Gamma \to \operatorname{Mod}\Lambda$ is called *coherent* if the composition $\widehat{F} = \operatorname{Hom}_\Lambda(\Lambda, -) \circ F$ with the forgetful functor is coherent, i.e., there exists an exact sequence

$$(**) \qquad \operatorname{Hom}_\Gamma(Y, -) \longrightarrow \operatorname{Hom}_\Gamma(X, -) \longrightarrow \widehat{F} \longrightarrow 0$$

with X and Y finitely presented. We denote by $\operatorname{Coh}(\Gamma, \Lambda)$ the category of coherent functors $\operatorname{Mod}\Gamma \to \operatorname{Mod}\Lambda$. For example, $\operatorname{Coh}(\Gamma, \mathbb{Z}) = \operatorname{Coh}\Gamma$. Note that for any $F \in \operatorname{Coh}(\Gamma, \Lambda)$, the Λ-action on $\widehat{F}(\Gamma)$ induces a ring homomorphism

$$\phi_F\colon \Lambda \longrightarrow \operatorname{End}_\mathbb{Z}(\widehat{F}(\Gamma))^{\mathrm{op}} \cong \operatorname{End}(\widehat{F})^{\mathrm{op}}.$$

It follows a list of basic properties of coherent functors.

(C1) *The functor*

$$\operatorname{Coh}(\Gamma, \Lambda) \longrightarrow \{(G, \psi) \mid G \in \operatorname{Coh}\Gamma, \psi \in \operatorname{Hom}(\Lambda, \operatorname{End}(G)^{\mathrm{op}})\}$$

which sends F *to* $(\widehat{F}, \phi_F)$ *is an equivalence.*

Proof. We provide an inverse for this functor. Let $G \in \operatorname{Coh}\Gamma$ and $\psi\colon \Lambda \to \operatorname{End}(G)^{\mathrm{op}}$ be a ring homomorphism. The map ψ defines a Λ-action on $G(M)$ for all $M \in \operatorname{Mod}\Gamma$ and this gives rise to a coherent functor $G_\psi\colon \operatorname{Mod}\Gamma \to \operatorname{Mod}\Lambda$. Clearly, $\widehat{F}_{\phi_F} = F$ and $(\widehat{G_\psi}, \phi_{(G_\psi)}) = (G, \psi)$. □

(C2) *A functor is coherent if and only if it preserves direct limits and products.*

Proof. One implication is easy. In fact, a representable functor $\operatorname{Hom}_\Gamma(X, -)$ preserves direct limits and products provided that X is a finitely presented Γ-module. From this follows that every coherent functor preserves direct limits and products. For the converse, see Corollary 12.2 in [11]. □

(C3) *Let ${}_\Gamma B_\Lambda$ be a bimodule. Then* $\operatorname{Mod}\Gamma \to \operatorname{Mod}\Lambda$, $M \mapsto M \otimes_\Gamma B$, *is coherent if and only if B is a finitely presented Γ-module.*

Proof. Apply (C2). The tensor functor $- \otimes_\Gamma B$ always preserves direct limits; it preserves products if and only if B is a finitely presented Γ-module. □

(C4) *A coherent functor sends pure-injective modules to pure-injective modules, and pure-exact sequences to pure-exact sequences.*

Proof. The assertion follows immediately from (C2) if one uses the appropriate description of pure-injectivity and pure-exactness: a module M is pure-injective if and only if for every set I the summation map $M^{(I)} \to M$ factors through the canonical map $M^{(I)} \to M^I$, and a sequence $0 \to L \to M \to N \to 0$ is pure-exact if and only if it is a direct limit of split exact sequences $0 \to L_i \to M_i \to N_i \to 0$; see [7]. □

(C5) *Let $F\colon$* $\operatorname{Mod}\Gamma \to \operatorname{Mod}\Lambda$ *be coherent and $\mathcal{X}$ be a definable subcategory of* $\operatorname{Mod}\Lambda$. *Then* $F^{-1}(\mathcal{X}) = \{M \in \operatorname{Mod}\Gamma \mid F(M) \in \mathcal{X}\}$ *is a definable subcategory of* $\operatorname{Mod}\Gamma$.

Proof. It follows directly from (C2) that a composition of coherent functors is again coherent. If $\mathcal{X}$ is defined by the vanishing of the family $(F_i)_{i\in I}$ in $\operatorname{Coh}\Lambda$, then $F^{-1}(\mathcal{X})$ is defined by the vanishing of the family $(F_i \circ F)_{i\in I}$ in $\operatorname{Coh}\Gamma$ and is therefore definable. □

(C6) *Let $F\colon$* $\operatorname{Mod}\Gamma \to \operatorname{Mod}\Lambda$ *be coherent and $\mathbf{U}$ be a closed subset of* $\operatorname{Ind}\Gamma$. *Then the indecomposable direct factors of modules in* $\{F(M) \mid M \in \mathbf{U}\}$ *form a closed subset of* $\operatorname{Ind}\Lambda$.

Proof. See Theorem 7.8 in [10]. □

(C7) *Let $F\colon$* $\operatorname{Mod}\Gamma \to \operatorname{Mod}\Lambda$ *be coherent and suppose that $F(M)$ is indecomposable for all $M \in$* $\operatorname{Ind}\Gamma$. *Then* $\operatorname{Ind}\Gamma \to \operatorname{Ind}\Lambda$, $M \mapsto F(M)$, *is a continuous and closed map.*

Proof. Combine (C4) – (C6). □

Given a coherent functor $F\colon \operatorname{Mod}\Gamma \to \operatorname{Mod}\Lambda$, we denote by n_F the minimal number of elements which generate the Γ-module X in a presentation $(**)$ of $\operatorname{Hom}_\Lambda(\Lambda, -)\circ F$.

(C8) *Let* $F\colon \operatorname{Mod}\Gamma \to \operatorname{Mod}\Lambda$ *be a coherent functor. Then* $\operatorname{endol}(F(M)) \le n_F \cdot \operatorname{endol}(M)$ *for all* $M \in \operatorname{Mod}\Gamma$.

Proof. The length of $F(M)$ over $\operatorname{End}_\Gamma(M)$ is bounded by $n_F \cdot \operatorname{endol}(M)$. Using the ring homomorphism $\operatorname{End}_\Gamma(M) \to \operatorname{End}_\Lambda(F(M))$, it follows that $\operatorname{endol}(F(M)) \le n_F \cdot \operatorname{endol}(M)$. □

(C9) *Let* $F\colon \operatorname{Mod}\Gamma \to \operatorname{Mod}\Lambda$ *be a coherent functor. If* Γ *and* Λ *are* k*-algebras and* F *is* k*-linear, then* $\dim_k(F(M)) \le n_F \cdot \dim_k(M)$ *for all* $M \in \operatorname{Mod}\Gamma$.

Proof. Adapt the argument of (C8). □

4. Tameness

We fix again a finite dimensional algebra Λ over some algebraically closed field k. Let us recall Drozd's definition of tameness. A *one-parameter family* of Λ-modules of dimension n is the set of Λ-modules

$$\{k[T]/(T-\lambda)\otimes_{k[T]} B \mid \lambda \in k\}$$

where B is a $k[T]$-Λ-bimodule which is free of rank n over $k[T]$. The algebra is said to be of *tame representation type* provided that there is for every $n \in \mathbb{N}$ a finite number of such one-parameter families such that every indecomposable Λ-module of dimension n is isomorphic to a module in one of these families [5, 1]. The minimal number of one-parameter families which is needed to parametrize all but finitely many modules in $\operatorname{ind}_n \Lambda$ is denoted by $\mu_\Lambda(n)$. We give now an alternative description of tameness which is based on the topology defined on $\operatorname{Ind}\Lambda$.

Theorem 4.1. *A finite dimensional algebra* Λ *is of tame representation type if and only if* $\overline{\operatorname{ind}_n \Lambda} \setminus \operatorname{ind}_n \Lambda$ *is finite for all* $n \in \mathbb{N}$. *Moreover, in this case*

$$\mu_\Lambda(n) = \operatorname{card}(\overline{\operatorname{ind}_n \Lambda} \setminus \operatorname{ind}_n \Lambda)$$

for all $n \in \mathbb{N}$.

We need the following easy reformulation of the classical tameness definition.

Lemma 4.2. *The algebra* Λ *is tame if and only if there exists for every* $n \in \mathbb{N}$ *a finite product* $\Gamma = k[T]\times \ldots \times k[T]$ *of polynomial rings and a* k*-linear exact and coherent functor* $F\colon \operatorname{Mod}\Gamma \to \operatorname{Mod}\Lambda$ *such that* $\operatorname{ind}_n \Lambda \subseteq F(\operatorname{ind}_1 \Gamma)$.

Proof. Straightforward. □

The next lemma presents the information about $\operatorname{Ind} k[T]$ which is sufficient for our purpose.

Lemma 4.3. $\operatorname{Ind}_1 k[T] = \operatorname{ind}_1 k[T] \cup \{k(T)\} = \overline{\operatorname{ind}_1 k[T]}$.

Proof. The Ziegler spectrum of a Dedekind domain is computed in [15, Example 9.5]. □

Proof of Theorem 4.1. Suppose first that Λ has tame representation type and let $n \in \mathbb{N}$. Consider the corresponding coherent functor $F\colon \operatorname{Mod}\Gamma \to \operatorname{Mod}\Lambda$ such that $\operatorname{ind}_n \Lambda \subseteq F(\operatorname{ind}_1 \Gamma)$ and $\Gamma = k[T] \times \ldots \times k[T]$ which exists by Lemma 4.2. Note that $\overline{\operatorname{ind}_1 \Gamma} = \operatorname{ind}_1 \Gamma \cup \{Q_1, \ldots Q_r\}$ by Lemma 4.3, where r denotes the number of factors of Γ. It follows from (C6) that every module in $\overline{\operatorname{ind}_n \Lambda} \setminus \operatorname{ind}_n \Lambda$ is isomorphic to a direct factor of $F(Q_i)$ for some i. However, each $F(Q_i)$ is endofinite by (C8) and has therefore only finitely many non-isomorphic direct factors by (E2). We conclude that $\overline{\operatorname{ind}_n \Lambda} \setminus \operatorname{ind}_n \Lambda$ is finite.

To prove the converse we assume that the algebra Λ is not tame. We apply the Tame and Wild Theorem in a form which is due to Crawley-Boevey [2]. In fact, there exists a k-linear *representation embedding*

$$F\colon \operatorname{Mod} k\langle X, Y\rangle \longrightarrow \operatorname{Mod}\Lambda$$

which as far as we are concerned means that F is coherent and induces an injective map $\operatorname{Ind} k\langle X,Y\rangle \to \operatorname{Ind}\Lambda$. Now consider for every pair $\alpha, \beta \in k$ the $k\langle X,Y\rangle$-module

$$M_{\alpha,\beta} = k\langle X,Y\rangle/(X-\alpha, Y-\beta)$$

and let Q_α be the endofinite $k\langle X,Y\rangle$-module whose underlying space is $k(T)$ with X acting by multiplication with α and Y acting by multiplication with T. Using Lemma 4.3, it is not hard to see that for each $\alpha \in k$ the closure of $\mathbf{U}_\alpha = \{M_{\alpha,\beta} \mid \beta \in k\}$ is $\mathbf{U}_\alpha \cup \{Q_\alpha\}$, and therefore $\mathbf{Q} = \{Q_\alpha \mid \alpha \in k\} \subseteq \overline{\mathbf{U}}$ with $\mathbf{U} = \bigcup_{\alpha\in k} \mathbf{U}_\alpha \subseteq \operatorname{Ind}_1 k\langle X,Y\rangle$. Now let $n_F \in \mathbb{N}$ such that $\operatorname{endol}(F(M)) \leq n_F \cdot \operatorname{endol}(M)$ for all $M \in \operatorname{Mod} k\langle X,Y\rangle$ which exists by (C8). It follows that

$$F(\mathbf{U}) \subseteq \operatorname{ind}_1 \Lambda \cup \ldots \cup \operatorname{ind}_{n_F} \Lambda$$

since F sends finite dimensional to finite dimensional modules by (C9). This implies

$$F(\mathbf{Q}) \subseteq F(\overline{\mathbf{U}}) = \overline{F(\mathbf{U})} \subseteq \overline{\operatorname{ind}_1 \Lambda} \cup \ldots \cup \overline{\operatorname{ind}_{n_F} \Lambda}$$

since the map $\operatorname{Ind} k\langle X,Y\rangle \to \operatorname{Ind}\Lambda$ is continuous and closed by (C7). Moreover,

$$F(\mathbf{Q}) \subseteq F(\overline{\mathbf{U}} \setminus \mathbf{U}) = \overline{F(\mathbf{U})} \setminus F(\mathbf{U})$$

which is contained in $\operatorname{Ind}\Lambda \setminus \operatorname{ind}\Lambda$ by (E7). We conclude that $\overline{\operatorname{ind}_n \Lambda} \setminus \operatorname{ind}_n \Lambda$ is infinite for some $n \leq n_F$.

It remains to show that

$$\mu_\Lambda(n) = \operatorname{card}(\overline{\operatorname{ind}_n \Lambda} \setminus \operatorname{ind}_n \Lambda)$$

for all $n \in \mathbb{N}$. In [9, Corollary 9.7] it is shown that an indecomposable module M belongs to $\overline{\operatorname{ind}_n \Lambda} \setminus \operatorname{ind}_n \Lambda$ if and only if M is infinite dimensional and $\operatorname{endol}(M)$ divides n. There are precisely $\mu_\Lambda(n)$ indecomposable modules with this property by [2, Theorem 5.6], and this gives the equality. □

We continue with a description of the set $\overline{\operatorname{ind}_n \Lambda} \setminus \operatorname{ind}_n \Lambda$ in terms of fp-idempotent ideals in $\operatorname{mod} \Lambda$. To this end let

$$\operatorname{fpnil}_n \Lambda = \{\mathfrak{I} \subseteq \operatorname{mod} \Lambda \mid \mathfrak{I} \text{ is fp-idempotent, nilpotent, and } 0 \neq \mathfrak{I} \subseteq [\operatorname{ind}_n \Lambda]\}.$$

Proposition 4.4. *Let $n \in \mathbb{N}$. Then $\operatorname{fpnil}_n \Lambda$ is finite if and only if $\overline{\operatorname{ind}_n \Lambda} \setminus \operatorname{ind}_n \Lambda$ is finite. Moreover, the map*

$$\overline{\operatorname{ind}_n \Lambda} \setminus \operatorname{ind}_n \Lambda \longrightarrow \{\mathfrak{I} \in \operatorname{fpnil}_n \Lambda \mid \mathfrak{I} \supseteq \mathfrak{J} \in \operatorname{fpnil}_n \Lambda \textit{ implies } \mathfrak{I} = \mathfrak{J}\}$$

which sends M to $[M]$ is bijective.

Proof. Let $\mathbf{U}_n = \operatorname{ind}_n \Lambda$. We use the bijection between closed subsets of $\operatorname{Ind} \Lambda$ and fp-idempotent ideals in $\operatorname{mod} \Lambda$. This correspondence sends $\overline{\mathbf{U}_n}$ to $[\mathbf{U}_n]$ by (E7), and induces therefore an injective map

$$\overline{\mathbf{U}_n} \longrightarrow \{\mathfrak{I} \subseteq \operatorname{mod} \Lambda \mid \mathfrak{I} \subseteq [\mathbf{U}_n] \text{ is fp-idempotent}\}, \quad M \mapsto [M],$$

because every module in $\overline{\mathbf{U}_n}$ is endofinite by (E5), hence a closed singleton by (E3). It follows from (E4) and (E7) that $[M]$ is nilpotent for every $M \in \overline{\mathbf{U}_n} \setminus \mathbf{U}_n$, and it is clear that the image of the restriction $\overline{\mathbf{U}_n} \setminus \mathbf{U}_n \to \operatorname{fpnil}_n \Lambda$ is precisely the set of minimal elements in $\operatorname{fpnil}_n \Lambda$. In particular

$$\operatorname{card}(\overline{\mathbf{U}_n} \setminus \mathbf{U}_n) \leq \operatorname{card}(\operatorname{fpnil}_n \Lambda).$$

On the other hand, every $\mathfrak{I} \in \operatorname{fpnil}_n \Lambda$ corresponds to a closed subset of $\overline{\mathbf{U}_n} \setminus \mathbf{U}_n$, and therefore $\operatorname{fpnil}_n \Lambda$ is finite if $\overline{\mathbf{U}_n} \setminus \mathbf{U}_n$ is finite. □

Combining Theorem 4.1 and Proposition 4.4, one obtains the following alternative description of tameness which only involves finite dimensional modules.

Corollary 4.5. *The algebra Λ is of tame representation type if and only if for every $n \in \mathbb{N}$ there are only finitely many non-zero fp-idempotent and nilpotent ideals in $\operatorname{mod} \Lambda$ which are contained in $[\operatorname{ind}_n \Lambda]$. Moreover, in this case $\mu_\Lambda(n)$ equals the number of minimal elements in the set of these ideals.*

5. One-parameter families and generic modules

The infinite dimensional modules occuring in the preceding characterizations of tame representation type are generic modules. Recall that an indecomposable Λ-module is *generic* if it is of finite endolength but of infinite length over Λ; see [2]. In this section we describe explicitly the relation between one-parameter families and generic modules over algebras of tame representation type. The results presented here are mainly due to Crawley-Boevey; those results which involve the Ziegler spectrum are due to the author of this note. We do not give proofs but refer to Section 5 in [2] and Section 9 in [9]. Throughout this section Λ is a finite dimensional algebra over some algebraically closed field.

Theorem 5.1. *Let Λ be of tame representation type and let $n \in \mathbb{N}$. Then there are one-parameter families $\mathbf{U}_1, \ldots, \mathbf{U}_{\mu_\Lambda(n)}$ of n-dimensional indecomposable Λ-modules having the following properties:*

(1) $\mathrm{ind}_n \Lambda \setminus (\mathbf{U}_1 \cup \ldots \cup \mathbf{U}_{\mu_\Lambda(n)})$ *is finite.*
(2) *The closure* $\overline{\mathbf{U}_i}$ *contains a unique generic* Λ*-module* G_i *for every* i.
(3) $G_i = G_j$ *if and only if* $i = j$.
(4) $\overline{\mathrm{ind}_n \Lambda} \setminus \mathrm{ind}_n \Lambda = \{G_1, \ldots, G_{\mu_\Lambda(n)}\}$.
(5) *A generic* Λ*-module belongs to* $\{G_1, \ldots, G_{\mu_\Lambda(n)}\}$ *if and only if its endolength divides* n.

In [1], Crawley-Boevey has shown that over a tame algebra almost all indecomposable modules of a fixed dimension belong to homogeneous tubes of the Auslander-Reiten quiver of Λ. Recall that a family $\mathbf{T} = (M_i)_{i \in \mathbb{N}}$ of Λ-modules forms a *homogeneous tube* if there are maps $M_i \to M_{i+1}$ and $M_{i+1} \to M_i$ for every $i \in \mathbb{N}$ which induce almost split sequences $0 \to M_i \to M_{i-1} \coprod M_{i+1} \to M_i \to 0$ for every $i \in \mathbb{N}$ where $M_0 = 0$. We say that $\mathbf{T}$ is *generic* if $\overline{\mathbf{T}} = \mathbf{T} \cup \{\varinjlim M_i, \varprojlim M_i, G\}$ for some generic module G.

Theorem 5.2. *Let* Λ *be of tame representation type and let* $n \in \mathbb{N}$. *Then all but finitely many* n*-dimensional indecomposable* Λ*-modules belong to generic homogeneous tubes.*

The occurence of homogeneous tubes is a characteristic phenomenon for tame algebras. This can be made precise as follows.

Theorem 5.3. *An algebra* Λ *is of tame representation type if and only if every generic* Λ*-module belongs to the closure of a (generic) homogeneous tube.*

References

[1] W.W. Crawley-Boevey, On tame algebras and bocses, Proc. London Math. Soc. **56** (1988), 451–483.
[2] W.W. Crawley-Boevey, Tame algebras and generic modules, Proc. London Math. Soc. **63** (1991), 241–264.
[3] W.W. Crawley-Boevey, Modules of finite length over their endomorphism ring, in: Representations of algebras and related topics, eds. S. Brenner and H. Tachikawa, London Math. Soc. Lec. Note Series **168** (1992), 127–184.
[4] W.W. Crawley-Boevey, Infinite dimensional modules in the representation theory of finite dimensional algebras, preprint (1996).
[5] Yu.A. Drozd, Tame and wild matrix problems, Representations and quadratic forms, Institute of Mathematics, Academy of Sciences, Ukrainian SSR, Kiev (1979), 39–74, Amer. Math. Soc. Transl. **128** (1986), 31–55.
[6] I. Herzog, The Ziegler spectrum of a locally coherent Grothendieck category, Proc. London Math. Soc. **74** (1997), 503–558.
[7] C.U. Jensen and H. Lenzing, Model theoretic algebra, Gordon and Breach, New York (1989).
[8] H. Krause, The spectrum of a locally coherent category, J. Pure Appl. Algebra **114** (1997), 259–271.
[9] H. Krause, Generic modules over artin algebras, Proc. London Math. Soc. **76** (1998), 276–306.
[10] H. Krause, Exactly definable categories, J. Algebra **201** (1998), 456–492.
[11] H. Krause, Functors on locally finitely presented categories, Colloq. Math. **75** (1998), 105–131.

[12] H. KRAUSE, The spectrum of a module category, Habilitationsschrift, Universität Bielefeld (1998), Mem. Amer. Math. Soc., to appear.
[13] M. PREST, Model theory and modules, London Math. Soc. Lec. Note Series **130** (1988).
[14] C.M. RINGEL, Recent advances in the representation theory of finite dimensional algebras, in: Representations of finite groups and finite dimensional algebras, Birkhäuser-Verlag, Basel (1991), 141–192.
[15] M. ZIEGLER, Model theory of modules, Ann. of Pure and Appl. Logic **26** (1984), 149–213.

FAKULTÄT FÜR MATHEMATIK, UNIVERSITÄT BIELEFELD, 33501 BIELEFELD, GERMANY
E-mail address: henning@mathematik.uni-bielefeld.de

Trends in Mathematics, © 2000 Birkhäuser Verlag Basel/Switzerland

INVARIANCE OF TAMENESS UNDER STABLE EQUIVALENCE: KRAUSE'S THEOREM

HELMUT LENZING

1. Introduction

This paper presents a streamlined proof of a theorem of H. Krause [8] regarding subsequent improvements by Krause and G. Zwara [10]. Our account is a revised version of a mini-course given at the Euroconference "Homological Invariants in Representation Theory" in Ioannina, March 1999. We include necessary background on infinite dimensional modules and functor categories centered around the concept of purity.

Theorem 1 (Krause). *For an algebraically closed field k let Λ and Γ be finite dimensional k-algebras whose stable categories $\overline{\text{mod}}\text{-}\Lambda$ and $\overline{\text{mod}}\text{-}\Gamma$ of finite dimensional modules modulo injectives (or modulo projectives) are equivalent. Then Λ is tame if and only if Γ is tame.*

The theorem is remarkable since through the passage from the module category mod-Λ to the stable category $\overline{\text{mod}}\text{-}\Lambda$ significant information is lost. At first sight, that is on the level of indecomposable objects, the information loss looks harmless since only finitely many indecomposable modules (the injectives, respectively the projectives) disappear. We note therefore that (a) the stable category $\overline{\text{mod}}\text{-}\Lambda$ is no longer abelian, (b) the morphism structure is destroyed, (c) the concept of dimension of modules, hence the geometric concept of families, essential for the definition of tameness, is lost. The roundabout to such difficulties comes from a theorem of W. Crawley-Boevey [2] characterizing tameness in terms of generic modules. Following [2], a Λ-module M is called *endofinite* if M has finite length as a (left) module over its endomorphism ring $\operatorname{End} M$, this length will be called the *endolength* of M. An equivalent but more flexible condition requests that $\operatorname{Hom}(E, M)$ has finite length over $\operatorname{End} M$ for each finitely presented module E. Further, M is called *generic* if M is endofinite, indecomposable and infinite dimensional over k.

Theorem 2 (Crawley-Boevey). *Let k be an algebraically closed field. For a finite dimensional k-algebra Λ the following assertions are equivalent:*

(i) *Λ is tame.*
(ii) *Λ is* generically tame, *that is, for each natural number d there are—up to isomorphism—only finitely many generic modules of endolength d.*

(iii) *For each generic module M its endomorphism algebra* $\operatorname{End} M$ *satisfies a polynomial identity.* □

Over an algebraically closed field, a finite dimensional algebra Λ is representation-finite if and only if Λ has no generic module. Thus all representation-finite algebras are generically tame. The proof of the implication "(iii) $\Longrightarrow$ (i)" uses Drozd's *tame-wild-dichotomy*, stating that a finite dimensional k-algebra is either tame or wild, whose proof uses heavily that the base field k is algebraically closed. In fact, for base fields that are not algebraically closed there exists, at present, no generally accepted definition of tameness in terms of families of modules. Property (ii) should therefore be taken as the *definition of tameness* for arbitrary base fields. With this convention in mind, Theorem 1 holds for arbitrary base fields.

The major tool in Krause's proof is a transfer mechanism, see Theorem 6.1, relating the pure-injective modules, in particular the generic modules, for finite dimensional algebras that are stably equivalent modulo injectives (or projectives). In more detail, one shows that each stable equivalence $f: \overline{\text{mod}}\text{-}\Lambda \to \overline{\text{mod}}\text{-}\Gamma$ induces a bijection from generic Λ-modules to generic Γ-modules, that allows to control the endolength. This fact is highly non-trivial; it is based on an analysis of the canonical functor $q: \text{Mod-}\Lambda \to \varinjlim \overline{\text{mod}}\text{-}\Lambda$ from the category Mod-Λ of all Λ-modules to the limit closure $\varinjlim \overline{\text{mod}}\text{-}\Lambda$ (under direct limits) of the stable category $\overline{\text{mod}}\text{-}\Lambda$, see Sections 4 and 5 for further details. Like Mod-A the category $\varinjlim \overline{\text{mod}}\text{-}A$ is locally finitely presented, so provides the proper framework for the study of pure-exact sequences, pure-injective and generic objects. Since the functor q commutes with direct limits and products it preserves purity and pure-injectivity. The core of Krause's proof, where all technical difficulties concentrate, is to establish that q induces a bijection between generic Λ-modules and generic objects of $\varinjlim \overline{\text{mod}}\text{-}\Lambda$ preserving radical factor rings.

Let X^I, respectively $X^{(I)}$, stand for the direct product, respectively the direct sum or coproduct, of a family of copies of X indexed by the set I.

2. Purity, flatness and pure-injectivity

Let R be a ring, associative with 1. Because of later generalizations of contents of this section it would be too restrictive to think of R as a finite dimensional algebra. Modules will always be right modules if not specified otherwise. An R-module E is called *finitely generated* (respectively *finitely presented*) if there is an exact sequence $R^n \to E \to 0$ (respectively $R^m \to R^n \to E \to 0$). By Mod-R, respectively by mod-R, we denote the category of all R-modules, respectively of all finitely presented R-modules. Finitely presented modules are important because each Λ-module M is a direct limit of finitely presented modules. In fact, the category $\mathcal{E}/M$ of finitely presented modules over M having objects (E, u), with E finitely presented and $u: E \to M$ R-linear, and morphisms $\alpha: (E', u') \to (E, u)$ given by R-linear maps $\alpha: E' \to E$ satisfying $u' = u\alpha$, is filtered and M is the direct limit of the forgetful functor $F: \mathcal{E}/M \to \text{Mod-}R$ sending (E, u) to E. As a useful consequence, a module E is finitely presented if and only if the functor $\mathrm{Hom}(E,\text{-})$ commutes with direct limits. Recall that a category $\mathcal{I}$ is *filtered* if for any two objects i, j there exists morphisms $u : i \to k$ and $v : j \to k$ into a third object k; moreover, for any two morphisms $u_1, u_2 : i \to j$ in $\mathcal{I}$ there exists $v : j \to k$ such that $vu_1 = vu_2$. In this paper the term direct limit refers to a colimit with respect to a (skeletally) small filtered category.

A short exact sequence $\eta : 0 \to X' \to X \to X'' \to 0$ is called *pure-exact* if every morphism $f : E \to X''$, with E finitely presented, lifts to X. Purity of η may thus be considered as a weak form of split-exactness, and in fact a short exact sequence η is pure-exact if and only if it is a direct limit $\eta = \varinjlim \eta_\alpha$ of split-exact sequences η_α. Another useful characterization of pure-exactness is in terms of tensor products: η is pure-exact if and only if $0 \to X' \otimes_R M \to X \otimes_R M \to X'' \otimes_R M \to 0$ is exact for each left R-module M. Since tensor products commute with direct limits it suffices to restrict this condition to finitely presented M's. Next we review the concepts of flat, fp-injective, pure-projective and pure-injective modules.

Flatness: A module M is called *flat* if each morphism $f : E \to M$, with E finitely presented, factors through some finitely generated free module R^n, or equivalently through some projective module. Flat modules are closed under direct limits, and flatness should be considered as a weak version of projectivity. In fact, a module is flat if and only if it is a direct limit of finitely generated projective, alternatively of finitely generated free, modules: If M is flat, then the category $\mathcal{F}/M$ of finitely generated free modules over M consisting of pairs (F, u), with $u : F \to M$ R-linear and F finitely generated free, is filtered and the direct limit of the forgetful functor $\mathcal{F}/M \to \text{Mod-}\Lambda$, $(F, u) \mapsto F$, is isomorphic to M. While each projective module is flat, the converse is not true in general, take for instance $\mathbb{Q}$ as a $\mathbb{Z}$-module.

It is an immediate consequence of the definitions that M is flat if and only if each exact sequence $0 \to X' \to X \to M \to 0$ is pure-exact. And further, M is flat if and only if $M \otimes_R \text{-}$ is an exact functor from left Λ-modules to abelian groups.

Fp-injectivity: A module Q is called *fp-injective* if $\mathrm{Ext}^1(F, Q) = 0$ for each finitely presented module F. To a certain extent this concept is dual to the concept of flatness. Exactly for the right noetherian rings fp-injectivity agrees with injectivity, otherwise it is weaker.

Pure-projectivity: A module P is called *pure-projective* if the functor $\mathrm{Hom}_R(P,\text{-})$ is exact on pure-exact sequences. Hence P is pure-projective if and only if each pure-exact sequence $0 \to X' \to X \to P \to 0$ splits.

Moreover, pure-projective modules are exactly the direct summands of, possibly infinite, direct sums of finitely presented R-modules. For a finite dimensional algebra, the pure-projective modules are just the direct sums of finitely presented modules.

Pure-injectivity: A module Q is called *pure-injective*, or algebraically compact, if the functor $\mathrm{Hom}_R(\text{-}, Q)$ is exact on pure-exact sequences, equivalently if each pure-exact sequence $0 \to Q \to Y \to Y'' \to 0$ splits. In particular, each injective module is pure-injective. On a formal level pure-injectivity is dual to pure-projectivity; however, the theory of pure-injective modules is much richer, equipped with many features without pure-projective counterpart. We illustrate this by a few examples:

Duals are always pure-injective. For instance, let R be a k-algebra, and $D = \mathrm{Hom}_k(\text{-}, k)$ the usual vector-space duality. Consider an arbitrary left R-module M, from the sequence of isomorphisms of functors $\mathrm{Hom}_R(\text{-}, DM) = \mathrm{Hom}_R(\text{-}, \mathrm{Hom}_k(M, k)) \cong \mathrm{Hom}_k(\text{-} \otimes_R M, k)$ we conclude that DM is pure-exact. It follows that a module over a finite dimensional algebra is pure-injective if and only if it is a direct summand of a direct product of finite dimensional modules.

Pure-injectivity is preserved under functors commuting with (infinite) direct sums and direct products. This follows from the fact [7] that a module is pure-injective if and only if the summation map $\Sigma : Q^{(I)} \to Q$, $(x_\alpha) \mapsto \sum_\alpha x_\alpha$, extends to an R-linear map $\bar{\Sigma} : Q^I \to Q$. In particular, pure-injectivity is preserved under various kinds of forgetful functors. The next proposition, recording a special case of a more general fact due to H. Bass [1], provides a good opportunity to verify how various aspects of purity fit together.

Proposition 2.1 (Bass). *Over a finite dimensional algebra Λ each flat module F is projective, moreover F decomposes into a direct sum of finite dimensional indecomposable projective modules.*

PROOF. The radical $J = \mathrm{rad}\,\Lambda$ of Λ is nilpotent, hence $J^n = 0$ for some integer n. Since A/J is a semisimple algebra, the module $F/F.J$ has a decomposition $F/F.J = \bigoplus_{\alpha \in I} S_\alpha$ into simple modules S_α. For each $\alpha \in I$ we choose a projective cover $\nu_\alpha : P_\alpha \to S_\alpha$ and put $P = \bigoplus_{\alpha \in I} P_\alpha$. By projectivity of P, the morphism $\oplus\nu_\alpha : \bigoplus P_\alpha \to \bigoplus S_\alpha = F/F.J$ lifts to a morphism $\pi : P \to F$ against the natural epimorphism $F \to F/F.J$. Let U be the image of π, then $U + F.J = F$, hence $F/U = (F/U).J$. By nilpotency of J this yields $F/U = 0$. Hence π is surjective, and we arrive at an exact sequence $0 \to N \xrightarrow{j} P \xrightarrow{\pi} F \to 0$, whose kernel term N sits in $P.J$. Moreover N is a pure submodule of P since F is flat. The aim is to show that $N = 0$. We claim that for each $x \in N$ there exists a linear map $\pi_x : P \to N$ with $\pi_x(x) = x$. Indeed, there exists a finite subset J of I such that $x.A \subseteq \bigoplus_{\alpha \in J} P_\alpha$, and we arrive at a commutative diagram

$$\begin{array}{ccccccccc} 0 & \to & N & \longrightarrow & \bigoplus_{\alpha\in I} P_\alpha & \longrightarrow & F & \to & 0 \\ & & i\uparrow & & j\uparrow & & \uparrow f & & \\ 0 & \to & x.A & \hookrightarrow & \bigoplus_{\alpha\in J} P_\alpha & \longrightarrow & E & \to & 0 \end{array}$$

with exact rows, where i and j are the respective inclusions. Since E is finitely presented, the mapping f lifts to $\bigoplus_{\alpha \in I} P_\alpha$, hence, by the homotopy-lemma, i

extends to $\bigoplus_{\alpha \in J} P_\alpha$, therefore to a linear map $\pi_x : P \to N$ with the required property. This implies in turn that $x = \pi_x(x)$ is contained in $\pi_x(P.J)$, hence in $N.J$, and therefore $N = N.J = N.J^2 = \cdots = N.J^n = 0$. □

3. The hierarchy of pure-injectives

In order to deal with generic modules, it is important to view their position in the world of pure-injective modules. We have the following hierarchy of pure-injectives:

pure-injective modules	$\supset$	Σ-pure-injective modules	$\supset$	endofinite modules	$\supset$	generic modules

A module Q is called *Σ-pure-injective* if the direct sum $Q^{(I)}$ is pure-injective for each index set I. Since direct sums of injective modules are injective for (right) noetherian rings, all such injectives are Σ-pure-injective.

As the next result will show, Σ-pure-injectivity is a strong finiteness condition. A subgroup of a Λ-module M is called *finitely definable* (alternatively pp-definable, or a subgroup of finite definition, or a finite matrix subgroup) if it arises as the image of the evaluation map $\operatorname{Hom}_\Lambda(E, M) \to M$, $u \mapsto u(e)$, for some finitely presented Λ-module E and some element $e \in E$. The next result is due to L. Gruson and C. U. Jensen [6]. For a proof and further characterizations of pure-injective modules we refer to [7].

Proposition 3.1 (Gruson–Jensen). *A module M is Σ-pure-injective if and only if M satisfies the descending chain condition for finitely definable subgroups.* □

Corollary 3.2. *Each endofinite Λ-module M is Σ-pure-injective.*

PROOF. Each finitely definable subgroup is an $\operatorname{End} M$-submodule of M. □

(i) Each finite dimensional Λ-module over a k-algebra, k a field, is endofinite, hence Σ-pure-injective. (ii) As follows from Proposition 3.1 a pure submodule of a Σ-pure-injective module is again Σ-pure-injective, hence is a direct factor. It further follows that each Σ-pure-injective module is a, typically infinite, direct sum of indecomposable Σ-pure-injectives. Note, that the corresponding assertions do not hold for pure-injective modules in general. (iii) The endomorphism ring E of an indecomposable pure-injective module is local, that is, $E/\operatorname{rad} E$ is a skew field. (iv) If M is an endofinite module over a finite dimensional algebra, then M has a decomposition $U_1^{(I_1)} \oplus \cdots \oplus U_n^{(I_n)}$, where each U_i is indecomposable and either finite dimensional or generic.

4. The limit closure $\varinjlim \mathcal{A}$

Let $\mathcal{A}$ be a small preadditive category. Then the category $(\mathcal{A}^{op}, \mathrm{Ab})$ of contravariant additive functors from $\mathcal{A}$ to abelian groups can—to a very large extend—be treated like a category of modules over a ring. In particular, if $\mathcal{A}$ consists of a single object $*$ with endomorphism ring R, then $(\mathcal{A}^{op}, \mathrm{Ab})$ is just the category

Mod-R of right R-modules. We may thus view a small preadditive category in general as a kind of multiobject ring and treat functors $M:\mathcal{A}^{op} \to \mathrm{Ab}$ as right modules over $\mathcal{A}$. The following (rudimentary) dictionary provides a largely automatic transfer of concepts from modules to functors. For technical convenience we assume from now on that $\mathcal{A}$ is *closed under finite direct sums and direct summands*, where the last request means that idempotents of $\mathcal{A}$ split.

A right module M over $\mathcal{A}$ is an additive functor $M:\mathcal{A}^{op} \to \mathcal{A}$. These functors form the category $(\mathcal{A}^{op}, \mathrm{Ab})$, to be interpreted as the category of right $\mathcal{A}$-modules. Exact sequences of functors are defined pointwise. In view of Yoneda's lemma $\mathrm{Hom}((\text{-},A),M) = M(A)$, the role of the finitely generated projective R-modules is taken by the representable functors $(\text{-},A)$ with A in $\mathcal{A}$. A functor E is called finitely generated (respectively finitely presented) if there is an exact sequence $(\text{-},A) \to M \to 0$ (respectively $(\text{-},A) \to (\text{-},B) \to M \to 0$) with A, $B \in \mathcal{A}$. Again, each M is a direct limit of finitely presented functors, and M is finitely presented if and only if $\mathrm{Hom}(M,\text{-})$ commutes with direct limits. A right ideal J in $\mathcal{A}$ is a system of subgroups $J(A,B) \subseteq (A,B)$ with $J(A,B) \circ (A',A) \subseteq (A',B)$. Like a module category $(\mathcal{A}^{op}, \mathrm{Ab})$ has exact direct limits with the objects $(-,A)$, $A \in \mathcal{A}$, forming a system of generators. It is therefore a *Grothendieck category*, in particular has injective envelopes. Finally, a functor M is flat if and only if each $f:E \to M$, E finitely presented, factors through some $(\text{-},A)$ with $A \in \mathcal{A}$.

We denote by $\mathrm{Flat}(\mathcal{A}^{op}, \mathrm{Ab})$ the full subcategory of $(\mathcal{A}^{op}, \mathrm{Ab})$ consisting of all flat functors. Like for modules, a direct limit of flat functors is flat; further each flat functor is a direct limit of finitely generated projective functors $(\text{-},A_\alpha)$, where $A_\alpha \in \mathcal{A}$. An object E is called *finitely presented* in $\mathrm{Flat}(\mathcal{A}^{op}, \mathrm{Ab})$ if $\mathrm{Hom}(E,\text{-})$ commutes with direct limits which happens exactly for the representable functors $(\text{-},A)$ with A in $\mathcal{A}$.

Proposition 4.1. *The category* $\mathrm{Flat}(\mathcal{A}^{op}, \mathrm{Ab})$ *is a category with direct limits. There is a full embedding* $j:\mathcal{A} \hookrightarrow \mathrm{Flat}(\mathcal{A}^{op}, \mathrm{Ab})$, $A \mapsto (\text{-},A)$, *yielding an identification between* $\mathcal{A}$ *and the full subcategory of finitely presented objects in* $\mathrm{Flat}(\mathcal{A}^{op}, \mathrm{Ab})$. *Moreover, each object of* $\mathrm{Flat}(\mathcal{A}^{op}, \mathrm{Ab})$ *is a direct limit of objects from* $\mathcal{A}$. □

Because of the Proposition, we call $\mathrm{Flat}(\mathcal{A}^{op}, \mathrm{Ab})$ the *limit closure* of $\mathcal{A}$ and use the notation $\varinjlim \mathcal{A}$. As an additive category with direct limits, where each object is a direct limit of finitely presented ones, $\varinjlim \mathcal{A}$ is a *locally finitely presented category*. As for module categories purity and the derived concepts like pure-injectivity make sense in such categories. As a general reference we recommend [3]. Next we determine the limit closure of mod-R.

Proposition 4.2. *Let* R *be a ring. An additive functor* $M:(\text{mod-}R)^{op} \to \mathrm{Ab}$ *is flat if and only if is left exact. Moreover, the full embedding* Mod-$R \to ((\text{mod-}R)^{op}, \mathrm{Ab})$, $M \mapsto (\text{-},M)$, *induces an equivalence between* Mod-R *and the category* $\mathrm{Flat}((\text{mod-}R)^{op}, \mathrm{Ab})$, *identifying* Mod-$R$ *with the limit closure* $\varinjlim$ mod-R *of* mod-R.

PROOF. Let $F:(\text{mod-}R)^{op} \to \text{Ab}$ be left exact. For any R-module X, and element $x \in X$ we denote by $\bar{x}: R \to X$ the R-linear mapping sending 1 to x. Then the mappings $\eta_X : FX \to \text{Hom}_R(X, FR)$, $u \mapsto [x \mapsto (F\bar{x})(u)]$, yields a natural transformation $\eta: F \to \text{Hom}(\text{-}, FR)$, such that η_R is an isomorphism. Therefore, by left exactness of $\text{Hom}(\text{-}, FR)$ and F, the mapping η_E is an isomorphism for each E with an exact sequence $R^m \to R^n \to E \to 0$. Next, let M be an R-module. Since $M = \varinjlim E_\alpha$, with $E_\alpha \in \text{mod-}R$, we get that $(\text{-}, M) = \varinjlim(\text{-}, E_\alpha)$ is flat, as a direct limit of projective functors. Conversely, assume that $F:(\text{mod-}R)^{op} \to \text{Ab}$ is flat, hence a direct limit of finitely generated free functors $(\text{-}, E_\alpha)$. It follows that F is left exact. □

From now on we assume Λ is a *finite dimensional algebra* over a field k. Let $\mathcal{C} = \text{add}(C)$ be the additive closure of some C in mod-Λ and let $\text{Add}(C)$ consist of all direct sums ob objects from $\mathcal{C}$.

Proposition 4.3. *Let* $\mathcal{C} = \text{add}(C)$ *with* $C \in \text{mod-}\Lambda$. *Then* $\varinjlim \mathcal{C}$ *is naturally equivalent to* $\text{Add}(C)$. *Moreover,* $\text{Add}(C)$ *is closed under the formation of direct products, direct limits and pure submodules.*

PROOF. We identify $\varinjlim \mathcal{C}$ with the full subcategory of right A-modules M such that each $f : E \to M$, $E \in \text{mod-}A$, factors through a member of $\mathcal{C}$. It is straightforward to check that $\varinjlim \mathcal{C}$ is closed under direct limits, direct products and pure submodules. To prove closedness under direct products, one uses that all $f : E \to C'$, $C' \in \mathcal{C}$, factor through a *fixed* morphism $\varphi : E \to C^n$, whose components $\varphi_1, \ldots, \varphi_n$ form a k-basis of $\text{Hom}(E, C)$. Denoting by $\{C\}$ the full subcategory of Mod-Λ consisting of the single object C, restriction of functors from $\mathcal{C}$ to $\{C\}$ yields natural equivalences $\varinjlim \mathcal{C} = \text{Flat}(\mathcal{C}^{op}, \text{Ab}) = \text{Flat}(\{C\}^{op}, \text{Ab}) = \text{Flat-End}\, C$. Since $\text{End}\, C$ is a finite dimensional algebra, the assertions follow from Proposition 2.1. □

5. Categories of exact functors

In this section we are going to convert pure-exact sequences into exact sequences and pure-injectives into injectives; further we represent Mod-Λ, and related categories, as categories of exact functors. The topic starts with a technique due to Gruson and Jensen [5].

Proposition 5.1 (Gruson–Jensen). *Let* Λ *be any ring. Then the functor* $j : \text{Mod-}\Lambda \to (\Lambda\text{-mod}, \text{Ab})$, $M \mapsto M \otimes_\Lambda -$ *is a full embedding, commuting with direct products and direct limits, that identifies* Mod-Λ *with the category* $\text{Fp-Inj}(\Lambda\text{-mod}, \text{Ab})$ *of fp-injective functors. Moreover a sequence* $0 \to M' \to M \to M'' \to 0$ *is pure-exact in* Mod-Λ *if and only if* $0 \to jM' \to jM \to jM'' \to 0$ *is exact in* $(\Lambda\text{-mod}, \text{Ab})$, *and* M *is pure-injective in* Mod-Λ *if and only if* jM *is injective in* $(\Lambda\text{-mod}, \text{Ab})$.

PROOF. We sketch the proof. The claim on pure-exact sequences follows from the characterization of pure-exactness by means of tensor products. Yoneda's lemma implies that a functor H is fp-injective if and only if it is right-exact, that is a functor of type $M \otimes$ -. Assume the module M pure-injective, and embed $M \otimes$ - into an injective functor $H = Q \otimes$-. Then M is pure in Q, hence a direct summand in Q. As a direct summand of $Q \otimes$ - the functor $M \otimes$ - is thus injective. □

One step further in the hierarchy of functor categories yields a presentation of Mod-Λ as a category of exact functors, a fact of central importance for the rest of the paper. We refer to [9] for a systematic treatment of such *exactly definable categories.*

Corollary 5.2. *Let* $\mathcal{E} = \mathrm{fp}(\Lambda\text{-mod}, \mathrm{Ab})$ *and* $\mathrm{Lex}(\mathcal{E}^{op}, \mathrm{Ab})$ *the Grothendieck category of left exact functors from* $\mathcal{E}^{op}$ *to abelian groups. Then the functor* $d : \text{Mod-}\Lambda \to \mathrm{Lex}(\mathcal{E}^{op}, \mathrm{Ab})$, $M \mapsto (\text{-}, M \otimes_\Lambda -)$, *is a full embedding, commuting with direct limits and direct products, which identifies pure-exact sequences with exact sequences and pure-injectives with injectives. Moreover d identifies* Mod-Λ *with the full subcategory* $\mathrm{Ex}(\mathcal{E}^{op}, \mathrm{Ab})$ *of exact functors.*

PROOF. Since $M \otimes_\Lambda -$ is fp-injective, the functor $(\text{-}, M \otimes_\Lambda -)$ is exact on finitely presented functors, hence on $\mathcal{E}$. By the functorial version of Proposition 4.1 each exact functor $H : \mathcal{E}^{op} \to \mathrm{Ab}$ has the form $(\text{-}, Q)$ for some Q, necessarily fp-injective, hence $Q = M \otimes$ -. □

Recall that a subcategory $\mathcal{S}$ of an abelian category $\mathcal{A}$ is called a *Serre category* if it is closed under subobjects, quotients and extensions. The objects of the *quotient category* $\mathcal{A}/\mathcal{S}$ are in bijection $A \mapsto qA$ with the objects of $\mathcal{A}$, moreover $\mathrm{Hom}(qA, qB) = \varinjlim \mathrm{Hom}(A', B/B')$, where A/A' and B' belong to $\mathcal{S}$. The quotient functor $q : \mathcal{A} \to \mathcal{A}/\mathcal{S}$ is exact and exact functors on $\mathcal{A}/\mathcal{S}$ are identified with exact functors on $\mathcal{A}$ annihilating $\mathcal{S}$, see [4]. Further $\mathcal{S}$ is called *localizing* if q admits a right adjoint (section functor) $s : \mathcal{A}/\mathcal{S} \to \mathcal{A}$.

Starting with the presentation Mod-$\Lambda = \mathrm{Ex}(\mathcal{E}^{op}, \mathrm{Ab})$, a subcategory $\mathcal{U}$ of Mod-Λ is called *definable* (by $\mathcal{S}$) if it consists of all functors H vanishing on a fixed subcategory $\mathcal{S}$ of $\mathcal{E}$. Clearly $\mathcal{S}$ can be assumed to be a Serre subcategory, hence $\mathcal{U} = \mathrm{Ex}((\mathcal{E}/\mathcal{S})^{op}, \mathrm{Ab})$. According to [9, Cor. 4.6] a subcategory of Mod-Λ is definable if and only if it is closed under direct limits, direct products and under pure submodules. By Proposition 4.3 the category $\mathrm{Add}(C)$ is hence definable in Mod-C for each $C \in$ mod-Λ. We provide a direct argument. A morphism $f : X \to Y$ in Λ-mod is called a *C-monomorphism* if $C \otimes f : C \otimes X \to C \otimes Y$ is an injective map.

Lemma 5.3. *A functor* $E \in \mathcal{E}$ *belongs to the Serre category* $\mathcal{S}$ *annihilated by* C *if and only if* E *has a presentation* $(Y, \text{-}) \xrightarrow{(f,\text{-})} (X, \text{-}) \to E \to 0$ *for some* C*-monomorphism* $f : X \to Y$. *It is equivalent that* $E(DC) = 0$. *In particular,* $\mathcal{E}/\mathcal{S} \cong \mathrm{End}\, DC\text{-mod} \cong \text{mod-End}\, C$.

PROOF. Choose a presentation $(Y, \text{-}) \xrightarrow{(f,\text{-})} (X, \text{-}) \to E \to 0$ for E with $f : X \to Y$ in Λ-mod. The functor $dC = (\text{-}, C \otimes_\Lambda -)$ annihilates E if and only if $C \otimes_\Lambda X \xrightarrow{C \otimes f} C \otimes_\Lambda$

Y is a monomorphism. Note also that the sequence $(Y, DC) \overset{(f,DC)}{\longrightarrow} (X, DC) \to E(DC) \to 0$ agrees with $D(C \otimes Y) \overset{D(C\otimes f)}{\longrightarrow} D(C \otimes X) \to E(DC) \to 0$. □

Proposition 5.4. *Let $C \in$ mod-Λ. For $M \in$ Mod-Λ the following are equivalent:*

(i) *M belongs to* Add(C).

(ii) *Each $M \to A$, with $A \in$ mod-Λ, factors through some C^n.*

(iii) *For each C-monomorphism f in Λ-mod, $M \otimes f$ is a monomorphism.*

(iv) *M belongs to the subcategory defined by $\mathcal{S}$, that is $d(M)(\mathcal{S}) = 0$.*

PROOF. (i) $\Longrightarrow$ (iv) is obvious. (iv) $\Longrightarrow$ (iii): the cokernel term of $(Y,\text{-}) \overset{(f,\text{-})}{\longrightarrow} (X,\text{-}) \to E \to 0$ is annihilated by $d(M)$, hence $M \otimes$ - is a monomorphism. (iii) $\Longrightarrow$ (ii): Let $\varphi: C^n \to A$ be C-universal such that $(C, \varphi): (C, C^n) \twoheadrightarrow (C, A)$ is an epimorphism. By duality $C \otimes D\varphi: C \otimes DA \to C \otimes DC^n$ is a monomorphism, hence also $M \otimes D\varphi: M \otimes DA \to M \otimes DC^n$. Passing to duals yields an epimorphism $(M, \varphi): (M, C^n) \to (M, A)$. (ii) $\Longrightarrow$ (i): We start with a pure embedding $M \hookrightarrow \prod_\alpha A_\alpha$ with $A_\alpha \in$ mod-Λ. By (ii) this yields a pure embedding $M \hookrightarrow \prod_\alpha C_\alpha^{n_\alpha}$, and $M \in$ Add(C) by Proposition 4.3. □

Lemma 5.5. *The categories $\mathcal{E}$ and $\mathcal{S}$ have sufficiently many injectives. Each $E \in \mathcal{E}$ ($S \in \mathcal{S}$) embeds into some $A \otimes$ - ($r(A \otimes \text{-})$, respectively) for some $A \in$ mod-A, where r denotes the right adjoint to inclusion $\mathcal{S} \hookrightarrow \mathcal{E}$.*

PROOF. Let $\bar{\mathcal{E}} =$ fp(mod-Λ, Ab), then $\bar{D}: \bar{\mathcal{E}} \to \mathcal{E}$, $F \mapsto DFD$, is a duality. [This uses that $\bar{D}(A,\text{-}) = A \otimes$ - is finitely presented for $A \in$ mod-Λ.] Since $\bar{\mathcal{E}}$ has sufficiently many projectives, $\mathcal{E}$ has sufficiently many injectives, all of the form $A \otimes$ -. We claim that inclusion $i: \mathcal{S} \to \mathcal{E}$ has the right adjoint $r: \mathcal{E} \to \mathcal{S}$, $E \mapsto E'$, where E' denotes the largest subobject of E belonging to $\mathcal{S}$. We choose a basis $\psi_1, \dots, \psi_n$ of $(E, C \otimes \text{-})$, yielding an exact sequence $0 \to E' \to E \overset{\psi}{\to} C^n \otimes$ -. By injectivity of $C \otimes$ - this implies exactness of $(C^n \otimes \text{-}, C \otimes \text{-}) \overset{\psi^*}{\to} (E, C \otimes \text{-}) \to (E', C \otimes \text{-}) \to 0$. By construction the map ψ^* induced by ψ is surjective, hence $(dC)(E') = 0$ and E' belongs to $\mathcal{S}$. Clearly $(S, E') = (S, E)$ for each $S \in \mathcal{S}$. Each $S \in \mathcal{S}$ embeds into some $A \otimes$ -, hence into $r(A \otimes \text{-})$ which is injective in $\mathcal{S}$. □

The *factor category* mod-$\Lambda/[\mathcal{C}]$ of mod-Λ modulo its two-sided ideal $[\mathcal{C}]$ of morphisms factoring through an object from $\mathcal{C}$ has the modules from mod-Λ as objects; morphisms are given by the factor spaces Hom$(X, Y)/(X, Y)[\mathcal{C}]$, where $(X, Y)[\mathcal{C}]$ consists of all morphisms factoring through an object from $\mathcal{C}$. As special cases for $C = \Lambda$ (respectively, $C = D\Lambda$) we obtain mod-$\Lambda/[\mathcal{C}] = \underline{\text{mod}}$-$\Lambda$ (respectively mod-$\Lambda/[\mathcal{C}] = \overline{\text{mod}}$-$\Lambda$) mod-$\Lambda/[\mathcal{C}] = \overline{\text{mod}}$-$\Lambda$) the stable category of Λ-modules modulo projectives (respectively injectives).

Proposition 5.6. *The restriction functor q:* Ex$(\mathcal{E}^{op}, \text{Ab}) \to$ Ex$(\mathcal{S}^{op}, \text{Ab})$, *$H \mapsto H|_{\mathcal{S}}$ induces an equivalence* $\varinjlim$ mod-$\Lambda/[\mathcal{C}] \overset{\sim}{\to}$ Ex$(\mathcal{S}^{op}, Ab)$.

PROOF. We first show that q induces an equivalence between the stable category mod-$\Lambda/[\mathcal{C}]$ and the category of finitely presented objects of Ex$(\mathcal{S}^{op}, \text{Ab})$.

For $S \in \mathcal{S}$ choose a presentation $(Y,\text{-}) \xrightarrow{(f,\text{-})} (X,\text{-}) \to S \to 0$ with $f: X \to Y$ a $\mathcal{C}$-monomorphism in Λ-mod. For $M \in$ Mod-Λ we have $qM = 0$ if and only if $0 = d(M)(S) = \ker(M \otimes f)$. By Proposition 5.4 this happens for all $S \in \mathcal{S}$ if and only if M belongs to Add(C). Next, consider $h: A \to B$ in mod-Λ with $q(h) = 0$. Choose $\varphi: C^n \to B$ C-universal such that $(C,\varphi): (C, C^n) \to (C, B)$ is an epimorphism, hence $C \otimes D\varphi: C \otimes DB \to C \otimes DC^n$ is a monomorphism. Let S be the member of $\mathcal{S}$ defined by the C-monomorphism $D\varphi$. From $q(h) = 0$ we get a factorization $[A \otimes DB \xrightarrow{h \otimes DB} B \otimes DB] = [A \otimes DB \xrightarrow{A \otimes D\varphi} A \otimes DC^n \xrightarrow{u} B \otimes DB]$ for some u. Passage to duals then yields a factorization $[(B,B) \xrightarrow{(h,B)} (A,B)] = [(B,B) \xrightarrow{u^*} (A, C^n) \xrightarrow{(A,\varphi)} (A,B)]$. Setting $\psi = u^*(1_B)$ we get $h = \varphi\psi \in (A,B)[\mathcal{C}]$.

Next we show that each $u \in (qA, qM)$ with $A \in$ mod-Λ and $M \in$ Mod-Λ is induced by some $h: A \to M$. By assumption u is defined on $\mathcal{S}$, that is for each $S \in \mathcal{S}$ we have a morphism $u_S: (S, A \otimes \text{-}) \to (S, M \otimes \text{-})$. Let $j: r(A \otimes \text{-}) \hookrightarrow A \otimes$ - be the inclusion of the maximal subobject of $A \otimes$ - belonging to $\mathcal{S}$ into $A \otimes$ -. By injectivity of $M \otimes$ -, see Proposition 5.1, $u_{r(A\otimes\text{-})}: r(A \otimes \text{-}) \to M \otimes$ - extends to a morphism $h \otimes \text{-}: A \otimes \text{-} \to M \otimes \text{-}$, hence $(h \otimes \text{-})j = u_{r(A\otimes\text{-})}(j)$. It is now straightforward to check that h induces u.

Finally, let $H \in \mathrm{Ex}(\mathcal{S}^{op}, \mathrm{Ab})$ be an exact functor. According to the functorial version of Proposition 4.2 as a left exact functor H is flat, hence is a direct limit of representable functors $(\text{-}, S_\alpha)$ with $S_\alpha \in \mathcal{S}$. We show that each morphism $h: (\text{-}, S) \to H$ factors through some $q(A)$ with $A \in$ mod-Λ, implying that H belongs to the limit closure of $q(\text{mod-}\Lambda) = \text{mod-}\Lambda/[\mathcal{C}]$. By Yoneda's lemma we identify h with a member of $H(S)$. Let $j: S \hookrightarrow I$ be the embedding in an injective object from $\mathcal{S}$, then $H(j): H(I) \to H(S)$ is surjective. We thus obtain $k \in H(I)$, corresponding to $k: (\text{-}, I) \to H$, with $H(j)(k) = h$, hence a factorization $[(\text{-}, S) \xrightarrow{h} H] = [(\text{-}, S) \xrightarrow{(\text{-},j)} (\text{-}, I) \xrightarrow{k} H]$ as required. □

6. The key result

The main technical tool to transfer tameness via stable equivalence is given by the next theorem. By Pinj-Λ we denote the category of all pure-injective Λ-modules; similarly Pinj$(\mathcal{A})$ denotes the pure-injectives of a locally finitely presented category $\mathcal{A}$. We fix $C \in$ mod-Λ and put $\mathcal{C} = \text{add}(C)$.

Theorem 6.1. *The natural functor* $q: \text{Mod-}\Lambda \to \varinjlim \text{mod-}\Lambda/[\mathcal{C}]$, $M \mapsto (\text{-}, M)/(\text{-}, C)[\mathcal{C}]$, *commutes with direct limits and direct products. Moreover, the following assertions hold:*

(i) *q induces an equivalence between the stable category* Pinj-Λ/[Add(C)] *and the category of pure-injective objects in* $\varinjlim \text{mod-}\Lambda/[\mathcal{C}]$.
(ii) *For* $E \in$ mod-Λ *and* $M \in$ Pinj-Λ *q induces an isomorphism* $(E, M)/(E, M)[\mathcal{C}] \xrightarrow{\sim} \mathrm{Hom}(qE, qM)$.
(iii) *$qM = 0$ if and only if* $M \in$ Add(C).

(iv) *For each* $M \in \text{Pinj-}\Lambda$ *the functor* q *induces an isomorphism* $\operatorname{End} M/\operatorname{rad}\operatorname{End} M \xrightarrow{\sim} \operatorname{End} qM/\operatorname{rad}\operatorname{End} qM$.

PROOF. For $M \in \text{Mod-}\Lambda$ the functor $(\text{-}, M)$ is flat in $(\text{mod-}\Lambda^{op}, \text{Ab})$, hence $(\text{-}, M)/(\text{-}, M)[\mathcal{C}]$ is flat in $((\text{mod-}\Lambda/[\mathcal{C}])^{op}, \text{Ab})$. It is obvious that q commutes with direct limits. To see that q also commutes with direct products we use that a morphism $f: E \to M$, with E finitely presented, factors through a module in $\mathcal{C}$ if and only if f factors through the *fixed* map $\varphi: E \hookrightarrow C^n$, where $\varphi_1, \ldots, \varphi_n$ form a basis of (E, C). (iii): We note that $qM = 0$ if and only if $(E, M)/(E, M)[\mathcal{C}] = 0$ for each $E \in \text{mod-}\Lambda$. This happens if and only if each $f: E \to M$ has a factorization $E \to C^n \to M$ for some integer n. It follows that M is a direct limit of modules C^{n_α}, hence by Proposition 4.3 belongs to $\text{Add}(C)$. The proof of (ii) is easy. (i): Since q commutes with direct sums and direct products it preserves pure-injectivity. Invoking (iii), q hence induces a functor $q: \text{Pinj-}\Lambda/[\text{Add}(C)] \to \text{Pinj}(\varinjlim \text{mod-}\Lambda/[\mathcal{C}])$, $M \mapsto (-, M)/(-, M)[\mathcal{C}]$. To show that q is an equivalence, and to establish (iv) we need some preparation.

We consider the sequence of categories $\text{Add}(C) \hookrightarrow \text{Mod-}\Lambda \xrightarrow{q} \varinjlim \text{mod-}\Lambda/[\mathcal{C}]$ associated with a subcategory $\mathcal{C} = \text{add}(C)$ of mod-Λ. Note that the three categories have direct limits and are the limit closures of a system of finitely presented objects given, respectively, by $\mathcal{C}$, mod-Λ and mod-$\Lambda/[\mathcal{C}]$. Each of the three categories is exactly definable. For the last two this follows from Corollary 5.2 and Proposition 5.6. Further, by Lemma 5.3 and Proposition 5.4 the category $\text{Add}(C)$ is equivalent to $\text{Ex}((\mathcal{E}/\mathcal{S})^{op}, \text{Ab})$. Starting with a presentation $\mathcal{A} = \text{Ex}(\mathcal{F}^{op}, \text{Ab})$, where $\mathcal{F}$ is small abelian, the category $\mathcal{D}(\mathcal{A}) = \text{Lex}(\mathcal{F}^{op}, \text{Ab})$ is a Grothendieck category. Moreover, the embedding $d: \mathcal{A} \hookrightarrow \mathcal{D}(\mathcal{A})$ identifies pure-exact sequences and pure-injective objects from $\mathcal{A}$ with exact sequences and injective objects from $\mathcal{D}(\mathcal{A})$. We then get a commutative diagram as follows (compare [9, Theorem 5.1])

$$\begin{array}{ccccc}
\text{Add}(C) & \overset{i}{\hookrightarrow} & \text{Mod-}\Lambda & \overset{q}{\to} & \varinjlim \text{mod-}\Lambda/[\mathcal{C}] \\
\downarrow d' & & \downarrow d & & \downarrow d'' \\
\mathcal{D}(\text{Add}(C)) & \underset{i^*}{\overset{i_*}{\rightleftarrows}} & \mathcal{D}(\text{Mod-}\Lambda) & \underset{q^*}{\overset{q_*}{\rightleftarrows}} & \mathcal{D}(\varinjlim \text{mod-}\Lambda/[\mathcal{C}]) \\
\| & & \| & & \| \\
\mathcal{A}/\mathcal{L} & \underset{i^*}{\overset{i_*}{\rightleftarrows}} & \mathcal{A} & \underset{q^*}{\overset{q_*}{\rightleftarrows}} & \mathcal{L}
\end{array}$$

with the following properties:

1. d', d and d'' are full embeddings, yielding equivalences with the full subcategories of pure-injectives of the respective source categories and injectives of the respective target categories.
2. i^* is an exact functor, whose kernel $\mathcal{L}$ is a localizing subcategory, equivalent to $\mathcal{D}(\varinjlim \text{mod-}\Lambda/[\mathcal{C}])$.
3. $\mathcal{D}(\text{Add}(C))$ is equivalent to the quotient category with respect to the quotient functor i^*.
4. i^*, q^* are exact functors and right adjoint to i_* and q_*, respectively.

Here $q_* : \mathrm{Lex}(\mathcal{E}^{op}, \mathrm{Ab}) \to \mathrm{Lex}(\mathcal{S}^{op}, \mathrm{Ab})$, is restriction to $\mathcal{S}$, and q^* is its left adjoint given by left Kan extension. Therefore q^* commutes with direct limits and preserves representable functors, that is $q^*(\mathcal{S}(-,S)) = \mathcal{E}(-,S)$ for each $S \in \mathcal{S}$. A similar definition yields i_* and i^*. Now, assume that N is pure-injective in $\varinjlim \text{mod-}\Lambda/[\mathcal{C}]$, therefore N is injective in $\mathcal{L}$. Then $\mathrm{E}(q^*N)$, the injective envelope of q^*N in $\mathcal{A}$, as an injective object from $\mathcal{A}$ yields a pure-injective object in $\varinjlim \text{mod-}\Lambda/[\mathcal{C}]$. This yields an inverse to $q : \text{Pinj-}\Lambda/[\mathrm{Add}(C)] \to \mathrm{Pinj}(\varinjlim \text{mod-}\Lambda/[\mathcal{C}])$.

Let $\mathrm{cl}(\mathcal{A})$ denote the full subcategory of *closed objects* Λ in $\mathcal{A}$ satisfying $\mathrm{Hom}(\mathcal{L}, A) = 0 = \mathrm{Ext}^1(\mathcal{L}, A)$. Since $\mathcal{L}$ is localizing in $\mathcal{A}$, restriction of i^* to $\mathrm{cl}(\mathcal{A})$ yields an equivalence $\mathrm{cl}(\mathcal{A}) \xrightarrow{\sim} \mathcal{A}/\mathcal{L}$ with inverse i_* [4]. Since i_* has an exact left adjoint, it sends $\mathrm{Inj}(\mathcal{A}/\mathcal{L})$ to $\mathrm{Inj}(\mathcal{A}) \cap \mathrm{cl}(\mathcal{A})$. Conversely, assume $j : i^*Q \hookrightarrow Y$ is a monomorphism in $\mathcal{A}$ for some $Q \in \mathrm{Inj}(\mathcal{A}) \cap \mathrm{cl}(\mathcal{A})$. Then the monomorphism $i_*(j) : Q \hookrightarrow i_*Y$ splits, hence $j = i^*i_*(j)$ splits. Hence i^*Q lies in $\mathrm{Inj}(\mathcal{A}/\mathcal{L})$. From now on we identify $\mathrm{Inj}(\mathcal{A}/\mathcal{L})$ with $\mathrm{Inj}(\mathcal{A}) \cap \mathrm{cl}(\mathcal{A})$.

We treat $\mathcal{L}$ as a full subcategory of $\mathcal{A}$ and q^* as the corresponding embedding. Note that in view of $\mathrm{Hom}(L, q_*A) = \mathrm{Hom}(L, \Lambda)$ the object q_*A is the largest subobject of Λ belonging to $\mathcal{L}$. In particular, q_* sends injectives to injectives, thus inducing by a functor, again called $q_* : \mathrm{Inj}(\mathcal{A}) \to \mathrm{Inj}(\mathcal{L})$.

(a) The functor q_* is dense. Let Q be injective in $\mathcal{L}$ and let $j : Q \hookrightarrow \mathrm{E}(Q)$ its injective envelope in $\mathcal{A}$. Then $q_*j : Q \hookrightarrow q_*\mathrm{E}(Q)$ splits. If this embedding is proper, there exists $0 \neq U \subseteq q_*\mathrm{E}(Q)$ with $U \cap Q = 0$. This implies $U \subseteq \mathrm{E}(Q)$ and hence $U = 0$ since Q is essential in $\mathrm{E}(Q)$. Thus $Q = q_*\mathrm{E}(Q)$ for each $Q \in \mathrm{Inj}(\mathcal{L})$.

(b) The functor q_* is full. Let Q_1, Q_2 be injective in $\mathcal{A}$ and let $f : q_*Q_1 \to q_*Q_2$. Recall that $q_*Q_i \subseteq Q_i$, hence by injectivity of Q_2 the morphism f extends to a morphism $\bar{f} : Q_1 \to Q_2$ and, clearly, $q_*\bar{f} = f$.

(c) We now determine the kernel of $q_* : \mathrm{Inj}(\mathcal{A}) \to \mathrm{Inj}(\mathcal{A}/\mathcal{L})$. Let $f : Q_1 \to Q_2$ be a morphism of injectives in $\mathcal{A}$ with $q_*f = 0$. Then $f(q_*Q_1) = 0$, hence f factors through $\mathrm{E}(Q_1/q_*Q_1)$. Since there are no non-zero morphisms from $\mathcal{L}$ to Q_1/q_*Q_1 there are no such morphisms from $\mathcal{L}$ to $\mathrm{E}(Q_1/q_*Q_1)$. Thus $\mathrm{E}(Q_1/q_*Q_1)$ belongs to $\mathrm{Inj}(\mathcal{A}) \cap \mathrm{cl}(\mathcal{A}) = \mathrm{Inj}(\mathcal{A}/\mathcal{L})$, and f factors through an object of $\mathrm{Inj}(\mathcal{A}/\mathcal{L})$. It easily follows that q_* induces an equivalence $\mathrm{Inj}(\mathcal{A})/\mathrm{Inj}(\mathcal{A}/\mathcal{L}) \xrightarrow{\sim} \mathrm{Inj}(\mathcal{L})$.

(d) Let $Q \in \mathrm{Inj}(\mathcal{A}) \backslash \mathrm{Inj}(\mathcal{A}/\mathcal{L})$ be indecomposable, then q_*Q is indecomposable in $\mathrm{Inj}(\mathcal{L})$. Moreover, $f \in \mathrm{End}\, Q$ is an isomorphism if and only if $q_*f \in \mathrm{End}\, q_*Q$ is an isomorphism.

Assume first that there exist non-zero objects Q_1 and Q_2 such that $Q_1 \oplus Q_2 = q_*Q$. Since $q_*Q \subseteq Q$ we get $\mathrm{E}(Q_1) \oplus \mathrm{E}(Q) = Q$, a contradiction. Concerning the second assertion assume that $q_*f = 0$, hence $\ker f \cap q_*Q = 0$. Since Q is indecomposable injective, this implies $\ker f = 0$. By injectivity of Q, f is then a split monomorphism, hence an isomorphism since Q is indecomposable.

(e) With the assumptions from (d) the surjection $\mathrm{End}\, Q \to \mathrm{End}\, q_*Q$ induced by q_* sends the radical of $\mathrm{End}\, Q$ to the radical of $\mathrm{End}\, q_*Q$, thus induces an isomorphism $\mathrm{End}\, Q/\mathrm{rad}\,\mathrm{End}\, Q \to \mathrm{End}\, q_*Q/\mathrm{rad}\,\mathrm{End}\, q_*Q$. □

7. Proof of Theorem 1

Since $DTr: \underline{\text{mod}}\text{-}\Lambda \to \overline{\text{mod}}\text{-}\Lambda$ is an equivalence, we may assume that there is an equivalence $f: \overline{\text{mod}}\text{-}\Lambda \to \overline{\text{mod}}\text{-}\Gamma$ modulo injectives. By Theorem 6.1 a stable equivalence $f: \overline{\text{mod}}\text{-}\Lambda \to \overline{\text{mod}}\text{-}\Gamma$ extends to an equivalence $F: \overline{\text{Pinj}}\text{-}\Lambda \to \overline{\text{Pinj}}\text{-}\Gamma$ between the stable categories modulo injective modules preserving radical factor rings and inducing isomorphisms $\overline{\text{Hom}}_\Lambda(E, M) \xrightarrow{\sim} \overline{\text{Hom}}_\Gamma(fE, FM)$ for each $E \in \overline{\text{mod}}\text{-}\Lambda$ and $M \in \text{Pinj-}\Lambda$. Since Λ is noetherian, direct limits of injective modules are injective, hence each Λ-module M has a maximal injective submodule. Each object in $\overline{\text{Pinj}}\text{-}\Lambda$ can thus be represented by a module M not having any non-zero injective submodule. In order to prove Theorem 1 we need to fix some notation: $[E, M]$ (respectively $[[E, M]]$) is the length of $\text{Hom}(E, M)$ over $\text{End}\, M$ (respectively the length of $\overline{\text{Hom}}(E, M)$ over $\overline{\text{End}}\, M$).

Step 1 *Let $S_1, \ldots, S_p$ be the system of simple Λ-modules. Then* $[[E, M]] \overset{(a)}{\leq} [E, M] \overset{(b)}{\leq} \sum_{i=1}^{p} m_i(E)[S_i, M]$, *where $m_i(E)$ is the multiplicity of S_i in a composition series of E.*

PROOF. Since $\text{End}\, M$ and $\overline{\text{End}}\, M$ have isomorphic radical factor rings, the module $\overline{\text{Hom}}(E, M)$ has the same length over $\text{End}\, M$ and over $\overline{\text{End}}\, M$, hence inequality (a) follows. For any short exact sequence $0 \to E' \to E \to E'' \to 0$ the sequence $0 \to (E'', M) \to (E, M) \to (E', M)$ is exact, hence $[E, M] \leq [E', M] + [E'', M]$, and assertion (b) follows.

Step 2 *Assume M is an endofinite module in* Mod-Λ *not having a non-zero injective submodule, then $[S, M] = [[S, M]]$ holds for each simple Λ-module S.*

PROOF. We show that $\text{Hom}(S, M) = \overline{\text{Hom}}(S, M)$. Assume that $0 \neq u: S \to M$ factors through an injective module, then f factors through the injective envelope $E(S)$ of S, yielding an embedding of $E(S)$ into M, contradiction.

Step 3 *There exists a constant c_f such that $e\ell\, FM \leq c_f \cdot e\ell\, M$ holds for each endofinite Λ-module M not having a non-zero injective submodule.*

PROOF. Let M be endofinite in Mod-Λ and put $N = FM$. Working in $\overline{\text{Pinj}}\text{-}\Gamma$ we may assume that N has no non-zero injective submodule. Next, we fix finite dimensional Λ-modules $U_1, \ldots, U_q$ such that $T_1 = FU_1, \ldots, T_q = FU_q$ form a complete system of simples in mod-Γ. Let $e\ell\, M$ denote the endolength of M. Now,

$$e\ell\, FM = [\Gamma, N] \overset{Step\ 1}{\leq} \sum_{j=1}^{q} m_j(\Gamma)[T_j, N] \overset{Step\ 2}{=} \sum_{j=1}^{q} m_j(\Gamma)[[T_j, N]].$$

By Theorem 6.1(ii) this equals $\sum_{j=1}^{q} m_j(\Gamma)[[U_j, M]]$, and we get

$$e\ell\, FM \overset{Step\ 1}{\leq} \sum_{j=1}^{q} m_j(\Gamma)[U_j, M] \overset{Step\ 1}{\leq} \sum_{i=1}^{p} \left(\sum_{j=1}^{q} m_j(\Gamma) m_i(U_j)) \right) [S_i, M]$$

$$\leq c_f \sum_{i=1}^{p} [S_i, M] \leq c_f\, [\Lambda/\text{rad}\,\Lambda, M] \leq c_f\, [\Lambda, M] = c_f \cdot e\ell\, M.$$

Here, we have defined c_f as the maximum of the expressions $\sum_{j=1}^{q} m_j(\Gamma)m_i(U_j)$ with i ranging from 1 to p. Invoking Crawley-Boevey's characterization of tameness (Theorem 2) in terms of numbers of generic modules of finite endolength, Theorem 1 now is an easy consequence. □

Krause's theorem [8] actually states a stronger result: Let $C \in$ mod-Λ, $D \in$ mod-Γ and $\mathcal{C} = \mathrm{add}(C)$, $\mathcal{D} = \mathrm{add}(D)$. Then tameness is preserved under equivalence mod-$\Lambda/[\mathcal{C}] \cong$ mod-$\Gamma/[\mathcal{D}]$. This may be deduced from Theorem 6.1, invoking a further result of Krause [9, Cor. 10.5] stating that an indecomposable pure-injective object M is endofinite if and only if each product M^I of copies of M is isomorphic to a direct product $M^{(J)}$. On the other hand the preceding argument $\ell\ell\, FM \leq c_f\, \ell\ell\, M$ from [10] allows to deduce that stable equivalence modulo injectives (respectively projectives) also preserves domestic type respectively polynomial growth [10].

References

[1] H. Bass. Finitistic dimension and a homological generalization of simiprimary rings. *Trans. Amer. Math. Soc.* **95**, 466–488 (1960).

[2] W. Crawley–Boevey. Tame algebras and generic modules. *Proc. London Math. Soc.* **63** (1991), 241–265.

[3] W. Crawley–Boevey. Locally finitely presented additive categories. *Commun. Algebra* **22** (1994), 1641–1674.

[4] P. Gabriel. Des catégories abéliennes. *Bull. Soc. Math. France* **90** (1962), 323–448.

[5] L. Gruson and C. U. Jensen. Modules algébriquement compacts et foncteures $\varprojlim^{(i)}$. *C. R. Acad. Sci. Paris Sér. A-B* **276** (1973), A1651-A1653.

[6] L. Gruson and C. U. Jensen. Deux applications de la notion de L-dimension. *C. R. Acad. Sci. Paris Sér. A-B* **282** (1976), A23-A24.

[7] C.U. Jensen and H. Lenzing. *Model Theoretic Algebra*, Gordon and Breach Science Publishers, 1989.

[8] H. Krause. Stable equivalence preserves representation type. *Comment. Math. Helv.* **72** (1997), 266–284.

[9] H. Krause. Exactly definable categories. *J. Algebra* **201** (1998), 456–492.

[10] H. Krause and G. Zwara. Stable equivalence and generic modules. *Preprint Series of SFB 343, Bielefeld* **98-119**, 1998.

Fachbereich Mathematik-Informatik, Universität-GH Paderborn, D-33095 Paderborn, Germany

E-mail address: `helmut@uni-paderborn.de`

Trends in Mathematics, © 2000 Birkhäuser Verlag Basel/Switzerland

THE KRULL-GABRIEL DIMENSION OF AN ALGEBRA - OPEN PROBLEMS AND CONJECTURES

JAN SCHRÖER

This note is a short introduction to the concept of the Krull-Gabriel dimension of an algebra. We mention some recent results and give a list of open problems and conjectures.

Throughout, let k be a fixed algebraically closed field. All algebras considered will be finite-dimensional associative k-algebras with unit element. Given an algebra A, we denote by mod-A the category of finite-dimensional right modules over A. Note that mod-A is an abelian category, and all objects in mod-A are of finite length. It turned out that there appear naturally other abelian categories related to mod-A which have objects of infinite length. The main example we have in mind is $\mathcal{C} = \mathcal{C}(A)$, the category of finitely presented covariant functors from mod-A to mod-k. The category $\mathcal{C}$ was intensively studied over the last 20 years, and is considered to be one of the central topics in modern representation theory. It is a hard problem to describe the category $\mathcal{C}$ even if the category mod-A is well understood. However, under a reasonable restriction, namely the existence of the Krull-Gabriel dimension of A, there is some hope to understand the structure of $\mathcal{C}$ as well. This concept is based on the following classical definition: If $\mathcal{T}$ is a Serre subcategory of an abelian category $\mathcal{A}$, then we denote by $\mathcal{A}/\mathcal{T}$ the corresponding abelian quotient category which was introduced by Grothendieck and Serre [Gr, 1.11], [Se, I,2]. Define $\mathcal{C}_{-1} = 0$. For each $n \geq 0$ let $\mathcal{C}_n$ be the subcategory of all functors $F \in \mathcal{C}$ such that $q_{n-1}(F)$ is of finite length, where q_{n-1} denotes the quotient functor from $\mathcal{C}$ to $\mathcal{C}/\mathcal{C}_{n-1}$. Note that $\mathcal{C}_n$ is a Serre subcategory of $\mathcal{C}$ for all n. Following Geigle define $\mathrm{KG}(A) = \min\{n \mid \mathcal{C}_n = \mathcal{C}\}$ if such a minimum exists, and $\mathrm{KG}(A) = \infty$, else [Ge1]. We call $\mathrm{KG}(A)$ the *Krull-Gabriel dimension* of A. This is a finitely presented version of a definition due to Gabriel [Ga, IV,1].

Remark. By n, d, $n(d)$ etc., we denote always natural numbers. For infinite ordinal numbers we use always greec letters.

1. Krull-Gabriel dimension and pure injectivity

There is a close relationship between the Krull-Gabriel dimension and the dimension of the Ziegler spectrum of an algebra. The points of this spectrum are the isomorphism classes of indecomposable pure injective modules. The dimension is again a reasonable finiteness condition. For this subject, we refer to the article of Prest in this conference proceedings.

2. Krull-Gabriel dimension and short exact sequences

Let $\eta : 0 \longrightarrow X \longrightarrow Y \longrightarrow Z \longrightarrow 0$ be a short exact sequence in mod-A. This induces an exact sequence $0 \longrightarrow \mathrm{Hom}(Z,-) \longrightarrow \mathrm{Hom}(Y,-) \longrightarrow \mathrm{Hom}(X,-) \longrightarrow F_\eta \longrightarrow 0$ of finitely presented functors [A1]. Define $m(\eta) = n$ if $q_{n-1}(F_\eta) \neq 0$ and $q_n(F_\eta) = 0$ if such an n exists, and let $m(\eta) = \infty$, else. Thus, the Krull-Gabriel dimension defines a hierarchy of short exact sequences in mod-A. It is a famous result of Auslander that the functors F_η where η is an almost split sequence are simple [A2, 4.4]. Thus, we have $m(\eta) = 0$ for all almost split sequences η. Geigle gave in [Ge1] a complete description of this hierarchy of short exact sequences for tame hereditary algebras. Unfortunately, these are the only known examples of this type. A sytematic study of other example classes is still missing.

3. Krull-Gabriel dimension and representation type

Recall that there exists a trichotomy of representation types: An algebra is representation-finite, tame or wild [CB], [Dr]. Auslander proved that an algebra A is representation-finite if and only if $\mathrm{KG}(A) = 0$ [A3, 3.14]. It is shown in [K2, 11.4] that there exists no algebra with Krull-Gabriel dimension 1. Every tame hereditary algebra has Krull-Gabriel dimension 2 [Ge1, 4.3]. If A is wild then $\mathrm{KG}(A) = \infty$, and there exist also tame algebras with Krull-Gabriel dimension ∞ [Ge2, 4.1]. Let A be a representation-infinite algebra. Assume that for each $d \geq 1$ there exist $n(d)$ affine lines in the affine variety of d-dimensional A-modules which parametrize all but finitely many d-dimensional indecomposable A-modules. Recall that in this case A is called *tame* [DS]. A tame algebra is called *domestic* if the numbers $n(d)$ have an upper bound. The following conjecture is due to Prest:

Conjecture. A is domestic if and only if $\mathrm{KG}(A) < \infty$.

4. Krull-Gabriel dimension and modular lattices

If $f : X \longrightarrow Y$ and $g : Y \longrightarrow Z$ are maps, then we denote the composition by $fg : X \longrightarrow Z$. Let A be an algebra and S a simple A-module. Let $f : S \longrightarrow X$ and $g : S \longrightarrow Y$ be two homomorphisms where X and Y are in mod-A. Define $f \geq g$ if there exists a homomorphism h such that $g = fh$. We call f and g equivalent if $f \geq g$ and $g \geq f$. Let L_S be the set of all equivalence classes of such homomorphisms. In the following we do not distinguish between equivalence classes and homomorphisms and work with representatives of equivalence classes. Define $f \vee g = [f,g] : S \longrightarrow X \oplus Y$ and $f \wedge g = h : S \longrightarrow Z$ where $\begin{bmatrix} h_1 \\ h_2 \end{bmatrix} : X \oplus Y \longrightarrow Z$ denotes the cokernel of $[f,g]$ and $h = fh_1$. With these two operations L_S becomes a modular lattice. We have $(\mathrm{Id}_S : S \longrightarrow S) \geq (f : S \longrightarrow X) \geq (f_S : S \longrightarrow I) > (0 : S \longrightarrow 0)$ for all $f \neq 0$.

Here f_S denotes the canonical embedding of S into its injective envelope I. Following [P1, 10.2] we define the dimension of some modular lattice L as follows: Given $x, y \in L$ define $x \sim y$ if the interval $\{z \in L \mid x \vee y \geq z \geq x \wedge y\}$ has finite length. This means that there exists a finite chain $x \vee y = z_0 > z_1 > \cdots z_{l-1} > z_l = x \wedge y$ such that for each i there is no $z \in L$ such that $z_i > z > z_{i+1}$. This defines a congruence relation on L. The set of congruence classes is denoted by $L/\sim$. Note that $L/\sim$ again carries the structure of a modular lattice. Let $L^{(-1)} = L$, and for $n \geq 0$ let $L^{(n)} = L^{(n-1)}/\sim$. Then define $\dim(L) = \min\{n \mid L^{(n)} = 0\}$ if such a minimum exists, and let $\dim(L) = \infty$, else.

Proposition [K2, 7.2]. *If A is an algebra, then* $\mathrm{KG}(A) = \max\{\dim(L_S) \mid$ *S simple A-module*$\}$.

A *linear poset* is a poset with the property that two given elements are always comparable. If C is a linear subposet of some poset L, then we call C *maximal chain*, if $C \cup \{x\}$ is not a linear subposet for all $x \notin C$.

Conjecture. If C is a maximal linear subposet of L_S, then $\dim(C) = \dim(L_S)$.

5. Krull-Gabriel dimension and the infinte radical of a module category

Let rad_A be the radical of mod-A. Recall that rad_A is the ideal of homomorphisms in mod-A which is generated by the non-invertible homomorphisms between indecomposable modules. Following [P2] we define rad_A^α for each ordinal number α as follows: If J is an ideal in mod-A and $n \geq 1$, then J^n is the n-fold product of the ideal J. For a limit number β define $\mathrm{rad}_A^\beta = \bigcap_{\alpha<\beta} \mathrm{rad}_A^\alpha$. Finally, let α be an arbitrary infinite ordinal number, thus $\alpha = \beta + n$ for some limit number β and some $n \geq 0$. In this case, define $\mathrm{rad}_A^\alpha = (\mathrm{rad}_A^\beta)^{n+1}$. We denote the first limit number by ω, the second by $\omega 2$, and so on. The minimal limit number, which is not of the form ωn for some $n \geq 1$, is denoted by ω^2.

Let $Q = (Q_0, Q_1)$ be a finite quiver with vertices Q_0 and arrows Q_1. For $\alpha \in Q_1$ let $s(\alpha)$ be its starting point and $e(\alpha)$ its end point. A *path* of length $n \geq 1$ in Q is of the form $\alpha_1 \alpha_2 \cdots \alpha_n$ where the α_i's are arrows with $e(\alpha_i) = s(\alpha_{i+1})$ for $1 \leq i \leq n-1$. To each $S \in Q_0$ we associate a path e_S of length 0. Let kQ be the (possibly infinite-dimensional) path algebra over k associated to Q. Recall that the paths in Q form a k-basis of kQ, and the multiplication is induced by the concatenation of paths. For $i \geq 0$ let $kQ^{(i)}$ the ideal generated by all paths of length $\geq i$. Let I be an ideal of kQ with $kQ^{(n)} \subseteq I \subseteq kQ^{(2)}$ for some $n \geq 2$. We call the finite-dimensional k-algebra kQ/I *special biserial* if the following hold:

1. Any vertex of Q is starting point of at most two arrows and also end point of at most two arrows;

2. Given an arrow β, there is at most one arrow α with $e(\alpha) = s(\beta)$ and $\alpha\beta \notin I$ and at most one arrow γ with $e(\beta) = s(\gamma)$ and $\beta\gamma \notin I$.

For cardinality reasons the radical series stabilizes for each finite-dimensional k-algebra A, i.e. there exists a minimal ordinal number $\mathrm{st}(A)$ such that $\mathrm{rad}_A^{\mathrm{st}(A)} = \mathrm{rad}_A^{\alpha}$ for all $\alpha \geq \mathrm{st}(A)$.

Theorem [Sch2]. *Let A be a special biserial algebra. Then the following are equivalent:*

(i) *A is domestic;*
(ii) $\mathrm{rad}_A^{\omega^2} = 0$*;*
(iii) $\mathrm{rad}_A^{\mathrm{st}(A)} = 0$.

Let $A = kQ/I$ be a special biserial algebra, and let ρ be the set of paths p in Q such that $p \in I$ or $p - \lambda q \in I$ for some path $q \neq p$ in Q which has the same start point and end point as p, $\lambda \in k \setminus \{0\}$. Given an arrow α in Q_1 denote by α^- a formal inverse where $s(\alpha^-) = e(\alpha)$ and $e(\alpha^-) = s(\alpha)$. Also let $(\alpha^-)^- = \alpha$. The set of formal inverses of arrows is denoted by Q_1^-. A *string* of length $n \geq 1$ is a sequence $c_1 \cdots c_n$ of elements in $Q_1 \cup Q_1^-$ such that $e(c_i) = s(c_{i+1})$ and $c_i \neq c_{i+1}^-$ for $1 \leq i \leq n-1$, and neither $c_i c_{i+1} \cdots c_{i+t}$ nor $c_{i+t}^- \cdots c_{i+1}^- c_i^-$ belong to ρ for $1 \leq i < i+t \leq n$. A string $C = c_1 \cdots c_n$ of length ≥ 2 with $c_1 \in Q_1^-$ and $c_n \in Q_1$ is called *band* if all powers C^m are defined, and if C is not the power of a string of smaller length. For bands B_1 and $B_2 = b_1 \cdots b_n$ let $B_1 \sim_r B_2$ if $B_1 = B_2$ or $B_1 = b_i \cdots b_n b_1 \cdots b_{i-1}$ for some $2 \leq i \leq n$. Let $\mathcal{B}$ be a set of representatives of $\sim_r$-equivalence classes of bands. Let G_A be the graph with vertices the elements of $\mathcal{B}$. For vertices $v \neq w$ we draw an arrow $v \longrightarrow w$ if there exists a string b of length ≥ 1 such that vbw is a string, too.

Theorem [Sch2]. *Let A be a special biserial algebra. Then the following hold:*

(i) *If there are no bands, then* $\mathrm{rad}_A^{\omega} = 0$*;*
(ii) *If there are only finitely many bands, then G_A has no oriented cycles, and we have $\mathrm{rad}_A^{\omega(n+1)} \neq 0$ and $\mathrm{rad}_A^{\omega(n+2)} = 0$ where n is the maximal length of a path in G_A;*
(iii) *If there are infinitely many bands then* $\mathrm{rad}_A^{\mathrm{st}(A)} \neq 0$.

Example. Let $A = kQ/I$ with Q of the form

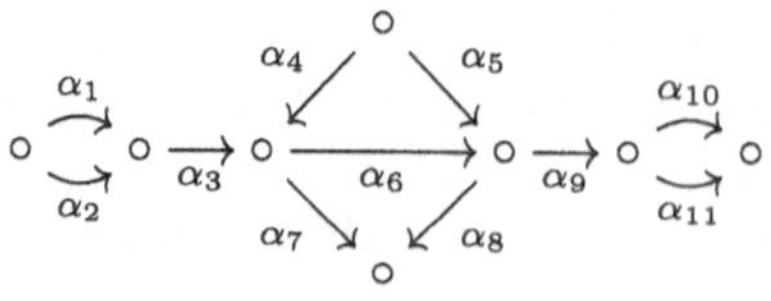

and I generated by

$$\{\alpha_2\alpha_3,\ \alpha_3\alpha_6,\ \alpha_4\alpha_7,\ \alpha_5\alpha_8,\ \alpha_4\alpha_6\alpha_8,\ \alpha_6\alpha_9,\ \alpha_9\alpha_{11}\}.$$

The graph G_A looks as follows:

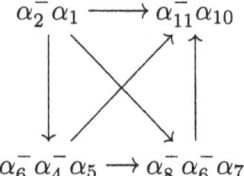

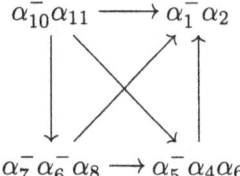

This implies $\mathrm{rad}_A^{\omega 4} \neq 0$ and $\mathrm{rad}_A^{\omega 5} = 0$.

Next, we present some results which relate the above results about the infinite radical with the Krull-Gabriel dimension. The following theorem is a consequence of the considerations in [Sch1].

Theorem. *Let A be a special biserial algebra with $\mathrm{rad}_A^{\omega n} \neq 0$ and $\mathrm{rad}_A^{\omega(n+1)} = 0$. Then there exists some simple A-module S and a maximal chain C in L_S with $\dim(C) = n + 1$.*

Theorem [BP], [Sch1]. *For every $n \neq 1$ there exists an algebra with Krull-Gabriel dimension n.*

This disproves a conjecture of Prest [P1, p.350, Conjecture 2].

For $t \geq 2$ we denote by $\Lambda_t = kQ/I$ the following special biserial algebra:

$$\circ \overset{\alpha_1}{\underset{\beta_1}{\rightrightarrows}} \circ \overset{\gamma_1}{\longrightarrow} \circ \overset{\alpha_2}{\underset{\beta_2}{\rightrightarrows}} \circ \overset{\gamma_2}{\longrightarrow} \cdots \overset{\gamma_{t-1}}{\longrightarrow} \circ \overset{\alpha_t}{\underset{\beta_t}{\rightrightarrows}} \circ$$

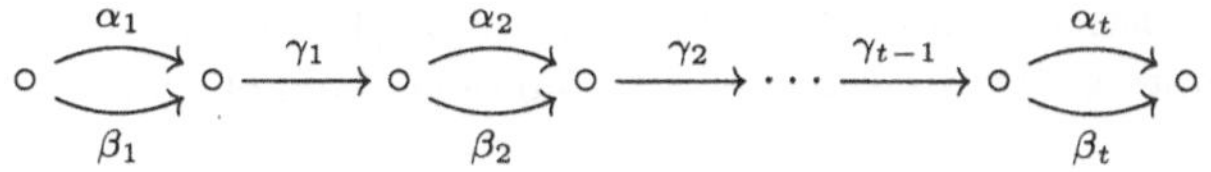

with I generated by $\{\beta_i\gamma_i, \gamma_i\beta_{i+1} \mid 1 \leq i \leq t-1\}$. It is shown independently in [BP] and [Sch1] that the Krull-Gabriel dimension of Λ_t is $t + 1$. Furthermore, we have $\mathrm{rad}_{\Lambda_t}^{\omega t} \neq 0$ and $\mathrm{rad}_{\Lambda_t}^{\omega(t+1)} = 0$.

It is proved in [K2] that $KG(A) = n$ implies $\mathrm{rad}_A^{\omega(n+1)} = 0$.

Conjecture. We have $\mathrm{rad}_A^{\omega(n-1)} \neq 0$ and $\mathrm{rad}_A^{\omega n} = 0$ if and only if $KG(A) = n$.

References

[A1] M. Auslander, Coherent functors, In: Proceedings of the conference on categorial algebra, La Jolla (1966), 189-231.

[A2] M. Auslander, Functors and morphisms determined by objects, In: Representation theory of algebras. Proc. Conf. Philadelphia 1976, (ed. R. Gordon), Dekker, New York (1978), 1-244.

[A3] M. Auslander, A functorial approach to representation theory, In: Representations of algebras (ed. M. Auslander, E. Luis), Springer Lecture Notes in Math. 944 (1980), 105-179.

[BP] K. Burke, M. Prest, The Ziegler and Zariski spectra of some domestic string algebras, Preprint (1998).

[CB] W.W. Crawley-Boevey, Tame algebras and bocses, Proc. London Math. Soc. **56** (1988), 451-483.

[Dr] Yu. A. Drozd, Tame and wild matrix problems, In: Representation theory II, Springer Lecture Notes in Math. 832 (1980), 242-258.

[Ga] P. Gabriel, Des catégories abéliennes, Bull. Soc. math. France **90** (1962), 323-448.

[Ge1] W. Geigle, The Krull-Gabriel dimension of the representation theory of a tame hereditary artin algebra and applications to the structure of exact sequences, Manuscripta Math. **54** (1985), 83-106.

[Ge2] W. Geigle, Krull dimension and artin algebras, In: Representation theory I, Finite dimensional algebras (ed. V. Dlab, P. Gabriel, G. Michler), Springer Lecture Notes in Math. 1177 (1984).

[Gr] A. Grothendieck, Sur quelques poins d'algèbre homologique, Tôhoku Math. J., séries 2, **9** (1957), 119-221.

[K1] H. Krause, Generic modules over artin algebras, Proc. London Math. Soc. **76** (1998), 276-306.

[K2] H. Krause, The spectrum of a module category, Habilitationsschrift, Universität Bielefeld (1998).

[P1] M. Prest, Model theory and modules, London Math. Soc. Lec. Note Series **130** (1988).

[P2] M. Prest, *Morphisms between finitely presented modules and infinite-dimensional representations*, Preprint (1996).

[Sch1] J. Schröer, On the Krull-Gabriel dimension of an algebra, Math. Z. (to appear), 19pp.

[Sch2] J. Schröer, On the infinite radical of a module category, Preprint (1998).

[Se] J.-P. Serre, Classes de groupes abéliens et groupes d'homotopie, Ann. of Math. **58** (1953), 258-294.

Universität Bielefeld, Fakultät für Mathematik, Postfach 100131, 33501 Bielefeld, Germany

E-mail address: jschroe@mathematik.uni-bielefeld.de

Trends in Mathematics, © 2000 Birkhäuser Verlag Basel/Switzerland

HOMOLOGICAL DIFFERENCES BETWEEN FINITE AND INFINITE DIMENSIONAL REPRESENTATIONS OF ALGEBRAS

SVERRE O. SMALØ

Meinem Freund Helmut Lenzing zum 60. Geburtstag

INTRODUCTION

The objective of this note is threefold. Firstly, to give a brief historical account of the finitistic dimension conjectures for finite dimensional algebras. Secondly, to give a considerably simpler example, refuting the conjecture that the little and big finitistic projective dimensions coincide, than the original one given by Birge Huisgen-Zimmermann in 1992. Thirdly, to give an account of the present state of art concerning the remaining part of the finitistic dimension conjecture.

Recall that the left little finitistic dimension of an algebra Λ is the supremum of the projective dimensions of the finitely generated left Λ-modules having finite projective dimension, and that the left big finitistic dimension of an algebra Λ is the supremum of the projective dimensions of all left Λ-modules having finite projective dimension. For a finite dimensional algebra Λ these dimensions will be denoted by $\ell\text{fin.dim}\,\Lambda$ and $\ell\text{Fin.dim}\,\Lambda$, respectively.

The notion of the (little) finitistic dimension of an algebra was introduced by Maurice Auslander and David Buchsbaum in [AB] for commutative local noetherian rings where they proved that for a commutative local noetherian ring the little finitistic dimension is equal to maximal length of a regular sequence in the maximal ideal. Then Irving Kaplansky considered the supremum of the projective dimensions over all modules of finite projective dimension, arriving at the notion of the big finitistic dimension. For a commutative noetherian ring this dimension coincide with the Krull dimension of the ring. That the big finitistic dimension is at least the Krull dimension for this class of rings was proved by Bass in 1962 [B2] and the proof of the equality was completed by Gruson and Raynaud in 1971 [GR]. Kaplansky also characterized at a seminar at the university of Chicago in 1958 those commutative rings where the big finitistic dimension vanishes. This was the background for the paper [B1] by Bass where the finitistic dimension conjectures were first formulated as questions raised by Rosenberg and Zelinsky. Question one: Is it true that the little and big finitistic dimension coincide? Question two: Is the little finitistic dimension finite?

1991 *Mathematics Subject Classification.* 16E10, 16G10, 16G20,16P10.

For a commutative noetherian local ring one has that the two finitistic dimensions coincide if and only if the depth and the Krull dimension coincide, which is usually expressed by saying that the ring is Cohen-Macauley. However, not all commutative noetherian local rings are Cohen-Macauley and therefore the two finitistic dimensions do not coincide for commutative noetherian local rings in general.

The first example of a finite dimensional algebra Λ where the two dimensions do not coincide was given in 1992 by Birge Huisgen-Zimmermann [HZ1]. The examples given in [HZ1] consists of a family of monomial relation algebras, are rather complicated compared to the algebras given in [S]. The examples from [S] will be recalled in this note. In [HZ1] it was also proved that for monomial relation algebras, the difference between the two dimensions is at most 1, and in the examples the left little finitistic dimension was at least 2. For monomial relation algebras it was also established that the two dimensions have to coincide if $\mathfrak{r}^3 = 0$ where $\mathfrak{r}$ denotes the radical of the algebra. The algebras given in Section 1 of this note have the property that $\mathfrak{r}^3 = 0$.

The little finitistic dimension of a commutative noetherian ring is not necessarily finite. Here one can use Nagata's first example of a commutative noetherian ring with infinite Krull dimension [N].

For finite dimensional algebras Λ with $\mathfrak{r}^3 = 0$, Green and Huisgen-Zimmermann proved that the little finitistic dimension is finite [GHZ]. A very nice proof of this fact was later given by Igusa and Todorov in an unpublished paper [IT]. In a short note Dräxler and Happel [DH] showed that the generalized Nakayama conjecture holds for a finite dimensional algebra Λ with $\mathfrak{r}^{2n+1} = 0$ such that $\Lambda/\mathfrak{r}^n$ is of finite representation type. Using the key result in the unpublished preprint by Igusa and Todorov, Wang was able to prove that the finitistic dimension conjecture holds for those algebras satisfying the condition Dräxler and Happel considered [W].

Going back to the comparison between the little and big finitistic dimensions of a finite dimensional algebra, an alternative approach of making the difference between them arbitrarily large which was pointed out to me by Jeremy Rickard, is useing the first example constructed by Birge Huisgen-Zimmermann together with properties of tensor products of algebras. For finite dimensional algebras Λ_1 and Λ_2 over an algebraically closed field k, the global dimension of $\Lambda_1 \otimes \Lambda_2$ is the sum of the global dimension of Λ_1 and the global dimension of Λ_2. For finite dimensional algebras Λ_1 and Λ_2 over an algebraically closed field one also gets that $\ell\text{fin.dim}(\Lambda_1 \otimes \Lambda_2) = \ell\text{fin.dim}\,\Lambda_1 + \ell\text{fin.dim}\,\Lambda_2$ and $\ell\text{Fin.dim}(\Lambda_1 \otimes \Lambda_2) = \ell\text{Fin.dim}\,\Lambda_1 + \ell\text{Fin.dim}\,\Lambda_2$. From these two equalities, just taking repeated tensor products of the algebra mentioned above, one obtains an arbitrary difference between the two finitistic dimensions. Observe that using the tensor product construction one also increases the Loewy lengths of the algebras involved. One is thus prompted to ask if the two equalities involving the finitistic dimensions are valid also for nonseparable finite dimensional algebras over a field?

For a finite dimensional algebra Λ, let $\operatorname{mod}\Lambda$ be the category of all finitely generated left Λ-modules and $\operatorname{Mod}\Lambda$ the category of all left Λ-modules. The full subcategories consisting of the objects of finite projective dimension in modΛ and ModΛ will be denoted by $\mathcal{P}^\infty(\text{mod}\,\Lambda)$ and $\mathcal{P}^\infty(\text{Mod}\,\Lambda)$, respectively. The left little finitistic dimension, $\ell\,\text{fin.dim}\,\Lambda$, is then the supremum of the projective dimensions of the objects in $\mathcal{P}^\infty(\text{mod}\,\Lambda)$, and the left big finitistic dimension, $\ell\,\text{Fin.dim}\,\Lambda$, is the supremum of the projective dimensions of the objects in $\mathcal{P}^\infty(\text{Mod}\,\Lambda)$.

Now, properties of the embedding of the subcategory $\mathcal{P}^\infty(\text{mod}\,\Lambda)$ into mod Λ turn out to have a remarkable influence on the two finitistic dimensions. The first observation of this was obtained by Auslander and Reiten in 1991 [AR] when they showed that ℓfin.dim Λ is finite if this embedding satisfied some frequently occurring condition which will now be described. Recall that for a full subcategory $\mathcal{C}$ in mod Λ, an object A in mod Λ is said to have a *right $\mathcal{C}$-approximation* in case the functor $\text{Hom}_\Lambda(\ ,A)$ restricted to $\mathcal{C}$ is finitely generated, and that $\mathcal{C}$ is said to be *contravariant finite* in mod Λ if each object A in mod Λ has a right $\mathcal{C}$-approximation. This notion was introduced by Auslander and the author in [AS]. It is easy to see that one have a right minimal version of any right approximation in mod Λ making this minimal right $\mathcal{C}$-approximation unique up to isomorphism. That $\mathcal{P}^\infty(\text{mod}\,\Lambda)$ is contravariantly finite in mod Λ is exactly what Auslander and Reiten need in order to obtain their result. In fact what they prove is that if $\mathcal{P}^\infty(\text{mod}\,\Lambda)$ is contravariantly finite in mod Λ, then each finitely generated module M of finite projective dimension is a direct summand of a module N having a filtration $0 = N_0 \subset N_1 \subset \cdots \subset N_m = N$ such that all subfactors N_{i+1}/N_i, $i = 1,\ldots,m-1$ are isomorphic to modules from the finite list of the right minimal $\mathcal{P}^\infty(\text{mod}\,\Lambda)$-approximation of one representative from each isomorphism class of the simple Λ-modules. Letting the simples be $S_1,\cdots,S_n$ and the minimal right $\mathcal{P}^\infty(\text{mod}\,\Lambda)$-approximations of these modules be $A_1,\ldots,A_n$, then one obtains that ℓfin.dim Λ is the supremum of the projective dimension of the A_i, $i = 1,\ldots,n$.

In this note I will give an exposition of how contravariant finiteness of $\mathcal{P}^\infty(\text{mod}\,\Lambda)$ in mod Λ not only forces finitely generated modules of finite projective dimension to have such filtrations, but forces all modules of finite projective dimension to be direct limits of modules having such finite filtrations. As a byproduct one then obtains not only that ℓfin.dim Λ is finite in case $\mathcal{P}^\infty(\text{mod}\,\Lambda)$ is contravariantly finite in mod Λ, but also that in this case the little and the big finitistic dimensions coincide. This result was obtained jointly with Birge Huisgen-Zimmermann [HZS].

Some of the tools used in the joint paper with Huisgen-Zimmermann dealing with infinite dimensional modules are included since they should be of independent interest. In particular it is shown that, over a left noetherian ring, each left module M of projective dimension n is a directed union of countably generated submodules which have projective dimensions bounded above by n. As a consequence, the left big finitistic dimension of a left artinian ring equals the supremum of the projective dimensions of the countably generated left modules of finite projective dimension.

For a more complete historical account of the finitistic dimension conjectures, we refer to the paper [HZ2].

1. The Example

The algebra Λ_n will be given as a path algebra of a quiver modulo relations and it will then be explained how it is obtained by means of repeated one point extensions using basically injective modules except in the first step. This point of view will then serve as basis for the argument leading to the announced conclusion.

Let Γ_n be the quiver

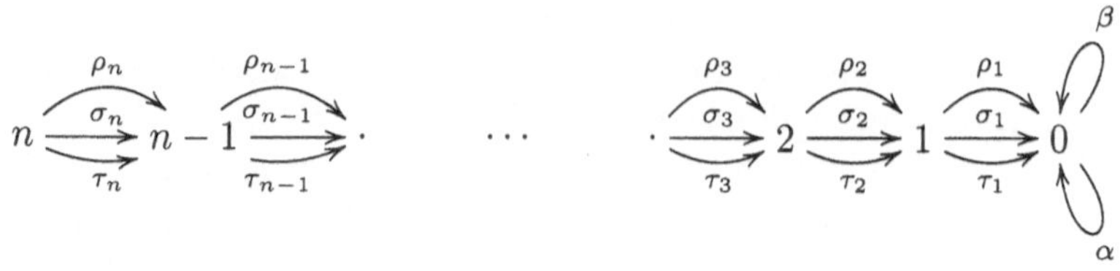

,

let k be any field, and let Λ be the path algebra of Γ over k modulo the ideal generated by the following relations: $\alpha^2, \beta^2, \alpha\beta, \beta\alpha, \alpha\rho_1, \alpha\sigma_1, \beta\tau_1, x_iy_{i+1}$ for $i = 1, 2, \cdots, n-1$ and $x \neq y; x, y \in \{\rho, \sigma, \tau\}$ and $x_ix_{i+1} - y_iy_{i+1}, x, y \in \{\rho, \sigma, \tau\}, i = 1, 2, \cdots, n-1$.

Theorem 1.1. *For each natural number $n \geq 1$, the algebra Λ_n given above has left little finitistic dimension 1 and left big finitistic dimension n.*

Proof: For each vertex $i = 1, 2, \cdots, n$ in the quiver Γ_n, let e_i be the corresponding idempotent in Λ_n, and let $P_i = \Lambda e_i$ and $S_i = P_i/\mathfrak{r}P_i$ be representatives of the indecomposable projective and simple Λ_n-modules respectively. Here $\mathfrak{r}$ denotes the radical of the algebra Λ_n and is the ideal generated by the residues of the arrows. Since the Loewy length of P_0 is 2 and all the indecomposable projective Λ_n-modules except P_0 have Loewy length 3, there can be no inclusion from a projective P into the radical of another projective Q if P contains an indecomposable direct summand not isomorphic to P_0. So, in order to have an inclusion of a nonzero projective module P into the radical of another projective module Q, the projective P has to be isomorphic to a direct sum of copies of P_0. However, P_0 has nonzero morphisms only to P_1, P_2 and itself. So if P is a finitely generated projective module which is a submodule of the radical of a finitely generated projective module, we have the situation $f : P_0^m \to P_0^{m_0} \amalg P_1^{m_1} \amalg P_2^{m_2}$ where the image of f is in the radical. But then, since the image of p_0f, where p_0 is the projection of $P_0^{m_0} \amalg P_1^{m_1} \amalg P_2^{m_2}$ onto $P_0^{m_0}$, and the image of p_2f, where p_2 is the projection of $P_0^{m_0} \amalg P_1^{m_1} \amalg P_2^{m_2}$ onto $P_2^{m_2}$, are both semisimple, we obtain an inclusion $P_0^m \to P_1^{m_1}$ by composing the given inclusion f with the projection p_1 of $P_0^{m_0} \amalg P_1^{m_1} \amalg P_2^{m_2}$ onto $P_1^{m_1}$. However, letting α and β also denote the residues of α and β in Λ_n, one have that the dimension of $\alpha \cdot P_0^m$ is m and that the dimension of $\alpha \cdot P_1^{m_1}$ is m_1, and therefore such an inclusion is possible only if $m_1 \geq m$. But then $\mathfrak{r}^2 P_1^{m_1}$, which has dimension $3m_1$ as a k-vector space, is not contained in the image of P_0^m which has an intersection with $\mathfrak{r}^2 P_1^{m_1}$ of dimension $2m$. Therefore the cokernel of any inclusion $P \to Q$ of finitely generated projective Λ_n-modules P and Q with the image in the radical, has Loewy length three and can therefore not be embedded in the radical of any projective Λ_n-module. Hence each finitely generated module of projective dimension 1 has Loewy length 3. Therefore the syzygy of any finitely generated Λ_n-module cannot have projective dimension 1 since the Loewy length of $\mathfrak{r}$ is 2, which shows that the left little finitistic dimension of Λ_n is 1 for all $n \geq 1$.

Next, let $e_{0,i}$ be the coordinates of $P_0^{(\mathbb{N})}$ with e_0 in the i'th place and 0 otherwise, and let analogously $e_{1,i}$ be the coordinates of $P_1^{(\mathbb{N})}$ with e_1 in the i'th place and 0 otherwise. Letting $\phi : P_0^{(\mathbb{N})} \to P_1^{(\mathbb{N})}$ be given by $\phi(e_{0,2i-1}) = \tau_1 e_{1,2i-1} + \sigma_1 e_{1,i}$ and $\phi(e_{0,2i}) = \tau_1 e_{1,2i} + \rho_1 e_{1,i}$, where τ_1, σ_1 and ρ_1 also denote the residues of τ_1, σ_1 and ρ_1 in Λ_n respectively, it is not hard to verify that ϕ is an inclusion $P_0^{(\mathbb{N})} \to P_1^{(\mathbb{N})}$ such that $\operatorname{Coker}\phi$ is annihilated by the residue classes of α and β and has socle equal to $S_0^{(\mathbb{N})}$. If we now go back and look at the definition of the algebra Λ_n, it follows that for $n \geq 3$ one has that Λ_n is a one point extension of Λ_{n-1} by the injective envelope of S_{n-2} considered as a Λ_{n-1}-module. Λ_2 is the one point extension of Λ_1 by the injective $\Lambda_1/(\alpha,\beta)$-envelope of S_0 considered as a Λ_1-module, where we also let α and β denote the residue classes of α and β in Λ_1. Denoting $\operatorname{Coker}\phi$ by X_1, we know that we can embed X_1 into $P_2^{(\mathbb{N})}$ in such a way that their socles coincide, yielding a quotient module which we denote by X_2 having Loewy length two with socle isomorphic to $S_1^{(\mathbb{N})}$. This can now be repeated leading to an exact sequence of modules

$$0 \to P_0^{(\mathbb{N})} \to P_1^{(\mathbb{N})} \to P_2^{(\mathbb{N})} \to \cdots \to P_n^{(\mathbb{N})} \to X_n \to 0$$

where each $P_i^{(\mathbb{N})}$ is projective. This shows that the big finitistic dimension of Λ_n is at least n.

That the big finitistic dimension cannot exceed n follows from the fact that the vertices in Γ_n corresponding to all indecomposable projective Λ_n-modules P_i except P_0 do not belong to any oriented cycle in Γ_n . This finishes the proof of the theorem.

2. Categories of countably generated modules of finite projective dimension

The primary goal of this section is to show that, for each left artinian ring R, the big finitistic dimension, $\ell\,\mathrm{Fin.dim}\,R$, coincides with the supremum of the projective dimensions of the countably generated left R-modules of finite projective dimension. The argument is in the spirit of Kaplansky's classical theorem [K] stating that every projective R-module is a direct sum of countably generated components.

Recall that a module X is said to be countably presented if there exists an exact sequence

$$R^{(K)} \to R^{(L)} \to X \to 0$$

with $|K|$, $|L| \leq \aleph_0$. Clearly, every left noetherian ring has the property that each countably generated left module is countably presented; the same is true if the base ring R is countable or a countably generated algebra over a field.

Proposition 2.1. *Let R be a ring with the property that each countably generated left R-module is countably presented. Then each left R-module M of finite projective dimension is the directed union of countably generated submodules which have projective dimensions bounded above by that of M.*

Proof: Let $\operatorname{pdim} M = m < \infty$ where $\operatorname{pdim} M$ denotes the projective dimension of the left Λ-module M. It clearly suffices to show that each countable subset of M is contained in a countably generated submodule M' with $\operatorname{pdim} M' \leq m$. Let

$$0 \longrightarrow P_m \xrightarrow{f_m} P_{m-1} \xrightarrow{f_{m-1}} \cdots \longrightarrow P_1 \xrightarrow{f_1} P_0 \xrightarrow{f_0} M \longrightarrow 0$$

be a projective resolution of M. By Kaplansky's theorem [K], each P_i is of the form $\amalg_{j\in A_i} P_{ij}$, where all of the P_{ij}'s are countably generated.

Given a countable subset U of M, choose a countable subset $B_0^{(1)} \subseteq A_0$ such that $f_0(\amalg_{j\in B_0^{(1)}} P_{0j})$ contains U. Then the kernel of the restriction of f_0 to $\amalg_{j\in B_0^{(1)}} P_{0j}$ is in turn countably generated by hypothesis, which permits us to pick a countable set $B_1^{(1)} \subseteq A_1$, with the property that

$$\operatorname{Ker}(f_0) \cap \left(\amalg_{j\in B_0^{(1)}} P_{0j}\right) \subseteq f_1\left(\amalg_{j\in B_1^{(1)}} P_{ij}\right).$$

An induction further yields countable subsets $B_i^{(1)} \subseteq A_i$ for $1 \leq i \leq m$ such that

$$\operatorname{Ker}(f_{i-1}) \cap \left(\amalg_{j\in B_{i-1}^{(1)}} P_{i-1,j}\right) \subseteq f_i\left(\amalg_{j\in B_i^{(1)}} P_{ij}\right)$$

for $1 \leq i \leq m$.

Now set $B_m^{(2)} = B_m^{(1)}$. In view of the fact that the image

$$f_m\left(\amalg_{j\in B_m^{(2)}} P_{mj}\right) \subseteq \operatorname{Ker} f_{m-1}$$

is countably generated, we can find a countable subset $B_{m-1}^{(2)} \subseteq A_{m-1}$ containing $B_{m-1}^{(1)}$ such that

$$f_m\left(\amalg_{j\in B_m^{(2)}} P_{mj}\right) \subseteq \operatorname{Ker}(f_{m-1}) \cap \left(\amalg_{j\in B_{m-1}^{(2)}} P_{m-1,j}\right).$$

An induction analogous to the preceding one then yields countable subsets $B_i^{(2)} \subseteq A_i$, $0 \leq i \leq m$ such that $B_i^{(1)} \subseteq B_i^{(2)}$ and

$$f_{i+1}\left(\amalg_{j\in B_{i+1}^{(2)}} P_{ij}\right) \subseteq \operatorname{Ker}(f_i) \cap \left(\amalg_{j\in B_i^{(2)}} P_{ij}\right)$$

for $0 < i < m - 1$.

Next set $B_0^{(3)} = B_0^{(2)}$, and continue. We iterate these m-step inductions moving back and forth along the given projective resolution of M, and an induction on this level will supply us, for each $k \in \mathbb{N}$, with countable sets $B_0^{(k)}, \cdots, B_m^{(k)}$ such that $B_i^{(k-1)} \subseteq B_i^{(k)} \subseteq A_i$ for $0 \le i \le m$, having the additional properties that

$$\mathrm{Ker}(f_{i-1}) \cap \left(\amalg_{j \in B_{i-1}^{(k)}} P_{i-1,j}\right) \subseteq f_i\left(\amalg_{j \in B_i^{(k)}} P_{ij}\right)$$

for $1 \le i \le m$ and odd k, as well as

$$f_{i+1}\left(\amalg_{j \in B_{i+1}^{(k)}} P_{i+1,j}\right) \subseteq \mathrm{Ker}(f_i) \cap \left(\amalg_{j \in B_i^{(k)}} P_{i,j}\right)$$

for $0 \le i \le m-1$ and even k.

Finally, we set $B_i = \cup_{k \in \mathbf{N}} B_i^{(k)} \subseteq A_i$ for $0 \le i \le m$, define M' to be the countably generated submodule $f_0(\amalg_{j \in B_0} P_{0j})$ of M, and observe that

$$0 \longrightarrow \amalg_{j \in B_m} P_{mj} \xrightarrow{f_m} \cdots \longrightarrow \amalg_{j \in B_1} P_{1j} \xrightarrow{f_1} \amalg_{j \in B_0} P_{0j} \xrightarrow{f_0} M' \longrightarrow 0$$

is a projective resolution of M'. This guarantees that $\mathrm{pdim}\, M' \le m$ as required. $\square$

Corollary 2.2. *If R is a left artinian ring, then $\ell\,\mathrm{Fin.dim}\, R$ equals*

$$\sup\{\mathrm{pdim}\, M \mid M \text{ a countably generated left } R\text{-module with } \mathrm{pdim}\, M < \infty\}.$$

Proof: Let N be any left R-module of finite projective dimension. Since R satisfies the hypothesis of Proposition 2.1, we have $N = \varinjlim_{i \in I} N_i$, where $(N_i)_{i \in I}$ is a directed family of countably generated modules with $\mathrm{pdim}\, N_i \le \mathrm{pdim}\, N$ for $i \in I$. In view of the fact that the functor Tor commutes with direct limits, this implies that the flat dimension of N is bounded above by the supremum of the flat dimensions of the N_i. But since R is left artinian, the flat dimension is equal to the projective dimension which completes the argument. $\square$

Observe that, in the conclusion of Proposition 2.1, the attribute 'countably generated' cannot be replaced by 'finitely generated' since this would mean that the big finitistic dimension and the little finitistic dimension coincide for finite dimensional algebras, which is not the case as the examples in Section 1 show.

3. Fin.dim Λ EQUALS fin.dim Λ WHEN $\mathcal{P}^\infty(\mathrm{mod}\,\Lambda)$ IS CONTRAVARIANTLY FINITE IN mod Λ

This section is devoted to proving the statement in its title, namely that, for any finite dimensional algebra Λ (or more generally any Artin algebra Λ), the left big and little finitistic dimensions coincide, provided that the subcategory $\mathcal{P}^\infty(\mathrm{mod}\,\Lambda)$ of finitely generated left Λ-modules of finite projective dimension is contravariantly finite in mod Λ.

Recall that $\mathcal{P}^\infty(\operatorname{mod}\Lambda)$ is contravariantly finite in modΛ if, for each finitely generated left Λ-module M, there exists an object A in $\mathcal{P}^\infty(\operatorname{mod}\Lambda)$ and a homomorphism $f : A \to M$ such that the induced sequence of functors from $\mathcal{P}^\infty(\operatorname{mod}\Lambda)$ to the category of abelian groups,

$$\operatorname{Hom}_\Lambda(-, A)|_{\mathcal{P}^\infty(\operatorname{mod}\Lambda)} \to \operatorname{Hom}_\Lambda(-, M)|_{\mathcal{P}^\infty(\operatorname{mod}\Lambda)} \to 0,$$

is exact. In that case, A is called a right $\mathcal{P}^\infty(\operatorname{mod}\Lambda)$-approximation of M. As already pointed out in the introduction when right $\mathcal{P}^\infty(\operatorname{mod}\Lambda)$-approximations exist for a module M, the ones of minimal length are all isomorphic, and hence it makes sense to refer to the minimal right $\mathcal{P}^\infty(\operatorname{mod}\Lambda)$-approximation of M in that case.

It is known that for any left serial Artin algebra Λ, the category $\mathcal{P}^\infty(\operatorname{mod}\Lambda)$ is contravariantly finite in modΛ [BHZ]; moreover, the minimal right $\mathcal{P}^\infty(\operatorname{mod}\Lambda)$-approximations of the simple left Λ-modules can be explicitly described in that case. Initial criteria for contravariant finiteness in more general situations were developed in [HHZ].

Throughout this section, one abbreviates $\ell\,\text{fin.dim}\,\Lambda$ by $\text{fin.dim}\,\Lambda$ and $\ell\,\text{Fin.dim}\,\Lambda$ by $\text{Fin.dim}\,\Lambda$.

Theorem 3.1. *Suppose that* $\mathcal{P}^\infty(\operatorname{mod}\Lambda)$ *is contravariantly finite in* modΛ. *Then*

$$\text{fin.dim}\,\Lambda = \text{Fin.dim}\,\Lambda.$$

Proof: Since $\text{Fin.dim}\,\Lambda$ is the supremum of the projective dimension of countably generated left Λ-modules of finite projective dimension by Corollary 2.2, it is enough to focus on countably generated modules M with $\operatorname{pdim} M < \infty$. Let M be such a module with $\operatorname{pdim} M = n < \infty$. Let $M_1 \subseteq M_2 \subseteq M_3 \subseteq \cdots \subseteq M_r \subseteq \cdots \subseteq M$ be a chain of finitely generated submodules such that $M = \cup_{i\geq 1} M_i$. For each i, one fixes the beginning of a finitely generated projective resolution of M_i, say

$$0 \to \Omega_{n,i} \to P_{n-1,i} \to \cdots \to P_{1,i} \to P_{0,i} \to M_i \to 0$$

and, over each of the inclusions $M_i \to M_{i+1}$, one chooses a chain morphism consisting of maps $f_{k,i,i+1} : P_{k,i} \to P_{k,i+1}$ for $0 \leq k \leq n-1$, and $f_{n,i,i+1} : \Omega_{n,i} \to \Omega_{n,i+1}$. Defining $f_{k,i,j} = f_{k,j-1,j} \circ \cdots \circ f_{k,i,i+1}$ for all $j > i$ and $0 \leq k \leq n$, one thus obtains a directed system of exact sequences indexed by the natural numbers. Passing to the direct limit of this system gives a projective resolution of the module M which is labeled

$$0 \to \varinjlim \Omega_{n,i} \to P_{n-1} \to \cdots \to P_1 \to P_0 \to M \to 0$$

In particular, $P_n := \varinjlim \Omega_{n,i}$ is projective since $\operatorname{pdim} M = n$. Each of the projective modules P_k is countably generated since the directed system is indexed by $\mathbf{N}$. Therefore, one can decompose P_n in the form $P_n = \amalg_{j\in\mathbf{N}} Q_{n,j}$, where all the $Q_{n,j}$ are finitely generated. Denoting by $g_i : \Omega_{n,i} \to P_n$ the canonical morphisms, one obtains an increasing sequence $i_1, i_2, i_3, \ldots$ of natural numbers such that $\amalg_{j\leq m} Q_{n,j} \subseteq \operatorname{Im} g_{i_m}$ for $m \in \mathbf{N}$. It is clearly harmless to pass to a suitable cofinal subsystem of the $\Omega_{n,i}$, which permits one to assume that $i_k = k$ for all k.

Next observe that each $\Omega_{n,i}$ contains a submodule $P_{n,i} = \amalg_{j\le i} P_{n,j,i}$ isomorphic to the finite direct sum $\amalg_{j\le i} Q_{n,j}$ with the property that g_i restricts to an isomorphism $P_{n,j,i} \to Q_{n,j}$ for all $j \le i$; just keep in mind that the restriction of the map g_i to the preimage of $\amalg_{j\le i} Q_{n,j}$ splits. Since $g_{i+1} f_{n,i,i+1}(P_{n,i}) = g_i(P_{n,i})$, the map $f_{n,i,i+1}$ induces a split monomorphism $P_{n,i} \to g_{i+1}^{-1}(\amalg_{j\le i+1} Q_{n,j})$. Consequently, an induction on i allows one to choose the $P_{n,j,i}$ in such a way that the squares

$$\begin{array}{ccc} P_{n,i+1} & \xrightarrow{g_{i+1}} & \amalg_{j\le i+1} Q_{nj} \\ {\scriptstyle f_{n,i,i+1}}\uparrow & & \uparrow \\ P_{n,i} & \xrightarrow[\cong]{g_i} & \amalg_{j\le i} Q_{nj} \end{array}$$

commute. In other words, this process yields a directed subsystem

$$\begin{array}{ccc} \vdots & & \vdots \\ \uparrow & & \uparrow \\ P_{n,i+1} & \hookrightarrow & \Omega_{n,i+1} \\ \uparrow & & \uparrow \\ P_{n,i} & \hookrightarrow & \Omega_{n,i} \\ \uparrow & & \uparrow \\ \vdots & & \vdots \end{array}$$

of the system $(\Omega_{n,i}, f_{n,i,j})_{i,j\in\mathbf{N}, i\le j}$ such that $\varinjlim P_{n,i} = \varinjlim \Omega_{n,i} = P_n$.

So far the assumption that $\mathcal{P}^\infty(\operatorname{mod}\Lambda)$ is contravariantly finite in $\operatorname{mod}\Lambda$ has not been used. To invoke this property, the chain of arguments is interrupt by a lemma which will allow one to supplement the original directed system of resolutions

$$(\mathbf{S_i}) \qquad P_{n-1,i} \to P_{n-2,i} \to \cdots \to P_{0,i} \to M_i \to 0$$

by a system

$$(\mathbf{T_i}) \qquad P'_{n-1,i} \to P'_{n-2,i} \to \cdots \to P'_{0,i} \to N_i \to 0,$$

where $N_i \in \mathcal{P}^\infty(\operatorname{mod}\Lambda)$, together with an epimorphism $(\mathbf{T_i}) \to (\mathbf{S_i})$ of directed systems, where $N = \varinjlim N_i$ has projective dimension bounded above by fin.dim Λ and the kernel of the induced epimorphism $N \to M$ is 'under control'.

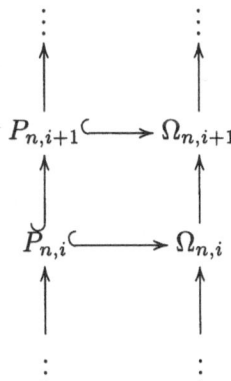

The following lemma is based on the fact that, if $\mathcal{P}^\infty(\operatorname{mod}\Lambda)$ is contravariantly finite in modΛ, then there exists an injective cogenerator inside the category $\mathcal{P}^\infty(\operatorname{mod}\Lambda)$ (see [AS]). Indeed, if I is the minimal right $\mathcal{P}^\infty(\operatorname{mod}\Lambda)$-approximation of a minimal injective cogenerator for modΛ, then each object of $\mathcal{P}^\infty(\operatorname{mod}\Lambda)$ embeds into an object of add I where add(I) denotes the full subcategory of mod Λ consisting of direct summands of finite direct sums of copies of I. Further, every inclusion $I' \hookrightarrow X$ with I' in add(I) and X in $\mathcal{P}^\infty(\operatorname{mod}\Lambda)$ is a split monomorphism. In particular, every monomorphism in add(I) is a split monomorphism.

Lemma 3.2. *Let I be a relative injective cogenerator for $\mathcal{P}^\infty(\operatorname{mod}\Lambda)$ as above, and suppose that there is given an exact commutative diagram of the form*

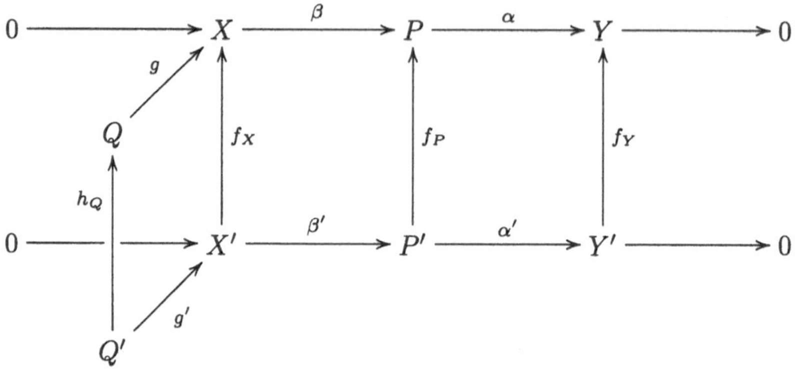

with Q, Q', P and $P' \in \mathcal{P}^\infty(\operatorname{mod}\Lambda)$. Then there exist modules I_0 and I_0' in add I, *together with homomorphisms $\gamma : Q \to I_0$ and $\gamma' : Q' \to I_0'$, as well as a homomorphism $h : I_0' \to I_0$ such that the following diagram has exact rows and commutes.*

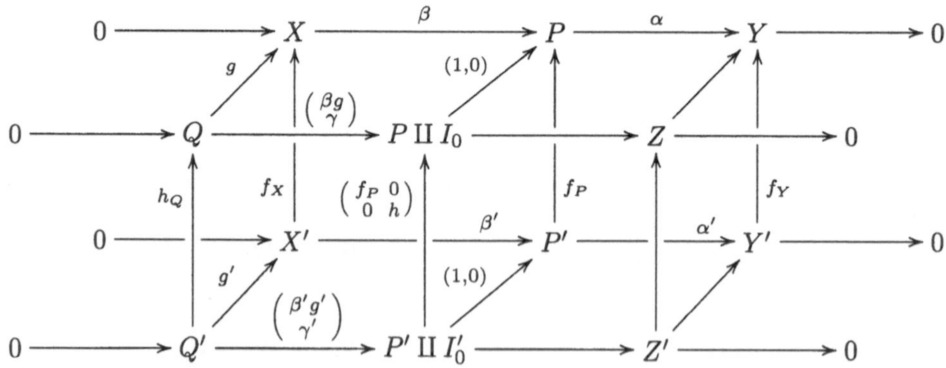

Here $Z = $ Coker $\binom{\beta g}{\gamma}$, *$Z' = $* Coker $\binom{\beta' g'}{\gamma'}$ *are in $\mathcal{P}^\infty(\operatorname{mod}\Lambda)$, and the remaining maps are induced by the cokernels.*

Proof of the Lemma: Since I is a cogenerator for $\mathcal{P}^\infty(\operatorname{mod}\Lambda)$, one can choose monomorphisms $\gamma : Q \to I_0$ and $\gamma' : Q' \to I_0'$. So if one introduces maps $Q \to P \oplus I_0$ and $Q' \to P \oplus I_0'$ as in the above diagram and denotes by Z and Z'

their cokernels, respectively, the two squares in each of the top and bottom planes commute. Moreover, Z, Z' belong to $\mathcal{P}^\infty(\text{mod}\,\Lambda)$ by the hypothesis on Q and Q'. Next, one can use the relative injectivity of I to obtain $h : I_0' \to I_0$ such that $\gamma h_Q = h\gamma'$. It is now straightforward to check that the entire diagram commutes, which completes the proof of the lemma. □

Now return to the proof of Theorem 4.1. Label the maps in the projective resolutions of the M_i, say $g_{k,i} : P_{k,i} \to P_{k-1,i}$. By applying the lemma, first to the diagrams

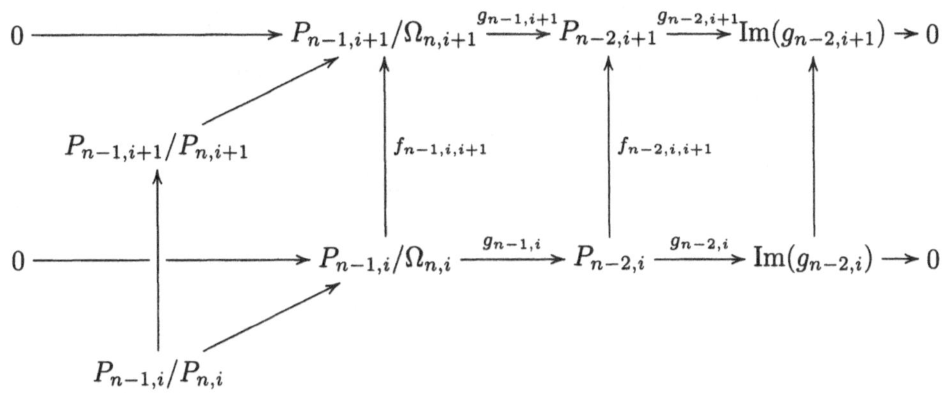

for $i \in \mathbf{N}$, and by then moving inductively along the sequences

$$P_{n-1,i} \to P_{n-2,i} \to \cdots \to P_{0,i} \to M_i \to 0,$$

one obtains the following directed system of diagrams in $\text{mod}\Lambda$:

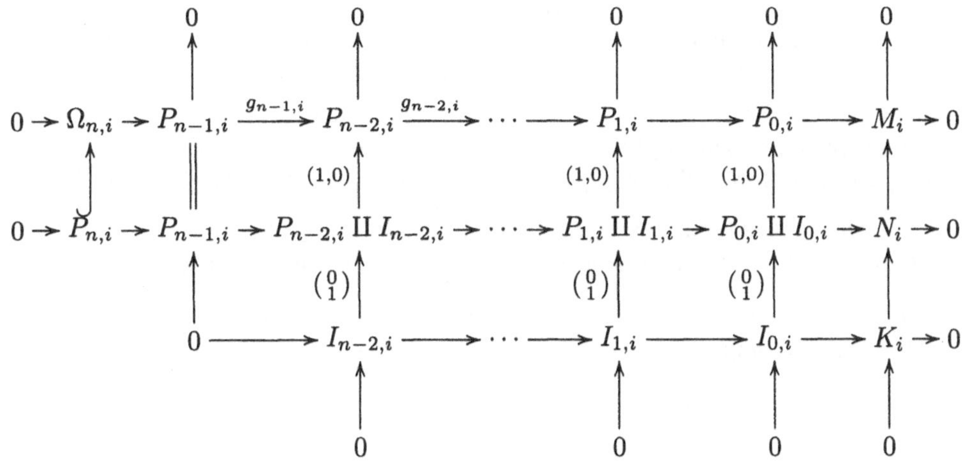

Here the objects $I_{k,i}$ all belong to $\text{add}(I)$. The upper two horizontal sequences of each of these diagrams are exact by construction, whereas the induced kernel sequence in the third row will not be exact in general; in fact, the homology in the term labeled $n-2$ of that sequence is isomorphic to $\Omega_{n,i}/P_{n,i}$. On the other hand, the direct limit of the inclusions $P_{n,i} \hookrightarrow \Omega_{n,i}$ is the identity $P_n \to P_n$, and hence the snake lemma ensures that, after taking direct limits, one arrives at an *exact* commutative diagram of the form

$$
\begin{array}{ccccccccccccccc}
 & & & & 0 & & 0 & & & & 0 & & 0 & & 0 \\
 & & & & \uparrow & & \uparrow & & & & \uparrow & & \uparrow & & \uparrow \\
0 & \to & P_n & \to & P_{n-1} & \longrightarrow & P_{n-2} & \longrightarrow & \cdots & \longrightarrow & P_1 & \longrightarrow & P_0 & \longrightarrow & M \to 0 \\
 & & \Vert & & \Vert & & \uparrow{\scriptstyle (1,0)} & & & & \uparrow{\scriptstyle (1,0)} & & \uparrow{\scriptstyle (1,0)} & & \uparrow \\
0 & \to & P_n & \to & P_{n-1} & \to & P_{n-2} \amalg I_{n-2} & \to & \cdots & \to & P_1 \amalg I_1 & \to & P_0 \amalg I_0 & \to & N \to 0 \\
 & & & & \uparrow & & \uparrow{\scriptstyle \binom{0}{1}} & & & & \uparrow{\scriptstyle \binom{0}{1}} & & \uparrow{\scriptstyle \binom{0}{1}} & & \uparrow \\
 & & & & 0 & \longrightarrow & I_{n-2} & \longrightarrow & \cdots & \longrightarrow & I_1 & \longrightarrow & I_0 & \longrightarrow & K \to 0 \\
 & & & & & & \uparrow & & & & \uparrow & & \uparrow & & \uparrow \\
 & & & & & & 0 & & & & 0 & & 0 & & 0
\end{array}
$$

Now using some facts about finitely generated modules over a finite dimensional algebra (see for example [L] or [Z]), one gets that each I_j, being a direct limit of modules in $\mathrm{add}(I)$, is itself a direct sum of modules in $\mathrm{add}(I)$. Also since all monomorphisms in $\mathrm{add}(I)$ are split monomorphisms, the same applies to the category consisting of arbitrary direct sums of copies of modules from $\mathrm{add}(I)$. As a consequence one obtains that K is a direct sum of modules in $\mathrm{add}(I)$. Since the N_i belong to $\mathcal{P}^\infty(\mathrm{mod}\,\Lambda)$ by construction, the relative cogenerating property of I furthermore permits one to choose a directed system of embeddings $N_i \hookrightarrow I_i'$ with I_i' in $\mathrm{add}(I)$, which shows that $N = \varinjlim N_i$ embeds into a module X which is a direct sum of modules from $\mathrm{add}(I)$. Then the composed inclusion $K \hookrightarrow N \hookrightarrow X$ is a split monomorphism which implies that the exact sequence $0 \to K \to N \to M \to 0$ splits, and therefore $\mathrm{pdim}\, M \leq \mathrm{pdim}\, N \leq \sup\{\mathrm{pdim}\, N_i \mid i \in \mathbf{N}\} \leq \mathrm{fin.dim}\,\Lambda$. □

In the light of the main Theorem in this section, the following is an immediate consequence of the result of Auslander and Reiten quoted in the introduction.

Corollary 3.3. *If* $\mathcal{P}^\infty(\mathrm{mod}\,\Lambda)$ *is contravariantly finite in* $\mathrm{mod}\Lambda$*, and* $A_1, \ldots, A_n$ *are* $\mathcal{P}^\infty(\mathrm{mod}\,\Lambda)$*-approximations of the simple left* Λ*-modules, then*

$$\mathrm{Fin.dim}\,\Lambda = \sup\{\mathrm{pdim}\, A_i \mid 1 \leq i \leq n\}.$$

□

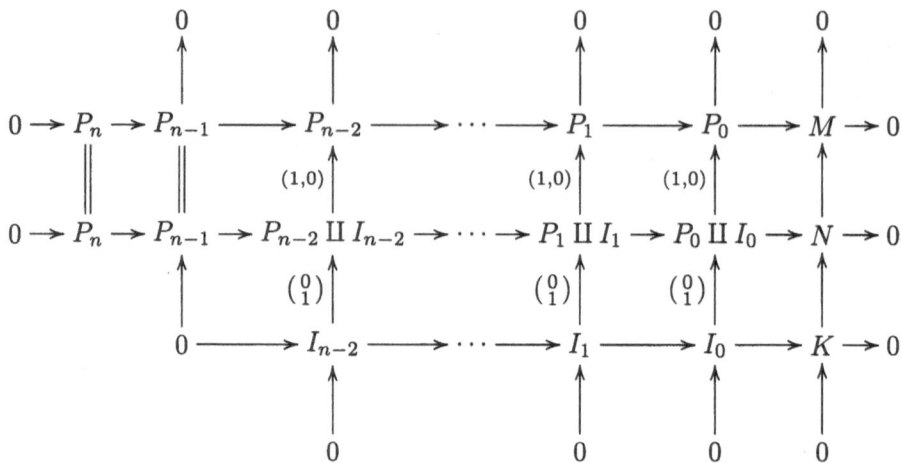

One can actually strengthen the conclusion of Theorem 4.1, so as to provide information on the structure of the non-finitely generated objects in $\mathcal{P}^\infty(\mathrm{Mod}\,\Lambda)$ as follows: Given modules $M_1, \cdots, M_n$ in modΛ one denotes by $\mathrm{filt}(M_1, \cdots, M_n)$ the full subcategory of modΛ having as objects all finitely generated modules X that possess filtrations $X = X_0 \supseteq X_1 \supseteq \cdots \supseteq X_r = 0$ such that each factor X_i/X_{i+1} is isomorphic to some M_j. Further, $\overrightarrow{\mathrm{filt}}(M_1, \cdots, M_n)$ denotes the full subcategory of ModΛ the objects of which are the direct limits of modules in $\mathrm{filt}(M_1, \cdots, M_n)$. The result of Auslander and Reiten states that in case $\mathcal{P}^\infty(\mathrm{mod}\,\Lambda)$ is contravariantly finite in modΛ and $A_1, \cdots, A_n$ are the minimal $\mathcal{P}^\infty(\mathrm{mod}\,\Lambda)$-approximations of the simple left Λ-modules, $\mathcal{P}^\infty(\mathrm{mod}\,\Lambda)$ consists of the direct summands of modules in $\mathrm{filt}(A_1, \cdots, A_n)$. In view of the proof of the main theorem in this section, this description of the finitely generated modules of finite projective dimension extends to non-finitely generated candidates as follows.

Theorem 3.4. *If* $\mathcal{P}^\infty(\mathrm{mod}\,\Lambda)$ *is contravariantly finite in* modΛ *and* $A_1, \cdots, A_n$ *are as above, then*

$$\mathcal{P}^\infty(\mathrm{Mod}\,\Lambda) = \overrightarrow{\mathrm{filt}}(A_1, \cdots, A_n).$$

In particular, each object of $\mathcal{P}^\infty(\mathrm{Mod}\,\Lambda)$ *is a direct limit of modules in* $\mathcal{P}^\infty(\mathrm{mod}\,\Lambda)$.

Proof: That $\overrightarrow{\mathrm{filt}}(A_1, \ldots, A_n)$ is contained in $\mathcal{P}^\infty(\mathrm{Mod}\,\Lambda)$ is clear. For the other inclusion, start by noting that each full subcategory $\mathcal{C}$ of ModΛ which is closed under direct limits is also closed under direct summands. This is well known and the argument goes as follows: If B is a direct summand of an object C in $\mathcal{C}$ and $\pi : C \to C$ a projection onto B, then B is the direct limit of the system

$$C \xrightarrow{\pi} C \xrightarrow{\pi} C \xrightarrow{\pi} \cdots$$

By Proposition 2.1, each object in $\mathcal{P}^\infty(\mathrm{Mod}\,\Lambda)$ is the direct union of countably generated objects in $\mathcal{P}^\infty(\mathrm{Mod}\,\Lambda)$, whence it suffices to show that each countably generated module M of finite projective dimension belongs to $\overrightarrow{\mathrm{filt}}(A_1, \ldots, A_n)$. But the proof of Theorem 3.1 shows that M is a direct summand of a direct limit of modules in $\mathcal{P}^\infty(\mathrm{mod}\,\Lambda)$ and thus belongs to the closure of $\mathcal{P}^\infty(\mathrm{mod}\,\Lambda)$ under direct limits as explained above. That the latter category is contained in $\overrightarrow{\mathrm{filt}}(A_1, \ldots, A_n)$, finally, follows from [AR]. □

Note that the arguments used for proving Theorems 3.1 and 3.4 only use the existence of a $\mathcal{P}^\infty(\mathrm{mod}\,\Lambda)$-approximation of an injective cogenerator for modΛ. This observation does not lead to any significant generalization of the results since Happel and Unger have proved that if $\mathrm{fin.dim}\,\Lambda < \infty$ and there is a $\mathcal{P}^\infty(\mathrm{mod}\,\Lambda)$-approximation of an injective cogenerator in mod Λ then $\mathcal{P}^\infty(\mathrm{mod}\,\Lambda)$ is in fact contravariantly finite in mod Λ [HU].

The final consequence of the main results was observed by Henning Krause, who pointed out that a result of Crawley-Boevey applies to this context. Here one needs the notion of a covariantly finite subcategory of mod Λ which is the dual notion of a contravariantly finite subcategory. In short, a subcategory $\mathcal{C}$ in mod Λ is covariantly finite if the restriction to $\mathcal{C}$ of each covariant representable functor $\mathrm{Hom}(A, \ )$ with A in mod Λ, is finitely generated as a functor on $\mathcal{C}$.

Corollary 3.5. *If the subcategory* $\mathcal{P}^\infty(\mathrm{mod}\,\Lambda)$ *of* modΛ *is contravariantly finite, then it is also covariantly finite. In particular,* $\mathcal{P}^\infty(\mathrm{mod}\,\Lambda)$ *has almost split sequences in that case.*

Proof: Suppose that $\mathcal{P}^\infty(\mathrm{mod}\,\Lambda)$ is contravariantly finite in modΛ. Then ℓFin.dim Λ is finite. In particular, $\mathcal{P}^\infty(\mathrm{Mod}\,\Lambda)$ is closed under direct products since direct products of projectives are projective for finite dimensional algebras. By Theorem 3.4, this is the same as saying that the closure of $\mathcal{P}^\infty(\mathrm{mod}\,\Lambda)$ under direct limits is also closed under direct products. Hence Theorem 4.2 of [CB] tells one that $\mathcal{P}^\infty(\mathrm{mod}\,\Lambda)$ is indeed covariantly finite. □

On the other hand, it is not true in general that covariant finiteness of $\mathcal{P}^\infty(\mathrm{mod}\,\Lambda)$ in modΛ implies contravariant finiteness. Since the category of modules of projective dimension ≤ 1 is always covariantly finite in modΛ by [AR], the examples in Section 1 (as well as the example in [IST]) exhibit covariantly finite subcategories $\mathcal{P}^\infty(\mathrm{mod}\,\Lambda)$ in mod Λ which fail to be contravariantly finite in mod Λ.

References

[AB] M. Auslander and D. Buchsbaum, *Homological Dimensions in Local Rings,* Trans. Amer. Math. Soc. vol. 85 no 2 (1957) 390-405.

[AR] M. Auslander and I. Reiten, *Applications of contravariantly finite subcategories,* Advances in Math. 86 (1991) 111-152.

[AS] M. Auslander and S. O. Smalø, *Preprojective modules over Artin algebras,* J. Algebra 66 (1980) 61-122.

[B1] H. Bass, *Finitistic Dimension and a Homological Generalization of Semi-Primary Rings,* Trans. Amer. Math. Soc. 95 (1960) 466-488.

[B2] H. Bass, *Injective Dimension in Noetherian Rings,* Trans. Amer. Math. Soc. 102 (1962) 18-29.

[BHZ] W. D. Burgess and B. Huisgen-Zimmermann, *Approximating modules by modules of finite projective dimension,* J. Algebra 178 (1995) 48-91.

[CB] W. Crawley-Boevey, *Locally finitely presented additive categories,* Comm. Algebra 22 (1994) 1641-1674.

[DH] P. Dräxler and D. Happel, *A proof of the generalized Nakayama conjecture for algebras with* $J^{2l+1} = 0$ *and* A/J^l *representation finite,* J. Pure and App. Alg. 78 (1992) 161-164.

[GHZ] E. Green and B. Huisgen-Zimmermann, *Finitistic dimension of artinian rings with vanishing radical cube* Math. Z. 206 (1991) 505-526.

[GR] L. Gruson and L. Raynaud, *Critères de platitude et de projectivité. Techniques de "platification" d'un module,* Invent.Math. 13 (1971) 1-89.

[HHZ] D. Happel and B. Huisgen-Zimmermann, *Viewing finite dimensional representations through infinite dimensional ones,* Pacific J. Math., (to appear).

[HU] D. Happel and L. Unger, *Partial tilting modules and covariantly finite subcategories,* Comm. Algebra 22 (1994) 1723-1727.

[HZ1] B. Huisgen-Zimmermann, *Homological domino effects and the first finitistic dimension conjecture,* Invent. Math. 108 (1992) 369-383.

[HZ2] ———, *The Finitistic Dimension Conjectures- A Tale of 3.5 Decades,* Abelian Groups and Modules, (Padova 1994) Kluwer Acad. Publishers, Dordrecht 1995.

[HZS] B. Huisgen-Zimmermann and S. O. Smalø, *A Homological Bridge Between Finite and Infinite Dimensional Representations* Journal of Representation Theory, to appear.

[IST] K. Igusa, S. O. Smalø, and G. Todorov, *Finite projectivity and contravariant finiteness,* Proc. Amer. Math. Soc. 109 (1990) 937-941.

[IT] K. Igusa and G. Todorov, *On the Finitistic Dimension Conjecture for Artin Algebras,* Unpublished preprint.

[K] I. Kaplansky, *Projective modules,* Annals of Math. 68 (1958) 372-377.

[L] H. Lenzing, *Homological transfer from finitely presented to infinite modules,* in Abelian Group Theory, Proc. Honolulu 1982/83, Lecture Notes in Math. 1006, Berlin (1983) Springer-Verlag, pp. 734-761.

[N] M. Nagata, *Local Rings*, John Wiley & Sons, Inc 1962.

[S] S. O. Smalø, *The supremum of the difference between the big and little finitistic dimensions is infinite,* Proc. Amer. Math. Soc., (to appear).

[W] Y. Wang, *A Note on the Finitistic Dimension Conjecture,* Comm. Algebra, 22 (7)(1994) 2525-2528.

[Z] W. Zimmermann, *Rein-injektive direkte Summen von Moduln,* Comm. Algebra 5 (1977) 1083-1117.

DEPARTMENT OF MATHEMATICAL SCIENCES, THE NORWEGIAN UNIVERSITY FOR SCIENCE AND TECHNOLOGY, 7491 TRONDHEIM, NORWAY

E-mail address: sverresm@math.ntnu.no